**Handbook of Asymmetric
Heterogeneous Catalysis**

*Edited by
Kuiling Ding and
Yasuhiro Uozumi*

Further Reading

Ertl, G., Knözinger, H., Schüth, F., Weitkamp, J. (Eds.)

Handbook of Heterogeneous Catalysis

8 Volumes

2008

ISBN: 978-3-527-31241-2

Maruoka, K. (Ed.)

Asymmetric Phase-Transfer Catalysis

2008

ISBN: 978-3-527-31842-1

Mikami, K

New Frontiers in Asymmetric Catalysis

2007

ISBN: 978-0-471-68026-0

van Santen, R. A., Neurock, M.

Molecular Heterogeneous Catalysis

A Conceptual and Computational Approach

2006

ISBN: 978-3-527-29662-0

Sheldon, R. A., Arends, I., Hanefeld, U.

Green Chemistry and Catalysis

2007

ISBN: 978-3-527-30715-9

Cornils, B., Herrmann, W. A., Muhler, M., Wong, C.-H. (Eds.)

Catalysis from A to Z

A Concise Encyclopedia

2007

ISBN: 978-3-527-31438-6

Handbook of
Asymmetric Heterogeneous Catalysis

Edited by
Kuiling Ding and Yasuhiro Uozumi

WILEY-VCH

WILEY-VCH Verlag GmbH & Co. KGaA

The Editors

Prof. Dr. Kuiling Ding
Chinese Academy of Sciences
Shanghai Institute of Organic Chemistry
Fenglin Road
Shanghai 200032
China

Prof. Dr. Yasuhiro Uozumi
Institute for Molecular Science (IMS)
Myodaiji
Okazaki 444-8787
Japan

All books published by Wiley-VCH are carefully produced. Nevertheless, authors, editors, and publisher do not warrant the information contained in these books, including this book, to be free of errors. Readers are advised to keep in mind that statements, data, illustrations, procedural details or other items may inadvertently be inaccurate.

Library of Congress Card No.: applied for

British Library Cataloguing-in-Publication Data
A catalogue record for this book is available from the British Library.

Bibliographic information published by the Deutsche Nationalbibliothek
Die Deutsche Nationalbibliothek lists this publication in the Deutsche Nationalbibliografie; detailed bibliographic data are available on the Internet at <http://dnb.d-nb.de>.

© 2008 WILEY-VCH Verlag GmbH & Co. KGaA, Weinheim

All rights reserved (including those of translation into other languages). No part of this book may be reproduced in any form – by photoprinting, microfilm, or any other means – nor transmitted or translated into a machine language without written permission from the publishers. Registered names, trademarks, etc. used in this book, even when not specifically marked as such, are not to be considered unprotected by law.

Composition SNP Best-set Typesetter Ltd., Hong Kong
Printing betz-druck GmbH, Darmstadt
Bookbinding Litges & Dopf GmbH, Heppenheim

Printed in the Federal Republic of Germany
Printed on acid-free paper

ISBN: 978-3-527-31913-8

Contents

Preface *XIII*
List of Contributors *XV*

1 **An Overview of Heterogeneous Asymmetric Catalysis** *1*
Zheng Wang, Kuiling Ding, and Yasuhiro Uozumi
1.1 Introduction *1*
1.2 Common Techniques for Immobilization of Homogeneous Asymmetric Catalysts *4*
1.2.1 Chiral Catalyst Immobilization on Inorganic Materials *4*
1.2.2 Chiral Catalyst Immobilization Using Organic Polymers *7*
1.2.3 Dendrimer-Supported Chiral Catalysts *8*
1.2.4 Self-Supported Chiral Catalysts in Asymmetric Reactions *10*
1.2.5 Chiral Catalyst Immobilization Using Nonconventional Media *12*
1.2.5.1 Catalyst Immobilization in Water *12*
1.2.5.2 Fluorous Phase-Separation Techniques in Catalysis *13*
1.2.5.3 Catalytic Reactions in Ionic Liquids *14*
1.2.5.4 Enantioselective Catalysis in Supercritical Carbon Dioxide *15*
1.2.6 Phase-Transfer Catalysis *16*
1.2.7 Immobilization of Chiral Organic Catalysts *16*
1.3 Chirally Modified Metal Surface for Heterogeneous Asymmetric Hydrogenation *17*
1.4 Heterogeneous Enantioselective Catalysts in Industrial Research and Application *18*
References *19*

2 **Heterogeneous Enantioselective Catalysis Using Inorganic Supports** *25*
Santosh Singh Thakur, Jae Eun Lee, Seo Hwan Lee, Ji Man Kim, and Choong Eui Song
2.1 Introduction *25*
2.2 Asymmetric Reduction *27*
2.2.1 Immobilization via a Covalent Link *31*
2.2.2 Immobilization via Hydrogen Bonding, Ionic and Other Interactions *34*

Handbook of Asymmetric Heterogeneous Catalysis. Edited by K. Ding and Y. Uozumi
Copyright © 2008 WILEY-VCH Verlag GmbH & Co. KGaA, Weinheim
ISBN: 978-3-527-31913-8

2.3	Asymmetric Oxidation *40*	
2.3.1	Asymmetric Epoxidation (AE) of Unfunctionalized Olefins *40*	
2.3.1.1	Immobilization via a Covalent Link *45*	
2.3.1.2	Immobilization via Coordination, Ionic and Other Interactions *45*	
2.3.1.3	Immobilization by the 'Ship-in-a-Bottle' Approach *47*	
2.3.1.4	Supported Ionic Liquid Catalysis (SILC) *48*	
2.3.2	AE of Allylic Alcohol *49*	
2.3.3	Enone Epoxidation *51*	
2.3.4	Asymmetric Dihydroxylation (AD) and Asymmetric Aminohydroxylation (AA) of Olefins *52*	
2.3.5	Asymmetric Aziridination of Olefin *53*	
2.4	Asymmetric Carbon–Carbon and Carbon–Heteroatom Bond Formation *54*	
2.4.1	Pd-Catalyzed Asymmetric Allylic Substitution *54*	
2.4.2	Enantioselective Addition of Dialkylzincs to Aldehydes *58*	
2.4.3	Asymmetric Diels–Alder Reaction *59*	
2.4.4	Ene Reactions *61*	
2.4.5	Asymmetric Conjugate Addition *61*	
2.4.6	Aldol and Nitroaldol Reactions *63*	
2.4.7	Asymmetric Cyclopropanation *64*	
2.4.8	Friedel–Craft Hydroxyalkylation *65*	
2.4.9	Si–H Insertion *65*	
2.5	Conclusions *66*	
	Acknowledgments *67*	
	References *67*	

3 **Heterogeneous Enantioselective Catalysis Using Organic Polymeric Supports** *73*
Shinichi Itsuno and Naoki Haraguchi

3.1	Introduction *73*
3.2	Asymmetric Alkylation of Carbonyl Compounds *74*
3.3	Asymmetric Phenylation *79*
3.4	Asymmetric Addition of Phenylacetylene *80*
3.5	Asymmetric Addition to Imine Derivatives *81*
3.6	Asymmetric Silylcyanation of Aldehyde *81*
3.7	Asymmetric Synthesis of α-Amino Acid *82*
3.8	Asymmetric Aldol Reaction *84*
3.9	Enantioselective Carbonyl-Ene Reaction *87*
3.10	Asymmetric Michael Reaction *87*
3.11	Asymmetric Deprotonation *91*
3.12	Enantioselective Diels–Alder Cycloaddition *92*
3.13	Enantioselective 1,3-Dipolar Cycloaddition *94*
3.14	Asymmetric Sharpless Dihydroxylation *94*
3.15	Asymmetric Epoxidation *95*
3.16	Hydrolytic Kinetic Resolution of Terminal Epoxide *100*
3.17	Enantioselective Borane Reduction of Ketone *101*

3.18	Asymmetric Transfer Hydrogenation	*104*
3.19	Enantioselective Hydrogenation of Ketones	*107*
3.20	Asymmetric Hydrogenation of Enamine	*110*
3.21	Enantioselective Hydrogenation of C=C Double Bonds	*110*
3.22	Enantioselective Hydrogenation of C=N Double Bonds	*111*
3.23	Asymmetric Allylic Alkylation	*111*
3.24	Asymmetric Allylic Nitromethylation	*115*
3.25	Asymmetric Cyclopropanation	*116*
3.26	Enantioselective Olefin Metathesis	*118*
3.27	Asymmetric Ring-Closing Metathesis	*120*
3.28	Enantioselective Reissert-Type Reaction	*120*
3.29	Asymmetric Wacker-Type Cyclization	*121*
3.30	Enantioselective Hydrolysis	*121*
3.31	Asymmetric Hydroformylation	*123*
3.32	Summary and Outlooks	*123*
	References	*124*
4	**Enantioselective Catalysis Using Dendrimer Supports**	*131*
	Qing-Hua Fan, Guo-Jun Deng, Yu Feng, and Yan-Mei He	
4.1	General Introduction	*131*
4.2	Core-Functionalized Dendrimers in Asymmetric Catalysis	*134*
4.2.1	Asymmetric Hydrogenation	*135*
4.2.2	Asymmetric Transfer Hydrogenation	*145*
4.2.3	Asymmetric Borane Reduction of Ketones	*147*
4.2.4	Asymmetric Addition of Organometallic Compounds to Aldehydes	*149*
4.2.5	Asymmetric Michael Addition	*151*
4.2.6	Asymmetric Allylic Substitution	*152*
4.2.7	Asymmetric Aldol Reaction	*153*
4.2.8	Asymmetric Hetero-Diels–Alder Reaction	*155*
4.3	Peripherally Modified Dendrimers in Asymmetric Catalysis	*156*
4.3.1	Asymmetric Hydrogenation	*157*
4.3.2	Asymmetric Transfer Hydrogenation	*159*
4.3.3	Asymmetric Borane Reduction of Ketones	*160*
4.3.4	Asymmetric Ring Opening of Epoxides	*161*
4.3.5	Asymmetric Addition of Dialkylzincs to Aldehydes and Imine Derivatives	*162*
4.3.6	Asymmetric Allylic Amination and Alkylation	*163*
4.3.7	Asymmetric Michael Addition	*166*
4.3.8	Asymmetric Diels–Alder Reaction	*167*
4.3.9	Asymmetric Aldol Reaction	*168*
4.3.10	Asymmetric Hydrovinylation	*169*
4.4	Solid-Supported Chiral Dendrimer Catalysts for Asymmetric Catalysis	*170*
4.4.1	Solid-Supported Internally Functionalized Chiral Dendrimer Catalysts	*170*

4.4.2	Solid-Supported Peripherally Functionalized Chiral Dendrimer Catalysts *174*
4.5	Conclusion and Perspectives *175*
	References *177*

5 Enantioselective Fluorous Catalysis *181*
Gianluca Pozzi

5.1	Introduction *181*
5.2	Designing Fluorous Catalysts *182*
5.3	C—O Bond Formation *184*
5.3.1	Epoxidation *184*
5.3.2	Hydrolytic Kinetic Resolution *186*
5.3.3	Allylic Oxidation *187*
5.4	C—H Bond Formation *189*
5.4.1	Reduction of Ketones *189*
5.4.2	Reduction of C=N and C=C Bonds *191*
5.5	C—C Bond Formation *193*
5.5.1	Addition of Organometallic Reagents to Aldehydes *193*
5.5.2	Pd-Catalyzed Reactions *197*
5.5.3	Cyclopropanation of Styrene *200*
5.5.4	Metal-Free Catalytic Processes *203*
5.6	Conclusions *205*
	References *206*

6 Heterogeneous Asymmetric Catalysis in Aqueous Media *209*
Yasuhiro Uozumi

6.1	Introduction *209*
6.2	Chiral-Switching of Heterogeneous Aquacatalytic Process *210*
6.2.1	Combinatorial Approach *210*
6.2.2	Imidazoindole Phosphine *214*
6.2.2.1	Design and Preparation *214*
6.2.2.2	Allylic Alkylation *214*
6.2.2.3	Allylic Amination *217*
6.2.2.4	Allylic Etherification *218*
6.2.2.5	Synthetic Application *219*
6.3	Heterogeneous- and Aqueous-Switching of Asymmetric Catalysis *222*
6.3.1	BINAP Catalysts *222*
6.3.2	DPEN Catalysts *224*
6.3.3	Miscellaneous *226*
	References *229*

7 Enantioselective Catalysis in Ionic Liquids and Supercritical CO_2 *233*
Sang-gi Lee and Yong Jian Zhang

7.1	Introduction *233*

7.2	Enantioselective Catalysis in Ionic Liquids 234
7.2.1	Asymmetric Reductions in Ionic Liquids 235
7.2.1.1	Asymmetric Hydrogenations of the C=C Bond 235
7.2.1.2	Asymmetric Hydrogenations of the C=O Bond 239
7.2.1.3	Asymmetric Transfer Hydrogenations 244
7.2.2	Asymmetric Oxidations in Ionic Liquids 245
7.2.2.1	Asymmetric Dihydroxylation 245
7.2.2.2	Asymmetric Epoxidations 248
7.2.3	Asymmetric Carbon–Carbon and Carbon–Heteroatom Bond Formation in Ionic Liquids 250
7.2.3.1	Asymmetric Diels–Alder Reactions 250
7.2.3.2	Asymmetric Ring Opening of Epoxides 252
7.2.3.3	Hydrolytic Kinetic Resolution of Epoxides 254
7.2.3.4	Asymmetric Cyanosilylation of Aldehydes 255
7.2.3.5	Asymmetric Allylic Substitution 256
7.2.3.6	Asymmetric Allylic Addition 258
7.2.3.7	Asymmetric Cyclopropanation 259
7.2.3.8	Asymmetric Sulfimidation 262
7.2.3.9	Asymmetric Diethylzinc Addition 262
7.2.3.10	Asymmetric Fluorination 262
7.2.4	Enantioselective Organocatalysis 264
7.2.4.1	Asymmetric Aldol Reactions 265
7.2.4.2	Asymmetric Michael Addition 269
7.2.4.3	Asymmetric Mannich, α-Aminoxylation, and Diels–Alder Reaction 270
7.3	Enantioselective Catalysis in Supercritical Carbon Dioxide (scCO$_2$) 274
7.3.1	Asymmetric Hydrogenation 274
7.3.2	Asymmetric Hydroformylation 279
7.3.3	Asymmetric Carbon–Carbon Bond Formation 282
7.4	Enantioselective Catalysis in the Combined Use of Ionic Liquids and Supercritical CO$_2$ 283
7.5	Summary and Outlook 285
	References 286
8	**Heterogenized Organocatalysts for Asymmetric Transformations** 293
	Maurizio Benaglia
8.1	Introduction 293
8.2	General Considerations on the Immobilization Process 295
8.3	Phase-Transfer Catalysts 299
8.4	Nonionic Cinchona-Derived Catalysts 302
8.5	Lewis Base Catalysts 305
8.6	Catalysts Derived from Amino Acids 307
8.6.1	Proline Derivatives 307
8.6.2	Amino Acid-Derived Imidazolinones 312

8.6.3	Other Amino Acids 314
8.7	Miscellaneous Catalysts 317
8.8	Outlook and Perspectives 319
	References 320

9 Homochiral Metal–Organic Coordination Polymers for Heterogeneous Enantioselective Catalysis: Self-Supporting Strategy 323
Kuiling Ding and Zheng Wang

9.1	Introduction 323
9.2	A Historical Account of Catalytic Applications of MOCPs 326
9.3	General Considerations on the Design and Construction of Homochiral MOCP Catalysts 330
9.4	Type I Homochiral MOCP Catalysts in Heterogeneous Asymmetric Reactions 333
9.4.1	Enantioselective C—C Bond-Forming Reactions 333
9.4.1.1	Carbonyl-Ene Reaction 333
9.4.1.2	Michael Addition 335
9.4.1.3	Diethylzinc Addition to Aldehydes 336
9.4.2	Enantioselective Oxidation Reactions 337
9.4.2.1	Epoxidation of α,β-Unsaturated Ketones 337
9.4.2.2	Sulfoxidation of Aryl Alkyl Sulfides 339
9.4.3	Asymmetric Hydrogenations 340
9.4.3.1	Hydrogenation of Dehydro-α-Amino Acids and Enamides 340
9.4.3.2	Hydrogenation of Ketones 342
9.5	Type II Homochiral MOCP Catalysts in Heterogeneous Asymmetric Reactions 343
9.5.1	Enantioselective Hydrogenations 343
9.5.2	Asymmetric C—C Bond-Forming Reactions 346
9.5.3	Epoxidation 349
9.5.4	Miscellaneous 350
9.6	Concluding Remarks and Outlook 351
	References 352

10 Heterogeneous Enantioselective Hydrogenation on Metal Surface Modified by Chiral Molecules 357
Takashi Sugimura

10.1	Introduction 357
10.2	History of the Chiral Modification of Metal Catalysts 358
10.3	Cinchona Alkaloid-Modified Platinum Catalysis [1, 4] 359
10.4	Tartaric Acid-Modified Nickel Catalysis [49–52] 363
10.5	Cinchona Alkaloid-Modified Palladium Catalysis [83] 368
10.5.1	Aromatic α,β-Unsubstituted Carboxylic Acids 369
10.5.2	Aliphatic α,β-Unsubstituted Carboxylic Acids 374
10.5.3	2-Pyrone Derivatives 376
10.6	Conclusions 377

Acknowledgments *377*
References *377*

11 Asymmetric Phase-Transfer Catalysis *383*
Xisheng Wang, Quan Lan, and Keiji Maruoka
11.1 Introduction *383*
11.2 Alkylation *384*
11.2.1 Pioneering Study *384*
11.2.2 Asymmetric Synthesis of α-Amino Acids and Their Derivatives *384*
11.2.2.1 Monoalkylation of Schiff Bases Derived from Glycine *384*
11.2.2.2 Dialkylation of Schiff Bases Derived from α-Alkyl-α-Amino Acids *391*
11.2.2.3 Alkylation of Schiff Base-Activated Peptides *394*
11.2.3 Other Alkylations and Aromatic or Vinylic Substitutions *396*
11.3 Michael Addition *398*
11.4 Aldol and Related Reactions *399*
11.5 Neber Rearrangement *403*
11.6 Epoxidation *404*
11.7 Strecker Reaction *406*
11.8 Conclusions *408*
Acknowledgments *408*
References *408*

12 The Industrial Application of Heterogeneous Enantioselective Catalysts *413*
Hans-Ulrich Blaser and Benoît Pugin
12.1 Introduction *413*
12.2 Industrial Requirements for Applying Catalysts *414*
12.2.1 Characteristics of the Manufacture of Enantiomerically Pure Products *414*
12.2.2 Process Development: Critical Factors for the Application of (Heterogeneous) Enantioselective Catalysts *414*
12.2.3 Requirements for Practically Useful Heterogeneous Catalysts *415*
12.2.3.1 Preparation Methods *415*
12.2.3.2 Catalysts *416*
12.2.3.3 Catalytic Properties and Handling *417*
12.2.4 Practically Useful Types of Heterogeneous Enantioselective Catalyst *417*
12.3 Chirally Modified Heterogeneous Hydrogenation Catalysts *418*
12.3.1 Nickel Catalysts Modified with Tartaric Acid *418*
12.3.1.1 Background *418*
12.3.1.2 Synthetic and Industrial Applications *419*
12.3.2 Catalysts Modified with Cinchona Alkaloids *420*
12.3.2.1 Background *420*
12.3.2.2 Industrial Applications *422*
12.4 Immobilized Chiral Metal Complexes *428*

12.4.1 Background *428*
12.4.2 Complexes Adsorbed on Solid Supports *430*
12.4.3 Toolbox for Covalent Immobilization *431*
12.5 Conclusions and Outlook *435*
References *435*

Index *439*

Preface

Stereoselective molecular transformation is a central theme in modern organic synthetic chemistry. Among the many strategies applied to prepare desired organic molecules in their optically active forms, catalytic asymmetric synthesis has been widely recognized as a powerful reaction class. Compared to the impressive developments of homogeneous asymmetric catalysis – the most splendid of which were awarded the Nobel Prize of 2001 – enantioselective organic synthesis with *heterogeneous* chiral catalysts has received only scant attention, in spite of the historical appearance of a silk–palladium composite as the first heterogeneous enantioselective catalyst in 1956.[1] During the past decade, however, in response to increasing environmental concerns about the harmful and resource-consuming waste of the heavy and/or rare metals that are frequently used as the catalytic centers of homogeneous catalysts, the importance of heterogeneous systems has again been realized and this area is presently undergoing very rapid growth.[2] In addition, the development of heterogeneous chiral catalysts has been attracting significant interest for their practical advantages: the safe and simple manipulation of work-up; reduced contamination by catalyst residues in the products; and the recovery and reuse of the costly chiral and/or metal resources. There is good reason to believe that the development of asymmetric heterogeneous catalysis – including the heterogeneous-switching of given homogeneous asymmetric processes and the chiral-switching of heterogeneous nonasymmetric processes – offers a practical, 'green', clean and safe alternative to more conventional methods of accomplishing many asymmetric processes.

This book, which comprises 12 review-type chapters, is intended to provide an overview of the main research areas of asymmetric heterogeneous catalysis, although the arrangement of the chapters is somewhat arbitrary. Chapter 1 (Z. Wang, K. Ding and Y. Uozumi) provides an introductory review to outline this volume, which should give guidance to the broad readership. Chapters 2 (S.S. Thakur, J.E. Lee, S.H. Lee, J.M. Kim and C.E. Song) and 3 (S. Itsuno and N. Haraguchi) mainly describe the development of chiral catalysts bound to

---

[1] For a review, see: Izumi, Y. (1971) *Angewandte Chemie – International Edition* 10, 871–948.

[2] For a review, see: De Vos, D.E., Vankelecom, I.F.J. and Jacobs, P. A. (eds) (2000) *Chiral Catalyst Immobilization and Recycling*, Wiley-VCH, Weinheim.

inorganic and organic (polymeric) supports, respectively, via the heterogenization of chiral organometallic and organic catalysts originally designed for homogeneous counterparts. In Chapter 4, Q.-H. Fan, G.-J. Deng, Y. Feng and Y.M. He present details of dendritic chiral catalysts, where the dendrimer moiety—a new class of highly branched polymer—often provides unique physical as well as chemical properties. Asymmetric heterogeneous catalysis in exotic liquid media is addressed in Chapters 5, 6 and 7. Thus, fluorous liquid–liquid biphasic systems with fluorophilic-modified chiral catalysts are reviewed in Chapter 5 (G. Pozzi), while water-based reactions with hydrophilic (or amphiphilic) polymeric catalysts are described in Chapter 6 (Y. Uozumi). The recent growth of chemistry with ionic liquids and supercritical carbon dioxide is also significant, and nowadays this is applied to heterogeneous asymmetric catalysis, which is detailed in Chapter 7 (S.-G. Lee and Y.J. Zhang). Although organocatalysis is a well-established class of organic transformations, chiral-switching is an up-to-date topic in the area of asymmetric catalysis. M. Benaglia introduces the heterogeneous-switching of asymmetric organocatalysis, mainly with cinchona- and amino acid-derivatives, in Chapter 8. The metal crosslinked assembly of chiral organic ligands forming chiral coordination polymers realized self-supporting systems of chiral complex catalysts where the catalytic activity and heterogeneous property are obtained in a single step (Chapter 9, K. Ding and Z. Wang). The chiral-switching of metal catalysts is a classic, yet immature, approach to asymmetric heterogeneous catalysis. Clearly, while pioneering strides have been made, additional studies on the chiral modification of metal surfaces are warranted, and this topic is reviewed by T. Sugimura in Chapter 10. The chiral-switching of phase-transfer catalysis (PTC) has been another eagerly awaited subject, and a breakthrough in asymmetric PTC was recently brought about by K. Maruoka, who contributes Chapter 11 together with his colleagues, X. Wang and Q. Lan. The final chapter of this volume is provided by H.-U. Blaser and B. Pugin, who introduce the industrial application of heterogeneous asymmetric catalysis.

We gratefully acknowledge the work of all authors in presenting up-to-date and well-referenced contributions; indeed, without their efforts this volume would not have been possible. Furthermore, it was a pleasure to collaborate with the Wiley-VCH 'crew' in Weinheim, who not only did an excellent job in producing the book but also helped us in a competent manner in all phases of its preparation. We are also grateful to Dr. Zheng Wang of Shanghai Institute of Organic Chemistry, who put a lot of effort into editing this volume. The collaborative studies of K.D. and Y.U. were partially supported by the Asian Core Program, sponsored by the Japan Society for the Promotion of Science.

Kuiling Ding *Shanghai and Okazaki*
Yasuhiro Uozumi *July 2008*

List of Contributors

Maurizio Benaglia
Università degli Studi di Milano
Dipartimento di Chimica
  Organica e Industriale
via Golgi 19
20133 Milan
Italy

Hans-Ulrich Blaser
Solvias AG
P.O. Box
4002 Basel
Switzerland

Guo-Jun Deng
Chinese Academy of Sciences
Institute of Chemistry
P.R. Beijing 100190
China

Kuiling Ding
Chinese Academy of Sciences
Shanghai Institute of Organic
  Chemistry
354 Fenglin Road
Shanghai 200032
P. R. China

Qing-Hua Fan
Chinese Academy of Sciences
Institute of Chemistry
Beijing 100190
P. R. China

Yu Feng
Chinese Academy of Sciences
Institute of Chemistry
Beijing 100190
China

Naoki Haraguchi
Toyohashi University of Technology
Department of Materials Science
Tempaku-cho, Toyohashi, 441-8580
Japan

Yan-Mei He
Chinese Academy of Sciences
Institute of Chemistry
Beijing 100190
P. R. China

Shinichi Itsuno
Toyohashi University of Technology
Department of Materials Science
Tempaku-cho, Toyohashi, 441-8580
Japan

Ji Man Kim
Sungkyunkwan University
Department of Chemistry
Cheoncheon-dong 300, Jangan-gu
440-746 Suwon
Korea

Handbook of Asymmetric Heterogeneous Catalysis. Edited by K. Ding and Y. Uozumi
Copyright © 2008 WILEY-VCH Verlag GmbH & Co. KGaA, Weinheim
ISBN: 978-3-527-31913-8

Quan Lan
Kyoto University
Department of Chemistry
Graduate School of Science
Sakyo
Kyoto 606-8502
Japan

Jae Eun Lee
Sungkyunkwan University
Department of Chemistry
Cheoncheon-dong 300,
  Jangan-gu
440-746 Suwon
Korea

Sang-gi Lee
Ewha Womans University
Department of Chemistry and
  Nano Science
Seoul 120-750
Korea

Seo Hwan Lee
Sungkyunkwan University
Department of Chemistry
Cheoncheon-dong 300,
  Jangan-gu
440-746 Suwon
Korea

Keiji Maruoka
Kyoto University
Department of Chemistry
Graduate School of Science
Sakyo
Kyoto 606-8502
Japan

Gianluca Pozzi
CNR-Istituto di Scienze e
  Tecnologie Molecolari
via Golgi 19
20133 Milan
Italy

Benoît Pugin
Solvias AG
P.O. Box
4002 Basel
Switzerland

Choong Eui Song
Sungkyunkwan University
Department of Chemistry
Cheoncheon-dong 300, Jangan-gu
440-746 Suwon
Korea

Takashi Sugimura
University of Hyogo
Graduate School of Material Science
Himeji Institute of Technology
3-2-1 Kohto
Kamigori
Ako-gun
Hyogo 678-1297
Japan

Santosh Singh Thakur
Sungkyunkwan University
Department of Chemistry
Cheoncheon-dong 300, Jangan-gu
440-746 Suwon
Korea

Yasuhiro Uozumi
Institute for Molecular Science (IMS)
Myodaiji
Okazaki 444-8787
Japan

Xisheng Wang
The Scripps Research Institute
Department of Chemistry
10550 North Torrey Pines Road
La Jolla, CA, 92037
USA

Zheng Wang
Chinese Academy of Sciences
Shanghai Institute of Organic
  Chemistry
354 Fenglin Road
Shanghai 200032
P. R. China

Yong Jian Zhang
Shanghai Jiao Tong University
School of Chemistry and Chemical
  Technology
800 Dongchuan Road
Shanghai 200240
P. R. China

1
An Overview of Heterogeneous Asymmetric Catalysis
Zheng Wang, Kuiling Ding, and Yasuhiro Uozumi

1.1
Introduction

Driven by the ever-increasing demand for nonracemic chiral chemicals, the development of efficient methods to provide enantiomerically enriched products is of great current interest to both academia and industry [1–3]. Among the various approaches employed for this purpose, asymmetric catalysis constitutes one of the most general and appealing strategies in terms of chiral economy and environmental considerations [4–9]. Over the past few decades, intense research in this field has greatly expanded the scope of catalytic reactions that can be performed with high enantioselectivity and efficiency. Consequently, thousands of chiral ligands and their transition metal complexes have been developed for the homogeneous asymmetric catalysis of various organic transformations. Despite this remarkable success, however, only a few examples of asymmetric catalysis have been developed into industrial processes, and today most chiral chemicals are still produced from natural chiral building blocks or through the resolution of racemic mixtures. The main concern for this situation is the need for reusable chiral catalysts for industrial implementation. Due to the high cost of both the metal and the chiral ligands, systems that allow the straightforward separation of expensive chiral catalysts from reaction mixtures and efficient recycling are highly desirable. Whilst this is particularly important for large-scale productions, unfortunately it is usually very difficult to achieve for homogeneous catalytic processes. Another major drawback often associated with homogeneous catalytic processes is that of product contamination by metal leaching; this is particularly unacceptable for the production of fine chemicals and pharmaceuticals. Heterogeneous asymmetric catalysis – including the use of immobilized homogeneous asymmetric catalysts and chirally modified heterogeneous metal catalysts for enantioselective reactions – provides a good way to resolve such problems and has recently attracted a great deal of interest [10, 11]. In this chapter we will briefly survey the field of heterogeneous asymmetric catalysis by summarizing the main features of some typical techniques at an introductory level. When relevant, we will present our

Handbook of Asymmetric Heterogeneous Catalysis. Edited by K. Ding and Y. Uozumi
Copyright © 2008 WILEY-VCH Verlag GmbH & Co. KGaA, Weinheim
ISBN: 978-3-527-31913-8

personal comments on weighing up their strengths and limitations, though without delving into too much detail. For in-depth discussions and a comprehensive elaboration of each technique, the reader is referred to excellent recent reviews [12–16] and the ensuing chapters in this handbook, all of which have been written by scientists with expertise in these areas of research. The application of immobilized biocatalysts (including enzymes) in enantioselective organic synthesis, although representing an important field of catalytic research, is beyond the scope of this book.

The term catalyst immobilization can be defined as 'the transformation of a homogeneous catalyst into a heterogeneous one, which is able to be separated from the reaction mixture and preferably be reused for multiple times'. The main goal for the development of an immobilized chiral catalyst is to combine the positive aspects of a homogeneous catalyst (e.g. high activity, high enantioselectivity, good reproducibility) with those of a heterogeneous catalyst (e.g. ease of separation, stability, reusability). Over the past few decades, a number of strategies have been developed for this purpose. Depending on whether the modifications are made on the catalyst structure or on the reaction medium, the immobilization techniques can be categorized into two general classes, namely heterogenized enantioselective catalysts and multiphase (or monophase) catalysis in nonconventional media (Figure 1.1). The immobilized chiral catalysts can be further subdivided into several types:

- Insoluble chiral catalysts bearing stationary supports such as inorganic materials or organic crosslinked polymers, or homochiral organic–inorganic coordination polymeric catalysts without using any external support.
- Soluble chiral catalysts bearing linear polymer supports or dendritic ligands.
- Chiral catalysts with some form of nonconventional reaction medium as the 'mobile carrier', such as aqueous phase, fluorous phase, ionic liquid or supercritical carbon dioxide ($scCO_2$).

In the latter case, these liquids can form biphasic systems with the immiscible organic product liquid, thus giving rise to the possibility of an easy isolation and recovery of the chiral catalysts by phase separation. An important option to the

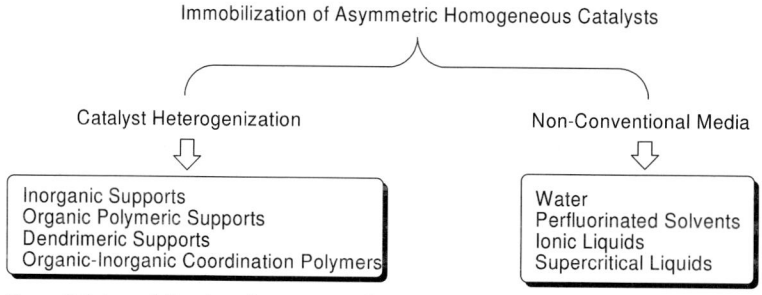

Immobilization of Asymmetric Homogeneous Catalysts

Catalyst Heterogenization	Non-Conventional Media
Inorganic Supports Organic Polymeric Supports Dendrimeric Supports Organic-Inorganic Coordination Polymers	Water Perfluorinated Solvents Ionic Liquids Supercritical Liquids

Figure 1.1 Immobilization of asymmetric homogeneous catalysts.

biphasic catalysis is asymmetric phase-transfer catalysis (PTC), where a nonracemic chiral additive is applied to increase the mobility of a given catalyst or reactant into a favored phase and furthermore control the stereochemical outcome. Although, in general, catalyst recycling is somewhat difficult in PTC systems, it represents a distinct type of heterogeneous asymmetric catalysis.

In order to apply an immobilized chiral catalyst to a chemical process, it is often necessary to make a critical evaluation in terms of its activity, productivity, enantioselectivity, stability, ease of recovery and reusability, and so on. An ideal immobilized catalyst should not only exhibit activity and selectivity comparable or superior to its homogeneous counterpart, but also be easily recoverable from the reaction stream without metal leaching, and reusable for many runs without any loss of catalytic performance. Unfortunately, this is seldom found in real-world cases, and numerous problems can occur during the immobilization of a homogeneous catalyst. For example, in a supported chiral catalyst, one often-observed negative effect is the lower catalytic activity (even complete deactivation) compared to a homogeneous catalyst, as a result of the poor accessibility of the active sites in the solid matrix. On the other hand, the geometry of an optimized homogeneous catalyst can be unintentionally disturbed by interactions with the support, and this often leads to a negative change in enantioselectivity. For these reasons, it is a common practice to use a linker of sufficient length to connect the complex and the support, so that the complex can move far away from the solid surface and into the liquid phase.

One general requirement for the reusability of any recoverable catalyst is that both the support material and the catalytic sites must be sufficiently stable to maintain the catalytic activity during the recycling process. Any type of poor stability in the catalytic moiety or the linker part and/or incompatibility of the support with the solvent may result in leaching of the metal and/or ligand. For this reason, the support material and the linkage for immobilization should have good mechanical, thermal and chemical stabilities in order to withstand the reaction conditions used in the catalytic process. In addition, the issue of the robustness of the complex itself can be nontrivial. Immobilization is sometimes found to decrease complex degradation by virtue of steric constraints imposed by the supporting matrix, and thus may improve the stability of an immobilized catalyst relative to its homogeneous analogues. Nevertheless, this beneficial effect of immobilization on catalyst stability cannot be taken for granted, as other factors – such as the presence of strong acids or oxidizing or reducing reagents or other harsh conditions – may lead to demetallation or ligand degradation of the complex. For these reasons, the stability issue of an immobilized catalyst must be addressed on a case-by-case basis, by including the data on catalyst leaching into the product phase for assessing the potential degree of catalyst decomposition. Despite these difficulties, major efforts continue to be made to develop more efficient and practical immobilization methods for homogeneous chiral catalysts. In this regard, numerous immobilized asymmetric catalytic systems have been examined over a broad range of reactions, and a number of innovative techniques for chiral catalyst immobilization have emerged during the past two decades. In favorable cases, higher enantioselectivities and/or improved efficiencies were observed with a heterogeneous than with its homogeneous analogue [17–19].

1.2
Common Techniques for Immobilization of Homogeneous Asymmetric Catalysts

1.2.1
Chiral Catalyst Immobilization on Inorganic Materials

By far the most commonly used immobilization method is to support the active chiral catalyst onto or into an insoluble solid, which can be either an inorganic solid or a organic polymer. Several distinct types of strategy, featuring covalent bonds or noncovalent interactions (e.g. physisorption, electrostatic interactions, H-bonding), have been employed for linking the complex to the solid support, either onto the external surface or into the interior pores (Figure 1.2). Each of these immobilization strategies has advantages and limitations with respect to the others. Covalent bonding linkage (Figure 1.2a) is the most frequently used strategy by far, and is generally assumed to furnish the strongest binding between the complex and a support. However, it is synthetically demanding since generally some special functionalization of the ligand is required, either for grafting to a preformed support or for forming an organic polymer by copolymerization with a suitable monomer. In contrast, a major advantage of noncovalent immobilization in general is the ease of catalyst preparation, often without the need for prior functionalization of the ligand. Adsorption (Figure 1.2b) represents a very facile immobilization method, as a simple impregnation procedure can be sufficient to furnish the heterogenized catalyst. Nevertheless, catalysts immobilized via adsorption tend to be nonstable when only weak van der Waals interactions are present, and this often results in extensive catalyst leaching due to the competing interactions with solvents and/or substrates. Immobilization by electrostatic interaction (Figure 1.2c) is another common and conceptually simple technique, which is applicable to heterogenization of ionic catalytic species. Here, the solid support can be either anionic or cationic, and the catalyst is adsorbed by ion-pairing.

a) Covalent linkage   b) Adsorption   c) Electrostatic interactions   d) Entrapment

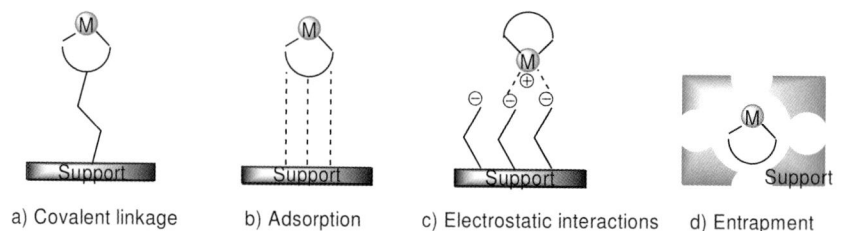

Figure 1.2 Schematic representation of the strategies for immobilizing homogeneous chiral catalysts with solid supports.

Various supports with ion-exchange capabilities have been used for this purpose, including organic or inorganic ion-exchange resins, inorganic clays and zeolites. Although this approach can provide relatively stable immobilized catalysts, it is still limited to the catalysts which can lend themselves to immobilization through electrostatic interaction. Furthermore, competition with ionic species (either present in or produced during the reaction) in solution may result in catalyst instability and leaching. Finally, the catalytic complex can also be entrapped within the pores of some solid matrix (Figure 1.2d). In this entrapment methodology, the size of the metal complex relative to that of the window or tunnel of a porous solid is the factor of paramount importance, leading to a mechanically immobilized catalyst. This can be accomplished by preparing the complex in well-defined cages of a porous support, or alternatively by building up a polymer network around a preformed catalyst. Although conceptually very elegant, the entrapment strategy is relatively complex to implement compared with the other methods, and the size of the substrate molecules may cause diffusion problems in the catalysis. In summary, it is difficult to predict whether a covalent or a noncovalent immobilization would be preferential for a given catalyst. Although covalent bonding remains the most popular approach to chiral catalyst immobilization (mainly due to stability advantages), examples illustrated in the following chapters in this Handbook have shown that noncovalent immobilizations are gaining increasing recognition as a feasible way to achieve good stability and reusability as well as high selectivity and activity of an immobilized chiral catalyst.

One major drawback of the insoluble solid-supported chiral catalysts is that, in many cases, lower reactivities and/or poorer enantioselectivities were obtained as compared with the corresponding homogeneous catalysts. However, recent results have shown that the reverse can also be true by correctly choosing the support/complex combination, even though this was largely achieved by trial-and-error rather than by rational design. A salient feature of the insoluble solid-supported catalysts is their easy recovery. In a batch operation, the solid-supported catalyst can be isolated from the reaction mixture by simple filtration, and in some cases can be reused for subsequent reaction cycles until deactivated. Alternatively, the heterogeneous catalysts can be employed in a continuous-flow reactor, with the advantages of easy automation and little or no reaction work-up [20–23].

Inorganic solids such as silicas, mesoporous solids (e.g. MCM-41, SBA-15), zeolites and clays have been widely used as supports for the immobilization of various homogeneous chiral catalysts [14, 24–30]. Depending on the properties of the complex and the structure of the support, the immobilization strategies can encompass the whole spectra of aforementioned interactions, from physical entrapment to covalent bonding. For example, zeolites are crystalline microporous aluminosilicates with interior cavities accessible to small reactants from the solution. A chiral metal complex with a suitable size can be assembled inside the zeolite cavity and entrapped snugly there for catalysis, as escape by diffusion through the small windows is very difficult. One advantage of the zeolite-entrapped catalyst is that the zeolite can impose shape selectivity to the catalytic system; that is, only those substrates with appropriate size and shape can reach the catalyst and react. Zeolite entrapment may also lead to a better catalyst stability as a result of protection of the inert framework. However, zeolite entrapment of a homogeneous chiral catalyst often leads to a decrease in enantioselectivity, presumably by a negative steric influence of the cage walls. Alternative inorganic solids were also examined as supports, for example, to immobilize the catalytic complex on the external surface of an nonporous solid such as amorphous silica, or in the interior of porous solids with void dimensions larger than zeolites. The immobilization of an electrically charged homogeneous complex by electrostatic interaction with the surface of the inorganic support is an attractive method owing to the experimental simplicity of the procedure. In this regard, lamellar solids bearing charged layers (clays such as montmorillonite K10, laponite or bentonite) have been used as supports to immobilize a variety of charged chiral metal complexes via simple anion-exchange procedures.

Usually, these immobilization approaches do not necessarily require special functionalization of the ligand part. In contrast, immobilization by covalent binding of a catalyst to an inorganic support is generally accomplished by reacting complementary functional groups – one located on the solid and the other at the complex moiety – to create a new covalent bond connecting the solid and the complex. Although a large variety of functional groups have been shown to be applicable for this purpose, and despite strong immobilization (not necessarily) being expected, the approach suffered from the major drawback of a need for extensive organic synthesis. Therefore, immobilization by covalent binding would be preferential only if the stability and reusability of the resulting catalyst were to be significantly improved relative to other methods.

The use of inorganic solids can demonstrate certain advantages over other types of support. In general, the rigid framework can prevent the aggregation of active catalysts which sometimes leads to the formation of inactive multinuclear species. The chemical and thermal stabilities of the inorganic supports are also superior, rendering them compatible with a broad range of reagents and relatively harsh reaction conditions. Compared with organic polymeric supports, inorganic solids are generally superior in their mechanical properties, which makes them less prone to attrition caused by stirring and solvent attack. One negative aspect of an inorganic solid-immobilized chiral catalyst is the extreme difficulty in the characterization of the catalytic species, apart from common problems suffered by heterogenized catalysts.

The use of an inorganic support for the immobilization of homogeneous chiral catalysts has been a steadily expanding area of research during recent years, as evidenced by the numerous examples described in Chapter 2. In some cases, the heterogenization of a chiral catalyst onto an inorganic material has not only provided a facile vehicle for catalyst recycling, but also has significantly improved the catalytic performance in terms of activity, stability and/or enantioselectivity by virtue of the site-isolation and confinement effect. Taken together, it is expected that this immobilization technique will continue to play an important role in the development of highly efficient heterogeneous chiral catalysts in the future.

1.2.2
Chiral Catalyst Immobilization Using Organic Polymers

The use of organic polymer supports (either soluble or insoluble) for (chiral) catalyst immobilization is an area of considerable research interest, and is also the subject of many excellent reviews (see also Chapter 3) [31–43]. To date, a wide variety of polymer-supported chiral complexes have been prepared and tested over a broad spectrum of synthetic reactions. Some systems have demonstrated catalytic performances (activity and selectivity) rivaling their homogeneous counterparts in certain model reactions, with the additional advantages of easy recovery and reusability. By virtue of its good chemical inertness, ready availability and ease of functionalization, polystyrene crosslinked with various amounts of *p*-divinylbenzene (DVB) is the most popular insoluble organic polymer support used in chiral catalyst immobilization. For the covalent immobilization of a homogeneous catalyst to an insoluble polymer support, two different approaches have been employed, depending on whether the polymer is formed in or before the process. One approach is to use a solid-phase synthesis, starting with a preformed functionalized polymer and anchoring the suitably derivatized chiral ligand or complex by stepwise assembly of the components. A wide array of commercially available functionalized polymers with a large variety of functional groups including chloromethyl, hydroxyl, amino, thiol or pyridine rings can be used for this purpose, and a large number of chiral ligands have been immobilized to the polystyrene support using this method. Alternatively, the insoluble polymer bearing the active complex can also be prepared by copolymerization of styryl derivatives of the chiral ligand or complex with styrene and divinylbenzene.

Compared to an inorganic support, the usually low surface area of an insoluble organic polymer may limit the interfacial contact between the supported complex and the substrate. Some polymer particles can be mechanically fragile or brittle, which constitutes some limitations encountered in the recycling of polymer-supported catalysts. It is generally accepted that the conformational influence, steric and polarity factors of the polymer backbones can provide a unique microenvironment for the reactants. The polymer supports have sometimes been found to exhibit beneficial impact on the catalytic reactions, leading to an enhanced selectivity and/or improved stability of the catalyst. However, in most cases, the exact role of the polymeric backbone (e.g. the linker or the degree of crosslinking) in the catalytic behavior of an immobilized metal complex is not clear and remained to be clarified.

Despite the well-known advantages of insoluble supports, there are several drawbacks in using these solids as supports for chiral catalysts due to the heterogeneous nature of the reaction conditions. The catalyst resides in the solid phase while reactants are in solution, which can often result in a decreased reaction rate owing to diffusion problems. Furthermore, the matrix effect of the solid support, though sometimes favorable, is difficult to predict and can often lead to lower enantioselectivities for the immobilized chiral catalysts than those for their homogeneous counterparts. Last, but not least, most of the insoluble catalysts developed so far still lack a full characterization, as the inherent heterogeneity of solid-phase systems precludes the use of many traditional analytical techniques. For these reasons, in recent years the use of soluble polymer-supported reagents and catalysts has gained significant attention as an alternative method to traditional insoluble-bound catalysts [44–52]. Soluble polymer-supported catalysts often have advantages of more facile characterization, allowing for (at least in principle) catalyst characterization on soluble polymer supports by routine analytical methods. Furthermore, the reactivity and selectivity of the soluble polymer-supported catalysts are somewhat (though not necessarily) predictable by virtue of the homogeneous nature of the solution-phase chemistry, and can be as high as those of their low-molecular-weight counterparts. Poly(ethylene glycol) (PEG) and non-crosslinked polystyrene are the most often-used polymer carriers in the preparation of soluble polymer-supported chiral catalysts. Although, anchoring of the chiral ligand/complex may be made directly to the polymer support with suitable functional groups, a linking group is often employed to impart anchor stability throughout synthesis and/or to improve accessibility to reagents. The immobilization is usually accomplished via coupling chemistry of the functionalized chiral ligands/complexes and the support. An interesting variation is that of chiral rigid-rod polymers containing binaphthyl groups as the catalyst supports, where the chiral groups are an integral part of the polymer's main chain; these have also been used as soluble and recoverable polymer-bound chiral ligands/catalysts [53, 54].

One potential problem with the use of soluble polymers as recoverable catalyst supports is how to isolate the polymer from the reaction mixture. This is usually accomplished by using the various macromolecular properties of the support. Most frequently, the homogeneous solution after the reaction is simply diluted with a poor solvent (i.e. it has a poor solvating power for the polymer support), and this can induce precipitation of the polymer catalyst. The resulting heterogeneous mixture is filtered to isolate the polymer-supported catalyst, which can be recharged with the reactants and solvent for subsequent cycles of reactions. A variety of separation techniques have been employed to recover a soluble polymer-bound catalyst or ligand, and the reader is referred to an elegant review on this topic [48].

1.2.3
Dendrimer-Supported Chiral Catalysts

Dendrimers are a class of macromolecules with highly branched and well-defined structures, and have recently attracted much attention as soluble supports for (chiral) catalyst immobilization [55–65]. As stated above, the catalysts anchored onto or into insoluble supports often possess an uneven catalytic site distribution and partly unknown structures, and generally suffer from diminished activity due to the mass transfer limitations. Dendrimers, on the other hand, allow for the precise construction of catalyst structures with uniformly distributed catalytic sites – a feature which is very valuable when performing an analysis of catalytic events and mechanistic studies. It is also possible to regulate the numbers and locations of the catalytic centers at the core or at the periphery, and to fine-tune the size, shape and solubility of dendrimers by ligand design, both of which are crucial for the catalytic performance of the system. Moreover, as soluble support dendrimers can, in principle, lead to catalytic systems with activities and selectivities similar to their monomeric analogues. Since the seminal publication by Brunner in 1994–1995 on the use of chiral dendritic catalysts in asymmetric catalysis [66, 67], research in this field has achieved remarkable progress. Indeed, excellent reviews are currently available with well-documented examples demonstrating the precise definition of catalytic sites and the possibility of recovering the dendritic catalysts [13, 68]. A variety of enantioselective dendritic catalysts are now known which, according to dendritic topology and/or the location/distribution of the catalytic sites, can be categorized into several general types, as summarized in Figure 1.3. Most of these dendritic catalysts are related to homogeneous catalysis (a, b, c and d), but some heterogeneously supported dendritic catalysts have also been developed (e). Typically, the dendrimer can adopt the shape of either a sphere (a and b) or a wedge (c and d), and the chiral catalytic site(s) can be located either at the periphery (a, c and e), at the core (b) or at the focal point (d) of the

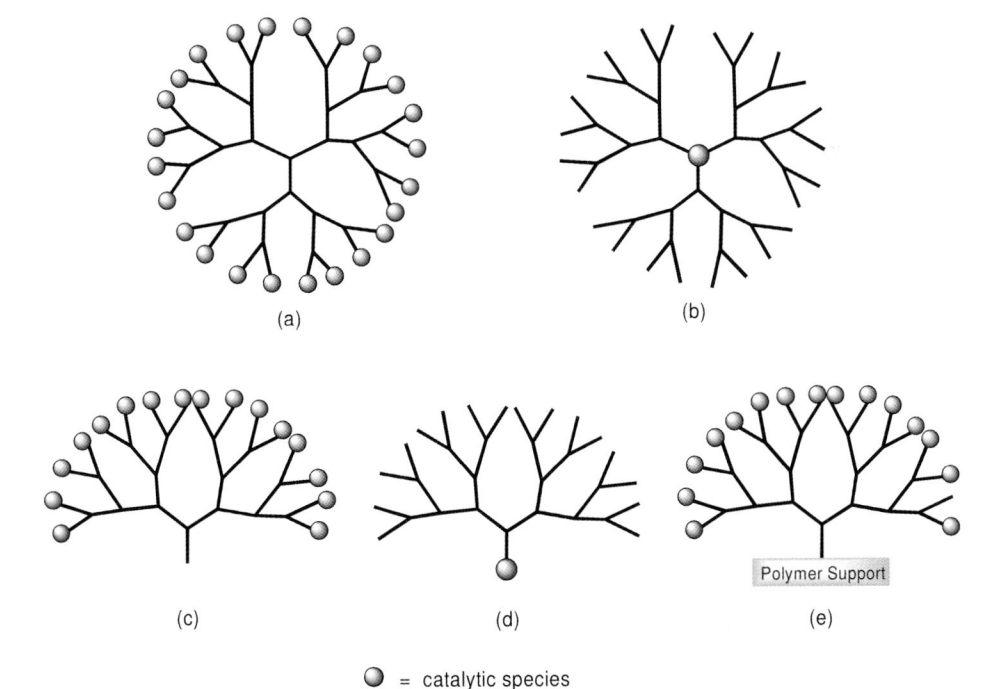

Figure 1.3 Various types of dendritic catalyst with catalytic species located at the periphery (a), core (b), periphery of a wedge (c), focal point of a wedge (d), and polymer-supported dendrimer with catalytic species at the periphery (e).

dendrimer. These diversified dendritic structures are very important for defining the catalytic performance, including the activity, enantioselectivity, stability and reusability of the catalyst system (dendritic effects). For example, the catalytic sites located at the surface of a periphery-functionalized dendrimer (a, c and e) would be easily accessible to the substrate, allowing reaction rates comparable to those of the homogeneous system. In addition, the multiple reaction sites in a periphery-functionalized dendrimer would result in high local catalyst concentrations, which can exhibit positive dendritic effects in cases where cooperative interactions between active sites are needed in the catalysis. On the other hand, for reactions that are deactivated by excess ligand, or in cases where a bimetallic deactivation mechanism is operative, core- or focal point-functionalized dendritic catalysts (b, d) would be especially beneficial owing to the site-isolation effect by the dendritic structure. The dendritic structure can also demonstrate a significant effect (negative or positive) on the enantioselectivity of an asymmetric catalysis. An elegant review of the many aspects involved in enantioselective catalysis using dendrimer supports is provided in Chapter 4 of this Handbook.

The dendritic catalysts can be recovered and reused using similar techniques as applied for other soluble polymeric catalysts discussed above, including solvent precipitation, column chromatography, two-phase catalysis or anchoring on insoluble supports. Moreover, the globular shapes and large size of the higher generations of dendrimers are well suited to catalyst–product separation by nanofiltration using a membrane, this being performed either batchwise or in a continuous-flow membrane reactor (CFMR).

It is clear that the attachment of chiral catalysts to dendrimer supports offers a potential combination of the advantages of homogeneous and heterogeneous asymmetric catalysis, and provides a very promising solution to the catalyst–product separation problem. However, one major problem which limits the practical application of these complicated macromolecules is their tedious synthesis. Thus, the development of more efficient ways to access enantioselective dendritic catalysts with high activity and reusability remains a major challenge in the near future.

1.2.4
Self-Supported Chiral Catalysts in Asymmetric Reactions

In the above-mentioned immobilization approaches, the chiral ligand is anchored either covalently or noncovalently to an insoluble or soluble support. Recently, another immobilization technique without the need for any external support was developed based on the assembly of multitopic chiral ligand and metals into homochiral metal–organic solids by coordination bonds [69–74]. The details of the story are described in Chapter 9 of this Handbook. In the simplest case, a ditopic or polytopic chiral ligand bearing two or more ligating units was used to link adjacent metal centers. Alternatively, chiral polyfunctional ligands were used as linkers and metal ions or clusters as nodes to construct homochiral metal–organic solids; however, the active sites were located as pendant auxiliaries on the main chain of the solids. In the former approach, the metal centers were incorporated into the main backbone of the resulting homochiral metal–organic assembly, where they played a dual role of structural binder and catalytically active site. Thus, it is essential that the metals used here should be capable of simultaneously bonding with at least two ligand moieties (same or different), but still have vacant or labile sites available for substrate and/or reagent coordination and activation. For the second approach, a prerequisite for the design is the orthogonal nature of the two types of functional group in the chiral bridging ligands. While the primary functional groups are responsible for connecting the network-forming metals to form extended structures, the secondary chiral groups are used to generate asymmetric catalytic sites with (or without) catalytic metals.

The modular nature of these solids allows for the fine-tuning of their catalytic performance through chemical modifications on the molecular building units. Once the chiral multitopic ligands are at hand, the synthetic protocols to assembly the homochiral solid catalysts can be quite simple. Combination of the ligand with an appropriate metal precursor in a compatible solvent would result in the spontaneous formation of polymeric structures (or oligomers). This polymerization process involves the successive formation of bonds between a multitopic ligand and a metal, and thus is facilitated by employing a ligand that shows high affinity towards the network metal. By virtue of the usually poor solubility of the solids in common organic solvents, as well as the chiral environment provided by the ligands, these self-supported chiral catalysts have demonstrated high enantioselectivities and efficiencies in a variety of heterogeneous catalytic asymmetric reactions. On completion of the reaction, the solid catalyst may be easily recovered by simple filtration and reused for several runs, without (in some reported cases) any significant loss of activity or enantioselectivity. In contrast to state-of-the-art heterogeneous catalysts which are mostly amorphous in nature, some homochiral metal–organic catalysts were even obtained in single crystal form with uniformly distributed catalytic sites, a salient feature which is highly desirable for a mechanistic understanding and structure–activity relationship (SAR) study of catalytic events which have occurred in or on a solid. The remarkable successes using self-supported metal–organic assemblies in heterogeneous (asymmetric) catalysis suggest a broader scope of application for these materials in the future.

Many challenges remain in this emerging field of heterogeneous asymmetric catalysis, however. Among other things, these solids typically maintain their structural integrity by metal–ligand coordination bonds, which encompass a broad spectrum of strengths. Therefore, beyond activity and selectivity, the stability and compatibility of the metal–organic catalysts with reaction systems is difficult to predict and should be addressed with care. To date, very few successful examples using homochiral metal–organic catalysts have been reported, and consequently further exploration of new types of homochiral assembly with a more critical assessment of their strengths and limitations in synthetic applications would be desirable.

1.2.5
Chiral Catalyst Immobilization Using Nonconventional Media

Changing the reaction medium from conventional volatile organic solvents to a form of nonconventional medium, with or without chemical manipulation on the catalyst, represents another type of general strategy used to recover and reuse a (chiral) catalytic system (see Figure 1.1) [75–83]. In general, an environmentally benign solvent is used as the nonconventional medium, as this can combine the advantages of catalyst recovery with those of 'green chemistry'. Most frequently, water, perfluorinated liquids, ionic liquids and supercritical fluids (usually supercritical CO_2) were used as the 'green' supplants for those environmentally hazardous organic solvents. Catalysis in such a medium can be carried out under either monophasic or biphasic conditions (e.g. aqueous/organic, fluorous/organic, or other immiscible multiphasic solvent combinations), with a variable degree of solvent effects (positive or negative) on the catalytic reactivity and/or selectivity. With regards to the catalyst recovery/reuse an idealized situation is that, upon completion of the reaction, the products (in organic phase) are effectively isolated by decantation, extraction, distillation or other routine phase-separation technique, while the catalyst remains in the reaction medium ready for reuse and without any decay in activity or selectivity. The potential advantages of these systems include easy catalyst separation and recovery, a reduced environmental impact of common organic solvents and increased safety – all of which are important concerns in the development of industrial catalytic asymmetric processes.

1.2.5.1 Catalyst Immobilization in Water

Water is abundant, cheap, nonflammable, nontoxic and safe to the environment, and is the solvent used by Nature for many biological organic functional group transformations catalyzed by enzymes. For these reasons, water as a reaction medium has attracted much interest in recent years. Indeed, a diverse range of synthetic transformations can be performed in water, and completely new reactivities have been identified in some previously inaccessible reactions. Several excellent reviews are available on this topic [84–89]. One additional advantage of water important for catalysis is that it allows for easy catalyst separation and is recycled via a biphasic catalysis, by virtue of its low miscibility with most organic compounds [90–93]. One classical aqueous/organic biphasic catalytic system which has been successfully commercialized is in the hydroformylation of propene using a sodium salt of sulfonated triphenylphosphine as ligand; this renders the rhodium-based catalyst soluble in water and facilitates its separation from the organic phase containing the product, butanal [94]. Given these facts, it is not surprising to see that asymmetric catalysis with a (recoverable) catalyst in water should represent one of the most attractive goals for synthetic chemists [95–99].

However, it is often more difficult to achieve high catalytic performance in water than in organic media, partly as a result of the generally poor water-compatibility of both the catalysts and/or the organic substrates. Several approaches have been invoked to achieve asymmetric catalysis in water, including hydrophilic modification of the chiral catalysts by sulfonation or quaternary cation salt formation, the use of chiral amphiphilic additives such as surfactants or phase-transfer catalysts, natural or synthetic water-soluble polymer-supported chiral catalysts, and so on. An insightful discussion on an alternative approach is provided in Chapter 6 of this Handbook, where insoluble polymer-supported chiral catalysts (chiral palladium phosphine complexes supported on TentaGel-type amphiphilic polymer bearing PEG chains) are used for heterogeneous asymmetric processes in water. As illustrated by the elegant examples, this approach combines the advantages of both aqueous reaction and ease of catalyst recovery in one system, with activity and selectivity rivaling that of their homogeneous counterparts. It should be noted that, despite the significant progresses made so far, the potential of polymer-supported chiral catalysts for aqueous reactions has not yet been fully exploited and the field is still far from mature. Hence, it remains an important issue – and a great challenge – to develop recoverable chiral catalyst systems in water, and the number of such systems will surely increase steadily in the future.

1.2.5.2 Fluorous Phase-Separation Techniques in Catalysis

Fluorous fluids such as perfluoroalkanes and perfluorodialkyl ethers are chemically inert and usually immiscible with most common organic solvents at ambient temperature. Their miscibility with organic solvents, however, improves remarkably with an increase in temperature. This special (thermotropic) solubility feature makes one-phase reaction and two-phase separation possible with a fluorous/organic solvent mixture. Pioneered by the seminal studies of Horváth and Rábai in 1994 for the use of fluorous biphase systems in rhodium-catalyzed hydroformylation [100], and which resulted in facile catalyst separation and reuse as well as high productivities, fluorous biphasic catalysis has attracted much attention as a powerful tool to facilitate catalyst/product separation [101–109]. In these systems, the reagent, substrate and product are dissolved in an organic solvent, which is in contact with a fluorous phase containing a fluorous-soluble catalyst. Analogous in principle to aqueous/organic biphasic catalyses based on water-soluble catalysts, the commonly used hydrocarbon-soluble catalysts must be modified to a 'fluorous-like' form in order to dissolve preferentially in the fluorous phase. This is usually accomplished by attaching perfluoralkyl chains ('fluorous ponytails') of appropriate size, shape and number to the ligand core of the catalyst, which would render the molecular catalyst fluorous-soluble. In order to diminish the potentially undesirable electron-withdrawing effects of the fluorous ponytails, it is often necessary to insert some spacer groups (such as $-CH_2CH_2CH_2-$) between the fluorous ponytail and the ligand backbone. The reaction involving the perfluoro-tagged catalyst in a fluorous and organic solvent mixture could proceed either heterogeneously with the catalyst in the fluorous phase and the substrate/reagents in the organic phase, or homogeneously since some fluorous biphasic systems can become single-phase at elevated reaction temperatures. On completion of the reaction, the mixture is cooled to room temperature and this results in a restoration of the two immiscible phases. The phases are then easily separated, with the product in the organic phase and the perfluoro-tagged catalyst in the fluorous layer; the catalyst can subsequently be reused for further runs in some cases. While fluorous biphase catalysis has received considerable interest since 1994, and has been the subject of several recent reviews, asymmetric fluorous catalysis has only been developed during the past decade as a novel technique that allows easy separation and recycling of the expensive chiral catalysts [110]. An overview of enantioselective fluorous catalysis, including both fluorous biphase catalysis and homogeneous enantioselective reactions promoted by fluorous catalysts is provided in Chapter 5. The chapter also outlines the remarkable achievements and research directions in this rapidly developing field.

1.2.5.3 Catalytic Reactions in Ionic Liquids

The use of ionic liquids as novel reaction media has generated considerable interest over the past decade, both as a convenient solution to environmental contamination problems caused by volatile organic solvents and as an efficient method for recycling homogeneous catalysts [111–123]. Ionic liquids are salts with a low melting point (<100 °C) and a negligible vapor pressure; they produce no atmospheric pollution and thus are regarded as 'green' solvents. Structurally, ionic liquids are generally composed of organic cations including alkylammonium, alkylphosphonium, N, N'-dialkylimidazolium and N-alkyl pyridinium cations, and weakly coordinating inorganic anions such as PF_6^- or BF_4^-, SbF_6^-, and so on. Some ionic liquids are commercially available, or can be prepared via direct quaternization of the appropriate amine or phosphane followed by subsequent anion exchange. It should be noted that the physical or chemical properties of an ionic liquid (e.g. melting behavior, polarity, hydrophilicity/lipophilicity, acidity, stability, etc.) are profoundly affected by the nature of its structural constituents, and as a result the properties can in principle be modulated and fine-tuned to the reaction by the suitable combination of cation and anion. For example, variation of the chain lengths of the alkyl groups on the cation has an effect on the lipophilicity of the ionic liquid, thus allowing its handling properties (miscibility with water or organic solvents) to be adjusted.

Ionic liquids have been used as solvents for a large number of catalytic reactions [111–123]. Ionic liquids represent a class of highly polar solvents; they can dissolve a wide range of organic, inorganic and organometallic compounds, and also tend to be immiscible with nonpolar solvents (i.e. hydrocarbons, diethyl ether). Ionic liquids have an especially high affinity for ionic or polar transition-metal complexes, which allows for the immobilization of many ionic catalysts within them, without modification of the ligands. The catalytic reactions in ionic liquids can be conducted either in a single phase by dissolving in them both substrate and catalyst, or in a biphasic system whereby the catalyst resides in the ionic liquid and the substrate/product in the second phase. Upon completion of the reaction, the product can be isolated by decantation, distillation or extraction with an organic solvent (or supercritical CO_2). In any cases, the catalyst should remain dissolved in the ionic liquid during the separation, and both the catalyst and the ionic liquid can be recovered and reused for subsequent runs.

Although the first example of homogeneous transition-metal catalysis in an ionic liquid was reported by Parshall in 1972 [124], the subject was largely neglected at the time; however, research into this area was revived about two decades later. The use of ionic liquids in asymmetric catalysis was reported even later, beginning with Chauvin's report on a catalytic asymmetric hydrogenation and hydroformylation of alkenes in 1995 [125]. Since then, enantioselective catalysis in ionic liquids has attracted remarkable interest as an approach to the facile recycling of expensive chiral ligands and catalysts, and a range of enantioselective catalytic transformations have been examined in ionic liquids [126]. In many cases, ionic liquids have a beneficial effect on the activities and enantioselectivities, and demonstrate facile recovery and reusability of the ionic solvent–catalyst systems. The reader is referred to Chapter 7 for an excellent review on the development of enantioselective catalysis in ionic liquids.

1.2.5.4 Enantioselective Catalysis in Supercritical Carbon Dioxide

Over the past decade, the use of supercritical fluids (SCF) as reaction media has begun to attract increasing attention in synthetic organic chemistry [127–134]. Both, homogeneously and heterogeneously catalyzed reactions have been performed in SCFs as reaction media, and for details on the properties and catalytic applications of SCFs the reader is referred to a thematic issue in *Chemical Reviews* [127]. Generally speaking, a SCF is a fluid at a state (conditions) above its critical point (the critical temperature T_c and pressure P_c), where upon it demonstrates both liquid-like (capable of dissolving many organic compounds) and gas-like (flow like gases and total miscibility with gaseous reactants such as H_2, CO, O_2, etc.) properties, but with no gas–liquid boundaries. The solvent properties of SCFs, such as dielectric constants, viscosity and solubility, can be easily fine-tuned by adjusting the temperature and pressure or addition of cosolvents, and this provides an attractive method for controlling the rates and selectivities of catalytic reactions. The excellent miscibility of SCFs with gases tends to make them beneficial for reactions involving gases, such as hydrogenation, hydroformylation and oxidation with oxygen, where the solubility of the gaseous reactant can be rate-limiting. Among the various SCFs that have been used as reaction media to date, supercritical carbon dioxide (scCO_2) is especially attractive in that it is nontoxic, nonflammable, chemically inert to a wide range of reaction conditions, and becomes supercritical at mild conditions (T_c = 31.1 °C and P_c = 73.8 bar). Today, scCO_2 is widely recognized as an environmentally benign solvent, and much attention has been focused on its use in various applications as an alternative to the environmentally damaging organic solvents. Since the first report of asymmetric hydrogenation reactions in scCO_2 in 1995 [135], some progress has been made in the catalytic asymmetric reactions in SCFs (predominantly scCO_2), as summarized in an excellent recent review [136]. Several types of catalytic asymmetric reaction (mainly hydrogenation and hydroformylation) have been investigated in scCO_2, with excellent results being obtained in some cases. The use of perfluoro tags on the ligands can considerably increase the otherwise often poor solubility of a polar chiral catalyst in the nonpolar $scCO_2$, allowing the reaction to be carried out in a single supercritical phase. A degree of success has also been achieved in the development of new protocols that allow for the separation and reuse of the catalyst in $scCO_2$ media, by use of the so-called catalysis and extraction using supercritical solutions (CESS), or biphasic systems wherein the catalysts are immobilized in an ionic liquid with the substrates and products being dissolved in $scCO_2$. The subject of enantioselective catalysis in $scCO_2$ and the advantages of combined use of ionic liquids and $scCO_2$ are discussed in Chapter 7.

1.2.6
Phase-Transfer Catalysis

For the above-discussed biphasic catalysis in liquid–liquid (e.g. aqueous-organic) systems, no additives are used, as the catalyst preferentially resides in a phase which is immiscible with that phase containing the product. One option to biphasic catalysis is to use amphiphilic additives such as phase-transfer catalysts which, by some interaction mechanism, can increase the mobility of a reactant or an intermediate into a favored phase, and thus provide an effective tool for accelerating the reaction and also controlling the selectivity. Phase-transfer catalysis (PTC) has been recognized for a long time, and typically involves simple experimental operations and mild reaction conditions, as well as environmentally friendly reagents and solvents [137]. The development of asymmetric PTC was triggered by a pioneering study performed by a Merck research group in 1984 [138]. Since then, several classes of chiral phase-transfer catalysts have been developed for use in asymmetric synthesis, mainly including cinchona alkaloid derivatives, purely synthesized chiral quaternary ammonium salts, and some chiral crown ethers. The synthetic utility of asymmetric PTC has been demonstrated for a wide range of reactions (most frequently involving anionic intermediates), among which the enantioselective alkylation of glycine derivatives has been most extensively studied and has become a powerful approach for the preparation of various α-amino acids [139–146]. The reactions can be performed in either solid–organic or aqueous–organic biphasic systems, typically with moderate to relatively high catalyst loading (1 to 20 mol%). Although, in most cases it is difficult to recover the catalyst, recent efforts using polymer-supported cinchona alkaloid-derived ammonium salts or fluorous chiral phase-transfer catalyst have resulted in notable achievements towards this direction [146]. Several recent reviews summarizing the achievements and highlighting the future perspectives of asymmetric PTC are described in Chapter 11.

1.2.7
Immobilization of Chiral Organic Catalysts

Since the 1990s, organocatalysis has become established as an extremely important alternative methodology to enzymatic or organometallic catalysis in the field of asymmetric catalysis [147–150]. In general, an organic catalyst can be defined as a relatively simple low-molecular-weight organic compound which can promote a chemical transformation in substoichiometric quantity. By virtue of its metal-free nature an organocatalyst can, in principle, be advantageous in comparison with its metal-based counterpart in terms of reaction handling, catalyst stability and complete freedom of metal contamination in the product. Compared with naturally occurring catalysts (enzymes), organic catalysts are structurally simpler, more stable, less expensive, have a broader application scope and can demonstrate excellent selectivities in some cases. With regards to the catalyst immobilization, organic catalysts may also possess certain advantages over both enzymes (difficult to manipulate) and metal-based catalysts (generally suffer from metal leaching). Several types of achiral or chiral organic catalyst supported on either organic polymers (soluble or insoluble) or inorganic matrices have been developed for a range of reactions, and with varying degree of successes. Several reviews are available relevant to this subject [151–153], and these are detailed in Chapter 8. Although the development of immobilized chiral organic catalysts remains firmly in its infancy stage, the field of research is expected to expand greatly in the future in view of the present 'gold rush' in organocatalysis.

1.3
Chirally Modified Metal Surface for Heterogeneous Asymmetric Hydrogenation

Conceptually different from the above-discussed strategies that have been applied in the immobilization of homogeneous chiral catalysts, the modification of an active metal surface by an adsorbed chiral modifier has been shown to be very successful for heterogeneous asymmetric catalysis [154–166]. In this approach, an achiral solid (most frequently a metal surface) is used in combination with a suitable chiral organic compound (chiral modifier) as an inherently heterogeneous asymmetric catalysts. While the metal is responsible for catalytic activity, the surface-adsorbed chiral modifier can induce stereochemical control of the reaction via some type of mutual interaction with the substrate. Due to the many technical advantages associated with heterogeneous catalysis, such as easy separation and possible reuse of the catalyst and facile modification to continuous process operation, research in this field has been attracting a longstanding interest from both academia and industry. During its rather long history of development, this strategy has achieved prominent successes predominantly in the enantioselective hydrogenation of several types of prochiral substrate. A high substrate specificity is observed when using a chirally modified metal surface as a heterogeneous catalyst in hydrogenations, as by far only a limited number of metal–modifier–substrate combinations can provide enantioselectivities that are useful for synthetic applications. Three types of widely recognized efficient catalyst system have been developed based on the chiral modification of metal catalysts: (i) the nickel–tartrate–NaBr system for the hydrogenation of β-ketoesters, β-diketones and methyl ketones; (ii) the platinum–cinchona alkaloid system for the hydrogenation of α-ketoesters, α-ketoacids and lactones; and (iii) palladium catalysts modified with cinchona alkaloids for selected activated alkenes. Even though the substrate scope is still relatively narrow, high enantioselectivities which are useful for technical applications have been achieved in the hydrogenation of some specific reactants. The mechanism involved has also received much attention, the aim being to understand the adsorption of the modifier onto the metal surface and its interaction with the reactant, to interpret and predict the catalytic behaviors of the systems, as well as to facilitate the rational design of a suitable modifier. By using a combination of surface techniques, spectroscopic studies, computational modeling and kinetic studies, considerable mechanistic insight has been obtained, especially in the enantioselective pyruvate hydrogenation over Pt [162]. Nevertheless, a rational design of suitable metal/modifier catalytic systems still represents an extremely challenging goal at the present stage, and continuous endeavor in this direction should be of prime importance from both practical and theoretical viewpoints. Last, but not least, the use of an achiral heterogeneous catalyst and a chiral modifier as asymmetric heterogeneous catalyst is not limited to hydrogenations. For example, recently developed new catalyst-modifier systems based on NAP-MgO (nanocrystalline aerogel-prepared magnesium oxide) have proven very successful for a variety of asymmetric reactions [167, 168]. Chirally modified metal nanoparticles have also been successfully tested in the enantioselective hydrosilylation of styrene [169], allylic alkylation [170], and Pauson–Khand-type reactions [171]. Although the catalyst-modifier systems other than those for hydrogenations are not described in Chapter 10 of this Handbook, these encouraging discoveries nonetheless underline the enormous application potential of the present strategy for asymmetric heterogeneous catalysis.

1.4
Heterogeneous Enantioselective Catalysts in Industrial Research and Application

As evidenced by the numerous publications in the field of heterogeneous asymmetric catalysis, over the years a large number of variously immobilized chiral catalysts – as well as several types of chirally modified catalyst – have been developed for a broad range of enantioselective reactions. Although some of these catalysts have demonstrated excellent performances in terms of activity, enantioselectivity, stability and/or reusability, the number of recognized industrial processes that use heterogeneous enantioselective catalysts for the commercial production of chiral compounds remains extremely small [172, 173]. As summarized in Chapter 12, to date only a few chirally modified heterogeneous hydrogenation catalysts (cinchona/Pt and Ni/tartaric acid/NaBr) are used industrially in the production of chiral intermediates. None of the immobilized chiral metal complexes has been applied industrially for large-scale production, despite several types of immobilized catalysts having demonstrated good potential for technical applications. Clearly, with few exceptions, most of the heterogeneous catalytic asymmetric methodologies developed to date have not been sufficiently mature to compete with alternative industrial methods (e.g. homogeneous catalysis, racemic mixture resolution). But this is not so surprising as it first appears, given the following general considerations. First, the inherent complexity associated with most heterogeneous systems makes the prediction of their catalytic performances extremely demanding. The often cumbersome immobilization procedure may considerably increase development time/costs, rendering the catalyst screening for a targeted reaction an even more challenging task. To make things worse, most of the immobilized metal complexes have suffered from problems of reduced activity and/or degraded enantioselectivity compared to their homogeneous counterparts as a result of the unintentional impact of supports on the active sites.

Nevertheless, we believe that heterogeneous asymmetric catalysis is – and will continue to be – an important field in the future, in view of its undisputable advantages over homogeneous counterparts with regards to separation and economy. To date, chemists have developed, with varying degrees of success, a wide variety of heterogenization techniques which will steadily broaden their applicability in the future. Finally, the identification of more economic and reliable heterogeneous asymmetric catalytic systems with high activity, selectivity, stability and reusability remains a challenging, but very worthwhile, goal which calls for collaborative efforts from both industry and academia.

References

1 Collins, A.N., Sheldrake, G.N. and Crosby, J. (1992) *Chirality in Industry: The Commercial Manufacture and Applications of Optically Active Compounds*, John Wiley & Sons, Ltd, Chichester.

2 Collins, A.N., Sheldrake, G.N. and Crosby, J. (1997) *Chirality in Industry II: Developments in the Commercial Manufacture and Applications of Optically Active Compounds*, John Wiley & Sons, Ltd, Chichester.

3 Sheldon, R.A. (1993) *Chirotechnology: Industrial Synthesis of Optically Active Compounds*, Dekker, New York.

4 Noyori, R. (1994) *Asymmetric Catalysis in Organic synthesis*, Wiley-Interscience, New York.

5 Ojima, I. (2000) *Catalytic Asymmetric Synthesis*, 2nd edn, Wiley-VCH Verlag GmbH, New York.

6 Doyle, M. (1995) *Advances in Catalytic Processes: Asymmetric Chemical Transformations*, Vol. **1**, JAI, Greenwich.

7 Jacobsen, E.N., Pfaltz, A. and Yamamoto, H. (1999) *Comprehensive Asymmetric Catalysis*, Vol. **I–III**, Springer, Berlin.

8 Yamamoto, H. (2001) *Lewis Acids in Organic Synthesis*, Wiley-VCH Verlag GmbH, New York.

9 Brunner, H. and Zettlmeier, W. (1993) *Handbook of Enantioselective Catalysis*. Vol. **1–2**, Wiley-VCH Verlag GmbH, New York.

10 De Vos, D.E., Van-kelecom, I.F.J. and Jacobs, P.A. (2000) *Chiral Catalyst Immobilization and Recycling*, Wiley-VCH Verlag GmbH, Weinheim.

11 For a thematic issue on recoverable catalysts and reagents, see: Gladzsz, J.A., (2002) *Chemical Reviews*, **102**, 3215–892.

12 Heitbaum, M., Glorius, F. and Escher, I. (2006) *Angewandte Chemie – International Edition*, **45**, 4732–62.

13 Fan, Q.H., Li, Y.M. and Chan, A.S.C. (2002) *Chemical Reviews*, **102**, 3385–465.

14 McMorn, P. and Hutchings, G.J. (2004) *Chemical Society Reviews*, **33**, 108–22.

15 Baleizao, C. and Garcia, H. (2006) *Chemical Reviews*, **106**, 3987–4043.

16 Corma, A. and Garcia, H. (2003) *Chemical Reviews*, **103**, 4307–65.

17 Hutchings, G.J. (1999) *Chemical Communications*, 301–6.
18 Fan, Q.H., Wang, R. and Chan, A.S.C. (2002) *Bioorganic and Medicinal Chemistry Letters*, **12**, 1867–71.
19 Song, C.E. (2005) *Annual Reports on the Progress of Chemistry Section C Physical Chemistry*, **101**, 143–73.
20 Jas, G. and Kirschning, A. (2003) *Chemistry – A European Journal*, **9**, 5708–23.
21 Hodge, P. (2003) *Current Opinion in Chemical Biology*, **7**, 362–73.
22 Kirschning, A., Solodenko, W. and Mennecke, K. (2006) *Chemistry – A European Journal*, **12**, 5972–90.
23 Hodge, P. (2005) *Industrial and Engineering Chemistry Research*, **44**, 8542–53.
24 Vankelecom, I.F.J. and Jacobs, P.A. (2000) Catalyst immobilization on inorganic supports, in *Chiral Catalyst Immobilization and Recycling* (eds D.E. De Vos, I.F.J. Van-kelecom and P.A. Jacobs), Wiley-VCH Verlag GmbH, Weinheim, pp. 19–42.
25 Song, C.E. and Lee, S.G. (2002) *Chemical Reviews*, **102**, 3495–524.
26 Corma, A. and Garcia, H. (2006) *Advanced Synthesis Catalysis*, **348**, 1391–412.
27 De Vos, D.E., Dams, M., Sels, B.F. and Jacobs, P.A. (2002) *Chemical Reviews*, **102**, 3615–40.
28 Li, C. (2004) *Catalysis Reviews – Science and Engineering*, **46**, 419–92.
29 Song, C.E., Kim, D.H. and Choi, D.S. (2006) *European Journal of Inorganic Chemistry*, 2927–35.
30 Li, C., Zhang, H., Jiang, D. and Yang, Q. (2007) *Chemical Communications*, 547–58.
31 Bergbreiter, D.E. (2000) Organic polymers as a catalyst recovery vehicle, in *Chiral Catalyst Immobilization and Recycling* (eds D.E. De Vos, I.F.J. Van-kelecom and P.A. Jacobs), Wiley-VCH Verlag GmbH, Weinheim, pp. 43–80.
32 Shuttleworth, S.J., Allin, S.M. and Sharma, P.K. (1997) *Synthesis*, 1217–39.
33 Pu, L. (1998) *Tetrahedron: Asymmetry*, **9**, 1457–77.
34 Saluzzo, C., Touchard, R., ter Halle, F., Fache, F., Schulz, E. and Lemaire, M. (2000) *Journal of Organometallic Chemistry*, **603**, 30–9.
35 Clapham, B., Reger, T.S. and Janda, K.D. (2001) *Tetrahedron*, **57**, 4637–62.
36 Bergbreiter, D.E. (2001) *Current Opinion in Drug Discovery and Development*, **4**, 736–44.
37 Leadbeater, N.E. and Marco, M. (2002) *Chemical Reviews*, **102**, 3217–73.
38 McNamara, C.A., Dixon, M.J. and Bradley, M. (2002) *Chemical Reviews*, **102**, 3275–99.
39 Bräse, S., Lauterwasser, F. and Ziegert, R.E. (2003) *Advanced Synthesis Catalysis*, **345**, 869–929.
40 Haag, R. and Roller, S. (2004) *Topics in Current Chemistry*, **242**, 1–42.
41 Mastrorilli, P. and Nobile, C.F. (2004) *Coordination Chemical Reviews*, **248**, 377–95.
42 El-Shehawy, A.A. and Itsuno, S. (2005) *Current Topics in Polymer Research*, 1–69.
43 Dioos, B.M.L., Vankelecom, I.F.J. and Jacobs, P.A. (2006) *Advanced Synthesis Catalysis*, **348**, 1413–46.
44 Gravert, D.J. and Janda, K.D. (1997) *Chemical Reviews*, **97**, 489–509.
45 Bergbreiter, D.E. (1998) *Catalysis Today*, **42**, 389–97.
46 Wentworth, P. Jr and Janda, K.D. (1999) *Chemical Communications*, 1917–24.
47 Toy, P.H. and Janda, K.D. (2000) *Accounts of Chemical Research*, **33**, 546–54.
48 Bergbreiter, D.E. (2002) *Chemical Reviews*, **102**, 3345–83.
49 Dickerson, T.J., Reed, N.N. and Janda, K.D. (2002) *Chemical Reviews*, **102**, 3325–43.
50 Bergbreiter, D.E. and Li, J. (2004) *Topics in Current Chemistry*, **242**, 113–76.
51 van de Coevering, R., Klein Gebbink, R.J.M. and van Koten, G. (2005) *Progress in Polymer Science*, **30**, 474–90.
52 Bergbreiter, D.E. and Sung, S.D. (2006) *Advanced Synthesis Catalysis*, **348**, 1352–66.
53 Pu, L. (1998) *Chemical Reviews*, **98**, 2405–94.
54 Pu, L. (1999) *Chemistry – A European Journal*, **5**, 2227–32.
55 Astruc, D. and Chardac, F. (2001) *Chemical Reviews*, **101**, 2991–3023.

56 Kreiter, R., Kleij, A.W., Gebbink, R.J.M.K. and van Koten, G. (2001) *Topics in Current Chemistry*, **217**, 163–99.
57 Crooks, R.M., Zhao, M., Sun, L., Chechik, V. and Yeung, L.K. (2001) *Accounts of Chemical Research*, **34**, 181–90.
58 Oosterom, G.E., Reek, J.N.H., Kamer, P.C.J. and Van Leeuwen, P.W.N.M. (2001) *Angewandte Chemie – International Edition*, **40**, 1828–49.
59 Twyman, L.J., King, A.S.H. and Martin, I.K. (2002) *Chemical Society Reviews*, **31**, 69–82.
60 Van Heerbeek, R., Kamer, P.C.J., Van Leeuwen, P.W.N.M. and Reek, J.N.H. (2002) *Chemical Reviews*, **102**, 3717–56.
61 King, A.S.H. and Twyman, L.J. (2002) *Journal of the Chemical Society – Perkin Transactions 1*, 2209–18.
62 Berger, A., Gebbink, R.J.M.K. and van Koten, G. (2006) *Topics in Organometallic Chemistry*, **20**, 1–38.
63 Ribaudo, F., van Leeuwen, P.W.N.M. and Reek, J.N.H. (2006) *Topics in Organometallic Chemistry*, **20**, 39–59.
64 Mery, D. and Astruc, D. (2006) *Coordination Chemistry Reviews*, **250**, 1965–79.
65 Helms, B. and Frechet, J.M.J. (2006) *Advanced Synthesis Catalysis*, **348**, 1125–48.
66 Brunner, H. and Altmann, S. (1994) *Chemische Berichte*, **127**, 2285–96.
67 Brunner, H. (1995) *Journal of Organometallic Chemistry*, **500**, 39–46.
68 Kassube, J.K. and Gade, L.H. (2006) *Topics in Organometallic Chemistry*, **20**, 61–96.
69 Dai, L.X. (2004) *Angewandte Chemie – International Edition*, **43**, 5726–9.
70 Ding, K.-L., Wang, Z., Wang, X.-W., Liang, Y.-X. and Wang, X.-S. (2006) *Chemistry – A European Journal*, **12**, 5188–97.
71 Kesanli, B. and Lin, W. (2003) *Coordination Chemistry Reviews*, **246**, 305–26.
72 Ngo, H.L. and Lin, W. (2005) *Topics in Catalysis*, **34**, 85–92.
73 Lin, W. (2005) *Journal of Solid State Chemistry*, **178**, 2486–90.
74 Williams, K.A., Boydston, A.J. and Bielawski, C.W. (2007) *Chemical Society Reviews*, **36**, 729–44.
75 Cornils, B. (1995) *Angewandte Chemie – International Edition*, **34**, 1575–7.
76 Hanson, B.E. (2000) Liquid biphasic enantioselective catalysis, in *Chiral Catalyst Immobilization and Recycling* (eds D.E. De Vos, I.F.J. Van-kelecom and P.A. Jacobs), Wiley-VCH Verlag GmbH, Weinheim, pp. 81–96.
77 Tzschucke, C.C., Markert, C., Bannwarth, W., Roller, S., Hebel, A. and Haag, R. (2002) *Angewandte Chemie – International Edition*, **41**, 3964–4000.
78 Cole-Hamilton, D.J. (2003) *Science*, **299**, 1702–6.
79 Keim, W. (2003) *Green Chemistry*, **5**, 105–11.
80 Sheldon, R.A. (2005) *Green Chemistry*, **7**, 267–78.
81 Baker, R.T., Kobayashi, S. and Leitner, W. (2006) *Advanced Synthesis Catalysis, Special Issue 348 (12 + 13): Multiphase Catalysis, Green Solvents and Immobilization*, 1317–771.
82 Liu, S. and Xiao, J. (2007) *Journal of Molecular Catalysis A – Chemical*, **270**, 1–43.
83 Afonso, C.A.M., Branco, L.C., Candeias, N.R., Gois, P.M.P., Lourenco, N.M.T., Mateus, N.M.M. and Rosa, N. (2007) *Chemical Communications*, 2669–79.
84 Li, C.J. and Chan, T.-H. (2007) *Comprehensive Organic Reactions in Aqueous Media*, 2nd edn, John Wiley & Sons, Ltd, Hoboken.
85 Li, C.J. (1993) *Chemical Reviews*, 2023–35.
86 Lindstroem, U.M. (2002) *Chemical Reviews*, **102**, 2751–71.
87 Li, C.J. (2005) *Chemical Reviews*, **105**, 3095–165.
88 Li, C.J. and Chen, L. (2006) *Chemical Society Reviews*, **35**, 68–82.
89 Hailes, H.C. (2007) *Organic Process Research & Development*, **11**, 114–20.
90 Cornils, B., Herrmann, W.A. and Eckl, R.W. (1997) *Journal of Molecular Catalysis A – Chemical*, **116**, 27–33.
91 Cornils, B. and Herrmann, W.A. (2002) Immobilization by aqueous catalysts, in *Applied Homogeneous Catalysis with Organometallic Compounds*, 2nd edn,

Wiley-VCH, Verlag GmbH, Weinheim, pp. 603–33.
92 Okuhara, T. (2002) *Chemical Reviews*, **102**, 3641–65.
93 Vancheesan, S. and Jesudurai, D. (2002) *Catalysis*, 311–37.
94 Cornils, B. (1996) *Applied Homogeneous Catalysis with Organometallic Compounds*, Vol. 1 (eds B. Cornils and W. A. Herrmann), Wiley-VCH Verlag GmbH, Weinheim, Germany, pp. 577–600.
95 Kobayashi, S. and Manabe, K. (2002) *Accounts of Chemical Research*, **35**, 209–17.
96 Sinou, D. (2002) *Advanced Synthesis Catalysis*, **344**, 221–37.
97 Uozumi, Y. (2004) *Topics in Current Chemistry*, **242**, 77–112.
98 Uozumi, Y. (2005) *Catalysis Surveys from Asia*, **9**, 269–78.
99 Kobayashi, S. (2007) *Pure and Applied Chemistry*, **79**, 235–45.
100 Horváth, I.T. and Rábai, J. (1994) *Science*, **266**, 72–5.
101 Cornils, B. (1997) *Angewandte Chemie-International Edition*, **36**, 2057–9.
102 Horvath, I.T. (1998) *Accounts of Chemical Research*, **31**, 641–50.
103 Fish, R.H. (1999) *Chemistry–A European Journal*, **5**, 1677–80.
104 de Wolf, E., van Koten, G. and Deelman, B.J. (1999) *Chemical Society Reviews*, **28**, 37–41.
105 Cavazzini, M., Montanari, F., Pozzi, G. and Quici, S. (1999) *Journal of Fluorine Chemistry*, **94**, 183–93.
106 Barthel-Rosa, L.P. and Gladysz, J.A. (1999) *Coordination Chemistry Reviews*, **190–2**, 587–605.
107 Horvath, I.T. (2002) Immobilization by other liquids: fluorous phases, in *Applied Homogeneous Catalysis with Organometallic Compounds*, 2nd edn, Vol. 2, Wiley-VCH Verlag GmbH, Weinheim, pp. 634–9.
108 Dobbs, A.P. and Kimberley, M.R. (2002) *Journal of Fluorine Chemistry*, **118**, 3–17.
109 Gladysz, J.A. and Curran, D.P. (2002) *Tetrahedron*, **58**, 3823–5.
110 Pozzi, G. and Shepperson, I. (2003) *Coordination Chemistry Reviews*, **242**, 115–24.
111 Welton, T. (1999) *Chemical Reviews*, **99**, 2071–83.
112 Wasserscheid, P. and Keim, W. (2000) *Angewandte Chemie–International Edition*, **39**, 3772–89.
113 Sheldon, R. (2001) *Chemical Communications*, 2399–407.
114 Olivier-Bourbigou, H. and Magna, L. (2002) *Journal of Molecular Catalysis A–Chemical*, **182–183**, 419–37.
115 Zhao, D., Wu, M., Kou, Y. and Min, E. (2002) *Catalysis Today*, **74**, 157–89.
116 Dupont, J., de Souza, R.F. and Suarez, P.A.Z. (2002) *Chemical Reviews*, **102**, 3667–91.
117 Zhao, H. and Malhotra, S.V. (2002) *Aldrichimica Acta*, **35**, 75–83.
118 Gordon, C.M. (2001) *Applied Catalysis A: General*, **222**, 101–17.
119 Baudequin, C. et al. (2003) *Tetrahedron: Asymmetry*, **14**, 3081–93.
120 Welton, T. (2004) *Coordination Chemistry Reviews*, **248**, 2459–77.
121 Muzart, J. (2006) *Advanced Synthesis Catalysis*, **348**, 275–95.
122 Miao, W. and Chan, T.H. (2006) *Accounts of Chemical Research*, **39**, 897–908.
123 Parvulescu, V.I. and Hardacre, C. (2007) *Chemical Reviews*, **107**, 2615–65.
124 Parshall, G.W. (1972) *Journal of the American Chemical Society*, **94**, 8716–19.
125 Chauvin, Y., Mussmann, L. and Olivier, H. (1995) *Angewandte Chemie–International Edition*, **34**, 2698–700.
126 Song, C.E. (2004) *Chemical Communications*, 1033–43.
127 Noyori, R. (1999) Supercritical fluids. *Chemical Reviews Thematic Issue*, **99**, 353–634.
128 Baiker, A. (1999) *Chemical Reviews*, **99**, 453–73.
129 Jessop, P.G., Ikariya, T. and Noyori, R. (1999) *Chemical Reviews*, **99**, 475–93.
130 Oakes, R.S., Clifford, A.A. and Rayner, C.M. (2001) *Journal of the Chemical Society–Perkin Transactions 1*, 917–41.
131 Musie, G., Wei, M., Subramaniam, B. and Busch, D.H. (2001) *Coordination Chemistry Reviews*, **219–221**, 789–820.
132 Leitner, W. (2002) *Accounts of Chemical Research*, **35**, 746–56.
133 Licence, P., Ke, J., Sokolova, M., Ross, S.K. and Poliakoff, M. (2003) *Green Chemistry*, **5**, 99–104.
134 Campestrini, S. and Tonellato, U. (2005) *Current Organic Chemistry*, **9**, 31–47.

135 Burk, M.J., Feng, S.G., Gross, M.F. and Tumas, W. (1995) *Journal of the American Chemical Society*, **117**, 8277–8.
136 Cole-Hamilton, D.J. (2006) *Advanced Synthesis Catalysis*, **348**, 1341–51.
137 Makosza, M. (2000) *Pure and Applied Chemistry*, **72**, 1399–403.
138 Dolling, U.-H., Davis, P. and Grabowski, E.J.J. (1984) *Journal of the American Chemical Society*, **106**, 446 7.
139 Nelson, A. (1999) *Angewandte Chemie – International Edition*, **38**, 1583–5.
140 O'Donnell, M.J. (2001) *Aldrichimica Acta*, **34**, 3–15.
141 Maruoka, K. and Ooi, T. (2003) *Chemical Reviews*, **103**, 3013–28.
142 O'Donnell, M.J. (2004) *Accounts of Chemical Research*, **37**, 506–17.
143 Lygo, B. and Andrews, B.I. (2004) *Accounts of Chemical Research*, **37**, 518–25.
144 Vachon, J. and Lacour, J. (2006) *Chimia*, **60**, 266–75.
145 Lygo, B. and Beaumont, D.J. (2007) *Chimia*, **61**, 257–62.
146 Ooi, T. and Maruoka, K. (2007) *Angewandte Chemie – International Edition*, **46**, 4222–66.
147 List, B. (2005) *Organic and Biomolecular Chemistry*, **3**, 719.
148 Dalko, P.I. and Moisan, L. (2001) *Angewandte Chemie – International Edition*, **40**, 3726–48.
149 Dalko, P.I. and Moisan, L. (2004) *Angewandte Chemie – International Edition*, **43**, 5138–75.
150 Berkessel, A. and Gröger, H. (2004) *Asymmetric Organocatalysis*, Wiley-VCH Verlag GmbH, Weinheim.
151 Benaglia, M., Puglisi, A. and Cozzi, F. (2003) *Chemical Reviews*, **103**, 3401–29.
152 Cozzi, F. (2006) *Advanced Synthesis Catalysis*, **348**, 1367–90.
153 Benaglia, M. (2006) *New Journal of Chemistry*, **30**, 1525–33.
154 Blaser, H.-U., Jalett, H.P., Müller, M. and Studer, M. (1997) *Catalysis Today*, **37**, 441–63.
155 Wells, P. B. and Wilkinson, A.G. (1998) *Topics in Catalysis*, **5**, 39–50.
156 Studer, M., Blaser, H.-U. and Exner, C. (2003) *Advanced Synthesis Catalysis*, **345**, 45–65.
157 Wells, P. B. and Wells, R.P.K. (2000) Enantioselective hydrogenation catalyzed by platinum group metals modified by natural alkaloids, in *Chiral Catalyst Immobilization and Recycling* (eds D.E. De Vos, I.F.J. Vankelecom and P.A. Jacobs), Wiley-VCH Verlag GmbH, Weinheim, pp. 123–54.
158 Baiker, A. (2000) Design of new chiral modifiers for heterogeneous enantioselective hydrogenation: a combined experimental and theoretical approach, in *Chiral Catalyst Immobilization and Recycling* (eds D.E. De Vos, I.F.J. Vankelecom and P.A. Jacobs), Wiley-VCH Verlag GmbH, Weinheim, pp. 155–71.
159 Tai, A. and Sugimura, T. (2000) Modified Ni catalysts for enantio-differentiating hydrogenation, in *Chiral Catalyst Immobilization and Recycling* (eds D.E. De Vos, I.F.J. Vankelecom and P.A. Jacobs), Wiley-VCH Verlag GmbH, Weinheim, pp. 173–209.
160 Osawa, T., Harada, T. and Takayasu, O. (2000) *Topics in Catalysis*, **13**, 155–68.
161 von Arx, M., Mallat, T. and Baiker, A. (2002) *Topics in Catalysis*, **19**, 75–87.
162 Buergi, T. and Baiker, A. (2004) *Accounts of Chemical Research*, **37**, 909–17.
163 Osawa, T., Harada, T. and Takayasu, O. (2006) *Current Organic Chemistry*, **10**, 1513–31.
164 Bartok, M. (2006) *Current Organic Chemistry*, **10**, 1533–67.
165 Studer, M. and Blaser, H.U. (2006) Enantioselective hydrogenation of activated ketones using heterogeneous Pt catalysts modified with cinchona alkaloids, in *Handbook of Chiral Chemicals*, 2nd edn, CRC Press LLC, Boca Raton, pp. 345–57.
166 Mallat, T., Orglmeister, E. and Baiker, A. (2007) *Chemical Reviews*, **107**, 4863–90.
167 Choudary, B.M., Kantam, M.L., Ranganath, K.V.S., Mahendar, K. and Sreedhar, B. (2004) *Journal of the American Chemical Society*, **126**, 3396–7.
168 Choudary, B.M., Ranganath, K.V.S., Pal, U., Kantam, M.L. and Sreedhar, B. (2005) *Journal of the American Chemical Society*, **127**, 13167–71.
169 Tamura, M. and Fujihara, H. (2003) *Journal of the American Chemical Society*, **125**, 15742–3.
170 Jansat, S., Gómez, M., Philippot, K., Muller, G., Guiu, E., Claver, C., Castillón,

S. and Chaudret, B. (2004) *Journal of the American Chemical Society*, **126**, 1592–3.
171 Park, K.H. and Chung, Y.K. (2005) *Advanced Synthesis Catalysis*, **347**, 854–66.
172 Blaser, H.U., Pugin, B. and Studer, M. (2000) Enantioselective heterogeneous catalysis: academic and industrial challenges, in *Chiral Catalyst Immobilization and Recycling* (eds D.E. De Vos, I.F.J. Vankelecom and P.A. Jacobs), Wiley-VCH Verlag GmbH, Weinheim, pp. 1–17.
173 End, N. and Schoening, K.U. (2004) *Topics in Current Chemistry*, **242**, 241–71.

2
Heterogeneous Enantioselective Catalysis Using Inorganic Supports

Santosh Singh Thakur, Jae Eun Lee, Seo Hwan Lee, Ji Man Kim, and Choong Eui Song

2.1
Introduction

For economic, environmental and social reasons, the trend towards the application of optically pure compounds is undoubtedly increasing. Among the various methods for selectively producing a single enantiomer, asymmetric catalysis is the most attractive from the atom-economic point of view. Over the past 30 years, numerous catalytic reactions enabling the enantioselective formation of C—H, C—C, C—O, C—N and other bonds have been discovered [1, 2]. A number of homogeneous chiral catalysts have already gained wide acceptance in terms of efficiency and selectivity, some of which are even used on an industrial scale [3], and the chemists involved in the pioneering breakthroughs were recently awarded the Nobel Prize [4]. However, in spite of the huge amount of work devoted to this subject in both academic and industrial fields, the contribution of asymmetric catalysis in the overall production of chiral chemicals has been much lower than was originally expected. One of the major drawbacks of homogeneous catalysis is the need to separate the relatively expensive catalysts from the reaction mixture at the end of the process. The possible contamination by the metal catalysts of the product also restricts their use in industry. In order to overcome these drawbacks, much effort has been devoted to developing effective immobilized catalyst systems, mainly by immobilizing a homogeneous chiral catalyst onto or into an inorganic solid support via: (i) covalent bonding; (ii) adsorption; (iii) ion-pair formation; (iv) entrapment; or (v) ship-in-a-bottle [5–12]. However, for a long time, most examples of supported catalysts tended to have inferior catalytic properties relative to their homogeneous counterparts. However, recent studies – especially those pioneered by the groups of Corma and Thomas – have shown that by choosing a suitable support a heterogeneous catalyst can provide much better catalytic performances than its homogeneous analogue [9–12]. Even nonenantioselective catalysts showed significant asymmetric induction when they were anchored onto a suitable support [13, 14a,c, 15, 16b]. This change of fortune for the immobilized catalysts can be mainly attributed to two reasons: site-isolation; and confinement effects.

Handbook of Asymmetric Heterogeneous Catalysis. Edited by K. Ding and Y. Uozumi
Copyright © 2008 WILEY-VCH Verlag GmbH & Co. KGaA, Weinheim
ISBN: 978-3-527-31913-8

First, site isolation is achievable through the appropriate choice of supports. The instability of homogeneous catalysts which results in a low catalyst turnover is often caused by the formation of inactive multinuclear species. Colocalization of the catalytic sites can easily be prevented by immobilizing the catalysts in confined spaces, that is, by the site isolation of the catalytic sites [17], in such a way that they cannot interact with each other. Second, and more importantly, the confinement effect of support originating from the additional interactions (hydrogen bond, van der Waals force, adsorption, etc.) between catalyst/substrate and the support surfaces plays an important role when the reactions take place within the nanosized pores or layers of supports. In this case, either the immobilized catalyst is conformationally confined by the interior of the pore or the access of the substrates to the catalyst is restricted [13, 14]. Because these additional interactions between catalyst/substrate and pore surfaces are about the same level as the energy difference (< 3 kcal mol^{-1}) between the two transition states corresponding to the R- and S-products, the enantioselectivity can be very sensitively (positively or negatively) altered [9]. A difference of 2.68 and 1.78 kcal mol^{-1} suffices to afford an enantiomeric excess (ee) of 98% at 20 and −78 °C, respectively (Figure 2.1). This is comparable to the energy of a fairly respectable hydrogen bond. Enantioselectivity can therefore be increased, decreased, or even reversed. Therefore, the enantioselectivity could be improved by careful tuning of the confinement effect based on the molecular design of the pore/surface of supports and the catalytic moiety according to the requirements of chiral reactions [9–12].

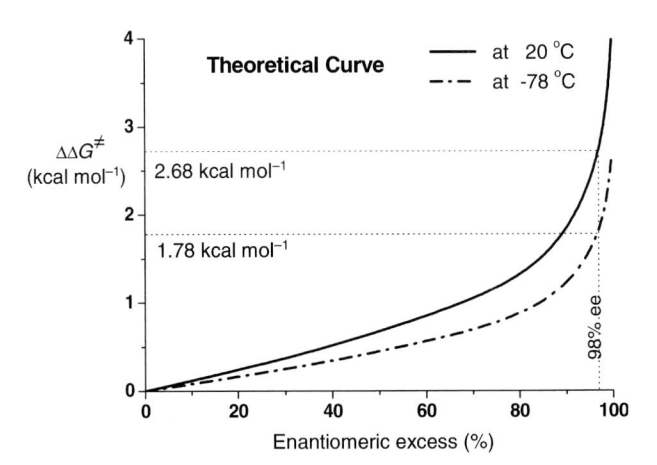

Figure 2.1 Difference in the Gibbs free energy change ($\Delta\Delta G^{\ddagger}$) for the R and S transition states as a function of product enantioselectivity for the reaction of a prochiral substrate for the R and S enantiomers. A subtle difference (1.78 kcal mol^{-1}) in the energy for the R and S transition states makes a big difference in the enantioselectivity (98% ee) at −78 °C.

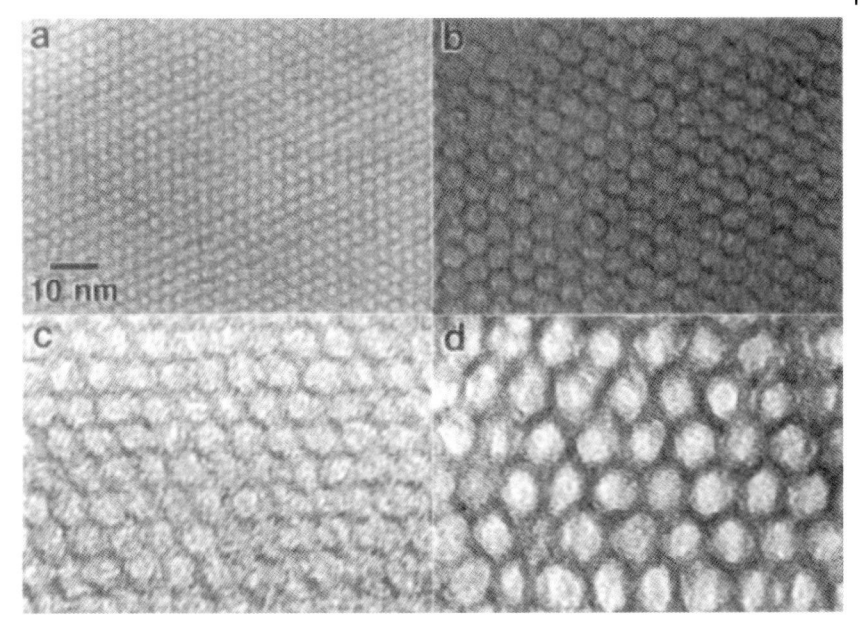

Figure 2.2 Transmission electron microscopy images of several MCM-41 materials having tunable pore sizes of (a) 2, (b) 4, (c) 6.5 and (d) 10 nm (from Ref. [18b]).

Considering the aforementioned site-isolation and confinement effects, the choice of a suitable support is decisive. In this respect, the application of inorganic materials as heterogeneous supports offers a number of advantages: their rigid pore or layered structure does not allow the aggregation of active catalysts. In addition, many of inorganic supports possess easily accessible pores/layers, the sizes and structures of which are often tunable (Figure 2.2). Among others, zeolites (e.g. 'ultrastabilized zeolite Y'), mesoporous materials (e.g. MCM-41, MCM-48 [18], SBA-15 [19], silica gel Grace332 [20]), and some mineral clays (e.g. hetorite, bentonite, montronite) [21] have been used successfully for the immobilization of asymmetric catalysts [8, 9a, 10–12].

It is the goal of this chapter to present an overview of asymmetric heterogeneous catalysis using inorganic supports, thereby portraying the state of the art and discussing its potential and limitations.

2.2
Asymmetric Reduction

Since the first successful report on heterogeneous asymmetric hydrogenation using inorganic materials by Kinting et al. in 1985 [22], numerous advances have been made in this area of research. Some representative examples of chiral reduction catalysts immobilized on inorganic support are summarized in Table 2.1. As

Table 2.1 Examples of heterogeneous asymmetric reduction using inorganic supports.

No.	Reaction/Catalyst	Support	Immobilization method	Substrate	ee (%)		Reference(s)
					Hetero[a]	Homo[b]	
Hydrogenation of C–C							
1	Rh(1) complex of (Eto$_3$)-Si(CH$_2$)$_n$P(menthyl)$_2$ (n = 1,3,5)	Silica (Kieselgel 100)	Covalent	(Z)-2-Aetamido-3-phenyl acrylic acid	87	67	[22]
2	Rh(COD)proline derivatives	Silica and modified USY zeolite	Covalent	(Z)-α-Acylcinnamic acid derivatives	84.1	97.9	[23]
3	Dichloro-(S)-6,6-dimethyl-2,2-diaminobiphenyl Ru Complex	MCM-41	Covalent	α-Acetamidocinnamate	97	69.8	[24]
4	1,1′-Bis(diphenylphosphino)-Ferrocene Pd complex	MCM-41	Covalent	Ethyl 1,4,5,6,-tetrahydronicotinate	17	racemic	[16]
5	[((R,R)-BDPBzSO$_3$)Rh(nbd)] OTf, [(+)-DIOP-Rh(nbd)]OTf, [(S)-BINAP-Rh(nbd)]OTf	Silica	Hydrogen bonding	Olefins	3–53	NA[c]	[25b]
6	[(R,R)-Me-(DuPhos) Rh-(COD)]OTf	MCM-41	Adsorption	Methyl 2-acetamido-3-methyl but-2-enoate	98	96.2	[26]
7	Rh complex of DiPamp, Prophos, Me-DuPhos, BPPM	Heteropolyacid-clays	Ionic interactions	Methyl 2-acetoamidoacrylate	93	76	[27]
8	Rhodium-MonoPhos	Phosphotungstic acid on alumina	Ionic interactions	Methyl 2-acetoamidoacrylate	97	97	[28]
9	Rh complex of (R,R)DIOP, (S,S)Me-DuPhos, (R)ProPhos, (S,S)Me-BPE, (S,S)ChiraPhos	Al–MCM-41, Al–MCM-48, Al–SBA-15	Impregnation	Methyl α-acetamidoacrylate	95	90	[29]

10	[Ru(II)(bpea)((S)-(BINAP))-Cl]BF$_4$	AlPO$_4$, Zeolite	Covalent	Dimethyl itaconate	>99	>99	[30a]
11	BPPM-Rh Complex	Silica, alumina, silica based ion-exchanger	Ionic interactions	Methyl (Z)-α-acetamidocinnamate	95	78	[30b]
12	[Ru-BINAP-(SO$_3$Na)$_4$(C$_6$H$_6$)Cl]Cl	Controlled pore glass	Hydrophilic interaction	2-(6′-Methoxy-2′-naphthalen-2-yl) acrylic acid	70	96	[30c–32g]
Hydrogenation of C=N bond							
13	Ir-BPPM complex	Silica	Covalent	N-(2-methyl-6-ethylphen-1-yl) methoxymethylmethylketimine	55	45	[31]
14	Ferroenyl diphosphine Ir Complex	Silica	Covalent	N-(2-methyl-6-ethylphen-1-yl) methoxymethylmethylketimine	79	79	[32a]
15	(Salen)Ni or Pd	MCM-41, delaminated Zeolite (ITQ-2, ITQ-6)	Covalent	(E)-N-Benzyl-(1-phenylethy-lidene)imine	10–15	NA^c	[32b]
Catalytic reduction of C=O bond							
16	[Rh(COD)AEP][BF$_4$]	MCM-41	Covalent	(E)-α-Phenylcinnamic acid	92	racemic	[14a]
17	[Rh(COD)(S)-PMP] CF$_3$SO$_3$	MCM-41	Hydrogen bonding	Methyl benzoylformate	82 racemic	racemic	[13a]
18	Ru-BINAP–DPEN	Porous Zr phosphate	Covalent	Acetophenone	96.3	80	[33a]
19	Ru-BINAP	Porous Zr phosphate	Covalent	Methyl 3-oxobutanoate	95	98.3	[33b]

Table 2.1 Continued

No.	Reaction/Catalyst	Support	Immobilization method	Substrate	ee (%) Hetero[a]	ee (%) Homo[b]	Reference(s)
20	Ru complexes of 4,4′-substituted BINAPs	SBA-15	Covalent	Ethyl 3-oxo-3-phenylpropanoate	91	99.2	[33c]
21	Chitosan Pd complex	Silica	Ionic interactions	3-Methyl-2-butanone	100	NA[c]	[34a]
22	Ru–BINAP	Poly (dimethylsiloxane)	Encapsulation	Methyl acetoacetate	70	98	[34b]
23	Rh(cod)diamine	Sol-gel matrix, Silica	Covalent	2-Acetylnaphthalene	98	98	[35a–c]
24	(+)-Norephedrine with [{RuCl$_2$(η^6-C$_6$Me$_6$)}$_2$]	MCM-41	Covalent	Acetophenone	81	92	[35d]
25	(+)-Norephedrine with [{RuCl$_2$(η^6-C$_6$Me$_6$)}$_2$]	SBA-15	Covalent	Acetophenone	81	85	[35e]
26	Ru-TsDPEN	Amorphous silica gel, MCM-41, SBA-15	Covalent	Acetophenone	97	97	[36a]
27	*Trans*-(1R,2R)-diaminocyclohexane [Rh(COD)Cl]$_2$	large pore mesoporous organosilicas	Covalent	2-Acetylnaphthalene	61	55	[36b]
28	Cu/R-(−)-(DTBM-segphos) bisphosphine ligand	Charcoal	Impregnation	Acetophenone	93	93	[36c]

[a] Hetero: Using heterogeneous catalyst.
[b] Homo: Using homogeneous catalyst.
[c] NA = data not available in the literature under identical experimental conditions.

a detailed description of each example is beyond the scope of this chapter, selected cases are described herein on the basis of the immobilization methods. These nicely demonstrate that a significant improvement in the activity and stereoselectivity of the chiral catalyst can be achieved by heterogenizing it into the confined nanospaces of porous or layered inorganic materials.

2.2.1
Immobilization via a Covalent Link

A spectacular, site-isolation effect in heterogeneous asymmetric catalysis was first reported by Pugin *et al.* The asymmetric hydrogenation of imine **1** is important for the commercial production of *(S)*-metolachlor, a herbicide presently produced at >10 000 tons per year. In this reaction, whereas homogeneous Ir-BPPM (**2**) catalyst prepared with $[Ir(COD)Cl]_2$ was deactivated after 26% conversion (turnover frequency (TOF) min^{-1} = ~0), the covalently immobilized Ir catalysts, Si-PPM (**3**)-Ir, were much more active and productive (TOF min^{-1} = up to 5.1; Scheme 2.1) [31]. It is known that Ir-complexes in the presence of hydrogen can form catalytically inactive hydrogen-bridged dimers. The finding that the activities of immobilized catalysts increase with decreasing catalyst loading (Scheme 2.1) indicates that the dimer formation can be prevented by a site-isolation effect.

Corma *et al.* [23] immobilized a number of proline derivatives covalently on silica and modified USY zeolite (pore diameter 12–30 Å) (**5a** and **5b** in Scheme 2.2), and used them for the hydrogenation of various *(Z)-N*-acylcinnamic acid derivatives

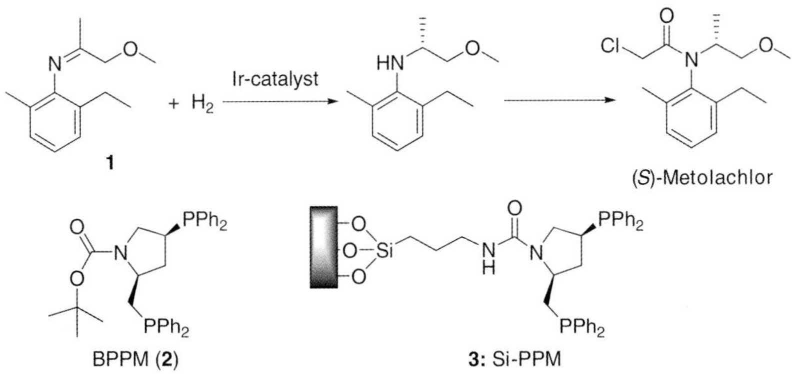

Ligand	Rate (TOF min^{-1})	ee (%)
BPPM (**2**)*	~0	45
Si-PPM (**3**) (0.016 mmol ligand g^{-1} support)	5.1	55
Si-PPM (**3**) (0.058 mmol ligand g^{-1} support)	2.4	55
Si-PPM (**3**) (0.19 mmol ligand g^{-1} support)	0.45	4.8

* Catalyst deactivated after 26% conversion

Scheme 2.1

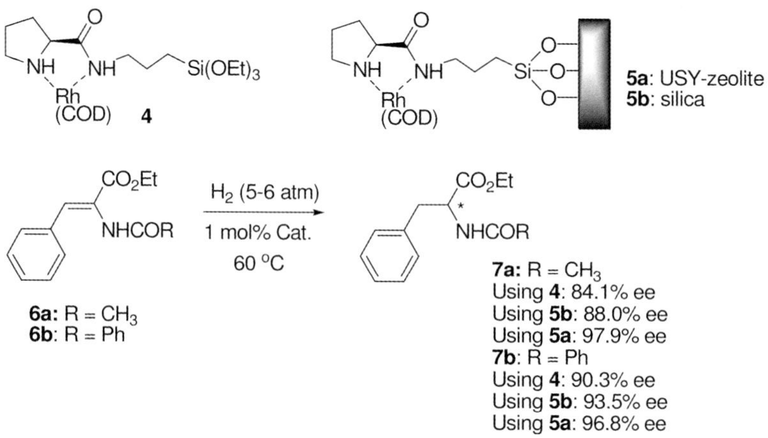

Scheme 2.2

6a: R = CH$_3$
6b: R = Ph

7a: R = CH$_3$
Using **4**: 84.1% ee
Using **5b**: 88.0% ee
Using **5a**: 97.9% ee
7b: R = Ph
Using **4**: 90.3% ee
Using **5b**: 93.5% ee
Using **5a**: 96.8% ee

5a: USY-zeolite
5b: silica

such as **6a** and **6b**. For all of the substrates tested using these catalysts, the ee-values of products **7** were higher with the zeolite-supported complex **5a** than with either the silica-supported **5b** or unsupported complex **4** (Scheme 2.2). This suggested that the steric constraints of the support play an important role, especially in the case of zeolite where the reaction must take place in the confined spaces of the supermicropores. This is the first example in which the steric constraints of the support (in this case, zeolite) were found to have a positive confinement effect on enantioselectivity [23a, b].

Recently, Pérez et al. also reported that the enantioselectivity of the supported complex was significantly increased compared to that obtained with the homogeneous complex [24]. (S)-MAB-Ru anchored covalently onto MCM-41 (**9-Ru**) catalyzed the asymmetric hydrogenation of itaconic acid (**10**) and α-acetamidocinnamic acid (**12**) to yield the (R)-products **11** and **13**, respectively, with 100% yield and circa 97% ee, while the homogeneous complex, (S)-MAB-Ru (**8-Ru**), gave only moderate ee-values of 80 and 69.8%, respectively (Scheme 2.3). The authors hypothesized that interaction of the ligands with the pendant hydroxy groups present on the pores of MCM-41 may increase the rigidity of the overall catalytic structure which, in turn, restricts the rotation of the transition state and favors formation of near-pure stereoisomers. In addition, the reaction yield remained at 100% while the enantioselectivity decreased only slightly, from 97 to 94%, after three runs.

The most striking examples for the positive confinement effect of supports on enantioselectivity have been provided by Thomas and Johnson's research group [13, 14, 16], who demonstrated that even nonenantioselective catalysts can show significant asymmetric induction when anchored onto a suitable support. The Pd catalyst **14** covalently supported within the pores of MCM-41 affected the hydrogenation of ethyl 1,4,5,6-tetrahydronicotinate (**16**) to afford nipecotic acid ethyl ester (**17**) at 17% ee (turnover number (TON) = 291), whereas the use of **15**, the

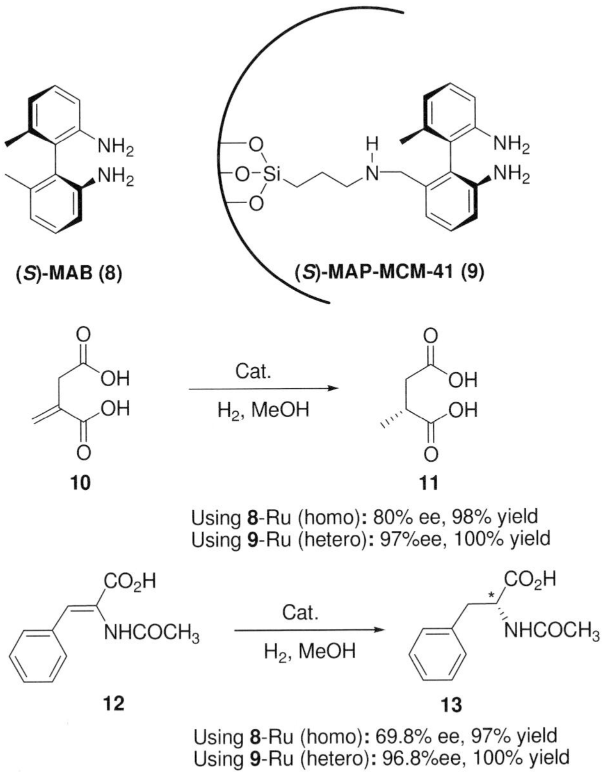

2.2 Asymmetric Reduction

(S)-MAB (8)

(S)-MAP-MCM-41 (9)

10 → **11**

Cat.
H$_2$, MeOH

Using **8**-Ru (homo): 80% ee, 98% yield
Using **9**-Ru (hetero): 97%ee, 100% yield

12 → **13**

Cat.
H$_2$, MeOH

Using **8**-Ru (homo): 69.8% ee, 97% yield
Using **9**-Ru (hetero): 96.8%ee, 100% yield

Scheme 2.3

soluble counterpart of **14**, resulted in a racemic mixture (TON = 98) (Scheme 2.4). This enhanced chiral induction was also explained in terms of the confinement effect of the ordered mesopore of MCM-41; that is, the active catalyst is conformationally confined by the interior of the pore and/or the access of the substrates to the catalyst is restricted [16]. The same research group [14a] covalently anchored (S)-2-aminomethyl-1-ethylpyrrolidine (AEP) or (1R,2R)-1,2-diphenylethylenediamine (DED) to the inner walls (concave surface) of porous MCM-41 (pore diameter 30 Å) and to the convex surface of nonporous silica (Cabosil). Catalytic reactions were carried out using heterogeneous Rh and Pd catalysts for the asymmetric hydrogenation of (E)-α-phenylcinnamic acid and methyl benzoylformate. Compared to the corresponding homogeneous catalyst or silica-supported catalyst, a significant enhancement of the ee-value was observed for the MCM-41-supported catalyst (with the concave surface). For example, quite surprisingly, the Rh(COD)- and Pd(allyl)-complex of AEP anchored on MCM-41, **18** and **19**, gave 92 and 87% ee, respectively, in the hydrogenation of methyl benzoylformate, while the corresponding homogeneous catalysts gave a racemic product (Scheme 2.5). This improvement was not observed with the catalyst immobilized on the convex surface of nonporous silica (Cabosil). This increase in the ee-values was attributed

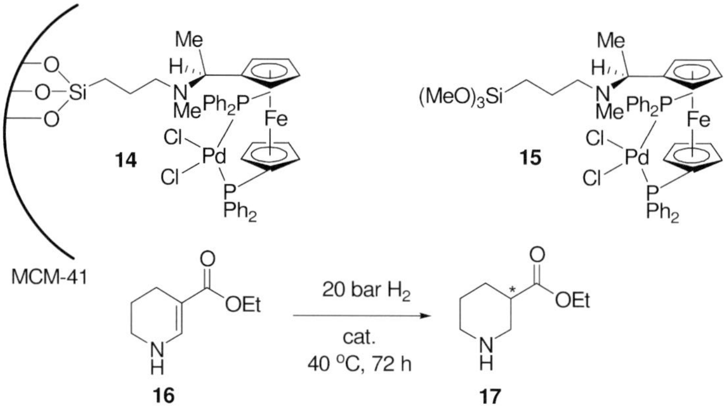

Scheme 2.4

Using **15** (homo): racemic, TON = 98
Using **14** (hetero): 17% ee, TON = 291

Scheme 2.5

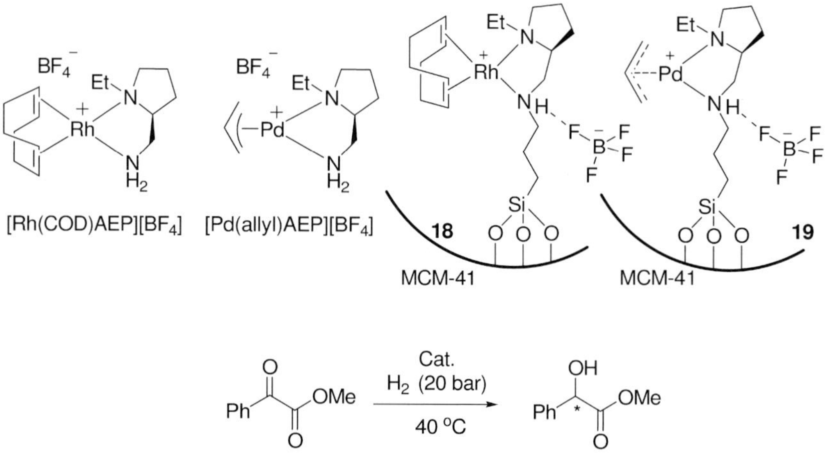

Using Rh(COD)AEP; racemic
Using heterogeneous catalyst **18**; 92% ee
Using Pd(allyl)AEP; racemic
Using heterogeneous catalyst **19**; 87% ee

mainly to the restricted access of the reactant to the active site generated by the pore's concavity.

2.2.2
Immobilization via Hydrogen Bonding, Ionic and Other Interactions

Bianchini et al. [25] disclosed a highly interesting immobilization approach which involves a hydrogen-bonding interaction between the silanol groups of silica and

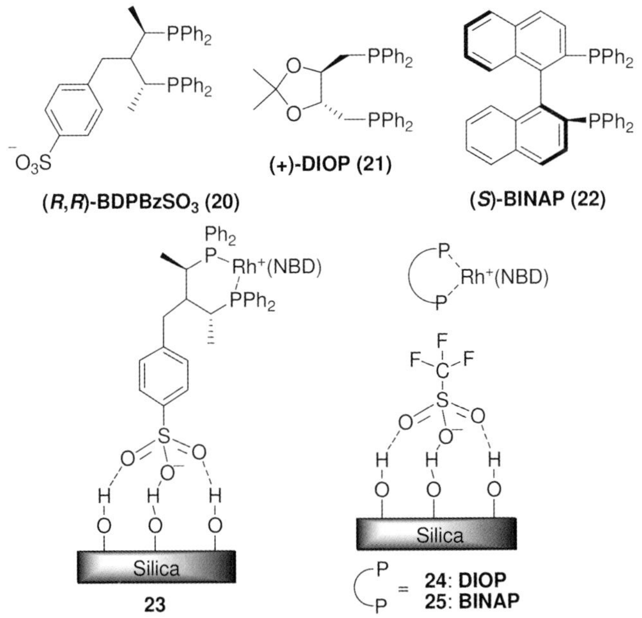

Scheme 2.6

sulfonate groups from the ligands and, also from the triflate counteranion of metal complexes, TfO⁻. By employing this immobilization approach, the Rh-complexes of the three optically pure bisphosphines **20**, **21** and **22**, [((R,R)-BDPBzSO₃)Rh(nbd)], [(+)-DIOP-Rh(nbd)]OTf and [(S)-BINAP-Rh(nbd)]OTf, were successfully immobilized on silica to give the supported hydrogen-bonded (SHB) catalysts **23**, **24** and **25**, respectively (Scheme 2.6). As expected, no immobilization was observed when the triflate counteranion in either DIOP– or BINAP–Rh complexes was replaced by other counteranions such as BPh₄⁻ that are not capable of hydrogen-bonding interactions with silica. Unfortunately, the enantioselectivities obtained for the hydrogenation of olefins with these immobilized catalysts were quite low (3–53% ee).

Similarly to Bianchini's approach, De Rege [26] also immobilized cationic [((R,R)-Me-duphos (**26**))Rh-(COD)]OTf complex noncovalently by the hydrogen-bonding interaction of triflate counterion with surface silanols of MCM-41 support. In contrast to the results obtained by Bianchini *et al.* [25c], the catalytic activity and selectivities of the immobilized **26**-Rh complex on MCM-41 were equal to or greater than the homogeneous counterparts (Scheme 2.7). Moreover, the catalysts were recyclable (up to four times, with no loss of activity) and did not leach. Here again, the counteranion was very important for the successful immobilization of the catalyst onto MCM-41. Whereas, the DuPhos–Rh complex with triflate anion was effectively immobilized (6.7 wt% based on Rh), the analogous complex with the lipophilic BAr$_F$ anion [BAr$_F$ = B(3,5-(CF₃)₂-C₆H₃)] was not loaded onto the support.

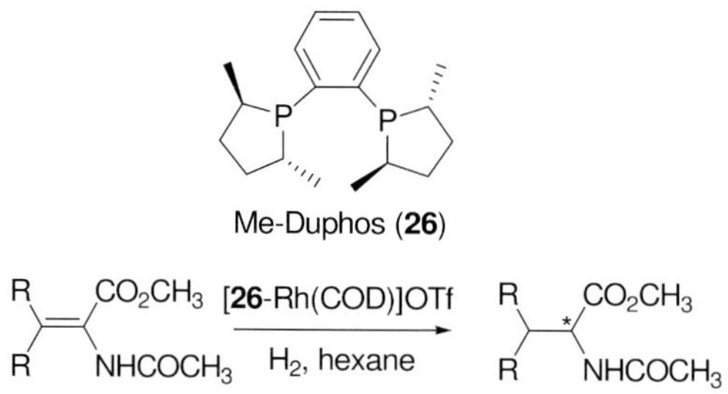

Me-Duphos (**26**)

$$\underset{R}{\overset{R}{\diagdown}}C=C\underset{NHCOCH_3}{\overset{CO_2CH_3}{\diagdown}} \xrightarrow[H_2, \text{hexane}]{[26\text{-Rh(COD)}]\text{OTf}} \underset{R}{\overset{R}{\diagdown}}CH-\overset{*}{C}H\underset{NHCOCH_3}{\overset{CO_2CH_3}{\diagdown}}$$

R = H
Using **26-Rh** (homo): 87% ee
Using MCM-41 supported **26-Rh**; 99% ee
R = Me
Using **26-Rh** (homo): 85% ee
Using MCM-41 supported **26-Rh**; 98% ee

Scheme 2.7

Recently, the same approach (i.e. noncovalent immobilization via H-bonding of the triflate ion, $CF_3SO_3^-$ to the silanol groups of inner surface of MCM-41) was adopted also by Thomas and Johnson's group [13a] to immobilize the Rh-complexes such as [Rh(COD)(S)-AEP][CF_3SO_3] and [Rh(COD)(1R,2R)-DED][CF_3SO_3] onto the inner surface of a set of silicas with narrowed pore size distributions (38, 60 and 250 Å). The catalytic performances of the heterogeneous catalysts were tested in the asymmetric hydrogenation of methyl benzoylformate to produce methyl mandelate. The noncovalently supported catalysts generally exhibited remarkably higher ee-values than their homogeneous counterparts. Surprisingly, the heterogeneous catalysts **27a** and **28a** gave up to 82 and 79% ee, respectively, while their two homogeneous counterparts, [Rh(COD)AEP][CF_3SO_3] and [Rh(COD)DED][CF_3SO_3], did not display any enantioselectivity (Scheme 2.8). The increase in the enantiocontrolling ability of the supported catalysts was observed in a manner that logically reflected the declining influence of spatial constraint in proceeding from the 38 Å pore diameter silica, through 60 Å and finally to 250 Å (Scheme 2.8). A similar enhancement of enantioselectivity in the hydrogenation of (E)-α-phenylcinnamic acid affording 2,3-diphenylpropanoic acid was also observed by heterogenizing [Rh(COD)(S)-AEP][CF_3SO_3] and [Rh(COD)(1R,2R)-DED][CF_3SO_3] [14b].

In a different approach, Augustine et al. [27] immobilized various Rh-complexes of several bisphosphine chiral ligands such as DiPamp (**29**) onto phosphotungstic acid (PTA)-modified inorganic supports such as montmorillonite K and alumina. The immobilization is based on the interaction of the ligand–metal and metal–PTA, as well as the support with an oxygen atom or hydroxyl group of the PTA. Therefore, a synthetic modification of the ligand is not necessary for the immobilization. Somewhat surprisingly, both reaction rates and enantioselectivities significantly increased with progressing catalyst recycling. For example, in the hydrogenation of methyl 2-acetoamidoacrylate using Rh(DiPamp, **29**)–PTA–

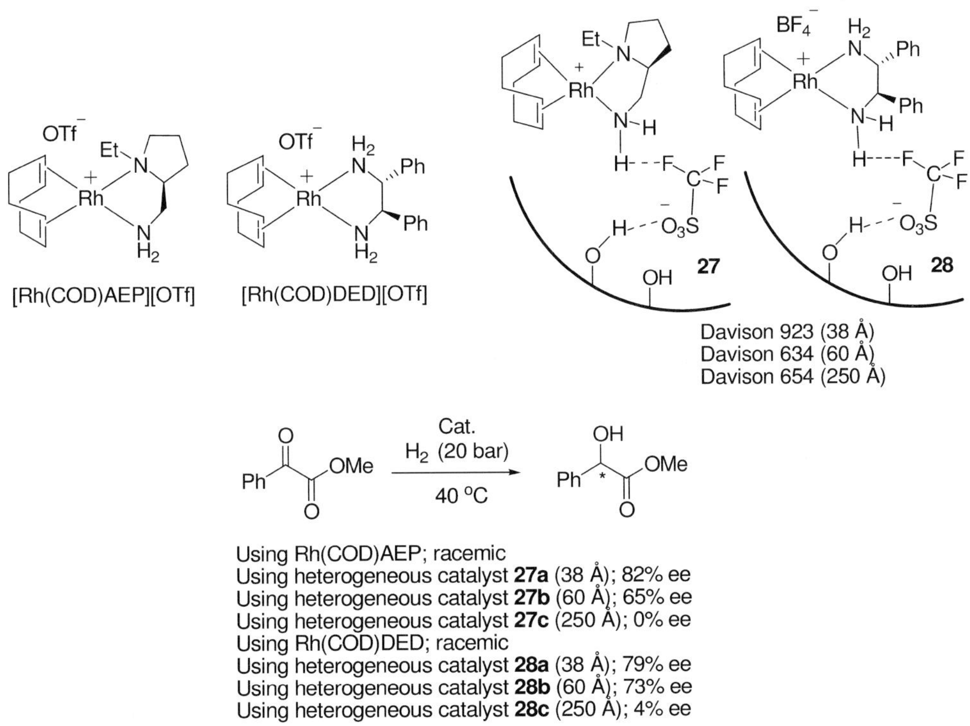

Scheme 2.8

montmorillonite K catalyst, the TOF and enantioselectivity (0.18 min^{-1}, 67% ee) for the first run increased significantly (1.29 min^{-1}, 97% ee) after nine recycles (Scheme 2.9). Simons et al. [28] also anchored the Rh-complex of MonoPhos (**30**) on PTA-modified alumina (Scheme 2.10). In the asymmetric hydrogenation of methyl-2-acetamidoacrylate, the supported catalyst **31** gave higher activity (TOF h^{-1} = 2300; 97% ee in EtOAc) than the homogeneous counter part (TOF h^{-1} = 1700; 97% ee in CH$_2$Cl$_2$). It was proposed that the superior anchoring ability was derived from the type of bonding between Rh and the PTA, which is thought to be partially covalent.

The immobilization of a chiral metal–ligand complex by ion exchange is another attractive strategy, of which no structural modification of the chiral ligand is required. Hoelderich et al. [29] anchored the Rh-complexes ([Rh(P–P)COD]Cl of various diphosphine ligands (P-P) such as *(R,R)*-DIOP (**21**), *(S,S)*-Me-DuPhos (**26**), *(R)*-ProPhos (**32**), *(S,S)*-Me-BPE (**33**), and *(S,S)*-ChiraPhos (**34**) over Al–MCM-41, Al–MCM-48, and Al–SBA-15 (Scheme 2.11). This heterogenization is based on an ionic interaction between the negatively charged Al-MCM-41, Al–MCM-48, and Al-SBA-15 framework and the cationic rhodium complex. In many cases, the heterogeneous catalysts **35** showed superior performance compared to its homogeneous counterpart, for example, the heterogeneous

2 Heterogeneous Enantioselective Catalysis Using Inorganic Supports

DiPamp (**29**)

Using homogeneous catalyst Rh-**29**; 76% ee, TOF(min^{-1})=0.25
Using PTA-montmorillonite-supported Rh-**29**; 67% ee, TOF(min^{-1})=0.18 (1st run)
92% ee, TOF(min^{-1})=1.20 (2nd run)
97% ee, TOF(min^{-1})=1.29 (9th run)
97% ee (15th run)

Scheme 2.9

MonoPhos (**30**)

PTA-modified alumina
31

Scheme 2.10

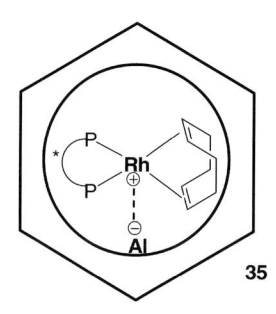

21   (S,S)-**26**   ProPhos (**32**)   (S,S)-Me-BPE (**33**)   (S,S)-Chiraphos (**34**)

35

Scheme 2.11

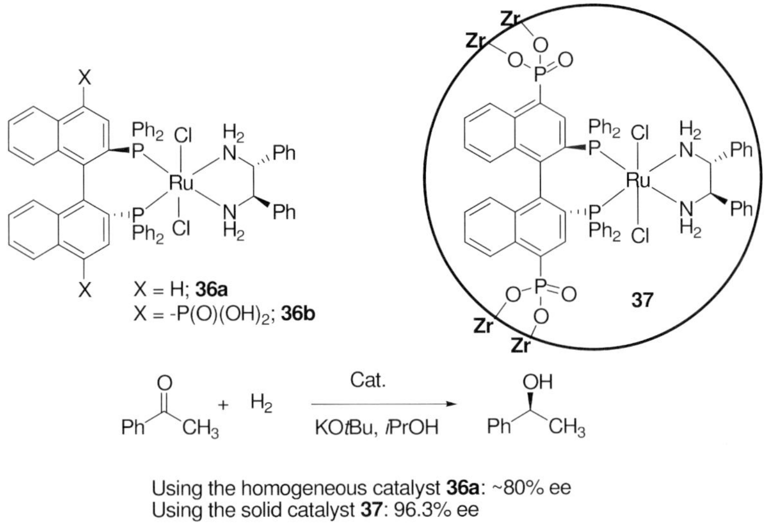

Scheme 2.12

Using the homogeneous catalyst **36a**: ~80% ee
Using the solid catalyst **37**: 96.3% ee

Rh-MeBPE(**33**)-Al-MCM-48 catalyst exhibited 95% ee for the hydrogenation of methyl α-acetamidoacrylate, while its homogeneous counterpart showed 90% ee. These chiral heterogeneous catalysts can be reused at least four times without any activity loss, with a TON in excess of than 4000.

In an another approach, Lin et al. [33a] immobilized the functionalized Ru–BINAP–DPEN derivative **36b** by self-assembly with one equivalent of soluble Zr(O*t*Bu)$_4$ to give chiral, porous, zirconium phosphonate **37**. This chiral, porous, hybrid solid **37** catalyzed the asymmetric hydrogenation of unfunctionalized aromatic ketones with activity and enantioselectivity remarkably higher than those of the parent homogeneous counterpart Ru–BINAP–DPEN system **36a**. For example, acetophenone was hydrogenated to 1-phenylethanol with complete conversion and 96.3% ee in 2-propanol with 0.1 mol% loading of the solid catalyst **37**. This level of enantioselectivity is significantly higher than that observed for the parent homogeneous catalyst **36a**, which typically gives ~80% ee for the same reaction (Scheme 2.12). A similar method has also been used to synthesize other chiral, porous, hybrid ruthenium solids [33b]. An ee-value of up to 95% was achieved in the heterogeneous asymmetric hydrogenation of β-ketoesters.

From the above examples, it is clear that the conventional heterogenization approach – that is, heterogenizing those catalysts designed for homogeneous processes – may not necessarily be the best solution for the development of efficient heterogenized chiral catalyst systems when it comes to immobilizing catalysts in the confined nanospaces of porous or layered support materials. Moreover, the above-mentioned results also suggest that the heterogenization of catalysts in confined spaces of inorganic materials may provide a new paradigm for the development of highly efficient chiral catalysts.

2.3
Asymmetric Oxidation

Various heterogeneous asymmetric oxidation reactions using inorganic supports are listed in Table 2.2. A few selected examples will be described herein on the basis of reaction type.

2.3.1
Asymmetric Epoxidation (AE) of Unfunctionalized Olefins

Asymmetric epoxidation (AE) of unfunctionalized alkenes catalyzed by chiral (salen)Mn(III) complex **38** (Scheme 2.13), developed by Jacobsen et al., is one of the most reliable methods [50]. As shown in Table 2.2, several different strategies have been formulated to immobilize Jacobsen's catalysts on inorganic supports [37–42]. Facilitation of catalyst separation, catalyst reuse, an increase in catalyst stability (e.g. minimization of the possibility of formation of inactive μ-oxo-manganese(IV) species [51a,b]) and sometimes improvement in enantioselectivity are the main objectives of such research. Heterogenized Mn(salen) systems have recently been reviewed by Salvadori et al. [51c] and Garcia et al. [51d]. Some selected cases are therefore described herein on the basis of the immobilization methods.

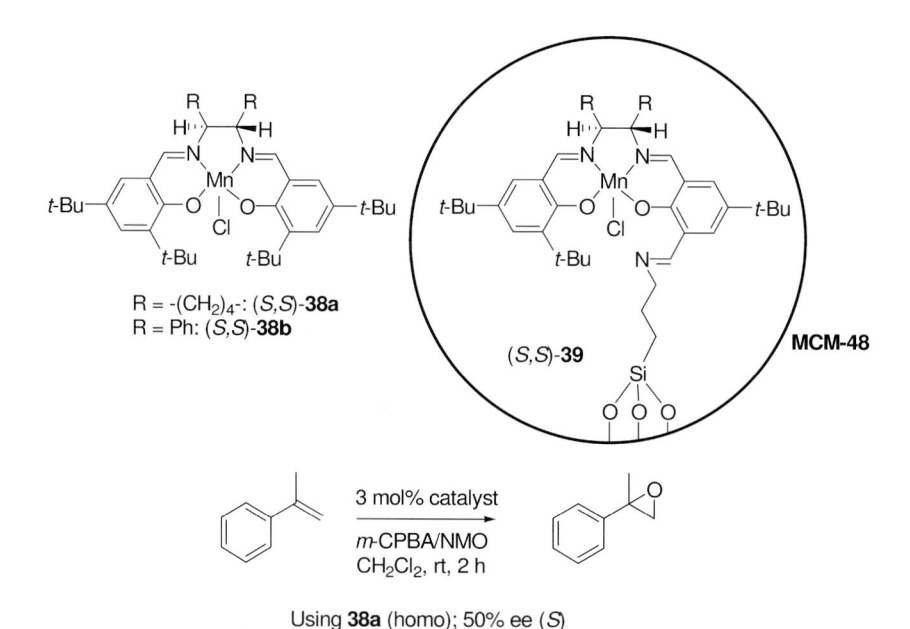

Scheme 2.13

2.3 Asymmetric Oxidation | 41

Table 2.2 Examples of heterogeneous asymmetric oxidation using inorganic supports.

No.	Reaction/Catalyst	Support	Method	Substrate	ee (%) Hetero[a]	ee (%) Homo[b]	Reference(s)
Asymmetric epoxidation (AE) of unfunctionalized olefins
1	Chiral (Salen)Mn complex	MCM-48	Covalent	α-Methylstyrene	>99	50	[37a]
2	Jacobsen [(Salen)MnCl]	Silica (Kieselgel 100)	Covalent	1-Phenylcyclohexene	58	>99	[37a]
3	Chiral (Salen)Mn complex	Amorphous silica, MCM-41	Covalent	1-Phenylcyclohexene	84	89	[37b]
4	Chiral (Salen)Mn complex	Modified MCM-41, SBA-15	Covalent	4-Chlorostyrene	73	46	[37c]
5	Chiral (Salen)Mn complex	MCM-48	Covalent	α-Methylstyrene	>99	50	[37d]
6	Chiral (Salen)Mn complex	MCM-41	Covalent	Styrene	70	65	[37e]
7	Chiral (Salen)Mn complex	MCM-41	Coordination	α-Methylstyrene	72	56	[38a]
8[d]	Chiral (Salen)Mn complex	MCM-41/SBA-15	Coordination	cis-β-Methylstyrene	94.8	54.8	[38b, c]
9	Chiral (Salen)Mn complex	Mesoporous MCM	Coordination	6-Cyano-2,2-dimethylchromene	90.6	80.1	[38d]
10	Chiral (Salen)Mn complex	Carbon	Coordination	α-Methylstyrene	57	45	[38e]
11	Binaphthyl Schiff base Cr	MCM-41	Coordination	cis-β-Methylstyrene	73	54	[38f]

Table 2.2 Continued

No.	Reaction/Catalyst	Support	Method	Substrate	ee (%) Hetero[a]	ee (%) Homo[b]	Reference(s)
12	Chiral (Salen)Mn complex	Montmorillonite	Encapsulation/electrostatic interaction	Styrene	70	45	[39a, b]
13	Jacobsen [(Salen)MnCl]	Al-MCM-41	Ion-exchange	cis-Stilbene	70	77	[39d]
14	Chiral (Salen)Mn complex	Al pillared clay	Encapsulation/electrostatic interaction	Styrene	14	6	[39e]
15	Chiral sulfanato Salen-Mn complex	Mg[II]-Al[III] layered double hydroxide	Encapsulation/electrostatic interaction	6-Cyanochromene	96	89	[39f]
16	Chiral (Salen)Mn complex	Synthetic Laponite	Ion-exchange	1,2-Dihydronaphthalene	32	46	[21]
17	(1R,2R)-(−)-diaminocyclohexane	Mn-exchanged Nanocrystalline MgO	Electrostatic interaction	Styrene	42	0	[40]
18	Chiral (Salen)Mn complex	MCM-22	Ship-in-a-bottle	α-Methylstyrene	91.3	51	[41a]
19	Dimeric Jacobsen (Salen)Mn Complex	Poly(dimethylsiloxane)	Entrapment	trans-β-Methylstyrene	18	0.5	[34b, 42d]
20	Chiral ruthenium porphyrin complex	MCM-41/MCM-48	Covalent	Styrene	75	69	[42e]

2.3 Asymmetric Oxidation

Asymmetric epoxidation (AE) of allylic alcohols							
21	Ti tartaric acid derivatives	Silica, MCM-41	Covalent	Allyl alcohol	86	83	[43a]
22	(2S,4R)-4-Hydroxyproline Mo(VI)	Modified USY-Zeolite	Covalent	Nerol	64	10.4	[43b]
23	Ta alkoxides/Ti alkoxides	Silica	Covalent	trans-2-Hexen-1-ol	90	96	[43c, d]
Asymmetric epoxidation (AE) of enones							
24	Polyamino acids (poly-L-(neo-pentyl)glycine (PLN))	Si	Adsorption	(E)-1-(2-aminophenyl)-3-phenyl prop-2-en-1-ones	97	97	[44]
25	(+)-Diethyl tartrate	Nanocrystalline MgO	Chemisorption	Chalcone	90	94	[45]
Asymmetric dihydroxylation (AD) of olefins							
26	Os$_3$(CO)$_{12}$	Al-MCM-41	Covalent	trans-Stilbene	90	Racemic	[15]
27	K$_2$OsO$_4$·2H$_2$O	Nanocrystalline MgO	Ion-exchange	trans-Stilbene	99	>99	[46a]
28	(DHQ)$_2$PYRD	Silica gel	Covalent	3-Hexene	45	93	[46b]
29	(DHQ)$_2$PHAL	Silica gel	Covalent	trans-Stilbene	99	>99	[46c]
30	(DHQ)$_2$PHAL	SBA-15	Covalent	trans-Stilbene	99.5	99.5	[46d]
31	(DHQD)$_2$PHAL	SBA-15, SiO$_2$	Covalent	trans-Stilbene	>99	>99	[46e]
32	(DHQD)$_2$PHAL with Titanosilicate-1	Silica gel	Covalent	trans-Stilbene	99	>99.5	[46f]

Table 2.2 Continued

No.	Reaction/Catalyst	Support	Method	Substrate	ee (%) Hetero[a]	ee (%) Homo[b]	Reference(s)
33	(DHQD)₂PHAL	Magnetic mesocellular mesoporous silica	Covalent	*trans*-Stilbene	>99.5	>99.5	[46g]
Asymmetric aminohydroxylation (AA) of olefins							
31	SGS-(QN)₂PHAL	Silica gel	Covalent	(E)-Isopropyl cinnamate	>99	99	[47a]
32	K₂OsO₄·2H₂O	Layered double hydroxides (LDH)	Ion-exchange	(E)-Methyl cinnamate	78	81	[47b]
Asymmetric aziridination of olefins							
33	Bis(oxazolines)	Copper-exchanged zeolite (CuHY)	Electrostatic interactions	Styrene	82	43	[48]
Asymmetric sulfoxidation							
34	Chiral (Salen)Mn/Cu complex	MCM-41, Silica	Covalent	Methyl phenyl sulfide	26	27	[49a]
35	K₂OsO₄·2H₂O	LDH	Ion-exchange	Methyl phenyl sulfide	51	NA[c]	[49b]

a Hetero: Using heterogeneous catalyst.
b Homo: Using homogeneous catalyst.
c NA = data not available in the literature under identical experimental conditions.
d ee given for *trans* product.

2.3.1.1 Immobilization via a Covalent Link

In an interesting example of this approach, Liu et al. [37a] recently immobilized chiral Mn(III) complexes **38** on the three-dimensional (3-D) cubic structure MCM-48. Very interestingly, in the epoxidation of α-methylstyrene in CH_2Cl_2 with m-CPBA/NMO as oxidant, the immobilized catalyst **39a** showed significantly improved enantioselectivity (>99% ee) compared to that (50% ee) of the homogeneous catalyst **38a** (Scheme 2.13). The authors proposed that, due to the 3-D cubic structure of the MCM-48, the diffusibilities of the reacting partners were more accessible to favor the desired specific geometry and energy of the transition state, thereby enhancing the enantioselectivity. The catalyst **39a** could be recycled for up to three successive runs for AE of α-methylstyrene while retaining its enantioselectivity.

2.3.1.2 Immobilization via Coordination, Ionic and Other Interactions

In addition to the above means of heterogenizing via a covalent link, Li et al. [38a] immobilized the chiral(salen)Mn complexes through the apical coordination of salen–manganese by the oxygen atoms of the phenoxyl groups grafted on the surface of MCM-41 (**40** in Scheme 2.14). Here again, heterogenization markedly increased the ee-values compared to those observed for the free complex in the AE of simple olefins. For example, in the epoxidation of α-methylstyrene in CH_2Cl_2 with NaOCl as the oxidant, the ee-value was notably increased from 56% for the homogeneous catalyst to 72% for the heterogeneous Mn(salen)/MCM-41 catalyst (**40**). Soon after, the same research group also reported that chiral manganese(salen) catalysts axially immobilized in nanopores via Ph sulfonic groups (**41**) also resulted in remarkably higher ee-values (up to 95%) for the AE of unfunctionalized olefins [38b–d]. In a similar manner, Freire et al. [38e] anchored the chiral (salen)Mn complexes by direct axial coordination of the Mn center onto the phenolate groups of a modified commercial activated carbon. It has also been reported by Zhou et al. [38f] that the ee-value for the enantioselective epoxidation of β-methylstyrene increased from 54 to 73% after the immobilization of Cr(salen) ((R)–Cr-binaphthyl Schiff base) through axial NH_2 complexation.

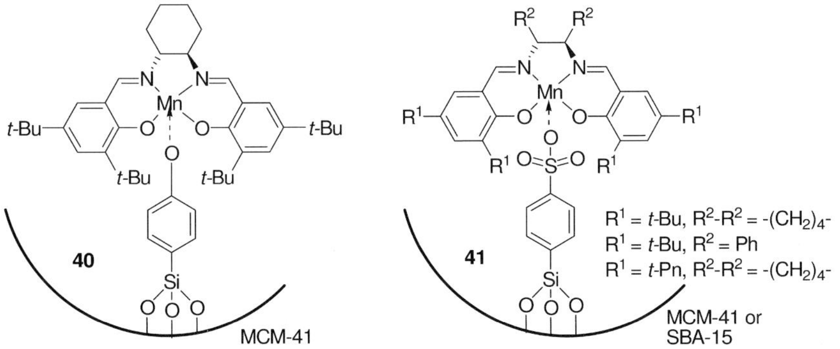

Scheme 2.14

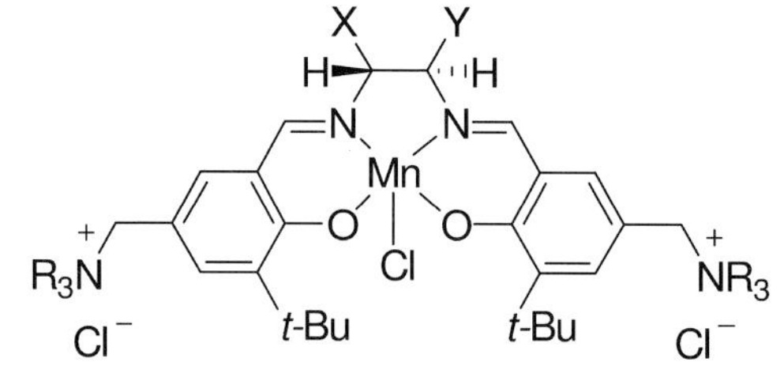

42

X = Y = Ph; R = ethyl, *i*-octyl
X--Y = -(CH$_2$)$_4$-; R = ethyl, *i*-octyl

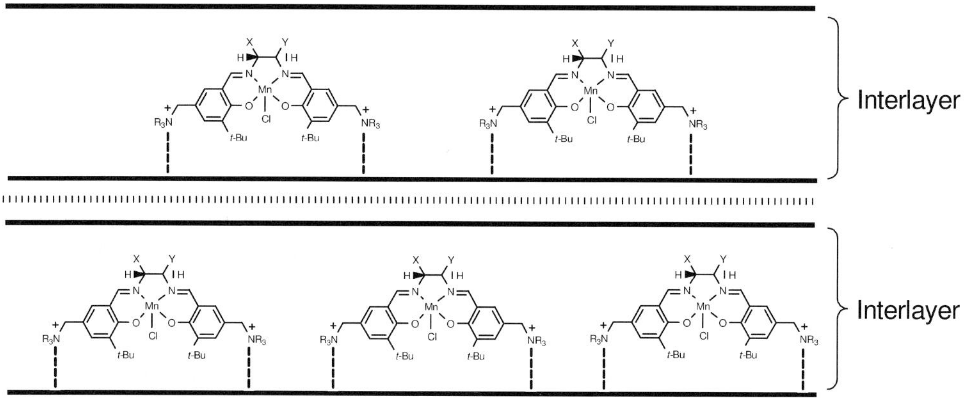

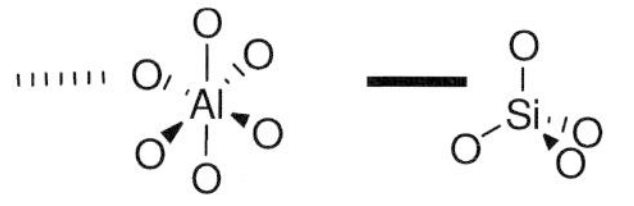

Scheme 2.15

Other than the aforementioned strategy, clay-supported chiral catalysts were also prepared by direct ion-exchange of the chiral (salen)Mn complexes. For example, Kureshy et al. [39a] immobilized the dicationic [(salen)Mn(III)] complexes **42** by cation exchange in the interlayers of montmorillonite (Scheme 2.15). In the AE of olefins in the presence of pyridine N-oxide as an axial base with NaOCl as an oxidant, the heterogeneous catalyst exhibited improved enantioselectivity. For example, in the case of styrene, significantly higher ee-values (69–70%) in the product epoxide were obtained with supported catalyst **43** than with their homogeneous analogues **42** (41–52% ee). This significant enhancement in ee was explained in terms of the unique spatial environment of the confined medium. Furthermore, the flexibility of the lamellar structure of the montmorillonite allowed for the selective conversion of sterically more demanding substrates such as 2,2-dimethyl-6-nitrochromene. The interlayer distance between cationic and anionic silicate sheets of mineral clays is usually increased (up to 10 nm) by swell-

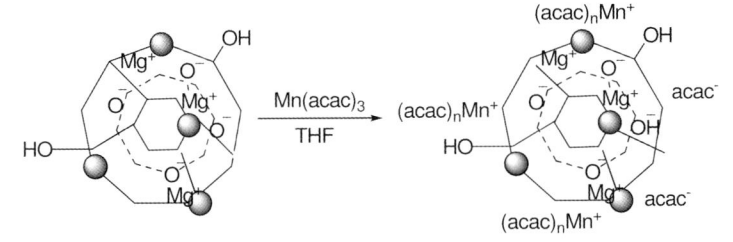

Scheme 2.16

ing in either water or an alcohol, to allow the ion-exchange of the large cationic complex (about 2 nm) into the inter-crystal layer [39c].

Choudary et al. [40] recently reported a new and quite useful approach of heterogenizing Mn(III) complexes. The Mn(acac)$_3$ complex was ion-exchanged successfully on aerogel-prepared nanocrystalline MgO, NAP-MgO (SSA: 590 m^2 g^{-1}), to obtain the heterogeneous catalyst **44** (Scheme 2.16). Catalyst **44** with (1R,2R)-(−)-diaminocyclohexane (DAC) as chiral ligand afforded good reactivity (90% yield, 18 h) and excellent enantioselectivity (91% ee) in the AE of 6-cyanochormene, taking tert-butyl hydroperoxide (TBHP) as an oxidant and tetrahydrofuran (THF) as a solvent of choice at a temperature of −20 °C. It was claimed that the catalyst could be recycled up to three times.

2.3.1.3 Immobilization by the 'Ship-in-a-Bottle' Approach

Alternatively, a 'ship-in-a-bottle' (SIB) approach to entrap the Jacobsen's catalyst **38** within the spacious supercages (7.1 × 18.2 Å) of zeolite MCM-22 (Scheme 2.17) has been developed by Gbery et al. [41a], by considering problems of a previously reported method on zeolite Y or EMT (hexagonal) which were plagued with the following two problems [41b–d]. First, the synthetic zeolites having FAU (cubic) or EMT topologies have been employed as host matrices for SIB complexes, but these structures cannot accommodate Jacobsen's catalyst **38**. The second problem was the method of encapsulation, which involved assembling the complex within the pores. Invariably, this strategy results in uncomplexed or partially complexed metal ions and ligands, as well as an uneven distribution of occluded species. In this view, MCM-22 has the advantage of encapsulating intact Jacobsen's catalyst **38** since MCM-22 has a very large cage (7.1 × 18.2 Å in diameter) and its synthesis proceeds through a layered precursor. That is, at this stage of synthesis, Jacobsen's catalyst can be intercalated between the layers. Thus, Gbery et al. [41a] successfully occluded Jacobsen's catalyst **38a** in MCM-22 during the synthesis of MCM-22. A significant enhancement of the enantioselectivity as a result of the confinement effect was observed in the epoxidation of α-methylstyrene with sodium hypochlorite (91.3% ee with **45** versus 51% ee with **38a**) (Scheme 2.17). Moreover, the encapsulated complex was threefold more active than the homogeneous counterpart **38a**, which was also attributed to site-isolation – that is, to the suppressed formation of dimeric and other oligomeric Mn oxo-complexes due to geometric

Scheme 2.17 (from Ref. [41a])

Supercage of MCM-22

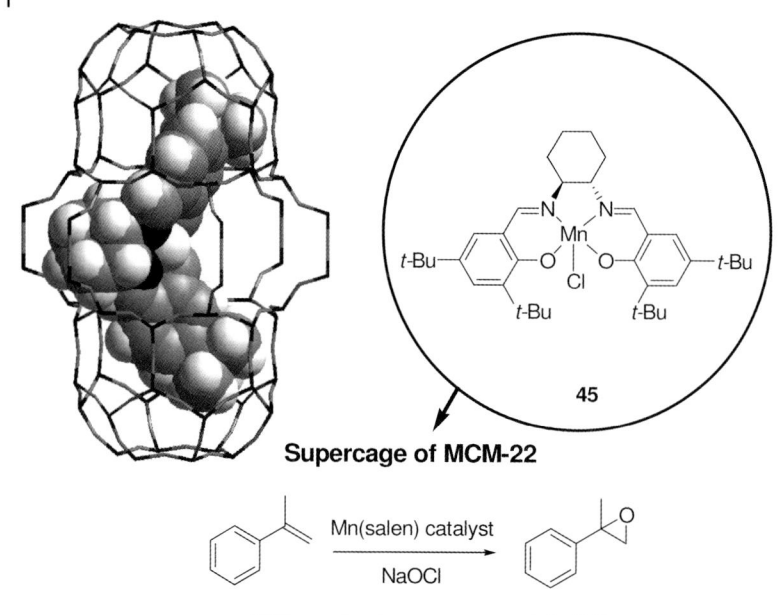

51% ee using homogeneous catalyst **38a**
91% ee using the entrapped catalyst **45**

constraints. A similar SIB strategy was also applied by Holderich et al. [41e–g] to encapsulate **38** and other (salen)metal-complexes into the X, Y and XY zeolites and the specially modified, faujasites host materials. The best result (100% conversion, 96% epoxide chemoselectivity and 91% diastereomeric excess (de)) was obtained in the diastereoselective epoxidation of (−)-α-pinene at room temperature and elevated pressure using O_2 as an oxidant and catalyzed by entrapped cobalt(II)(salen) complex without suffering the leaching problem [41g].

2.3.1.4 Supported Ionic Liquid Catalysis (SILC)

A relatively new concept of supported ionic liquid catalysis (SILC) has been developed by Mehnert et al. [42a,b]. Liu et al. [42c] recently adopted this methodology for immobilizing the chiral (salen)Mn(III) complexes **38**. The immobilized catalyst **46** ([bmim][PF$_6$]/**38a**-MCM-48) on a supported ionic liquid phase (MCM-48 was taken as a support) exhibited notably higher enantioselectivity and comparable catalytic activity (>99% ee; 99% conversion) than those of the homogeneous catalyst **38a** (50% ee; >99% conversion) in the α-methylstyrene epoxidation using m-CPBA/NMO oxidant (Scheme 2.18). Both, the obtained conversion and enantioselectivity were significantly higher than those already reported for inorganic supports [37e, 38a,b]. This may be due to the combination of solution dynamics (ionic liquid) and the spatial effect originated from the heterogeneous carrier media MCM-48, which possesses a well-defined pore size and 3-D topological structure. The heterogeneous catalysts **46** and **47** were recycled and reused without

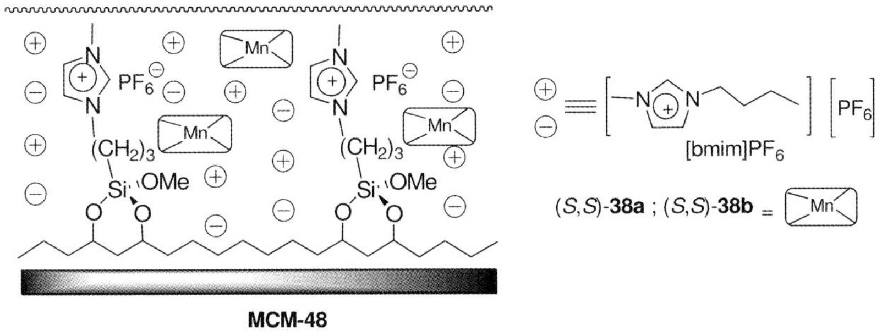

[bmim][PF₆] / [1-Methyl-3-(3-methoxysilylpropyl)-imidazolium][PF₆]-MCM48-**38a** = 46
[bmim][PF₆] / [1-Methyl-3-(3-methoxysilylpropyl)-imidazolium][PF₆]-MCM48-**38b** = 47

Scheme 2.18

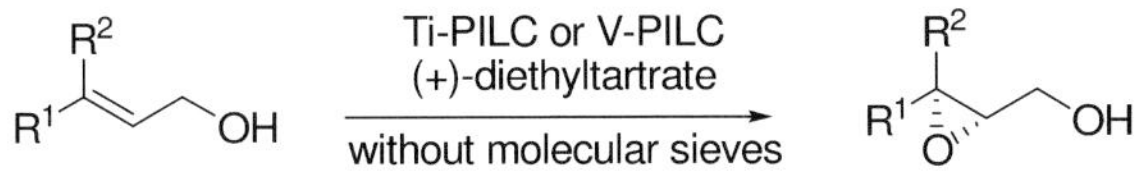

R^1 = Me, n-Pr, n-hex, Ph, Me$_2$C=CHCH$_2$CH$_2$
R^2 = H or Me

using Ti-PILC : 90–98% ee
using V-PILC : 20% ee (R^1=n-Pr, R^2=H)

Scheme 2.19

any obvious decrease in activity and enantioselectivity at least for three runs, in the AE of 1-phenylcyclohexene. The leaching of ionic liquid and (salen)Mn(III) complex have not been detected by nuclear magnetic resonance (NMR) and inductively coupled plasma (ICP) analyses.

2.3.2
AE of Allylic Alcohol

The Sharpless AE of allylic alcohols has become a benchmark classic in catalytic asymmetric synthesis [52a,b] and has found use in some industrial applications [52c]. Although this catalytic process seems to be well understood, a heterogeneous system would be advantageous as it could avoid a complicated separation of the product from the catalyst, which can lead to decomposition of the epoxide formed [52c]. However, only a few heterogeneous versions of this important reaction have been conducted successfully [43, 51c, 53].

A successful example for this class of reaction was reported by Choudary et al. [53a]. With the combination of a dialkyl tartrate and titanium-pillared montmorillonite (Ti-PILC), excellent ee-values in the range of 90–98% were achieved (Scheme 2.19). In contrast to the homogeneous conditions, this heterogeneous system was operational without the use of molecular sieves; however, no recycling experiment was reported. Distinct from Ti-PILC, the use of vanadium-pillared montmorillonite catalyst for the AE of (E)-hex-2-enol, however, led to only 20% enantiomeric excess [53b].

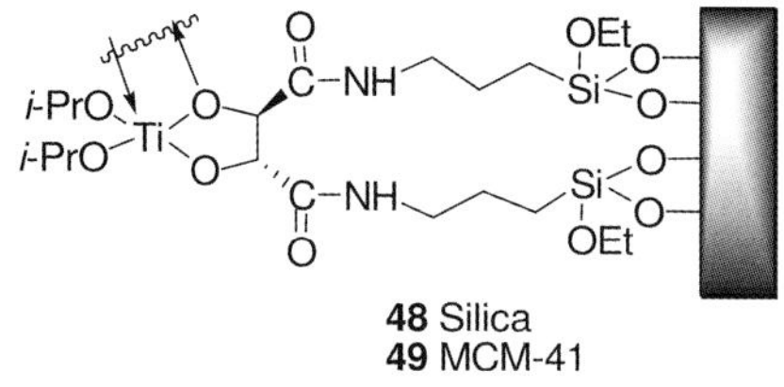

Scheme 2.20

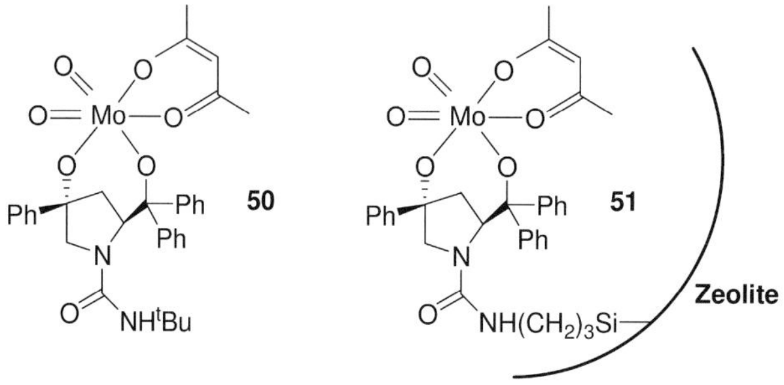

48 Silica
49 MCM-41

Scheme 2.21

Recently, Li *et al.* [43a] also reported a successful result in this reaction. A chiral tartaric acid derivative grafted onto the surface of silica and separately in the mesopores of MCM-41 material (**48** and **49**, respectively, in Scheme 2.20) exhibited similar activity and enantioselectivity (86% ee) compared to the homogeneous Sharpless system (83% ee) in the AE reaction of allyl alcohol. Although the product could be separated easily from the catalyst by simple filtration, no data were available for the recyclability of the heterogeneous catalyst.

Another example was provided by Corma *et al.* The use of Mo(VI) complexes of chiral ligands derived from (2S,4R)-4-hydroxyproline heterogenized onto a modified USY-zeolite by covalent bonding (Scheme 2.21) [43b], in the epoxidation of geraniol and nerol with TBHP as the oxygen source resulted in much higher enantioselectivities (47% ee for geraniol and 64% ee for nerol) than those (up to 27.6% ee and 10.4% ee, respectively) obtained with the homogeneous counterpart **50**. This enhanced enantioselectivity of the heterogeneous catalyst **51** can also be ascribed to the additional steric constraints imposed by the zeolite pores. The lifetime of the heterogenized catalyst was also examined by its repeated use, which showed similar rates and yields of epoxide, even after five runs. The enhanced stability of the heterogenized complex was ascribed to the stronger coordination of the dihydroxy-ligand to Mo than that of the acetylacetonate or hydroxyl group present on the substrate, *tert*-butyl alcohol or hydroperoxide. The site-isolated catalyst molecules on the mesopores could also avoid being deactivated by dimerization and oligomerization of the Mo species.

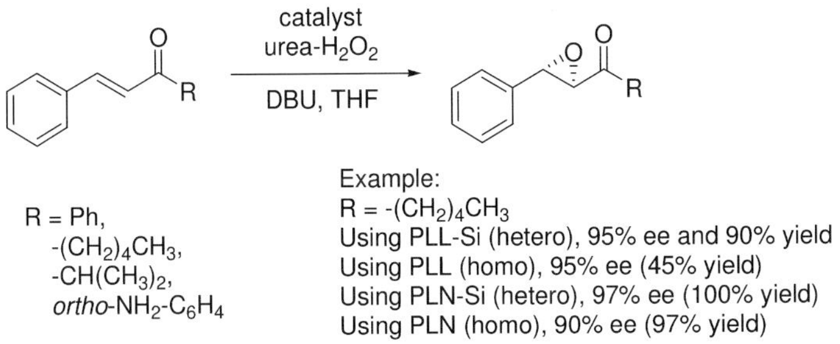

2.3 Asymmetric Oxidation

Scheme 2.22

R = Ph,
-(CH$_2$)$_4$CH$_3$,
-CH(CH$_3$)$_2$,
ortho-NH$_2$-C$_6$H$_4$

Example:
R = -(CH$_2$)$_4$CH$_3$
Using PLL-Si (hetero), 95% ee and 90% yield
Using PLL (homo), 95% ee (45% yield)
Using PLN-Si (hetero), 97% ee (100% yield)
Using PLN (homo), 90% ee (97% yield)

2.3.3
Enone Epoxidation

Polyamino acids such as poly-L-leucine or poly-L-alanine catalyze the AE of α,β-unsaturated ketones in organic solvent/alkaline hydrogen peroxide or in an organic solvent/peroxide donor (e.g. urea–hydrogen peroxide complex)/organic base such as diazabicycloundecene (DBU) [54]. However, the most serious problem of this reaction is the difficulty in the recovery of the gel- or paste-like catalyst. In order to overcome this problem, Roberts *et al.* [44] simply adsorbed polyamino acids (poly-L-(*neo*-pentyl)glycine (PLN), poly-L-leucine (PLL), etc.) onto several inorganic solid carriers, such as aluminum oxide, Celite, molecular sieve, zeolite TS1, silica gel 60, and so on. Among these immobilized materials, the silica-adsorbed catalysts (e.g. PLN-Si, PLL-Si) were outstanding. Besides demonstrating an increased catalytic activity and enantioselectivity compared to nonadsorbed polyamino acids (Scheme 2.22), the polyamino acid-on-silica catalyst was very easily recovered by filtration, thereby simplifying its recycling. The recovered silica-adsorbed polyamino acid catalyst retained its catalytic efficiency even after six recycling runs. In addition to these advantages the catalyst exhibited extreme robustness; for example, in contrast to unsupported polyamino acid (e.g. PLL), the catalytic activity of PLL-Si was fully retained even after heating at 150 °C for 12 h under vacuum.

In a continuation of their studies (Section 2.3.1.2, [40]) on the use of nanocrystalline MgO as a heterogeneous catalyst, Choudary *et al.* recently reported that the nanocrystalline MgO (NAP-MgO) can also act as a nanoheterogeneous catalyst for the Claisen–Schmidt condensation of benzaldehydes with acetophenones to yield chalcones, followed by AE to afford chiral epoxy ketones with moderate to good yields (52–70%) and impressive enantioselectivities (53–90% ee), in a two-pot reaction using (+)-diethyl tartrate (DET) as a chiral auxiliary (Scheme 2.23) [45]. These authors observed that the silylated MgO samples had longer reaction times than the corresponding MgO samples in Claisen–Schmidt condensation and, moreover, that there was essentially no epoxidation reaction. These results strongly indicated that Brønsted hydroxyls present on NAP-MgO are sole contributors to the epoxidation reaction, while they add on to the Claisen–Schmidt condensation, which is largely driven by Lewis-basic O^{2-} sites. When (+)-diethyl-2,3-*O*-isopropyledene-*R,R*-

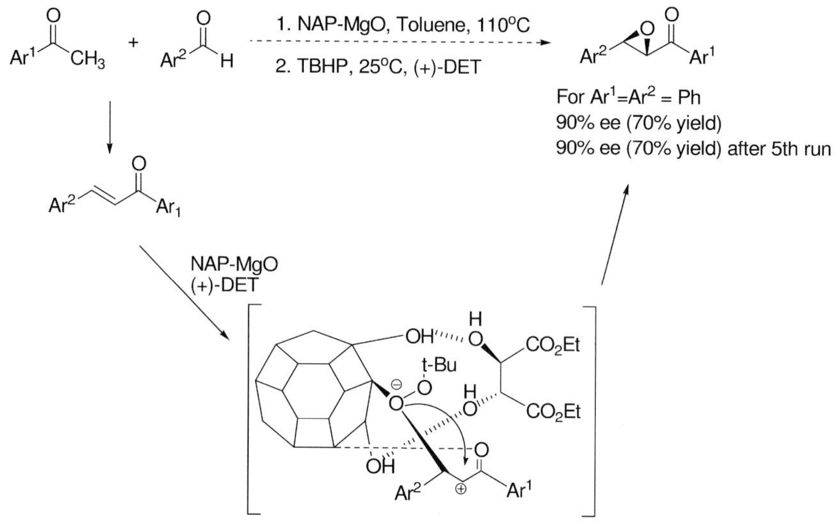

Scheme 2.23

tartrate, which has no free –OH groups, was used instead of DET, no enantiomeric excess was observed, which establishes that the hydrogen-bond interactions between the –OH groups of DET and MgO are essential for the induction of enantioselectivity (Scheme 2.23). This system conceptually evolved the single-site chiral catalyst by the successful transfer of molecular chemistry to surface metal–organic chemistry during the heterogenization of homogeneous chiral catalysts.

2.3.4
Asymmetric Dihydroxylation (AD) and Asymmetric Aminohydroxylation (AA) of Olefins

The Sharpless Os-catalyzed asymmetric dihydroxylation (AD) and aminohydroxylation (AA) of olefins are undoubtedly highly efficient methods of synthesizing chiral vicinal diols and aminoalcohols, respectively [55]. Although these reactions offer a number of processes that can be applied to the synthesis of chiral drugs, natural products and fine chemicals, and so on, the cost, toxicity and contamination of products with osmium restrict their use in industry. Therefore, a major effort has been directed at immobilizing the catalyst system (ligand or the osmium catalyst itself) [46]. Unfortunately, however, most examples of supported catalysts reported to date have exhibited inferior catalytic properties compared to their homogeneous counterparts. Some recent reviews on this topic are available [51c, 56], but in this chapter we will present only one exciting example of the immobilization of Os catalyst recently provided by Caps et al. [15]. This demonstrates that the heterogenization of catalysts in confined spaces can be used not only to improve the stereoselectivity but also to generate a new chiral species *in situ* from achiral catalyst precursors (Scheme 2.24). The heterogenization of the achiral cluster, $Os_3(CO)_{12}$, on the internal space of MCM-41 using simple chemical vapor

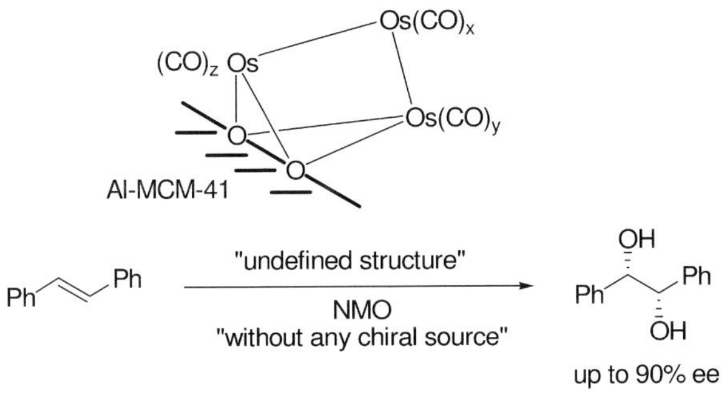

Scheme 2.24

deposition (CVD) gave superior enantioselectivity towards the *(S,S)*-configuration of the 1,2-diphenyl-1,2-ethanediol in the dihydroxylation of *trans*-stilbene using N-methylmorpholine N-oxide (NMO) as an oxidant, without adding any chiral ligand. This surprising effect was even more pronounced when surface Al sites were introduced into the silicate (90% ee for *(S,S)*-isomer) (Scheme 2.24). The 90% ee towards the *(S,S)* configuration was much higher than that reported using a homogeneous OsO_4–NMO–acetone/water system containing a chiral ligand dihydroquinidine *p*-chlorobenzoate (78%) [57]. One possible reason for this chiral induction might be the 'spontaneous symmetry breaking' [58] of achiral $Os_3(CO)_{12}$ during CVD on the MCM-41 or related surfaces, resulting in a new surface chiral catalytic species. However, despite many attempts to reproduce this highly interesting result we have not yet succeeded.

2.3.5
Asymmetric Aziridination of Olefin

Copper complexes of chiral bis(oxazoline)s are known to be highly effective enantioselective homogeneous catalysts for aziridination as well as for many C—C and C—X bond-forming reactions. Hutchings *et al.* immobilized chiral bisoxazoline, 52–55, on Cu-exchanged zeolite Y and employed the resulting solid copper catalysts for the asymmetric aziridination of alkenes using two nitrene sources, [N-(*p*-tolylsulfonyl)imino]phenyliodinane (PhI=NTs) and [N-(*p*-nitrophenylsulfonyl)imino]phenyliodinane (PhI=NNs) [48]. As mentioned above, this heterogenization is based on the electrostatic interactions of the copper cations with the anionic support. Interestingly, in the asymmetric aziridation of styrene derivatives, in general, the heterogeneous catalyst was found to produce an enhanced enantioselection for a range of bis(oxazolines) compared to the homogeneous catalyst (Scheme 2.25) [48a,b]. This was considered to be due to the confinement effect within the micropores of the zeolite. Electron paramagnetic resonance (EPR) spectroscopic studies indicated that the copper(bisoxazoline) complexes were located in the pores of the zeolite [59]. An interesting observation with the copper-bis(oxazoline)-catalyzed aziridination reaction was that the enantioselectivity increased with the alkene conversion, most likely due to interaction of the chiral

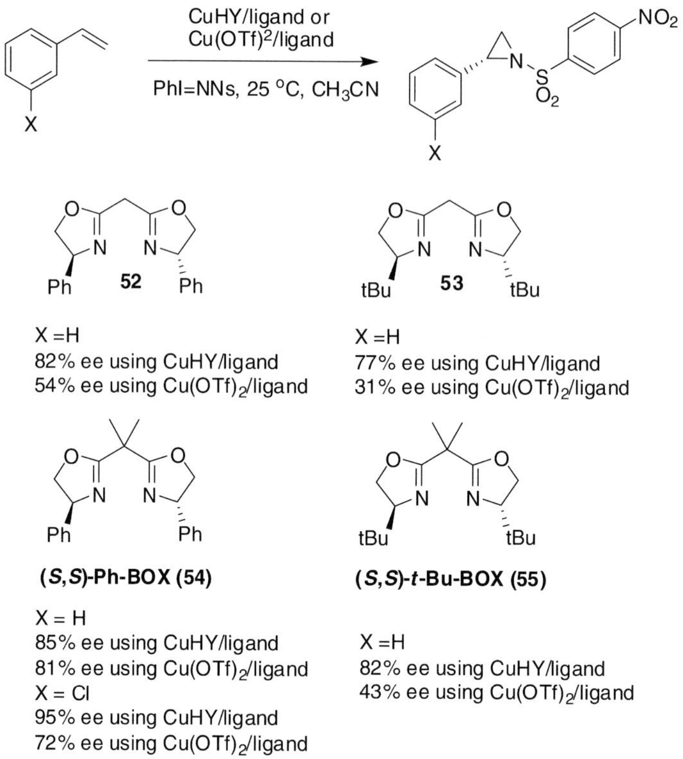

54 | *2 Heterogeneous Enantioselective Catalysis Using Inorganic Supports*

Scheme 2.25

Reaction: styrene derivative + PhI=NNs, 25 °C, CH₃CN, CuHY/ligand or Cu(OTf)₂/ligand → aziridine product (N-Ns, with NO_2 on sulfonyl aryl)

Ligand **52** (Ph-BOX, non-gem-dimethyl):
X = H
82% ee using CuHY/ligand
54% ee using Cu(OTf)₂/ligand

Ligand **53** (t-Bu-BOX, non-gem-dimethyl):
X = H
77% ee using CuHY/ligand
31% ee using Cu(OTf)₂/ligand

(S,S)-Ph-BOX (**54**)
X = H
85% ee using CuHY/ligand
81% ee using Cu(OTf)₂/ligand
X = Cl
95% ee using CuHY/ligand
72% ee using Cu(OTf)₂/ligand

(S,S)-t-Bu-BOX (**55**)
X = H
82% ee using CuHY/ligand
43% ee using Cu(OTf)₂/ligand

aziridine product with the active copper–nitrene intermediate [48c]. These heterogeneous catalysts also gave selectivities which were comparably high or even improved in relation to the homogeneous catalysts in the Diels–Alder [60] and carbonyl- and imino-ene reactions [61], both of which are discussed later in this chapter.

2.4
Asymmetric Carbon–Carbon and Carbon–Heteroatom Bond Formation

The enantioselective C–C and C–X bond-formation catalyzed by inorganic supported heterogeneous chiral catalysts is summarized in Table 2.3, and selected examples are further described in this section.

2.4.1
Pd-Catalyzed Asymmetric Allylic Substitution

The most exciting example in this reaction was provided by Thomas and Johnson *et al.* [14c]. In the case of the allylic amination of cinnamyl acetate, the Pd-catalyst

2.4 Asymmetric Carbon–Carbon and Carbon–Heteroatom Bond Formation

Table 2.3 Examples of heterogeneous asymmetric carbon–carbon and carbon–heteroatom bond formation using inorganic supports.

No.	Reaction/Catalyst	Support	Method	Substrate	ee (%) Homo[a]	ee (%) Hetero[b]	Reference(s)
	Pd-catalyzed asymmetric allylic substitution						
1	Ferrocenyl Pd-complex	MCM-41, Carbosil	Covalent	Cinnamyl acetate with benzylamine	>99	NA[c]	[14c]
	Asymmetric addition of dialkylzincs to aldehydes						
2	Chiral N-alkylnorephedrines	Silica gel, alumina	Covalent	Benzaldehyde with Et$_2$Zn	59	NA[c]	[62a]
3	Chiral ephedrine	Silica gel	Covalent	Pyrimidine-5-carbaldehyde with i-Pr$_2$Zn	97	98	[62b]
4	BINOL derivatives	Cd(NO$_3$)$_2$·4H$_2$O	Covalent	1-Naphthaldehyde with Et$_2$Zn	90	94	[63a]
5	Chiral β-amino alcohols	Silica gel	Covalent	Benzaldehydes with Et$_2$Zn	77	NA[c]	[63c]
6	(S)-BINOL	MCM-41, SBA-15	Covalent	Benzaldehydes with Et$_2$Zn	81	89	[63d]
7	Pyrrolidinemethanol	Amorphous silica, MCM-41, SBA-15	Covalent	Benzaldehydes with Et$_2$Zn	75	93	[63e,f]
8	TADDOL-Ti complex	Controlled-pore glass	Covalent	Benzaldehydes with Et$_2$Zn	96	92.5	[63g]

56 | *2 Heterogeneous Enantioselective Catalysis Using Inorganic Supports*

Table 2.3 Continued

No.	Reaction /Catalyst	Support	Method	Substrate	ee (%) Homo[a]	ee (%) Hetero[b]	Reference(s)
Asymmetric Diels–Alder reaction							
9	Bis(oxazoline)Cu(II) complex	Silica	Hydrogen bonding	Acryloxazolidinone and cyclopentadiene	93	88	[64a]
10	Bis(oxazoline)Cu(II) complex	MCM-41, Al-SBA-15, MSU-2, Zeolite Y	Electrostatic interactions	(E)-ethyl-2-oxo-3-pentenonate with vinyl ethyl ether	41	20	[60]
11	Bis(oxazoline)Cu(II) complex	Aerosil 200	Covalent	3-Acryloyl-2-oxazolidinone with cyclopentadiene	65	27	[64c]
12	TADDOL	Controlled pore glass	Covalent	Diphenylnitrone with N-crotonyl-1,3-oxazolidin-2-one	70	85	[63g, 65]
13	Imidazolidin-4-one	Siliceous mesocellular foams	Covalent	Cinnamaldehyde with cyclopentadiene	83	91	[66]
Ene reaction							
14	Chiral bis(oxazoline) Cu(II) complex	Zeolite Y	Electrostatic interactions	Methylene cyclopentane with ethyl glyoxalate	93	57	[61]
Asymmetric conjugate addition							
15	L-Proline amide	USY zeolites or silica	Covalent	4-Phenylbut-3-en-2-one with Et$_2$Zn	95	75	[67]

2.4 Asymmetric Carbon–Carbon and Carbon–Heteroatom Bond Formation

#	Catalyst	Support	Immobilization	Reaction/Substrate	Yield	ee	Ref
18	(1R,2R)-(−)-diaminocyclohexane	Nanocrystalline magnesium oxide NAP-MgO	Hydrogen bonding	Chalcone with nitromethane	96	NA[c]	[69a]
	Aldol/nitroaldol reaction						
19	L-Proline	Mesoporous support, MCM-41	Covalent	Benzaldehyde with hydroxyacetone	80	75	[70]
20	L-Proline	Silica gel supported ionic liquid	Covalent and Adsorption	Benzaldehyde with acetone	64	60	[71a]
21	(1S,2S)-(+)-1,2-diaminocyclohexane	Nanocrystalline MgO	Chemisorption	4-Fluorobenzaldehyde with acetone	60	NA[c]	[72a]
	Asymmetric cyclopropanation						
22[d]	Bisoxazoline Cu complex	Siliceous mesocellular foams	Covalent	Styrene with ethyl diazoacetate	95	86	[73a]
	Friedel–Craft hydroxylation						
23	Chiral bis(oxazoline)Cu(II) Complex	Silica, MCM-41	Covalent	1,3-Dimethoxybenzene with 3,3,3-trifluoropyruvate	92	72	[74]
	Si–H insertion						
24	[(4S)-BNOX]₃Rh₂	Aerosil 200	Ligand exchange, adsorption	Methyl phenyldiazoacetate with dimethylphenylsilane	28	2	[75]

[a] Hetero: Using heterogeneous catalyst.
[b] Homo: Using homogeneous catalyst.
[c] NA = data not available in the literature under identical experimental conditions.
[d] ee given for *trans* product.

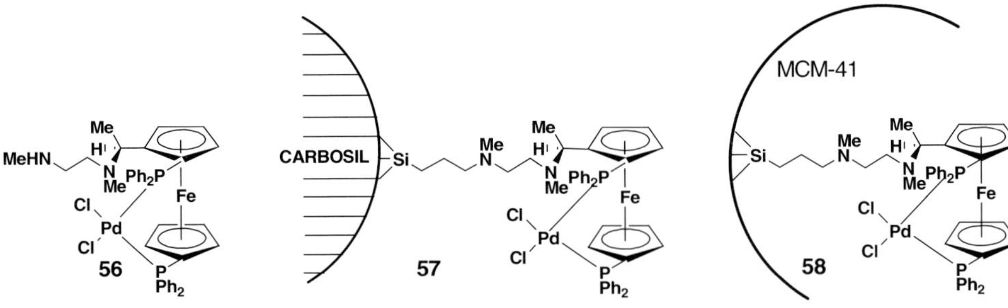

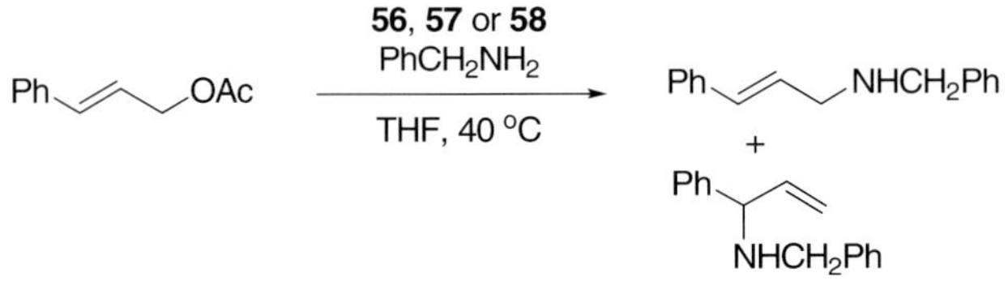

58 | 2 Heterogeneous Enantioselective Catalysis Using Inorganic Supports

Scheme 2.26

Catalyst	Conversion (%)	Straight chain (%)	Branched (%)	ee (%)
56	76	>99	-	-
57	98	98	2	43
58	>99	49	51	>99

58, when immobilized on the inner walls of mesoporous MCM-41, exhibited superior catalytic properties to catalyst 57 anchored on Carbosil (a nonporous, high-area silica) and to homogeneous catalyst 56. The MCM-41-immobilized catalyst 58 showed a degree of regioselectivity for the desirable branched product (51%) and extremely high enantioselectivity (>99% ee), whereas the Carbosil-immobilized catalysts 57 afforded the branched product in only 2% yield and with 43% ee. In the case of the homogeneous catalyst 56, the reaction produced exclusively the straight-chain product (Scheme 2.26). Although the regioselectivity of 58 is still not very high, the further careful design of active centers and control of the pore structure of supports would most likely lead to more unexpected selectivities.

2.4.2
Enantioselective Addition of Dialkylzincs to Aldehydes

Soai et al. [62a] first reported the use of silica gel or alumina as a heterogeneous support for chiral catalysts in the enantioselective addition of dialkylzincs to aldehydes. Chiral N-alkylnorephedrines (R = Me, Et, n-Pr) were immobilized covalently on (3-chloropropyl)silyl-functionalized alumina or silica gel via a nucleophilic substitution. However, the catalytic activities and enantioselectivities were only moderate (24–59% ee) in comparison with those of homogeneous and polymer-

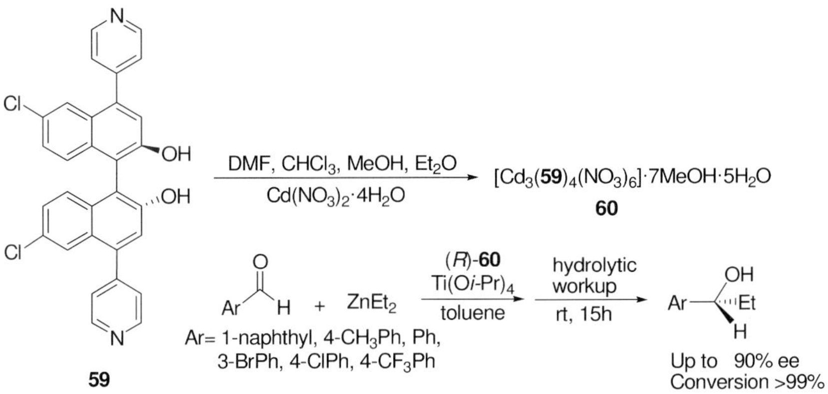

Scheme 2.27

supported counterparts. Since then, several research groups [63] have made advancements in this field (Table 2.3). Among these, a highly interesting example of this reaction was recently reported by Lin et al. [63a,b], who rationally designed the zeolite-mimicking, crystalline, 3-D, homochiral, porous, metal–organic frameworks (MOFs) **60** built from the chiral bridging ligand (L*∩*) **59** and metal-connecting point. Colorless crystals of $[Cd_3(L*\cap*)_4(NO_3)_6]\cdot 7MeOH \cdot 5H_2O$ **60** were synthesized by slow diffusion of diethyl ether into a mixture of $Cd(NO_3)_2\cdot 4H_2O$ and *(R)*-6,6′-dichloro-2,2′-dihydroxy-1,1′-binaphthyl-4,4′-bipyridine (**59**) in DMF/$CHCl_3$/MeOH at room temperature (Scheme 2.27). The approximate dimensions of the channels of MOF **60** were found to be $4.9\,\text{Å} \times 13.1\,\text{Å}^2$ along the *a* and *b* axes, and $13.5\,\text{Å} \times 13.5\,\text{Å}^2$ along the *c* axis [63b]. After treatment of the MOF **60** with excess $Ti(OiPr)_4$ in toluene, **60**-Ti proved to be a highly active and enantioselective heterogeneous catalyst for the addition of diethylzinc to aromatic aldehydes, to afford chiral secondary alcohols upon hydrolytic work-up. In particular, **60**-Ti catalyzed the addition of diethylzinc to 1-naphthaldehyde to afford *(R)*-1-(1-naphthyl)-propanol with complete conversion and 90.0% ee, which is similar to the homogeneous analogue (94% ee) [63b] (Scheme 2.27).

2.4.3
Asymmetric Diels–Alder Reaction

For heterogeneous asymmetric Diels–Alder/hetero-Diels–Alder reactions, several research groups have attempted to immobilize the chiral bis(oxazoline)(BOX)-copper complex [60, 64], chiral (salen)Cr-complex [65], TADDOL-Ti catalyst [63g, 76] and other chiral auxiliaries [66] onto silica, ordered mesoporous silica, siliceous mesocellular foams, clays, nafionsilica nanocomposites, zeolite and alumina. The reactions using these heterogeneous catalysts exhibited relatively good enantioselectivity and activity; the following selected examples demonstrate how the confinement effect of the porous supports leads to changes in catalytic properties. For example, chiral bis(oxazoline) complexes of Cu(II), Zn(II) and Mg(II) supported

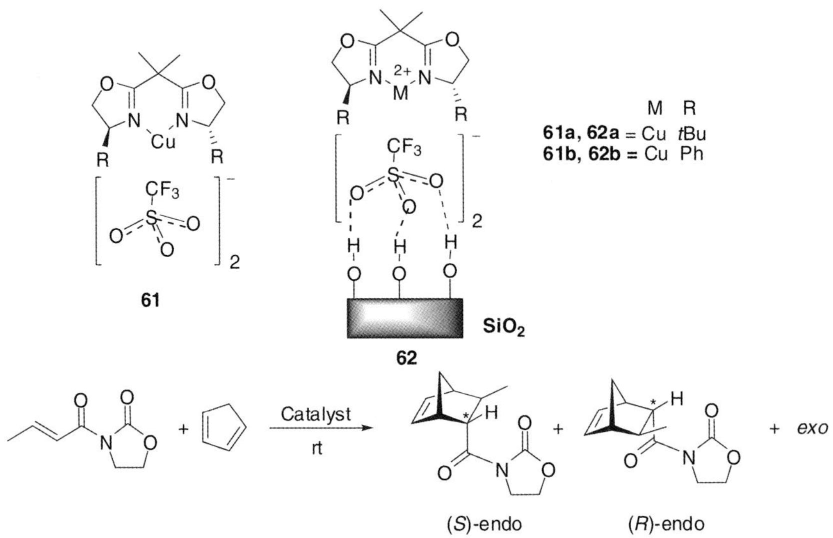

Scheme 2.28

on silica via hydrogen-bonding interactions between the triflate anions and surface silanol groups (Scheme 2.28) displayed excellent enantioselectivity and enhanced activity compared to those of homogeneous analogues **61**, when the less-polar toluene was used as the solvent, in the Diels–Alder reaction between acryloxazolidinone and cyclopentadiene at room temperature (Scheme 2.28) [64a]. The heterogeneous catalyst **62a** can be reused without losing enantioselectivity and decreasing the activity in the second run. Intriguingly, the supported catalyst **62b** provided a reversed product configuration with improved enantioselectivity compared to the homogeneous (S)-PhBOX-Cu(OTf)$_2$ (**61b**) in this reaction (Scheme 2.28). On the basis of experimental results and theoretical calculations, the authors explained that the configuration reversal upon immobilization is triggered by dissociation of the triflate anion from the metal cation due to hydrogen-bonding interactions with the surface silanols, resulting in a geometric change of the PhBOX (**54**)-Cu(II)-dienophile intermediate. In parallel, Gebbink *et al.* also reported the same strategy in immobilizing BOX complexes by hydrogen-bonding interactions between the triflate anions and surface silanol groups of silicas for the Diels–Alder reactions [64b].

Hutchings *et al.* [60] extended the use of the heterogeneous catalyst (**54**-CuHY) to the asymmetric hetero-Diels–Alder reaction. As mentioned above, this heterogenization is based on the electrostatic interactions of the copper cations with the anionic support. Here too, **54**-CuHY exhibited significantly higher enantioselectivity (41% ee) than that obtained with the homogeneous catalyst **54**-Cu(OTf)$_2$ (**61b**) (20% ee) in the reaction of (E)-ethyl-2-oxo-3-pentenoate (**63**) with ethyl vinyl ether, to afford the dihydropyran **64** (Scheme 2.29). Similar to the above example of the Diels–Alder reaction [60], the configuration of the product (2S,4R) obtained from the heterogeneous reaction was opposite to that (2R,4S) obtained from the homo-

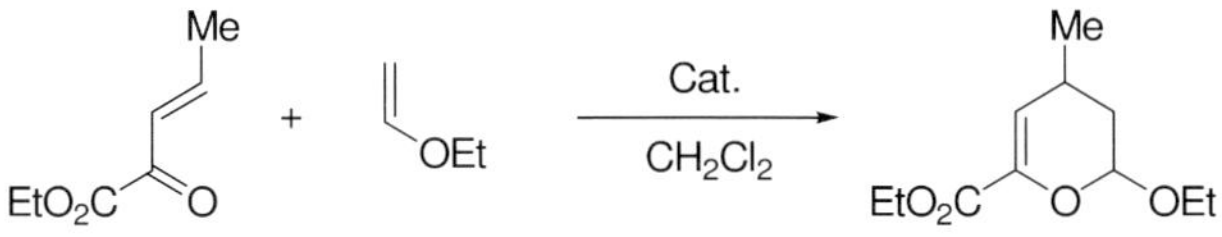

2.4 Asymmetric Carbon–Carbon and Carbon–Heteroatom Bond Formation

Scheme 2.29

Using **54**-Cu(OTf)$_2$ (homo): 20% ee (2R,4S)
Using **54**-CuHY (hetero): 41% ee (2S,4R)

Scheme 2.30

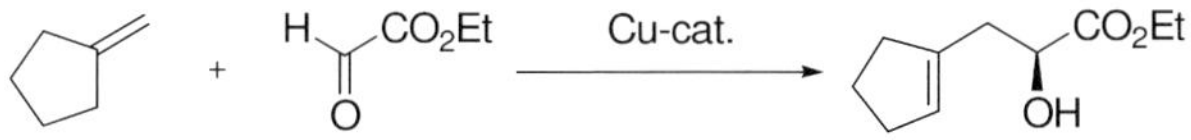

With **54**-CuHY (hetero): 93% ee
With **54**-Cu(OTf)$_2$ (homo): 57% ee

geneous reaction. As described above in Section 2.1, this reversed enantioselectivity could also be attributed to the confinement effect of the porous supports.

2.4.4
Ene Reactions

The zeolite Y-supported, heterogeneous catalyst (**54**-CuHY) was also used by Hutchings et al. [61] for the enantioselective carbonyl- and imino-ene reactions. In a carbonyl-ene reaction of methylene cyclopentane with ethyl glyoxylate, the heterogeneous catalyst **54**-CuHY exhibited superior enantioselectivity (93% ee) compared to the homogeneous catalyst **54**-Cu(OTf)$_2$ (57% ee) (Scheme 2.30), due to the confinement effect of the zeolite pores.

2.4.5
Asymmetric Conjugate Addition

Sanchez et al. [67] anchored the Ni complex of L-proline amide on USY zeolites (**66**) or silica (**67**) by covalent bonding. Although the use of insoluble heterogeneous catalysts **66** and **67** slowed down the conjugate addition of ZnEt$_2$ to enones **68**, the ee-values of the saturated ketone products **69** were much higher for the zeolite-supported Ni complex **66** (95%) than for either its homogeneous counterpart **65a** (75%) or the silica-supported analogue **67** (34%) (Scheme 2.31). The steric constraint imposed by the zeolite pore was thought to contribute to the enhanced enantioselectivities that were observed. The same research group also attached a series of chiral amines covalently to modified USY zeolite and MCM-41 zeolite supports and evaluated the ability of the obtained chiral base catalysts confined in porous hosts to improve the stereoselectivity in conjugate additions of nucleophiles to enones [68]. The reaction rate was enhanced and the catalysts showed higher selectivity than did the free amines.

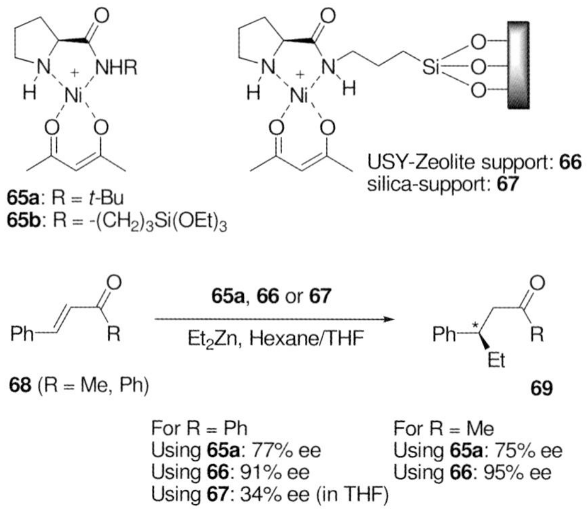

2 Heterogeneous Enantioselective Catalysis Using Inorganic Supports

65a: R = t-Bu
65b: R = -(CH$_2$)$_3$Si(OEt)$_3$

USY-Zeolite support: **66**
silica-support: **67**

68 (R = Me, Ph) → **69**

65a, 66 or 67, Et$_2$Zn, Hexane/THF

For R = Ph
Using **65a**: 77% ee
Using **66**: 91% ee
Using **67**: 34% ee (in THF)

For R = Me
Using **65a**: 75% ee
Using **66**: 95% ee

Scheme 2.31

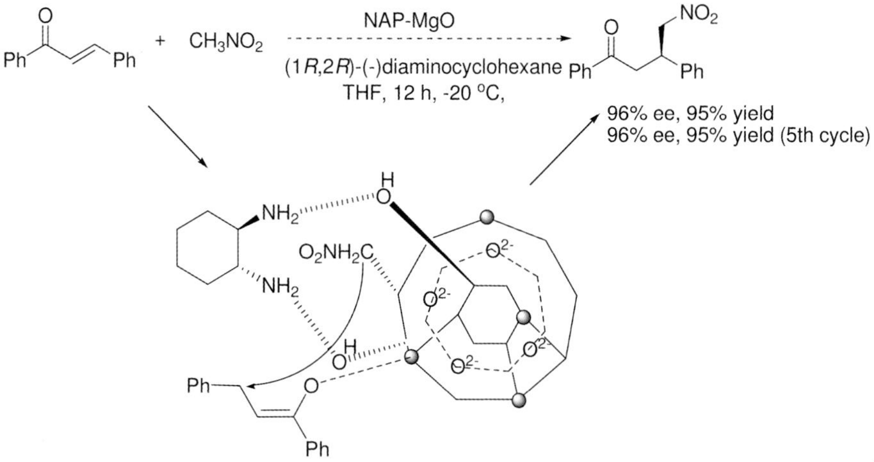

NAP-MgO
(1R,2R)-(−)-diaminocyclohexane
THF, 12 h, −20 °C,

96% ee, 95% yield
96% ee, 95% yield (5th cycle)

Scheme 2.32

As described in Sections 2.3.1.2 and 2.2.3, Choudary et al. recently revealed nanocrystalline magnesium oxide (NAP-MgO) as a recyclable heterogeneous catalyst [40, 45]. These authors extended the use of this new type of heterogeneous catalyst for the asymmetric Michael reaction of different acyclic enones with nitromethane and 2-nitropropane [69a]. In a Michael reaction of chalcone with nitromethane in THF solvent at −20 °C, NAP-MgO/(1R,2R)-(−)-diaminocyclohexane (DAC) was found to be an excellent catalyst system (96% ee, 95% yield) (Scheme 2.32). This Michael reaction proceeds via the dual activation of both substrates (nucleophiles and electrophiles) by NAP-MgO. The Lewis basic site (O^{2-}/O$^-$) of the NAP-MgO activates the nitroalkanes, while the Lewis acid moiety (Mg^{2+}/Mg$^+$)

2.4 Asymmetric Carbon–Carbon and Carbon–Heteroatom Bond Formation

activates the carbonyls of enones (Scheme 2.32). Moreover, the hydrogen-bond interactions between the –NH groups of chiral auxiliary and –OH groups of MgO are found to be essential for the induction of enantioselectivity (Scheme 2.32). The –OH groups present on the edge and corner sites on the NAP-MgO, which are stretched in 3-D space, are more isolated and accessible for the chiral ligand DAC for greater alignment, and consequently provided excellent ee-values. The NAP-MgO catalyst was reused five times, and showed consistent yields and ee-values in the Michael reaction. The same research group also reported [69b] that the use of heterogeneous layered double hydroxide (LDH)-proline and LDH-BINOL chiral catalysts in the Michael reaction of cyclohexenone with diethylmalonate afforded the corresponding adduct in moderate to good yields, albeit in racemic form.

2.4.6
Aldol and Nitroaldol Reactions

Mayoralas et al. [70] reported the aldol reaction of hydroxyacetone with different aldehydes catalyzed by immobilized L-proline on a mesoporous support. Heterogenized L-proline on MCM-41 showed higher enantioselectivity (80% ee) than its homogeneous counterpart (75% ee) in the aldol reaction of benzaldehyde with hydroxyacetone in dimethylsulfoxide (DMSO) solvent with the assistance of microwave heating.

Gruttadauria et al. [71a] utilized the previously described supported ionic liquid catalysis (SILC) concept for the L-proline-catalyzed aldol reaction (Scheme 2.33). The immobilized catalyst (SiO$_2$-[bmim]BF$_4$-L-proline, **70**) displayed higher enantioselectivity (64% ee) with acceptable activity (51% yield) than the homogeneous conditions (60% ee, 62% yield) in the aldol reaction of acetone and benzaldehyde [71b]. Additionally, this system was even more efficient than the PEG-proline-

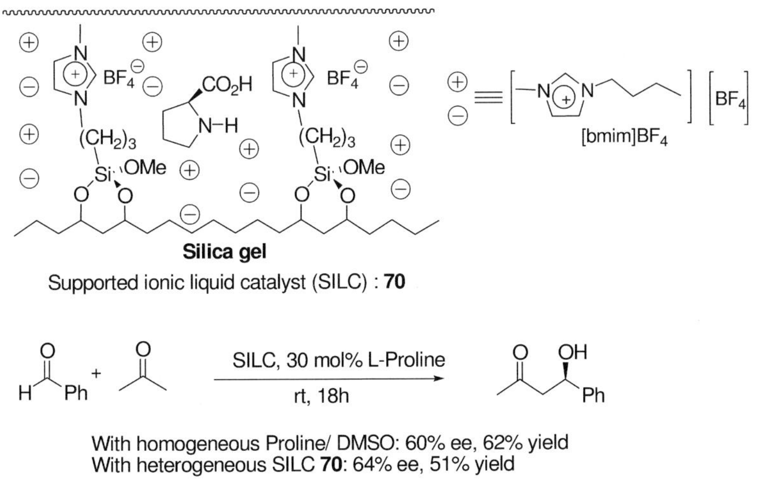

Scheme 2.33

catalyzed aldol reaction of acetone and benzaldehyde in dimethylformamide (DMF) (59% ee, 45% yield) [71c]. However, the catalytic performance was lower than that observed under the homogeneous ionic liquid conditions in the L-proline-catalyzed aldol reaction of acetone and benzaldehyde in [bmim]PF_6 (76% ee; 55% yield) [71d]. Other examples of aldol and nitroaldol (Henry) reactions catalyzed by heterogeneous chiral catalysts supported on nanocrystalline MgO [72a] or mesoporous supports, including MCM-41 [72b], have been reported by several research groups (Table 2.3) [72]. In particular, the catalyst NAP-MgO proved to be a very useful heterogeneous catalyst in the asymmetric Henry reaction of different substituted benzaldehydes with nitromethane using chiral *(S)*-BINOL ligand as a chiral auxiliary to afford the corresponding chiral nitro alcohols with up to 98% ee and 95% yield.

2.4.7
Asymmetric Cyclopropanation

A good example of heterogeneous asymmetric cyclopropanation was recently provided by Ying *et al.* [73], who immobilized the chiral *t*-butylbisoxazoles (BOX) on spherical siliceous mesocellular foams (MCF), the surface of which had been partially modified with trimethylsilyl (TMS) groups prior to the immobilization (Scheme 2.34). The resulting MCF-supported bisoxazoline **72**-Cu(I) catalyst showed superior enantioselectivity (up to 95% ee for the *trans* product, 92% ee for the *cis* isomer) in the benchmark asymmetric cyclopropanation of styrene with ethyl diazoacetate to that of the homogeneous analogue **71**-Cu(I) and the MCF-supported bisoxazoline without TMS precapping. The authors explained that the TMS precapping might have provided a better dispersion of BOX ligands on the silica support, by reducing their interactions with the surface silanol groups on the MCF a priori, and thereby increasing their enantioselectivity [73a].

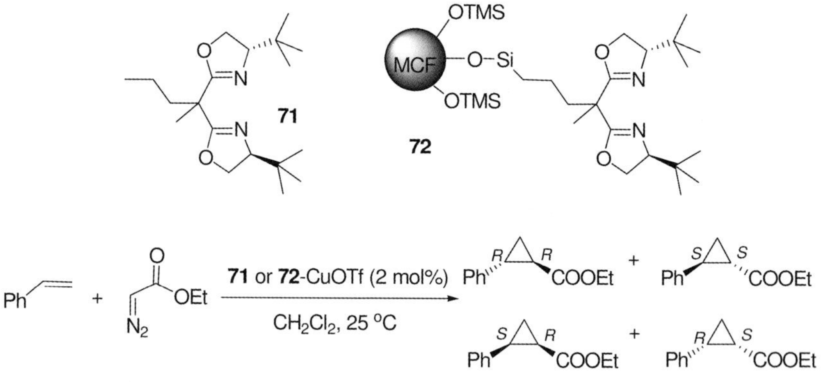

With **71**-Cu (homo): 80% yield, 68/32 (*cis/trans*), 86% ee (*trans*), 84% ee (*cis*)
With **72**-Cu (hetero): 80% yield, 65/35 (*cis/trans*), 95% ee (*trans*), 92% ee (*cis*)

Scheme 2.34

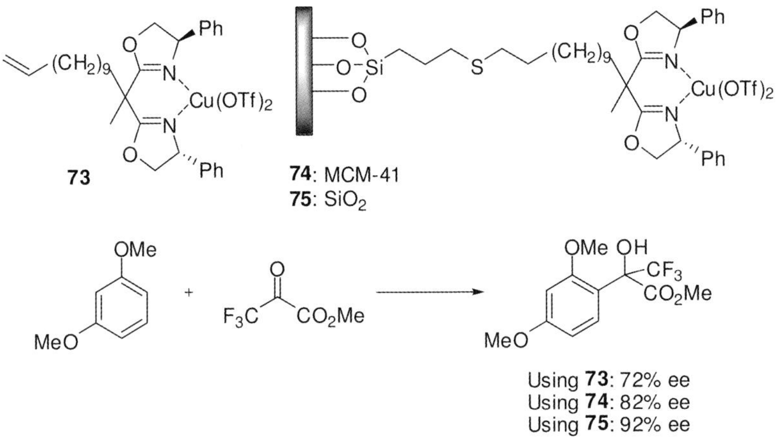

2.4 Asymmetric Carbon–Carbon and Carbon–Heteroatom Bond Formation

Scheme 2.35

2.4.8
Friedel–Craft Hydroxyalkylation

An interesting example of this class of reaction is the alkylation of 1,3-dimethoxybenzene with 3,3,3-trifluoropyruvate using a MCM-41(**74**)- and silica (**75**)-immobilized copper-bis(oxazoline) [74]. The heterogeneous catalysts **74** and **75** were immobilized on inorganic support through a long flexible linker to minimize spatial restrictions by the support. The homogeneous reaction using **73** gave an ee-value of 72% at 44% conversion, whereas the MCM-41-immobilized catalyst gave 82% ee at 77% conversion, and the silica-immobilized catalyst 92% ee at 72% conversion (Scheme 2.35).

2.4.9
Si–H Insertion

Recently, Maschmeyer et al. [75] immobilized the chiral dirhodium complex **76** on Aerosil 200 and observed a significant improvement in enantioselectivity in the Si–H insertion reactions of dimethylphenylsilane with methyl phenyl diazoacetate. This immobilized catalyst, SiO_2–$(CH_2)_2COO$–$Rh_2(4S$-$BNOX)_3$ **77**, showed a more than 10-fold increase in enantioselectivity (up to 28% ee), while the homogeneous analogue $Rh_2(4R$-$BNOX)_4$ (**76**) gave the near-racemic product (Scheme 2.36). In contrast to the catalysts immobilized on Aerosil 200, which has an average pore size of 50 nm, none of the catalysts immobilized inside the pores of MCM-41 showed any activity. The pores of MCM-41 may have lacked sufficient space for the reaction to take place. This result was felt to be reasonable, considering that the average pore diameter (19 Å) of MCM-41 is only slightly larger than the catalyst size (between 19 and 13 Å). A transition state requiring a space-demanding conformation of the catalyst might, therefore, be impossible under these circumstances.

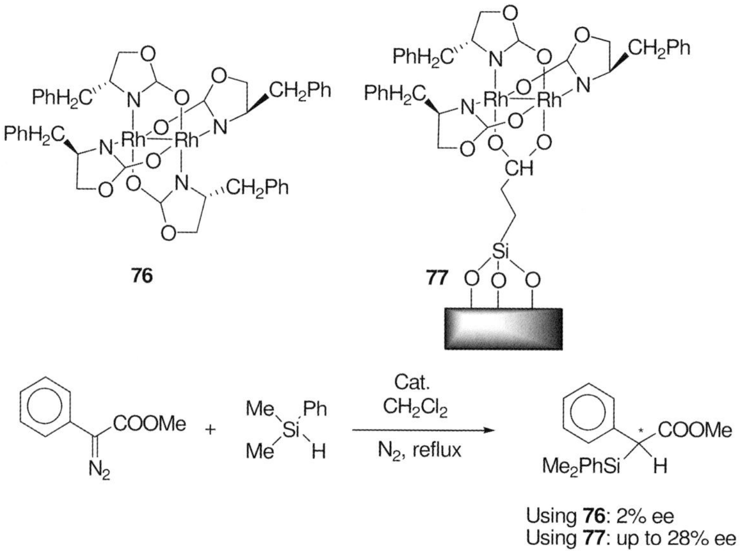

Scheme 2.36

Using **76**: 2% ee
Using **77**: up to 28% ee

2.5
Conclusions

As the many representative examples provided in this chapter demonstrate, the choice of inorganic support in the design of immobilized catalysts has regained favor in the multifarious research area of heterogeneous asymmetric catalysis, and the potential scope of this approach continues to expand. The heterogenization of a chiral catalyst onto an inorganic material provides not only straightforward catalyst recycling but also a significant improvement in the catalytic performances (activity, stability, enantioselectivity), due mainly to site-isolation and confinement effects. By careful tuning these effects, based on the design of the pore surface and the catalyst structure, even nonenantioselective catalysts have shown significant asymmetric induction when anchored into the confined spaces of inorganic materials [13, 14a,c, 16b]. Moreover, without any chiral source the asymmetric induction was, unexpectedly, achieved in confined nanospaces [15]. Taken together, these outcomes stimulate the concept that conventional approaches – that is, the heterogenization of catalysts that are well designed for homogeneous processes – may not represent the best solution for developing efficient heterogeneous chiral catalyst systems. Despite some negative effects occurring due to immobilization, we are convinced that the heterogenization of catalysts on inorganic materials will in the future play a crucial role as an essential technology for the discovery of highly efficient chiral catalysts. Moreover, this will be even more likely if advantage is taken of the special properties of inorganic materials, coupled with a combinatorial approach and high-throughput catalyst screening strategies.

Acknowledgments

These studies were supported by a Korea Research Foundation Grant (KRF-2005-005-J11901) funded by MOEHRD, and by grants R01-2006-000-10426-0 (KOSEF) and R11-2005-008-00000-0 (SRC program of MOST/KOSEF).

References

1. (a) Ojima, I. (2000) *Catalytic Asymmetric Synthesis*, 2nd edn, Wiley-VCH Verlag GmbH, New York.
   (b) Jacobsen, E.N., Pfaltz, A. and Yamamoto, H. (1999) *Comprehensive Asymmetric Catalysis I-III*, Springer Verlag, Berlin.
   (c) Noyori, R. (1994) *Asymmetric Catalysis in Organic Synthesis*, John Wiley & Sons, Ltd, New York, Chapter 2.
2. Yoon, T.P. and Jacobsen, E.N. (2003) *Science*, **299**, 1691–3.
3. (a) Blaser, H.U. and Schmidt, E. (2004) *Asymmetric Catalysis on Industrial Scale*, Wiley-VCH Verlag GmbH, Weinheim.
   (b) Collins, A.N., Sheldrake, G.N. and Crosby, J. (1992) *Chirality in Industry I*, John Wiley & Sons, Ltd, Chichester.
   (c) Collins, A.N., Sheldrake, G.N. and Crosby, J. (1996) *Chirality in Industry II*, John Wiley & Sons, Ltd, Chichester.
4. Knowles, W.S. For details see Nobel lectures: (a) (2002) *Angewandte Chemie – International Edition*, **41**, 1998–2007 and visit nobel_prizes/chemistry/laureates/2001/knowles-lecture.pdf.
   (b) Noyori, R. (2002) *Angewandte Chemie – International Edition*, **41**, 2008–22 and visit nobel_prizes/chemistry/laureates/2001/noyori-lecture.pdf.
   (c) Sharpless, K.B. (2002) *Angewandte Chemie – International Edition*, **41**, 2024–32 and visit nobel_prizes/chemistry/laureates/2001/sharpless-lecture.pdf.
5. De Vos, D.E., Vankelecom, I.F.J. and Jacobs, P.A. (2000) *Chiral Catalyst Immobilization and Recycling*, Wiley-VCH Verlag GmbH, New York.
6. Vankelecom, I.F.J. and Jacobs, P.A. (2000) Catalyst immobilization on inorganic supports, in *Chiral Catalyst Immobilization and Recycling* (eds D.E. De Vos, I.F.J. Vankelecom and P.A. Jacobs), Wiley-VCH Verlag GmbH, Weinheim, p. 19.
7. Song, C.E. and Lee, S.-G. (2002) *Chemical Reviews*, **102**, 3495–524.
8. (a) Fan, Q.H., Li, Y.M. and Chan, A.S.C. (2002) *Chemical Reviews*, **102**, 3385–466.
   (b) McMorn, P. and Hutchings, G.J. (2004) *Chemical Society Reviews*, **33**, 108–22.
   (c) Heitbaum, M., Glorius, F. and Escher, I. (2006) *Angewandte Chemie – International Edition*, **45**, 4732–62.
9. (a) Li, C. (2004) *Catalysis Reviews – Science and Engineering*, **46**, 419–92.
   (b) Li, C., Zhang, H., Jiang, D. and Yang, Q. (2007) *Chemical Communications*, 547–58.
   (c) Blackmond, D.G. (1998) *Cattech*, **3**, 17–32.
   (d) Ishihara, K. and Yamamoto, H. (1997) *Cattech*, **1**, 51–62.
10. Song, C.E., Kim, D.H. and Choi, D.S. (2006) *European Journal of Inorganic Chemistry*, **15**, 2927–35.
11. Song, C.E. and Park, I.S. (2005) *Advances in Organic Synthesis*, Bentham Publishers, Chap. 8, p. 233.
12. Song, C.E. (2005) *Annual Reports on the Progress of Chemistry C*, **101**, 143–73.
13. (a) Raja, R., Thomas, J.M., Jones, M.D., Johnson, B.F.G. and Vaughan, D.E.W. (2003) *Journal of the American Chemical Society*, **125**, 14982–3.
    (b) Jones, M.D., Raja, R., Thomas, J.M. and Johnson, B.F.G. (2003) *Topics in Catalysis*, **25**, 71–9.
14. (a) Jones, M.D., Raja, R., Thomas, J.M., Johnson, B.F.G., Lewis, D.W., Rouzaud, J. and Harris, K.D.M. (2003) *Angewandte Chemie – International Edition*, **42**, 4326–31.
    (b) Rouzaud, J., Jones, M.D., Raja, R., Johnson, B.F.G., Thomas, J.M. and Duer, M.J. (2003) *Helvetica Chimica Acta*, **86**, 1753–59.

(c) Johnson, B.F.G., Raynor, S.A., Shephard, D.S., Mashmeyer, T., Thomas, J.M., Sankar, G., Bromley, S., Oldroyd, R., Gladden, L. and Mantle, M.D. (1999) *Chemical Communications*, 1167–8.

15 Caps, V., Paraskevas, I. and Tsang, S.C. (2005) *Chemical Communications*, 1781–3.

16 (a) Thomas, J.M., Maschmeyer, T., Johnson, B.F.G. and Shephard, D.S. (1999) *Journal of Molecular Catalysis*, **141**, 139–44.
(b) Raynor, S.A., Thomas, J.M., Raja, R., Johnson, B.F.G., Bell, R.G. and Mantle, M.D. (2000) *Chemical Communications*, 1925–6.

17 (a) Pugin, B. (1996) *Journal of Molecular Catalysis A–Chemical*, **107**, 273–9.
(b) Meunier, D., Piechaczyk, A., Mallmann, A.De and Basset, J.-M. (1999) *Angewandte Chemie–International Edition*, **38**, 3540–2.

18 (a) Kresge, C.T., Leonowicz, M.E., Roth, W.J., Vartuli, J.C. and Beck, J.S. (1992) *Nature*, **359**, 710–12.
(b) Beck, J.S., Vartuli, J.C., Roth, W.J., Leonowicz, M.E., Kresge, C.T., Schmitt, K.D., Chu, C.T-W., Olson, D.H., Sheppard, E.W., McCullen, S.B., Higgins, J.B. and Schlenker, J.L. (1992) *Journal of the American Chemical Society*, **114**, 10834–43.

19 Zhao, D., Feng, J., Huo, Q., Melosh, N., Frederickson, G.H., Chmelka, B.F. and Stucky, G.D. (1998) *Science*, **279**, 548–52.

20 Pugin, B. and Müller, M. (1993) Heterogeneous catalysis and fine chemicals III, in *Studies in Surface Science and Catalysis*, Vol. 78 (eds M. Guisnet, J. Barrault, C. Bouchhoule, D. Duprez, G. Pérot and C. Montassier), Elsevier, pp. 107–14.

21 Fraile, J.M., García, J.I., Massam, J. and Mayoral, J.A. (1998) *Journal of Molecular Catalysis A–Chemical*, **136**, 47–57.

22 Kinting, A., Krause, H. and Capka, M. (1985) *Journal of Molecular Catalysis*, **33**, 215–23.

23 (a) Corma, A., Iglesias, M., del Pino, C. and Sánchez, F. (1991) *Journal of the Chemical Society D–Chemical Communications*, 1253–5.
(b) Corma, A., Iglesias, M., del Pino, C. and Sánchez, F. (1992) *Journal of Organometallic Chemistry*, **431**, 233–46.

(c) Carmona, A., Corma, A., Iglesias, M., José, A.S. and Sánchez, F. (1995) *Journal of Organometallic Chemistry*, **492**, 11–21.
(d) Corma, A., Iglesias, M., Mohino, F. and Sánchez, F. (1997) *Journal of Organometallic Chemistry*, **544**, 147–56.

24 Pérez, C., Pérez, S., Fuentes, G.A. and Corma, A. (2003) *Journal of Molecular Catalysis A–Chemical*, **197**, 275–81.

25 (a) Bianchini, C., Burnaby, D.G., Evans, J., Frediani, P., Meli, A., Oberhauser, W., Psaro, R., Sordelli, L. and Vizza, F. (1999) *Journal of the American Chemical Society*, **121**, 5961–71.
(b) Bianchini, C., Barbaro, P., Scapacci, G. and Zanobini, F. (2000) *Organometallics*, **19**, 2450–61.
(c) Bianchini, C., Barbaro, P., Dal Santo, V., Gobetto, R., Meli, A., Oberhauser, W., Psaro, R. and Vizza, F. (2001) *Advanced Synthesis Catalysis*, **343**, 41–5.

26 De Rege, F.M., Morita, D.K., Ott, K.C., Tumas, W. and Broene, R.D. (2000) *Chemical Communications*, 1797–8.

27 Augustine, R., Tanielyan, S., Anderson, S. and Yang, H. (1999) *Chemical Communications*, 1257–8.

28 Simons, C., Hanefeld, U., Arends, I.W.C.E., Maschmeyer, T. and Sheldon, R.A. (2006) *Journal of Catalysis*, **239**, 212–19.

29 (a) Crosman, A. and Hoelderich, W.F. (2007) *Catalysis Today*, **121**, 130–9.
(b) Crosman, A. and Hoelderich, W.F. (2005) *Journal of Catalysis*, **232**, 43–50.

30 (a) Bautista, F.M., Caballero, V., Campelo, J.M., Luna, D., Marinas, J.M., Romero, A.A., Romero, I., Serrano, I. and Llobet, A. (2006) *Topics in Catalysis*, **40**, 193–205.
(b) Flach, H.N., Grassert, I., Oehme, G. and Capka, M. (1996) *Colloid & Polymer Science*, **274**, 261–8.
(c) Wan, K.T. and Davis, M.E. (1994) *Journal of Catalysis*, **148**, 1–8.
(d) Wan, K.T. and Davis, M.E. (1994) *Nature*, **370**, 449–50.
(e) Wan, K.T. and Davis, M.E. (1995) *Journal of Catalysis*, **152**, 25–30.
(f) Davis, M.E. and Wan, K.T. (1998) US Patent 5,736,480. (g) Davis, M.E. and Wan, K.T. (1998) US Patent 5,827,794.

31 Pugin, B. (1996) *Journal of Molecular Catalysis A–Chemical*, **107**, 273–9.

32 (a) Pugin, B. (1996) WO 9,632,400 A1; (1997) WO 9,702,232 A1, (1996) EP 7,299,69.
(b) Rincon, V., Ayala, A., Corma, M., Iglesias, J.A. and Sanchez, F. (2004) *Journal of Catalysis*, **224**, 170–7.

33 (a) Hu, A., Ngo, H.L. and Lin, W. (2003) *Journal of the American Chemical Society*, **125**, 11490–1.
(b) Hu, A., Ngo, H.L. and Lin, W. (2003) *Angewandte Chemie – International Edition*, **42**, 6000–3.
(c) Kesanli, B. and Lin, W. (2004) *Chemical Communications*, 2284–5.

34 (a) Yin, M.-Y., Yuan, G.-L., Wu, Y.-Q., Huang, M.-Y. and Jiang, Y.-Y. (1999) *Journal of Molecular Catalysis A – Chemical*, **147**, 93–8.
(b) Vankelecom, I.F.J., Tas, D., Parton, R.F., Van de Vyver, V. and Jacobs, P.A. (1996) *Angewandte Chemie – International Edition in English*, **35**, 1346–8.

35 (a) Adima, A., Moreau, J.J.E. and Man, M.W.C. (2000) *Chirality*, **12**, 411–20.
(b) Bied, C., Gauthier, D., Moreau, J.J.E. and Man, M.W.C. (2001) *Journal of Sol-Gel Science and Technology*, **20**, 313–20.
(c) Moreau, J.J.E., Vellutini, L., Man, M.W.C. and Bied, C. (2001) *Journal of the American Chemical Society*, **123**, 1509–10.
(d) Jin, M.-Y., Kim, S.-H., Lee, S.-J. and Ahn, W.-S. (2003) *Studies in Surface Science and Catalysis*, **146**, 509–12.
(e) Jin, M.-Y., Sarkar, M.S., Takale, V.B. and Park, S.-E. (2005) *Bulletin of the Korean Chemical Society*, **26**, 1671–2.

36 (a) Liu, P.N., Gu, P.M., Wang, F. and Tu, Y.Q. (2004) *Organic Letters*, **6**, 169–72.
(b) Jiang, D., Yang, Q., Wang, H., Zhu, G., Yang, J. and Li, C. (2006) *Journal of Catalysis*, **239**, 65–73.
(c) Lipshutz, B.H., Frieman, B.A. and Tomaso, A.E. Jr (2006) *Angewandte Chemie – International Edition*, **45**, 1259–64.

37 (a) Yu, K., Lou, L.-L., Lai, C., Wang, S., Ding, F. and Liu, S. (2006) *Catalysis Communications* **7**, 1057–60.
(b) Bigi, F., Moroni, L., Maggi, R. and Sartori, G. (2002) *Chemical Communications*, 716–17.
(c) Kureshy, R.I., Ahmad, I., Khan, N.H., Abdi, S.H.R., Pathak, K. and Jasra, R.V. (2005) *Tetrahedron: Asymmetry*, **16**, 3562–9.
(d) Pini, D., Mandoli, A., Orlandi, S. and Salvadori, P. (1999) *Tetrahedron: Asymmetry*, **10**, 3883–6.
(e) Kim, G.-J. and Shin, J.-H. (1999) *Tetrahedron Letters*, **40**, 6827–30.

38 (a) Xiang, S., Zhang, Y., Xin, Q. and Li, C. (2002) *Chemical Communications*, 2696–7.
(b) Zhang, H. and Xiang, S. and Li, C. (2005) *Chemical Communications*, 1209–11.
(c) Zhang, H., Zhang, Y. and Li, C. (2006) *Journal of Catalysis*, **238**, 369–81.
(d) Zhang, H. and Li, C. (2006) *Tetrahedron*, **62**, 6640–9.
(e) Silva, A.R., Budarin, V., Clark, J.H., de Castro, B. and Freire, C. (2005) *Carbon*, **43**, 2096–105.
(f) Zhou, X.-G., Yu, X.-Q., Huang, J.-S., Che, C.-M., Li, S.-G. and Li, L.-S. (1999) *Chemical Communications*, 1789–90.

39 (a) Kureshy, R.I., Khan, N.H., Abdi, S.H.R., Ahmad, I., Singh, S. and Jasra, R.V. (2004) *Journal of Catalysis*, **221**, 234–40.
(b) Kureshy, R.I., Khan, N.H., Abdi, S.H.R., Ahmad, I., Singh, S. and Jasra, R.V. (2003) *Catalysis Letters*, **91**, 207–10.
(c) Nagel, U. and Kinzel, E. (1986) *Journal of the Chemical Society D – Chemical Communications*, 1098–9.
(d) Piaggio, P., McMorn, P., Murphy, D., Bethell, D., Page, P.C.B., Hancock, F.E., Sly, C., Kerton, O.J. and Hutchings, G.J. (2000) *Journal of the Chemical Society – Perkin Transactions 2*, 2008–15.
(e) Das, P., Kuzniarska-Biernacka, I., Silva, A.R., Carvalho, A.P., Pires, J. and Freire, C. (2006) *Journal of Molecular Catalysis A – Chemical*, **248**, 135–43.
(f) Choudary, B.M., Ramani, T., Maheswaran, H., Prasant, L., Ranganath, K.V.S. and Kumar, K.V. (2006) *Advanced Synthesis Catalysis*, **348**, 493–8.

40 Choudary, B.M., Pal, U., Kantam, M.L., Ranganath, K.V.S. and Sreedhar, B. (2006) *Advanced Synthesis Catalysis*, **348**, 1038–42.

41 (a) Gbery, G., Zsigmond, A. and Balkus, K.J.J. (2001) *Catalysis Letters*, **74**, 77–80.
(b) Ogunwumi, S.B. and Bein, T. (1997) *Chemical Communications*, 901–2.
(c) Sabater, M.J., Corma, A., Domenech, A., Fornes, V. and Garcia, H. (1997) *Chemical Communications*, 1285–6.

(d) Frunza, L., Kosslick, H., Landmesser, H., Hoft, E. and Fricke, R. (1997) *Journal of Molecular Catalysis A–Chemical*, **123**, 179–87.
(e) Heinrichs, C. and Holderich, W.F. (1999) *Catalysis Letters*, **58**, 75–80.
(f) Schuster, C. and Holderich, W.F. (2000) *Catalysis Today*, **60**, 193–207.
(g) Schuster, C., Mollmann, E., Tompos, A. and Holderich, W.F. (2001) *Catalysis Letters*, **74**, 69–75.

42 (a) Mehnert, C.P., Cook, R.A., Dispenziere, N.C. and Afeworki, M. (2002) *Journal of the American Chemical Society*, **124**, 12932–3.
(b) Mehnert, C.P. (2005) *Chemistry–A European Journal*, **11**, 50–6.
(c) Lou, L.L., Yu, K.Y., Diang, F., Zhou, W., Peng, X. and Liu, S. (2006) *Tetrahedron Letters*, **47**, 6513–16.
(d) Janssen, K.B.M., Laquiere, I., Dehaen, W., Parton, R.F., Vankelecom, I.F.J. and Jacobs, P.A. (1997) *Tetrahedron: Asymmetry*, **8**, 3481–7.
(e) Zhang, J.L., Liu, Y.-L. and Che, C.M. (2002) *Chemical Communications*, 2906–7.

43 (a) Xiang, S., Zhang, Y.-L., Xin, Q. and Li, C. (2002) *Angewandte Chemie–International Edition*, **41**, 821–4.
(b) Corma, A., Fuerte, A., Iglesias, M. and Sanchez, F. (1996) *Journal of Molecular Catalysis A–Chemical*, **107**, 225–34.
(c) Meunier, D., Piechaczyk, A., de Mallmann, A. and Basset, J.-M. (1999) *Angewandte Chemie–International Edition*, **38**, 3540–2.
(d) Meunier, D., Mallmann, A.D. and Basset, J.M. (2003) *Topics in Catalysis*, **23**, 183–9.

44 (a) Dhanda, A., Drauz, K.-H., Geller, T. and Roberts, S.M. (2000) *Chirality*, **12**, 313–17.
(b) Geller, T. and Roberts, S.M. (1999) *Journal of the Chemical Society–Perkin Transactions 1*, 1397–8.
(c) Carde, L., Davies, H., Geller, T.P. and Roberts, S.M. (1999) *Tetrahedron Letters*, **40**, 5421–4.

45 Choudary, B.M., Kantam, M.L., Ranganath, K.V.S., Mahendar, K. and Sreedhar, B. (2004) *Journal of the American Chemical Society*, **126**, 3396–7.

46 (a) Choudary, B.M., Jyothi, K., Kantam, M.L. and Sreedhar, B. (2004) *Advanced Synthesis Catalysis*, **346**, 1471–80.
(b) Lohray, B.B., Nandanan, E. and Bhushan, V. (1996) *Tetrahedron: Asymmetry*, **7**, 2805–8.
(c) Song, C.E., Yang, J.W. and Ha, H.-J. (1997) *Tetrahedron: Asymmetry*, **8**, 841–4.
(d) Lee, H.M., Kim, S.-W., Hyeon, T. and Kim, B.M. (2001) *Tetrahedron: Asymmetry*, **12**, 1537–41.
(e) Motorina, I. and Crudden, C.M. (2001) *Organic Letters*, **3**, 2325–8.
(f) Choudary, B.M., Chowdari, N.S., Jyothi, K. and Kantam, M.L. (2002) *Advanced Synthesis Catalysis*, **344**, 503–6.
(g) Lee, D., Lee, J., Lee, H., Jin, S., Hyeon, T. and Kim, B.M. (2006) *Advanced Synthesis Catalysis*, **348**, 41–6.

47 (a) Song, C.E., Oh, C.R., Lee, S.W., Canali, S.-G., Lee, L. and Sherrington, D.C. (1998) *Chemical Communications*, 2435–6.
(b) Choudary, B.M., Chowdari, N.S., Jyothi, K. and Kantam, M.L. (2003) *Journal of Molecular Catalysis A–Chemical*, **196**, 151–6.

48 (a) Taylor, S., Gullick, J., McMorn, P., Bethell, D., Page, P.C.B., Hancock, F.E., King, F. and Hutchings, G.J. (2001) *Journal of the Chemical Society–Perkin Transactions 2*, 1714–23.
(b) Taylor, S., Gullick, J., McMorn, P., Bethell, D., Page, P.C.B., Hancock, F.E., King, F. and Hutchings, G.J. (2001) *Journal of the Chemical Society–Perkin Transactions 2*, 1724–8.
(c) Ryan, D., McMorn, P., Bethell, D. and Hutchings, G.J. (2004) *Organic and Biomolecular Chemistry*, **2**, 3566–72.

49 (a) Ayala, V., Corma, A., Iglesiasc, M. and Sanchez, F. (2004) *Journal of Molecular Catalysis A–Chemical*, **221**, 201–8.
(b) Kantam, M.L., Prakash, B.V., Bharathi, B. and Reddy, C.V. (2005) *Journal of Molecular Catalysis A–Chemical*, **226**, 119–22.

50 Jacobsen, E.N. and Wu, M.H. (1999) Epoxidation of alkenes other than allylic alcohols, in *Comprehensive Asymmetric Catalysis*, Vol. **2** (eds E.N. Jacobsen, A. Pfaltz and H. Yamamoto), Springer Verlag, Berlin, p. 649.

51 (a) Srinvasan, K., Michaud, P. and Kochi, J.K. (1986) *Journal of the American Chemical Society*, **108**, 2309–20.

(b) De, B.B., Lohray, B.B., Sivaram, S. and Dhal, P.K. (1995) *Tetrahedron: Asymmetry*, **6**, 2105–8.
(c) Salvadori, P., Pini, D., Petri, A. and Mandoli, A. (2000) Catalytic heterogeneous enantioselective dihydroxylation and epoxidation, in *Chiral Catalyst Immobilization and Recycling* (eds D.E. De Vos, I.F.J. Vankelecom and P.A. Jacobs), Wiley-VCH Verlag GmbH, Weinheim, p. 235.
(d) Baleizao, C. and Garcia, H. (2006) *Chemical Reviews*, **106**, 3987–4043.

52 (a) Katsuki, T. (1999) Epoxidation of allylic alcohols, in *Comprehensive Asymmetric Catalysis*, Vol. 2 (eds E.N. Jacobsen, A. Pfaltz and H. Yamamoto), Springer Verlag, Berlin, p. 621.
(b) Johnson, R.A. and Sharpless, K.B. (2000) Catalytic asymmetric epoxidation of allylic alcohols, in *Catalytic Asymmetric Synthesis*, 2nd edn (ed. I. Ojima), Wiley-VCH Verlag GmbH, New York, p. 231.
(c) Shum, W.P. and Cannarsa, M.J. (1996) Sharpless asymmetric epoxidation: scale-up and industrial production, in *Chirality in Industry II* (eds A.N. Collins, G.N. Shedrake and J. Crosby), John Wiley & Sons, p. 363.

53 (a) Choudary, B.M., Valli, V.L.K. and Prasad, A.D. (1990) *Journal of the Chemical Society D – Chemical Communications*, 1186–7.
(b) Choudary, B.M., Valli, V.L.K. and Prasad, A.D. (1990) *Journal of the Chemical Society D – Chemical Communications*, 721–2.

54 (a) Aggarwal, V.K. (1999) Epoxide formation of enones and aldehydes, in *Comprehensive Asymmetric Catalysis*, Vol. 2 (eds E.N. Jacobsen, A. Pfaltz and H. Yamamoto), Springer Verlag, Berlin, p. 679.
(b) Banfi, S., Colonna, S., Molinari, H., Juliá, S. and Guixer, J. (1984) *Tetrahedron*, **40**, 5207–11.
(c) Carrea, G., Colonna, S., Meek, A.D., Ottolina, G. and Roberts, S.M. (2004) *Tetrahedron: Asymmetry*, **15**, 2945–9.
(d) Carrea, G., Stefano, C., Kelly, D.R., Lazcano, A., Ottolina, G. and Roberts, S.M. (2005) *Trends in Biotechnology*, **23**, 507–13.

55 (a) Johnson, R.A. and Sharpless, K.B. (2000) Catalytic asymmetric dihydroxylation- discovery and development, in *Catalytic Asymmetric Synthesis*, 2nd edn (ed. I. Ojima), Wiley-VCH Verlag GmbH, p. 357.
(b) Bolm, C., Hildebrand, J.P. and Muniz, K. (2000) Recent advances in asymmetric dihydroxylation and aminohydroxylation, in *Catalytic Asymmetric Synthesis*, 2nd edn (ed. I. Ojima), Wiley-VCH Verlag GmbH, p. 399.

56 Kobayashi, S. and Sugiura, M. (2006) *Advanced Synthesis Catalysis*, **348**, 1496–504 and references cited therein.

57 Kwong, H.-L., Sorato, C., Ogino, Y., Chen, H. and Sharpless, K.B. (1990) *Tetrahedron Letters*, **31**, 2999–3002.

58 (a) Gridnev, I.D. (2006) *Chemistry Letters*, **35**, 148–53.
(b) Mikami, K. and Yamanaka, M. (2003) *Chemical Reviews*, **103**, 3369–400.
(c) Konepudi, D.K.K. and Asakura (2001) *Accounts of Chemical Research*, **34**, 946–54.

59 Traa, Y., Murphy, D.M., Farley, R.D. and Hutchings, G.J. (2001) *Physical Chemistry Chemical Physics*, **3**, 1073–80.

60 Wan, Y., McMorn, P., Hancock, F.E. and Hutchings, G.J. (2003) *Catalysis Letters*, **91**, 145–8.

61 Caplan, N.A., Hancock, F.E., Bulman, P.P.C. and Hutchings, G.J. (2004) *Angewandte Chemie – International Edition*, **43**, 1685–8.

62 (a) Soai, K., Watanabe, M. and Yamamoto, A. (1990) *Journal of Organic Chemistry*, **55**, 4832–5.
(b) Sato, I., Shimizu, M., Kawasaki, T. and Soai, K. (2004) *Bulletin of the Chemical Society of Japan*, **77**, 1587–8.

63 (a) Wu, C.-D. and Lin, W. (2007) *Angewandte Chemie – International Edition*, **46**, 1075–8.
(b) Wu, C.-D., Hu, A., Zhang, L. and Lin, W. (2005) *Journal of the American Chemical Society*, **127**, 8940–1.
(c) Fraile, J.M., Mayoral, J.A., Serrano, J., Pericás, M.A., Solá, L. and Castellnou, D. (2003) *Organic Letters*, **5**, 4333–5.
(d) Pathak, K., Bhatt, A.P., Abdi, S.H.R., Kureshy, R.I., Khan, N.H., Ahmad, I. and Jasra, R.V. (2006) *Tetrahedron: Asymmetry*, **17**, 1506–13.

(e) Bae, S.J., Kim, B.M., Kim, S.-W. and Hyeon, T. (2000) *Chemical Communications*, 31–2.
(f) Kim, S.-W., Bae, S.J., Hyeon, T. and Kim, B.M. (2001) *Microporous and Mesoporous Materials*, **44-45**, 523–9.
(g) Heckel, A. and Seebach, D. (2000) *Angewandte Chemie – International Edition*, **39**, 163–5.

64 (a) Wang, H., Liu, X., Xia, H., Liu, P., Gao, J., Ying, P., Xiao, J. and Li, C. (2006) *Tetrahedron*, **62**, 1025–32.
(b) O'Leary, P., Krosveld, N.P., De Jong, K.P., van Koten, G. and Klein Gebbink, R.J.M. (2004) *Tetrahedron Letters*, **45**, 3177–80.
(c) Tanaka, S., Tada, M. and Iwasawa, Y. (2007) *Journal of Catalysis*, **245**, 173–83.
(d) Selkala, S.A., Tois, J., Pihko, P.M. and Koskinen, A.M.P. (2002) *Advanced Synthesis Catalysis*, **344**, 941–5.

65 Heckel, A. and Seebach, D. (2002) *Helvetica Chimica Acta*, **85**, 913–25.

66 Zhang, Y., Zhao, L., Lee, S.S. and Ying, J.Y. (2006) *Advanced Synthesis Catalysis*, **348**, 2027–32.

67 Corma, A., Iglesias, M., Martin, M.V., Rubio, J. and Sanchez, F. (1992) *Tetrahedron: Asymmetry*, **3**, 845–8.

68 Iglesias, M. and Sanchez, F. (2000) *Studies in Surface Science and Catalysis*, **130D**, 3393–8.

69 (a) Choudary, B.M., Ranganath, K.V.S., Pal, U., Kantam, M.L. and Sreedhar, B. (2005) *Journal of the American Chemical Society*, **127**, 13167–71.
(b) Choudary, B.M., Kavita, B., Chowdari, N.S., Sreedhar, B. and Kantam, M.L. (2002) *Catalysis Letters*, **78**, 373–7.

70 Calderón, F., Fernández, R., Sánchez, F. and Mayoralas, A.-F. (2005) *Advanced Synthesis Catalysis*, **347**, 1395–403.

71 (a) Gruttadauria, M., Riela, S., Meo, P.L., D'Anna, F. and Noto, R. (2004) *Tetrahedron Letters*, **45**, 6113–16.
(b) Sakthivel, K., Notz, W., Bui, T. and Barbas, C.F. III (2001) *Journal of the American Chemical Society*, **123**, 5260–7.
(c) Benaglia, M., Cinquini, M., Cozzi, F., Puglisi, A. and Celentano, G. (2002) *Advanced Synthesis Catalysis*, **344**, 533–42.
(d) Loh, T.-P., Feng, L.-C., Yang, H.-Y. and Yang, J.-Y. (2002) *Tetrahedron Letters*, **43**, 8741–3.

72 (a) Choudary, B.M., Chakrapani, L., Ramani, T., Kumar, K.V. and Kantam, M.L. (2006) *Tetrahedron*, **62**, 9571–6.
(b) Bhatt, A.P., Pathak, K., Jasra, R.V., Kureshy, R.I., Khan, N.H. and Abdi, S.H.R. (2006) *Journal of Molecular Catalysis A – Chemical*, **251**, 123–8.

73 (a) Lee, S.S., Hadinoto, S. and Ying, J.Y. (2006) *Advanced Synthesis Catalysis*, **348**, 1248–54.
(b) Lee, S.S. and Ying, J.Y. (2006) *Journal of Molecular Catalysis A – Chemical*, **256**, 219–24.

74 Corma, A., Garcia, H., Moussaif, A., Sabater, M.J., Zniber, R. and Redouane, A. (2002) *Chemical Communications*, 1058–9.

75 (a) Hultman, H.M., de Lang, M., Nowotny, M., Arends, I.W.C.E., Hanefeld, U., Sheldon, R.A. and Maschmeyer, T. (2003) *Journal of Catalysis*, **217**, 264–74.
(b) Hultman, H.M., de Lang, M., Arends, I.W.C.E., Hanefeld, U., Sheldon, R.A. and Maschmeyer, T. (2003) *Journal of Catalysis*, **217**, 275–83.

76 Heckel, A. and Seebach, D. (2002) *Chemistry – A European Journal*, **8**, 559–72.

3
Heterogeneous Enantioselective Catalysis Using Organic Polymeric Supports

Shinichi Itsuno and Naoki Haraguchi

3.1
Introduction

The development of polymer-immobilized ligands and catalysts for asymmetric synthesis is a rapidly growing field [1], which has great importance due mainly to the easy separation and recyclability of polymer-supported catalysts. The polymeric catalysts also represent one of the most powerful tools for 'green' sustainable chemistry, in the sense that they can be easily recovered and reused many times.

The use of a polymer support in heterogeneous catalysis sometimes results a lowering of the chemical yield of a synthetic reaction, due mainly to the heterogeneous nature of the catalyst, because efficient interactions between substrate in solution and a supported catalyst may not be achieved [1]. However, several highly reactive polymer-supported catalysts and reagents have recently been developed for a variety of reactions [1], and excellent enantioselectivities in many asymmetric reactions have been obtained by using polymer-supported chiral catalysts. In some cases, even higher enantioselectivities were attained by using polymeric catalysts than when using low-molecular-weight catalysts. Under some circumstances, a polymer network may provide a favorable microenvironment for the asymmetric reactions.

The most common polymer supports used for chiral catalyst immobilization are polystyrene-based crosslinked polymers, although poly(ethylene glycol) (PEG) represents an alternative choice of support. In fact, soluble PEG-supported catalysts show relatively high reactivities (in certain asymmetric reactions) [1e] which can on occasion be used in aqueous media [1e]. Methacrylates, polyethylene fibers, polymeric monoliths and polynorbornenes have been also utilized as efficient polymer supports for the heterogenization of a variety of homogeneous asymmetric catalysts.

3.2
Asymmetric Alkylation of Carbonyl Compounds

Many studies have been reported on the asymmetric alkylation of carbonyl compounds with organozinc reagents using polymeric chiral ligands [1a,b]. Some recent examples of this have involved the use of polymer-supported prolinol derivatives or other chiral amino alcohols [1f] Scheme 3.1 shows some typical examples of polymer-supported amino alcohol derivatives used for the asymmetric diethylzinc addition to benzaldehyde. Chiral α,α-diaryl-prolinols were attached to polystyrene containing phenylboronic acid moieties by means of Suzuki coupling reaction to give **1** [2]. When the polymeric catalyst **1** was used at 30 mol% relative to aldehyde in toluene at 20 °C, diethylzinc addition to the aromatic aldehyde occurred smoothly to give the expected alcohols in 78–94% enantiomeric excess (ee) and >90% yields in 20 h. The polymeric catalysts can be easily recycled and reused up to nine times in the diethylzinc addition reaction.

Degni *et al.* developed a new method for the anchoring of chiral prolinol derivatives, whereby N-(4-vinylbenzyl)-α,α-diphenyl-prolinol was immobilized on polyethylene fibers by electron beam-induced pre-irradiation grafting using styrene as a comonomer. The chirally modified polymer fibers **2** were used as a catalyst for the asymmetric ethylation of aldehyde, with moderate enantioselectivity [3].

A chiral diamino alcohol derived from bispidine was immobilized on the cross-linked polystyrene, to give **3** [4]. Asymmetric ethylation of benzaldehyde with

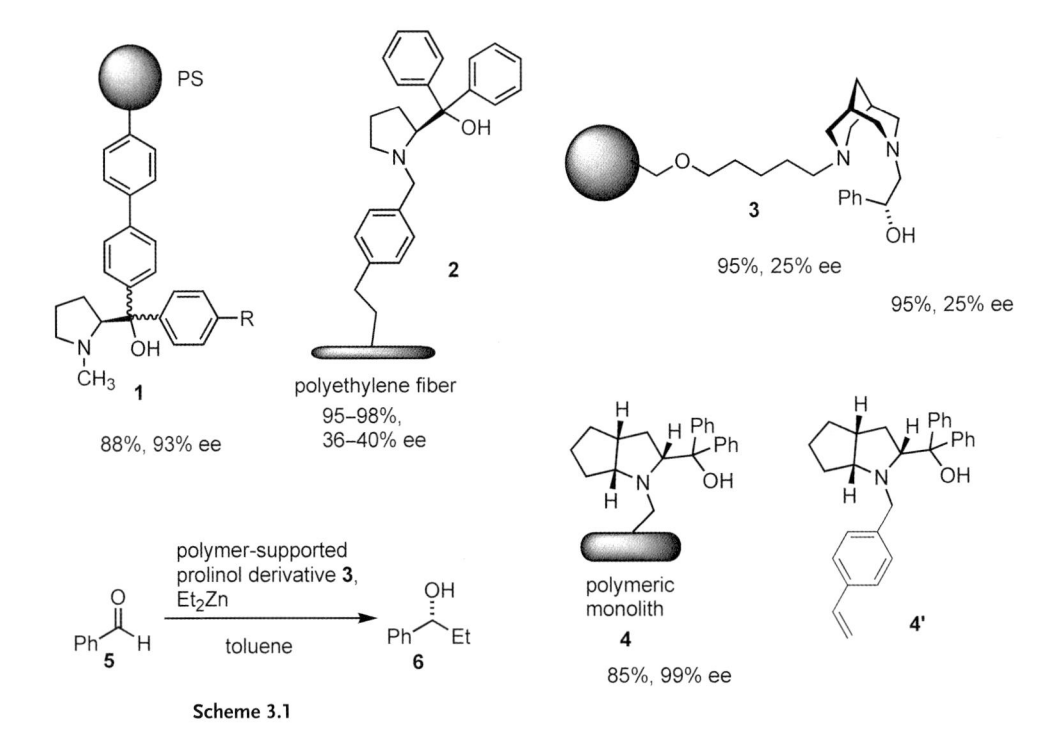

Scheme 3.1

diethylzinc was monitored in the presence of the polymer-supported diamino alcohol **3**. Although the corresponding low-molecular-weight chiral diamino alcohol ligand showed excellent enantioselectivity (96% ee), a drastic lowering of the ee-value was observed with the polymer-supported ligand [4].

Luis and coworkers prepared a chiral amino alcohol monomer **4′**. Polymerization of a mixture containing 40wt% monomers (10 mol% chiral monomer, 90 mol% divinylbenzene (DVB)) and 60wt% toluene-1-dodecanol as the porogenic mixture (10wt% toluene) gave monoliths **4** [5, 6]. The monolithic column allowed the design of a flow system, and under flow conditions benzaldehyde was asymmetrically ethylated with diethylzinc to give the chiral alcohol in 99% ee [5].

The ring-opening metathesis polymerization (ROMP) [7] of norbornenes bearing catalytically active prolinol units gave the highly functionalized polymers **9** [8]. Asymmetric diethylzinc addition to benzaldehyde in the presence of **9** proceeded to give the chiral alcohol in 75–85% yield and 87% ee (Scheme 3.2).

Nonracemic Ti-BINOLate (BINOL = 1,1′-bi-2-naphthol) and Ti-TADDOLate (TADDOL = α,α,α′,α′-tetraaryl-2,2-dimethyl-1,3-dioxolan-4,5-dimethanol) complexes are also effective chiral catalysts for the asymmetric alkylation of aldehydes [9–11]. Seebach developed polystyrene beads with dendritically embedded BINOL [9] or TADDOL derivatives **11** [10, 11]. As the chiral ligand is located in the core of the dendritic polymer, less steric congestion around the catalytic center was achieved after the treatment with Ti(OiPr)$_4$. This polymer-supported TiTADDOLate **14** was then used for the ZnEt$_2$ addition to benzaldehyde. Chiral 1-phenylpropanol was obtained in quantitative yield with 96% ee (Scheme 3.3), while the polymeric catalyst could be recycled many times.

One significant problem which is associated with the conventional polystyrene supports is that of low mechanical strength and often poor thermo-oxidative stability. The immobilization of TADDOL derivatives on mechanically stable and chemically inert polyethylene (PE) fibers by means of electron beam-induced grafting has also been investigated. The TADDOL-containing fibers **15** obtained were transformed into the titanium TADDOLate and used for the asymmetric ethylation of benzaldehyde in quantitative yield and 94% ee [12].

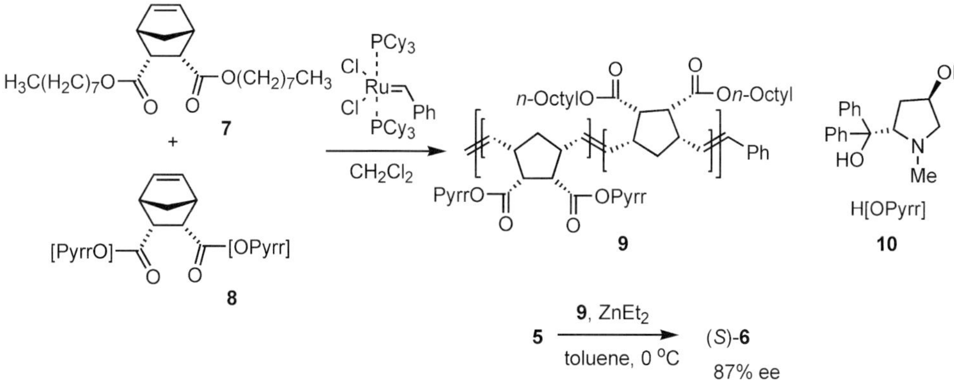

Scheme 3.2

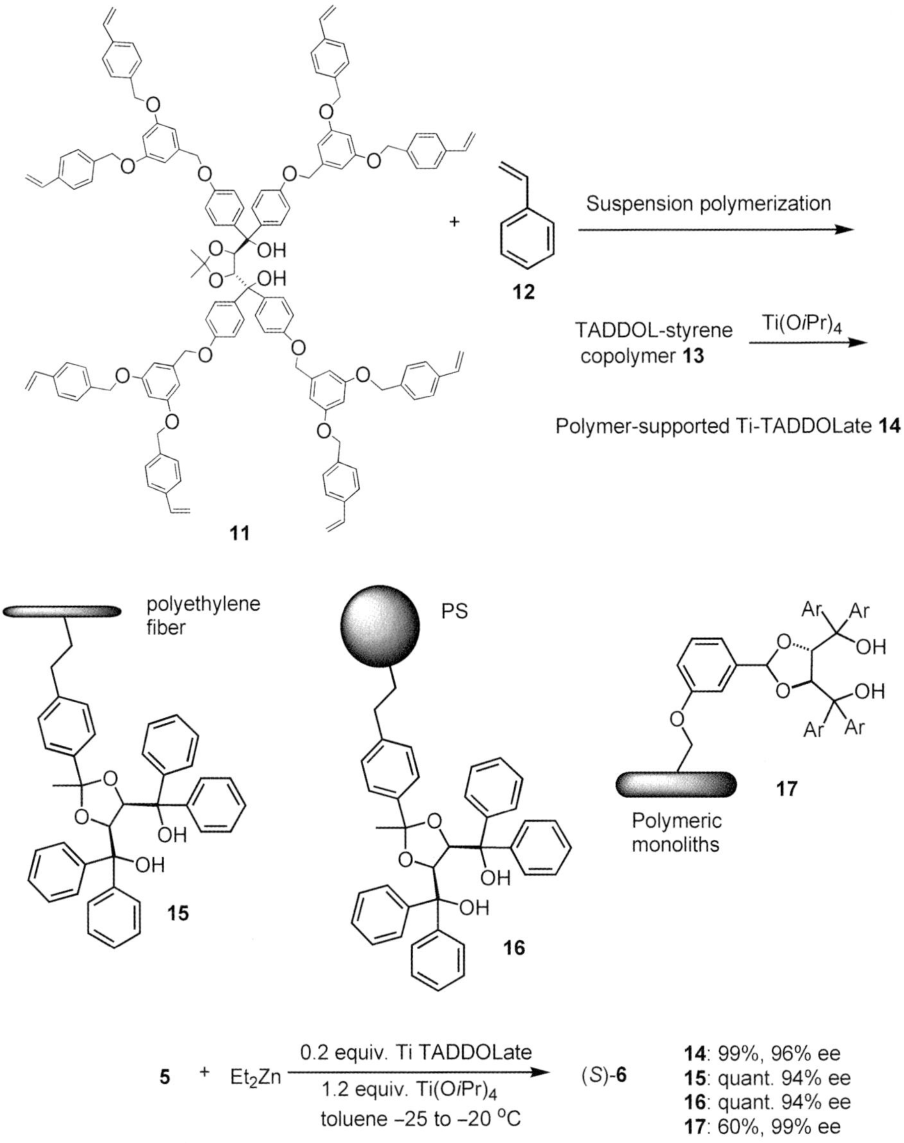

Scheme 3.3

Luis prepared polymeric monoliths **17** containing TADDOL subunits [13]; these were synthesized with a thermally induced radical solution polymerization of a mixture containing TADDOL monomer, styrene and DVB, using toluene/1-dodecanol as the precipitating porogenic mixture and azoisobutyronitrile (AIBN) as the radical initiator. The polymer-supported Ti-TADDOLates generated from **17** and Ti(O*i*Pr)$_4$ were then used for the asymmetric alkylation of benzaldehyde to give 1-phenylethanol in 60% yield and 99% ee [13].

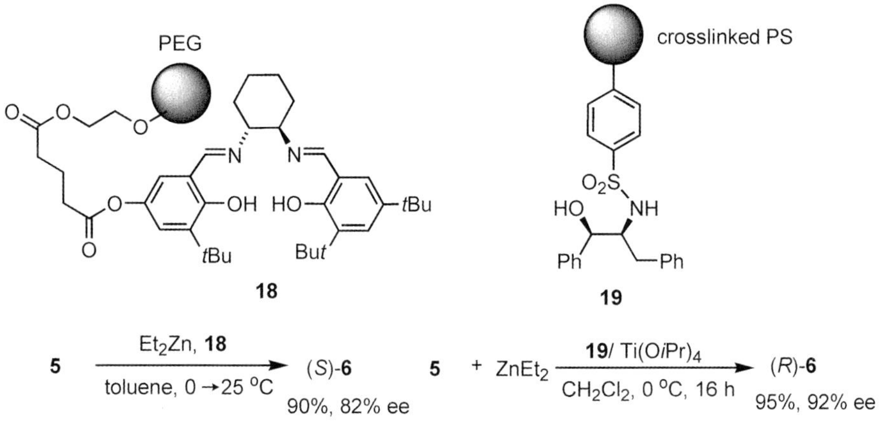

3.2 Asymmetric Alkylation of Carbonyl Compounds | 77

Scheme 3.4

Some chiral salen (*N,N*-ethylenebis(salicylimine)) ligands were attached as the end group of PEG. The soluble polymer-supported catalyst **18**, with a glutarate spacer between the ligand and PEG, performed well in toluene and provided 82% ee in the asymmetric ethylation of benzaldehyde (Scheme 3.4) [14].

Crosslinked polystyrene (PS)-supported *N*-sulfonylated amino alcohols can generally be prepared by either of the following two methods [15, 16]:

- The polymer reaction method, which involves chlorosulfonylation of a crosslinked PS, followed by the attachment of amino alcohol by *N*-sulfonylation to give **19a** [15].
- Polymerization of the monomer bearing the chiral *N*-sulfonylamino alcohol moiety with styrene in the presence of DVB as crosslinking agent, to give the similar polymer-supported *N*-sulfonylamino alcohol **19b**.

The polymers **19** were treated with Ti(O*i*Pr)$_4$ and the resulting polymeric complexes were used as a catalyst for the diethylzinc addition to aromatic aldehyde. In the diethylzinc addition to benzaldehyde, the titanium catalyst prepared from **19a** afforded 1-phenylethanol in only 62% yield and 44% ee; this poor performance was attributed to interference due to the close proximities of the active catalytic metal centers (98% ligand loading) in **19a** [17]. In contrast, **19b** with 20% ligand loading gave the product in 95% yield and 92% ee in the same reaction [17]. The polymeric Ti catalyst **19b** was reused five times, albeit with slightly decreasing enantioselectivities of the product.

Sasai developed micelle-derived polymer-supported catalysts for a variety of enantioselective reactions, including diethylzinc addition [18]. The surfactant monomer **20** having tetraethylene glycol chains formed micelles in water and, followed by copolymerization with styrene, gave the spherical polymer. A coupling reaction of the polymer with a BINOL derivative and deprotection of the methoxymethyl groups of the BINOL moiety afforded the desired chiral polymer **21**, as shown in Scheme 3.5. The catalytic asymmetric ethylation of benzaldehyde was

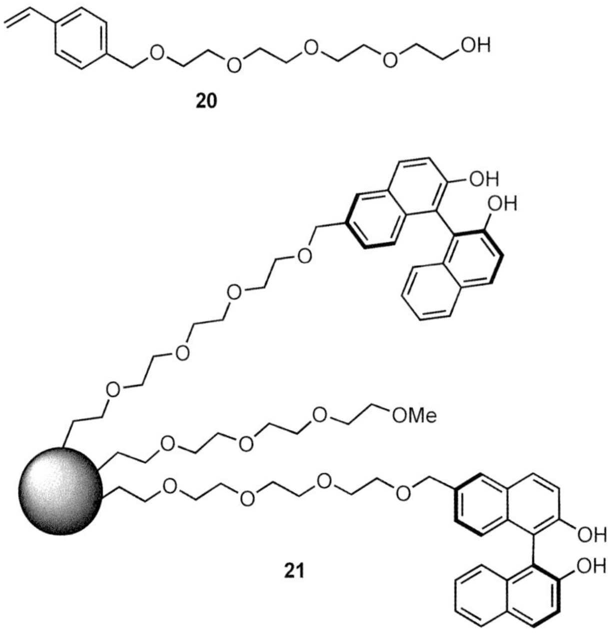

Scheme 3.5

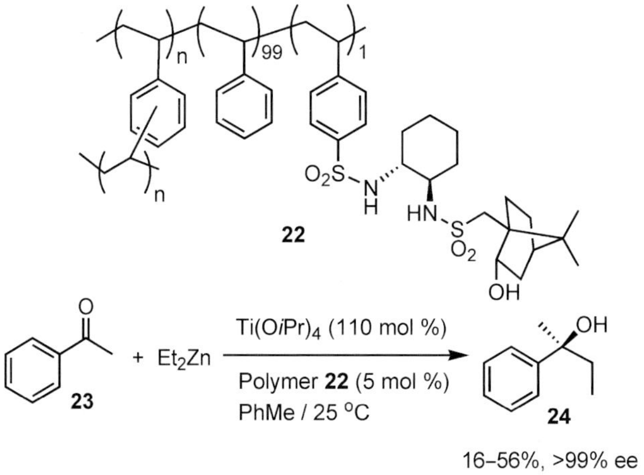

Scheme 3.6

performed over the polymeric Ti-BINOLate complex prepared from **21**, to give 1-phenylethanol in 96% yield with 84% ee.

The asymmetric alkylation of the ketone carbonyl groups occurred under selected reaction conditions [19]. One recent example of this involves the use of polymer-supported chiral disulfonamide **22** as a chiral ligand (Scheme 3.6) [20]. The polymer **22**-Ti(O*i*Pr)$_4$ complex was utilized for diethylzinc addition to simple

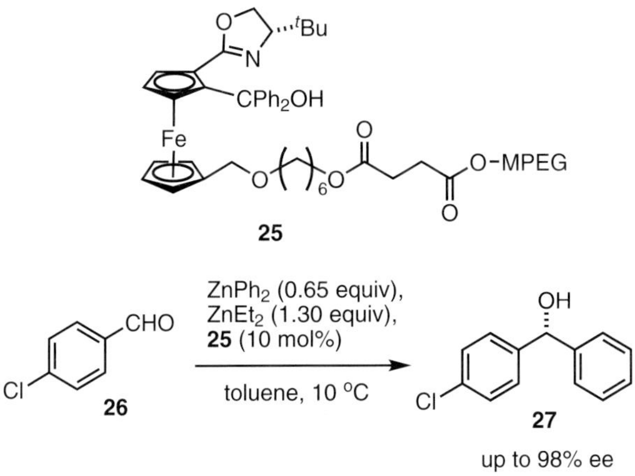

Scheme 3.7

ketones. Although the use of polymeric chiral ligand showed excellent enantioselectivity (>99% ee) in all cases, a lowering of the yield was observed which was dependent on the degree of crosslinkage. The use of a lightly crosslinked polymer (**22**; n = 1) gave a relatively high yield (56%) with very high enantioselectivity (>99% ee).

3.3
Asymmetric Phenylation

Chiral diarylmethanols are useful intermediates in the synthesis of biologically active substrates. One of several highly enantioselective catalytic systems utilized for the organometallic phenyl transfer reaction involves the use of organozinc reagents [21]. Bolm developed a ferrocenyl oxazoline ligand for catalysis of the phenyl transfer reaction [22]. A polymer-supported version of the ferrocenyl oxazoline **25** was also prepared and utilized for the same reaction by using $ZnPh_2$ and $ZnEt_2$ as the phenyl transfer agent [23]. When the polymeric catalyst prepared by **25** was used in an asymmetric phenyl addition to aromatic aldehyde, almost the same enantioselectivities as that of the low-molecular-weight counterpart in solution system were obtained. In the case of addition of phenylzinc to 4-chlorobenzaldehyde **26**, the corresponding chiral diarylmethanol (**27**) was obtained in 97% ee (Scheme 3.7).

Enantioselective phenylation of aromatic ketones also proceeded by means of a polymeric catalyst derived from **22** [20]; triphenylborane is the source of the phenyl transferring group in this case (Scheme 3.8). The linear polymer **22** (n = 0) gave a higher yield of **29** with high enantioselectivity than did the crosslinked polymer. Aliphatic ketones such as 2-hexanone gave a lower enantioselectivity (38% ee) than their aromatic counterparts, using the same polymeric catalyst.

Scheme 3.8

Scheme 3.9

Scheme 3.10

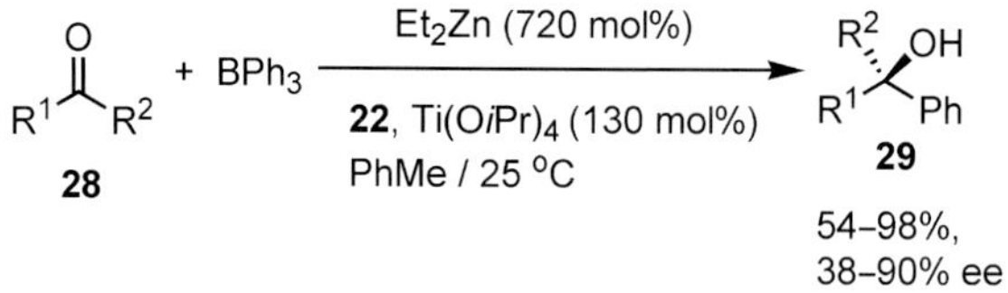

3.4
Asymmetric Addition of Phenylacetylene

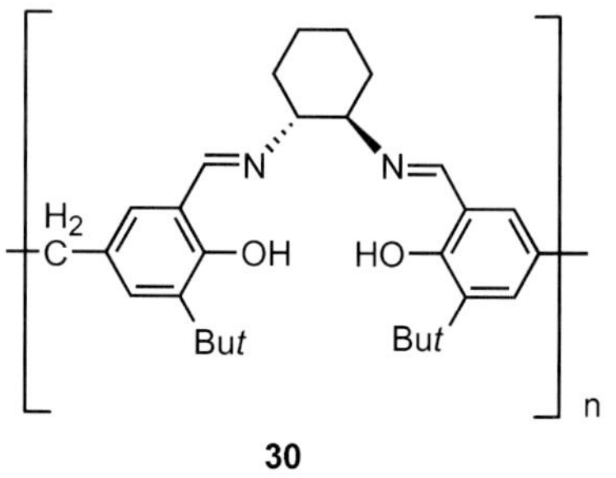

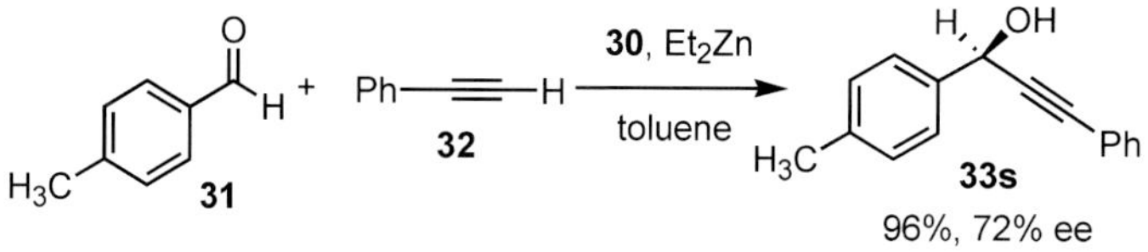

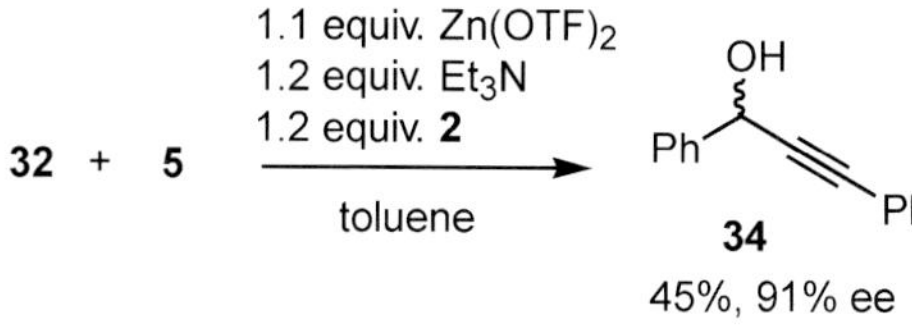

The enantioselective alkynylation of ketones catalyzed by Zn(salen) complexes has been reported [24]. Polymeric salen ligand **30** was prepared with a polycondensation reaction and subsequently used as a polymeric chiral ligand of Zn. The polymeric Zn(salen) complex (prepared by **30**) was then used as a catalyst of asymmetric addition of phenylacetylene to aldehyde in the presence of 2 equivalents of Et_2Zn. Subsequent asymmetric alkynylation of **31** gave **33** in 96% yield and 72% ee (Scheme 3.9) [25].

The polyethylene fiber-supported chiral amino alcohol **2**, which has been used for the asymmetric ethylation of aldehyde, can also be applied to an asymmetric addition of phenylacetylene to benzaldehyde, to give the corresponding propargylic alcohol **34** with 91% ee in 45% yield (Scheme 3.10) [3].

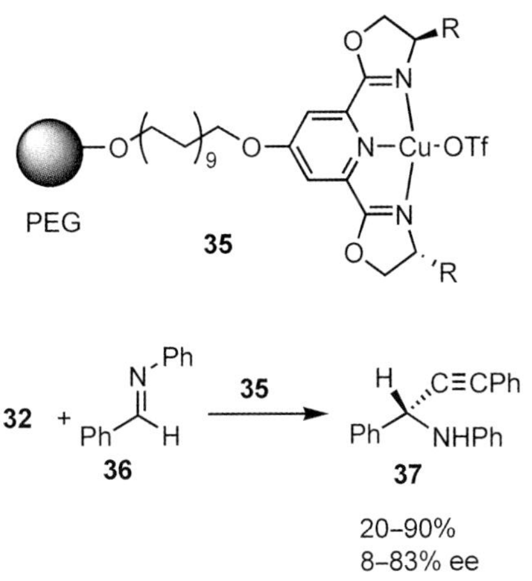

Scheme 3.11

3.5
Asymmetric Addition to Imine Derivatives

Copper complexes of chiral Pybox (pyridine-2,6-bis(oxazoline))-type ligands have been found to catalyze the enantioselective alkynylation of imines [26]. Moreover, the resultant optically active propargylamines are important intermediates for the synthesis of a variety of nitrogen compounds [27], as well as being a common structural feature of many biologically active compounds and natural products. Portnoy prepared PS-supported chiral Pybox-copper complex **35** via a five-step solid-phase synthetic sequence [28]. Cu(I) complexes of the polymeric Pybox ligands were then used as catalysts for the asymmetric addition of phenylacetylene to imine **36**, as shown in Scheme 3.11. tBu-Pybox gave the best enantioselectivity of 83% ee in the synthesis of **37**.

Scheme 3.12 illustrates the polymer-supported allylboron reagents derived from chiral N-sulfonylamino alcohols and used for the asymmetric synthesis of homoallylic alcohols and amines (see Scheme 3.12) [29]. All of these asymmetric allylborations were performed using the polymeric chiral allylboron reagent prepared from triallylborane and PS-supported N-sulfonylamino alcohols **38–41**. High levels of enantioselectivity were obtained in the asymmetric allylboration of imines with the polymeric reagent derived from norephedrine.

3.6
Asymmetric Silylcyanation of Aldehyde

The asymmetric addition of trimethylsilylcyanide to aldehyde was catalyzed by a chirally modified Lewis acid. Polymer-supported chiral bis(oxazoline)s (**46, 47**) were prepared and used as ligands of ytterbium chloride [30]. The polymeric ligands exhibited as high a reactivity in the asymmetric silylcyanation as did their

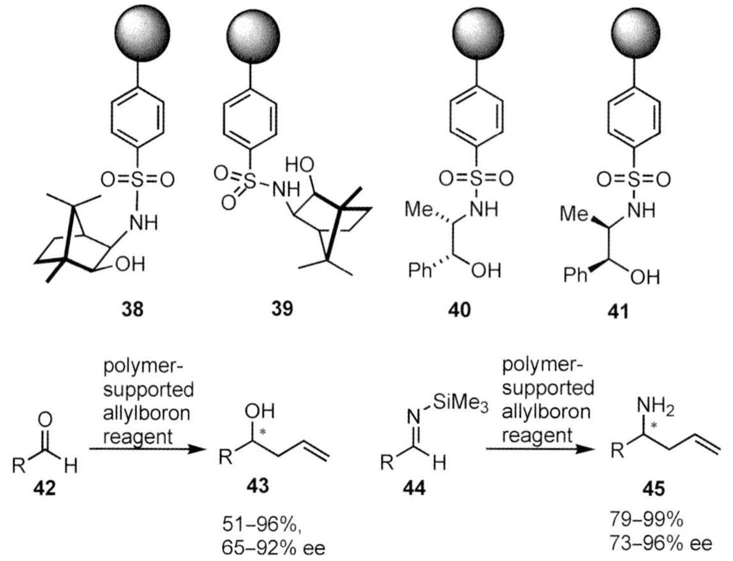

Scheme 3.12

homogeneous analogues. However, the enantioselectivities obtained by using these polymeric catalysts were marginally lower (Scheme 3.13).

3.7
Asymmetric Synthesis of α-Amino Acid

The asymmetric alkylation of glycine derivatives is one of the most simple methods by which to obtain optically active α-amino acids [31]. The enantioselective alkylation of glycine Schiff base **52** under phase-transfer catalysis (PTC) conditions and catalyzed by a quaternary cinchona alkaloid, as pioneered by O'Donnell [32], allowed impressive degrees of enantioselection to be achieved using only a very simple procedure. Some examples of polymer-supported cinchona alkaloids are shown in Scheme 3.14. Polymer-supported chiral quaternary ammonium salts **48** have been easily prepared from crosslinked chloromethylated polystyrene (Merrifield resin) with an excess of cinchona alkaloid in refluxing toluene [33]. The use of these polymer-supported quaternary ammonium salts allowed high enantioselectivities (up to 90% ee) to be obtained.

PEG-supported cinchona ammonium salts **54** were applied to the asymmetric alkylation of *tert*-butyl benzophenone Schiff base derivatives **52** [34]. The use of a water-soluble polymer support allowed the reaction to be conducted in a 1 M KOH aqueous solution to give the α-amino acid derivatives **53** in high chemical yields (up to 98%). Ten different types of electrophile have been tested for the reaction, with the best enantioselectivity being obtained with *o*-chlorobenzylchloride (97% ee) (Scheme 3.15).

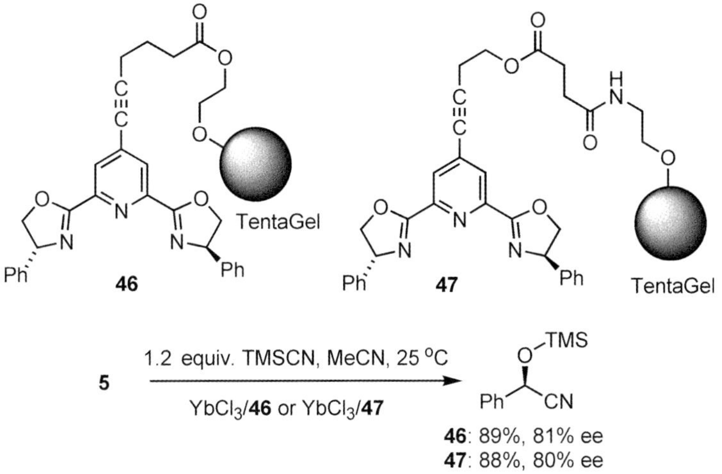

3.7 Asymmetric Synthesis of α-Amino Acid

Scheme 3.13

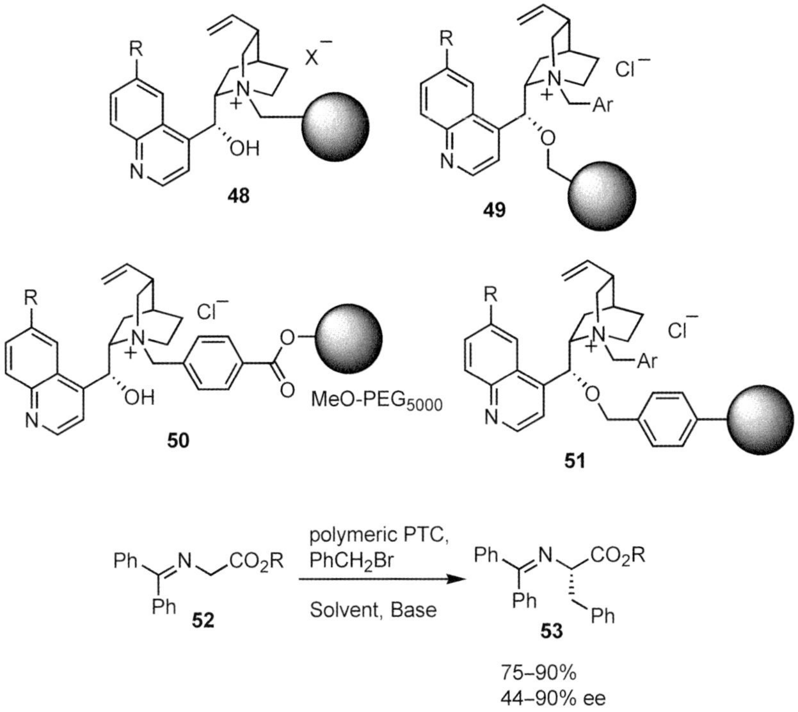

5 →(1.2 equiv. TMSCN, MeCN, 25 °C; YbCl$_3$/**46** or YbCl$_3$/**47**)→ Ph-CH(OTMS)-CN

46: 89%, 81% ee
47: 88%, 80% ee

Scheme 3.14

52 (Ph$_2$C=N-CH$_2$-CO$_2$R) →(polymeric PTC, PhCH$_2$Br, Solvent, Base)→ **53**

75–90%
44–90% ee

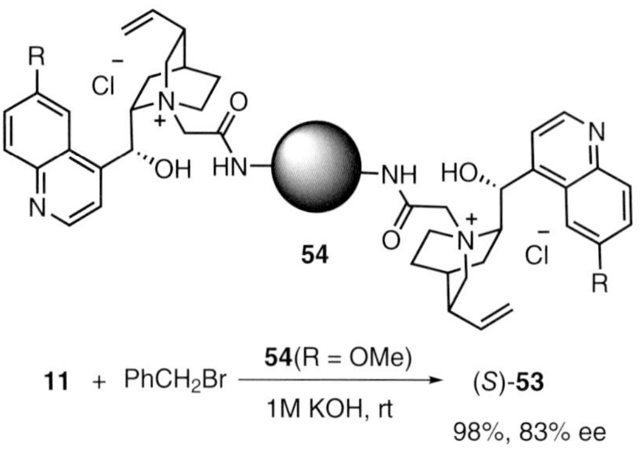

Scheme 3.15

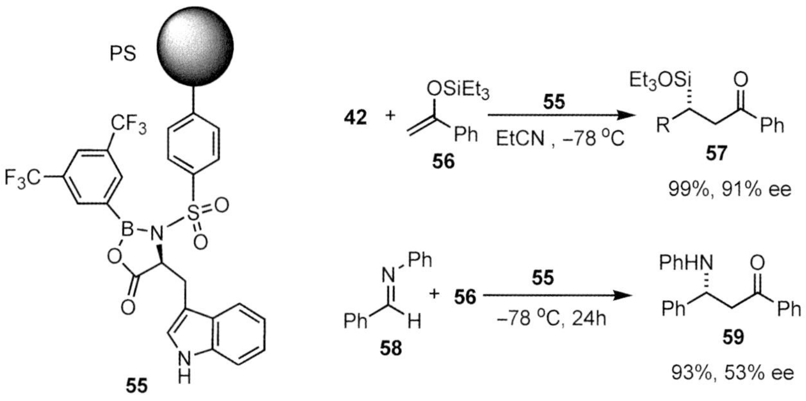

11 + PhCH$_2$Br $\xrightarrow{\text{54(R = OMe)}}$ (S)-53
           1M KOH, rt   98%, 83% ee

Scheme 3.16

42 + 56 (OSiEt$_3$, Ph) $\xrightarrow{\text{55, EtCN, −78 °C}}$ 57 (Et$_3$OSi, R, Ph) 99%, 91% ee

58 (N-Ph, Ph, H) + 56 $\xrightarrow{\text{55, −78 °C, 24h}}$ 59 (PhHN, Ph, Ph) 93%, 53% ee

3.8
Asymmetric Aldol Reaction

In recent years, catalytic asymmetric Mukaiyama aldol reactions have emerged as one of the most important C—C bond-forming reactions [35]. Among the various types of chiral Lewis acid catalysts used for the Mukaiyama aldol reactions, chirally modified boron derived from N-sulfonyl-(S)-tryptophan was effective for the reaction between aldehyde and silyl enol ether [36, 37]. By using polymer-supported N-sulfonyl-(S)-tryptophan synthesized by polymerization of the chiral monomer, the polymeric version of Yamamoto's oxazaborolidinone catalyst was prepared by treatment with 3,5-bis(trifluoromethyl)phenyl boron dichloride [38]. The polymeric chiral Lewis acid catalyst 55 worked well in the asymmetric aldol reaction of benzaldehyde with silyl enol ether derived from acetophenone to give β-hydroxyketone with up to 95% ee, as shown in Scheme 3.16. In addition to the Mukaiyama aldol reaction, a Mannich-type reaction and an allylation reaction of imine 58 were also asymmetrically catalyzed by the same polymeric catalyst [38].

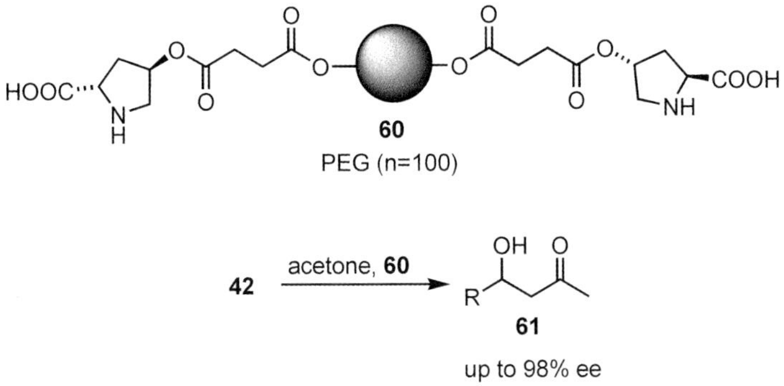

3.8 Asymmetric Aldol Reaction

Scheme 3.17

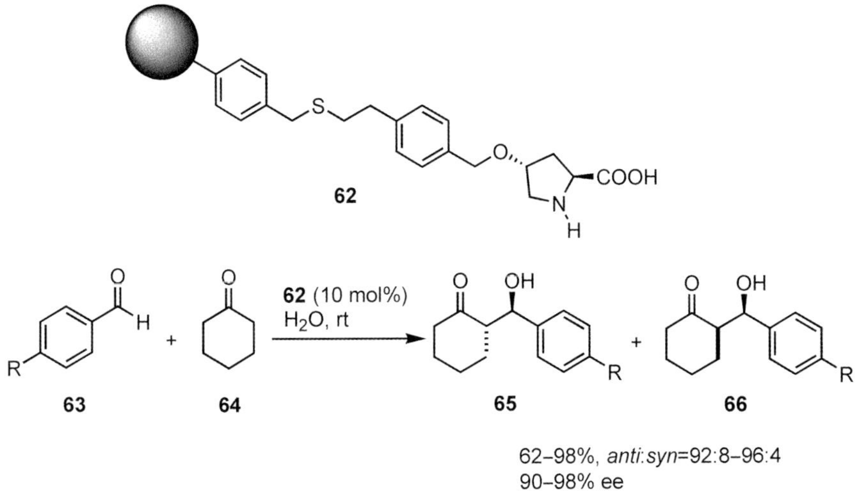

Scheme 3.18

Proline has recently been recognized as being capable of catalyzing asymmetric C—C bond-forming reactions [39], including aldol and iminoaldol reactions. (2S,4R)-4-Hydroxyproline was immobilized onto both ends of PEG to give **60**, and this was used as a catalyst for the aldol reaction of acetone and aldehyde to give β-hydroxyketones **61** (Scheme 3.17) [40]. In the case of the reaction with cyclohexanecarboaldehyde (**42**, R = cyclohexyl), an excellent enantioselectivity (>98% ee) was obtained with 81% yield. However, when aldimine was used instead of aldehyde, the asymmetric iminoaldol reaction also occurred with the same catalyst to give β-amino ketones with a high level of enantioselectivity [40].

Proline was also attached to crosslinked PS through a benzyl thioether linkage, as shown in Scheme 3.18. The PS-supported proline **62** was used as organocatalyst in the asymmetric aldol reaction between cyclohexanone **64** and substituted benzaldehyde **63** in water [41]. The reaction with p-cyanobenzaldehyde gave the aldol adduct in 98% conversion with a high level of enantioselectivity (98% ee). The

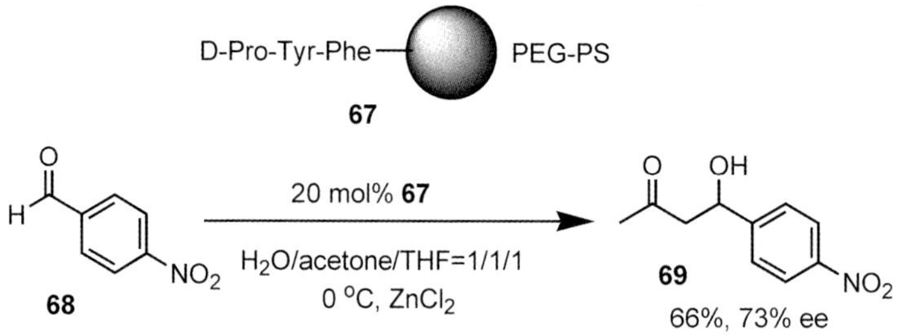

Scheme 3.19

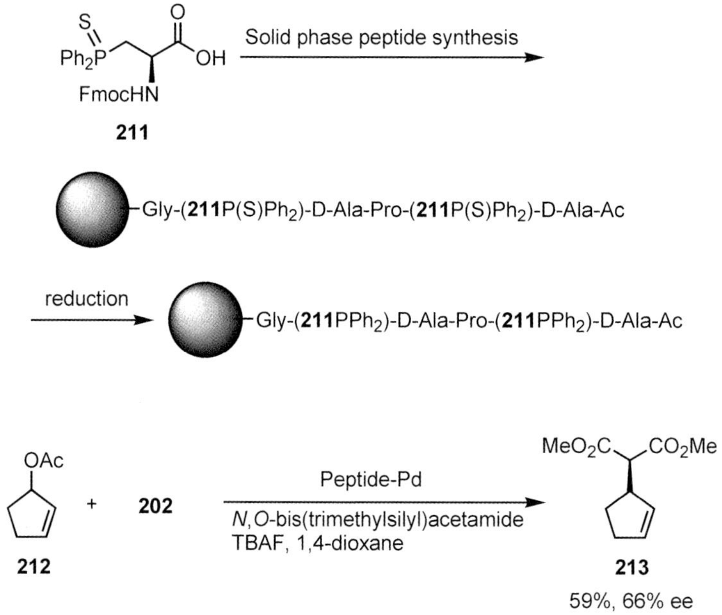

Scheme 3.20

same polymer-supported proline was also effective as an organocatalyst for the α-selenenylation of aldehyde.

Poly(ethylene glycol) grafted on crosslinked polystyrene (PEG-PS) resin has often been used as a polymer support for chiral catalysts of reactions performed in aqueous media. Peptides immobilized to PEG-PS resin have been developed and used as a catalyst for direct asymmetric aldol reactions in aqueous media (Scheme 3.19) [42]. When tripeptide-supported PEG-PS **67** was used as chiral catalyst in the reaction between **70** and acetone, the corresponding aldol product **69** was obtained with 73% ee. Kudo further developed the one-pot sequential reaction of acidic deacetalization and enantioselective aldol reaction by using an Amberlite and PEG-ST-supported peptide catalyst **67** [43]. The enantioenriched aldol product **72** was obtained in 74% isolated yield from acetal **70** in a one-pot reaction (Scheme 3.20).

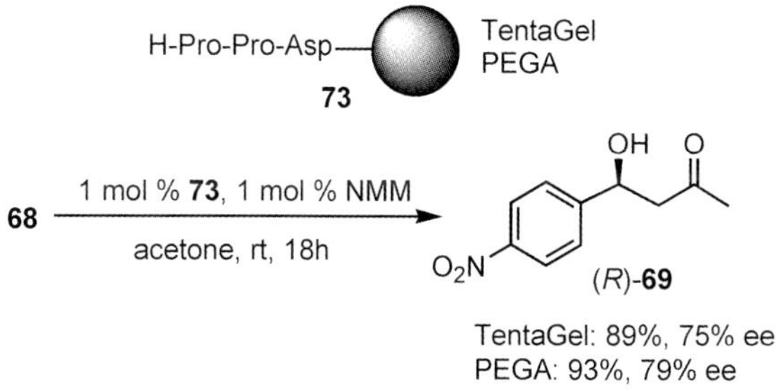

Scheme 3.21

Wennemers found that tripeptide H-Pro-Pro-Asp-NH$_2$ was a highly active and selective catalyst for asymmetric aldol reactions [44]. This peptide was immobilized to a polymer support and used as a catalyst for the aldol reaction of p-nitrobenzaldehyde 73 and acetone (Scheme 3.21). By using a TentaGel-supported peptide 73 the aldol adduct 69 was obtained in 89% yield with 75% ee, while a polyethylene glycol-polyacrylamide (PEGA)-supported peptide gave the same adduct in 93% yield and 79% ee [45].

3.9
Enantioselective Carbonyl-Ene Reaction

A bimetallic titanium complex of BINOL derivative can be used to catalyze the asymmetric carbonyl-ene reaction [46]. Insoluble polymeric catalyst 74 was prepared from a self-assembly of Ti(O*i*Pr)$_4$ and non-crosslinked copolymers with (R)-binaphthol pendant groups (Scheme 3.22) [47]. The self-assembled polymeric Ti complex is insoluble in organic solvent and catalyzed the carbonyl-ene reaction of glyoxylate 75 and α-methylstyrene 76. When the reaction of 75 and 76 was carried out with 20 mol% of 74 in CH$_2$Cl$_2$ at room temperature, an 85% yield of the product with 88% ee was obtained. Following its recovery by filtration, this catalyst was reused five times with full retention of its activity and enantioselectivity, without further treatment.

The polymer-supported bisBINOL–Ti complex 78 was also effective for the same reaction to give a high enantioselectivity (96% ee) of the product (Scheme 3.23) [48]. The catalyst was recovered and reused three times, maintaining high enantioselectivity.

3.10
Asymmetric Michael Reaction

The heterobimetallic catalyst prepared from (R,R)-3-aza-benzyl-1,5-dihydroxy-1,5-diphenylpentane was used for the asymmetric Michael addition reaction of malonates and thiophenols to enones [49]. The polymer-supported version of the chiral

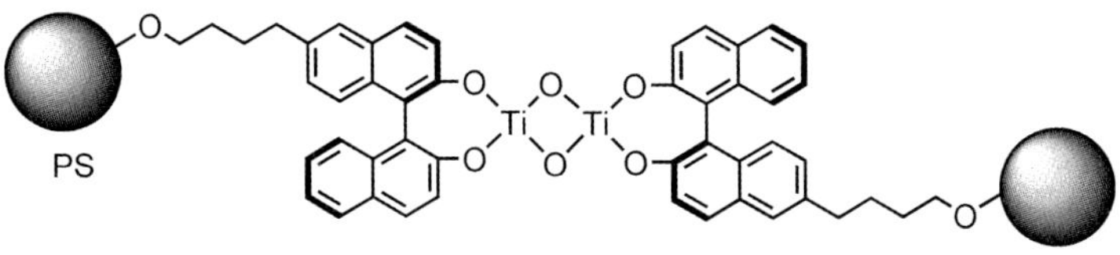

Scheme 3.22

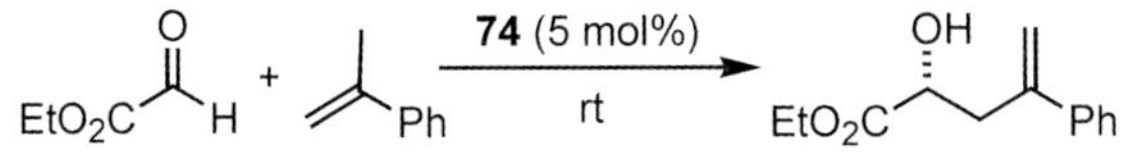

74 (5 mol%)

EtO$_2$C-CHO (**75**) + CH$_2$=CH-Ph (**76**) $\xrightarrow{\text{rt}}$ EtO$_2$C-CH(OH)-CH$_2$-C(=CH$_2$)-Ph (**77**)

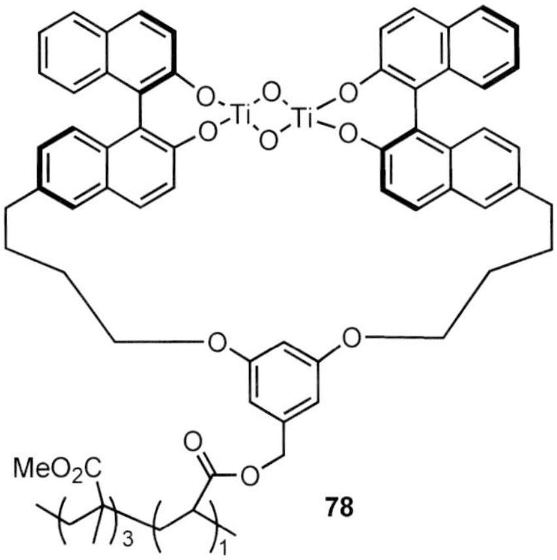

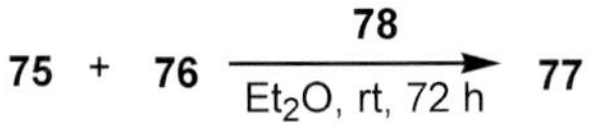

49–87%
49–84% ee

Scheme 3.23

75 + **76** $\xrightarrow[\text{Et}_2\text{O, rt, 72 h}]{\textbf{78}}$ **77**

ligand was synthesized and treated with LiAlH$_4$ to give the chiral polymeric catalyst **79** (Scheme 3.24) [50] for Michael addition of benzylamine to ethyl cinnamate. The product β-amino acid **85** was obtained in 60% yield with 81% ee.

Asymmetric Michael reactions are known to be well-catalyzed by the AlLi-bis(binaphthoxide) complex (ALB) [51]. In order to immobilize such a highly organized multicomponent asymmetric catalyst, Sasai developed a new strategy based on the use of a catalyst analogue [52]. Si-tethered binaphthol **86** as a stable catalyst analogue was copolymerized with methyl methacrylate and ethylene glycol

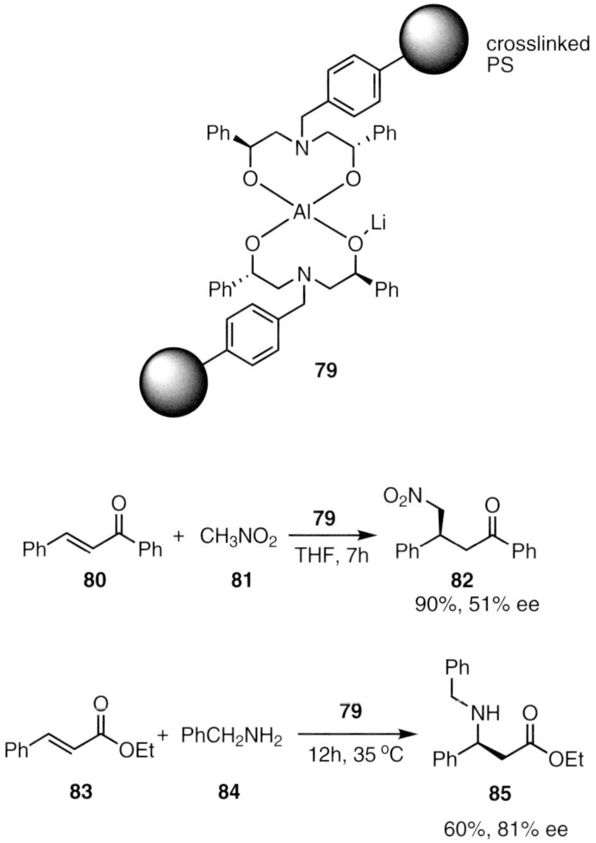

Scheme 3.24

dimethacrylate as a crosslinker. After removal of the Si by hydrolysis the polymer-supported AlLi-bis(binaphthoxide) catalyst **87** was generated by treatment with AlMe$_3$ and *t*BuLi (Scheme 3.25). This method of using a catalyst analogue allowed chiral ligands to be arranged at suitable positions along the polymer chains. The polymeric catalyst **87** promoted the Michael reaction of 2-cyclohexen-1-one with dibenzylmalonate to give the adduct in 73% yield with 91% ee.

A similar approach has been examined by using polymer-supported ALB **91** (Scheme 3.25). When this polymeric chiral ALB catalyst was used for the asymmetric Michael reaction, the corresponding chiral adduct was obtained in 91% yield with 96% ee [48].

PEG-supported proline **92** was designed to catalyze the asymmetric Michael addition of ketones to nitrostyrene (Scheme 3.26) [53]. Using 5 mol% of the polymeric catalyst, the Michael adduct **95** was obtained in good yields (up to 94%) and moderate to good enantioselectivities (up to 86% ee). A high level of diastereoselectivities (>98/2, *syn/anti*) was also observed. The enantiomeric excesses obtained

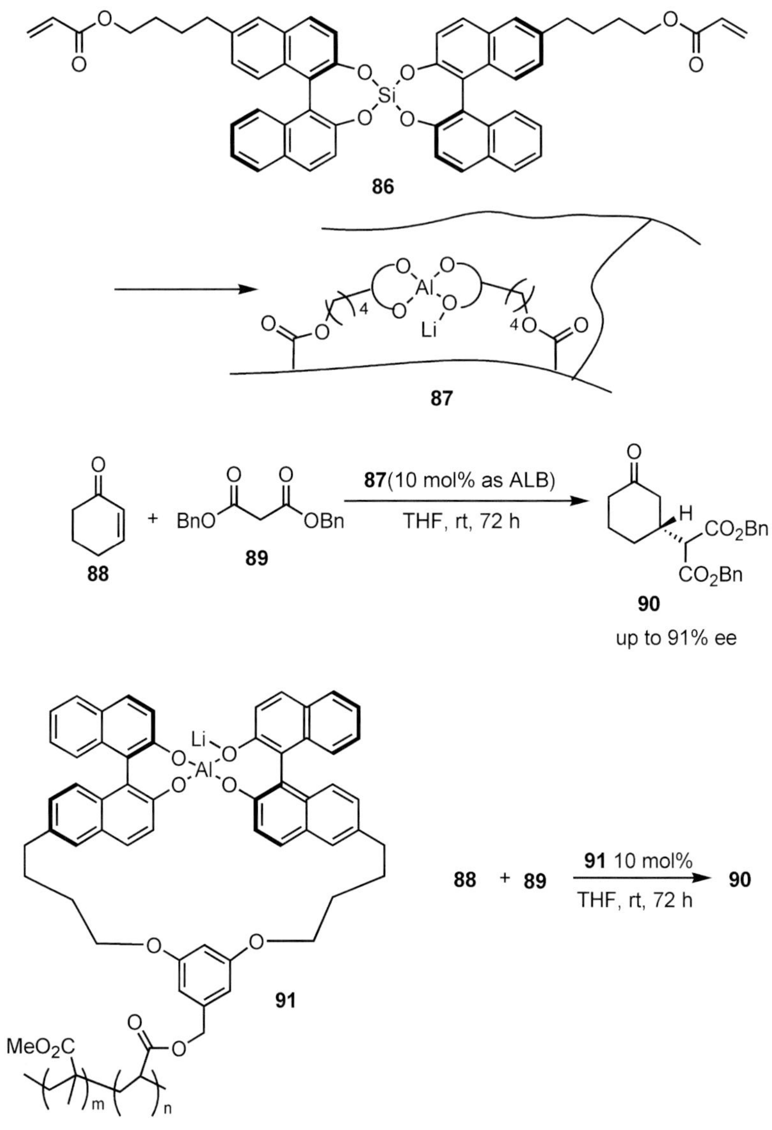

Scheme 3.25

by the polymeric proline were higher than those obtained from the nonsupported proline. Cinchona alkaloids are known to represent another chiral amine catalyst for Michael addition reactions, and a polymer-supported version of cinchona alkaloid **96** has been prepared and used for the asymmetric Michael reaction of **97** and **98**. The same polymer was also used in the continuous-flow system [54] (Scheme 3.27).

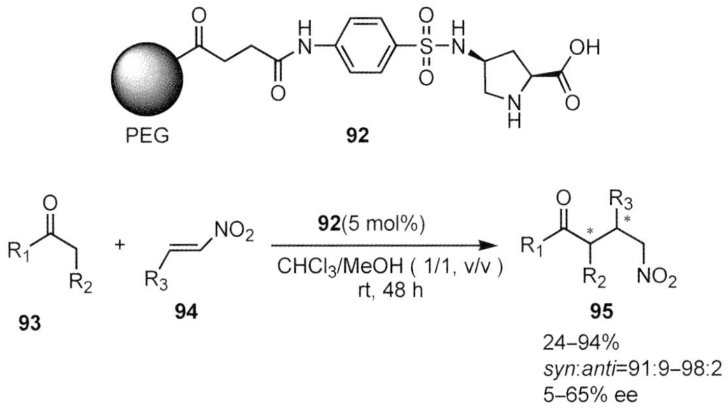

Scheme 3.26

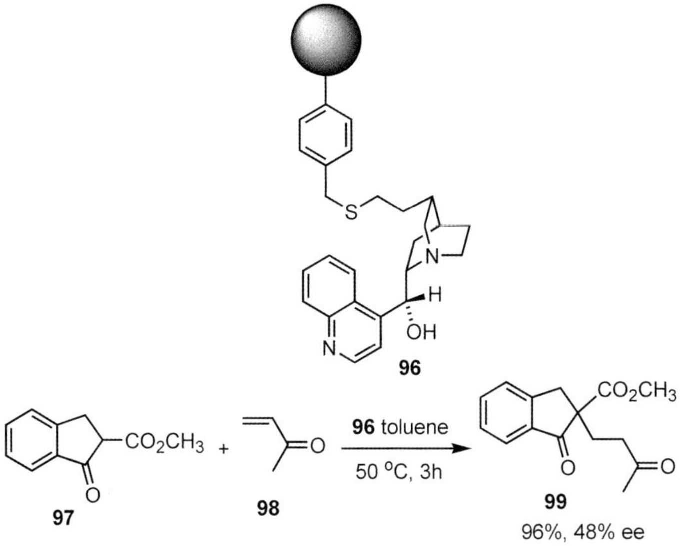

R_1–CO–R_2 (**93**) + R_3–CH=CH–NO_2 (**94**) → **92** (5 mol%), $CHCl_3$/MeOH (1/1, v/v), rt, 48 h → **95**

24–94%
syn:anti=91:9–98:2
5–65% ee

Scheme 3.27

97 + **98** → **96** toluene, 50 °C, 3h → **99**

96%, 48% ee

3.11
Asymmetric Deprotonation

Chiral lithium amide bases have been used successfully in the asymmetric deprotonation of prochiral ketones [55, 56]. Williard prepared polymer-supported chiral amines from amino acid derivatives and Merrifield resin [57]. The treatment of *cis*-2,6-dimethylcyclohexanone with the polymer-supported chiral lithium amide base, followed by the reaction with TMSCl, gave the chiral silyl enol ether. By using polymeric base **96**, asymmetric deprotonation occurred smoothly in tetrahydrofuran to give the chiral silyl enol ether *(S)*-**102** in 94% with 82% ee (Scheme 3.28).

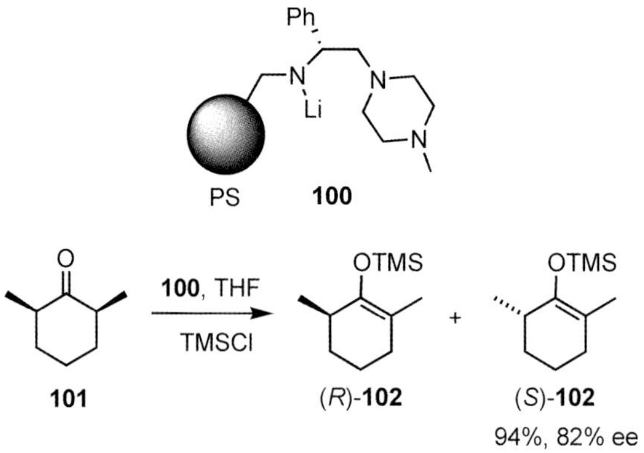

Scheme 3.28

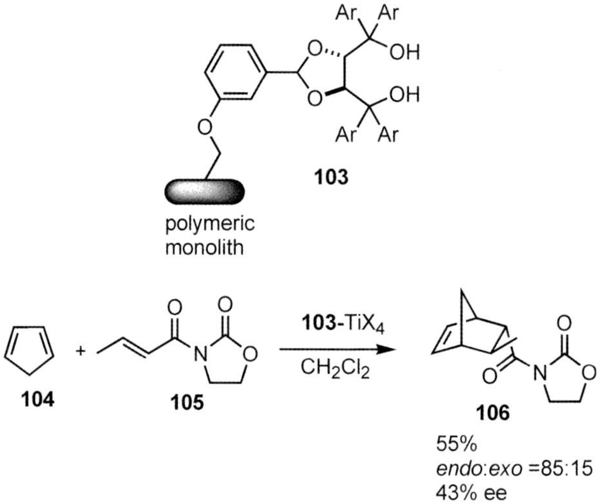

Figure 3.29

3.12
Enantioselective Diels–Alder Cycloaddition

Various types of chirally modified Lewis acids have been developed for asymmetric Diels–Alder cycloadditions. Some of these, including Ti-TADDOLates, have been attached to crosslinked polymers [11]. A recent example of this approach involved polymeric monoliths **103** containing TADDOL subunits (Scheme 3.29). The treatment of **103** with TiX$_4$ afforded Ti-TADDOLates, which were used for the asymmetric Diels–Alder reaction of cyclopentadiene **104** and **105**. The major product obtained in this reaction was the *endo* adduct with 43% ee [58]. The supported Ti-catalysts showed an extraordinary long-term stability, being active for at least one year.

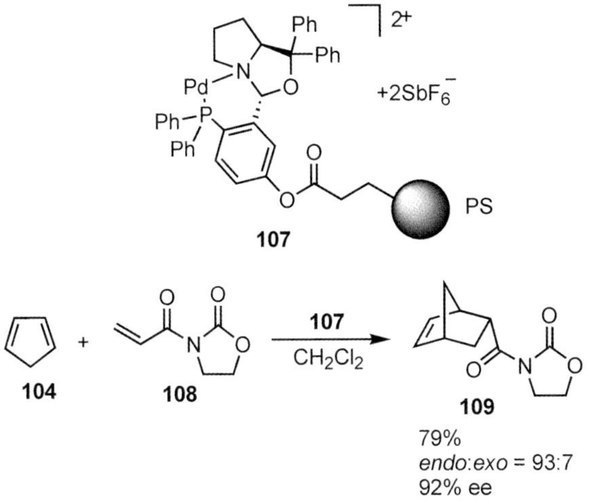

3.12 Enantioselective Diels–Alder Cycloaddition

Scheme 3.30

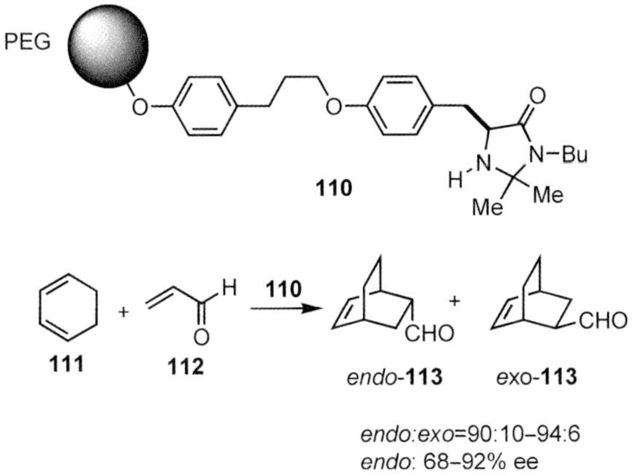

Scheme 3.31

Chiral cationic Pd-complexes with phosphinooxazolidine (POZ) represent another choice of catalyst for asymmetric Diels–Alder reaction [59]. Polymer-supported cationic POZ catalyst **107** effectively catalyzed the Diels–Alder reaction of cyclopentadiene **104** and acryloyl-1,3-oxazolidin-2-one **108** (Scheme 3.30).

Chiral secondary amines such as nonracemic imidazolidin-4-ones have been found to be effective asymmetric organocatalysts in the Diels–Alder cyclization of cyclopentadiene and α,β-unsaturated aldehydes [60]. A tyrosine-derived imidazolidin-4-one was immobilized on PEG to provide a soluble, polymer-supported catalyst **110**. In the presence of **110**, Diels–Alder cycloaddition of acrolein **112** to 1,3-cyclohexadiene **111** proceeded smoothly to afford the corresponding cycloadduct **113** with high *endo* selectivity and enantioselectivity up to 92% ee (Scheme 3.31) [61].

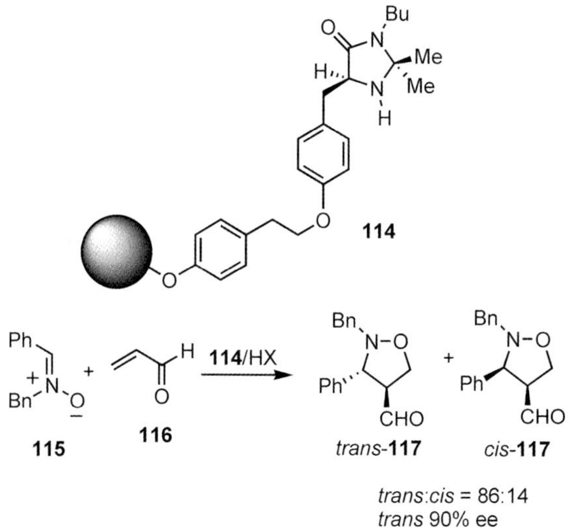

Scheme 3.32

3.13
Enantioselective 1,3-Dipolar Cycloaddition

A polymer-supported version of MacMillan's catalyst **114** has been developed by anchoring a tyrosine-derived imidazolidin-4-one via a spacer to the monomethyl ether of PEG (Scheme 3.32) [62]. The reaction of N-benzyl-C-phenyl nitrone **115** with acrolein **116** proceeded in the presence of PEG-supported chiral imidazolidinone and acid in wet nitromethane to give the isoxazolidine **117**. The supported catalysts behaved very similarly to their nonsupported counterparts in terms of enantioselectivity, but were somewhat less efficient in terms of chemical yield.

3.14
Asymmetric Sharpless Dihydroxylation

Ionic polymers **120** containing a mesylate anion were prepared by the quaternization of poly(4-vinylpyridine/styrene) **118** with tri(ethylene glycol) monomesylate monomethyl ether **119**. These polymers were used to immobilize OsO_4, as shown in Scheme 3.33. The resultant polymer **122** showed excellent catalytic performance in the Sharpless asymmetric dihydroxylation of styrene derivatives [63]. For example, **123** was dihydroxylated to styrene glycol **124** in 88% yield with 99% ee.

An alternative way to use the polymer support is to microcapsulate the catalyst. Kobayashi developed a microcapsulated osmium tetroxide using phenoxyethoxymethyl-polystyrene, and applied this to the asymmetric dihydroxylation of olefins in water (Scheme 3.34) [64]. The reaction proceeded smoothly in water with cata-

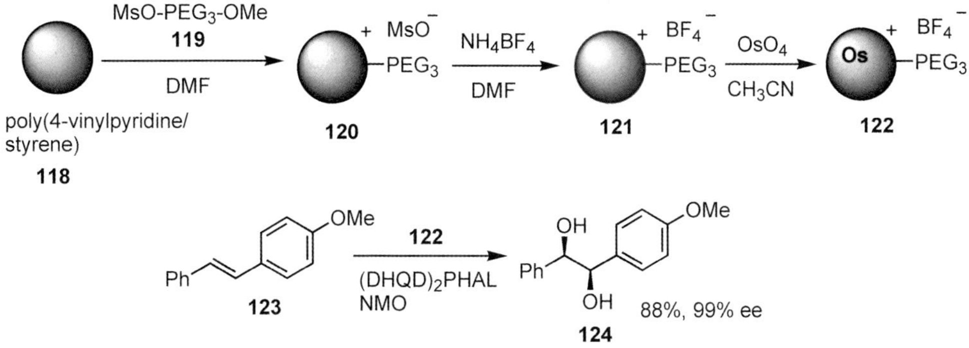

Scheme 3.33

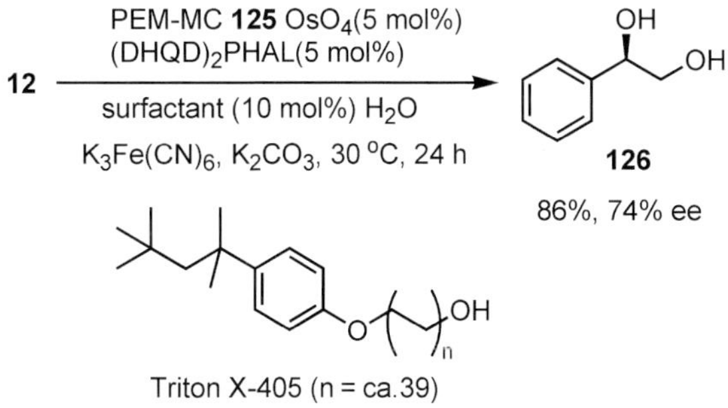

Scheme 3.34

lytic amounts of **125**, a chiral ligand ((DHQD)$_2$-PHAL), and a surfactant (Triton X-405).

3.15
Asymmetric Epoxidation

PTC methodology was applied to the asymmetric epoxidation of electrodeficient olefins, and is probably the simplest and most straightforward way to scale up such a process. Asymmetric PTC epoxidation using cinchona alkaloids was pioneered by Wynberg [65] during the mid-1970s and later improved by Lygo [66] and Corey [67]. Polymer-supported versions of cinchona alkaloids have been studied extensively for a variety of asymmetric PTC reactions. Wang and colleagues reported that the water-soluble, PEG-supported cinchona ammonium salts **54** could be successfully prepared and used as catalysts for the asymmetric epoxidation of chalcones [68]. Chalcone **80** was asymmetrically epoxidized with tBuOOH in the presence of **54** to give chiral epoxide **127** (Scheme 3.35).

Metalated chiral salen ligands were first introduced during the 1990s by Jacobsen and Katsuki as highly enantioselective catalysts for the asymmetric

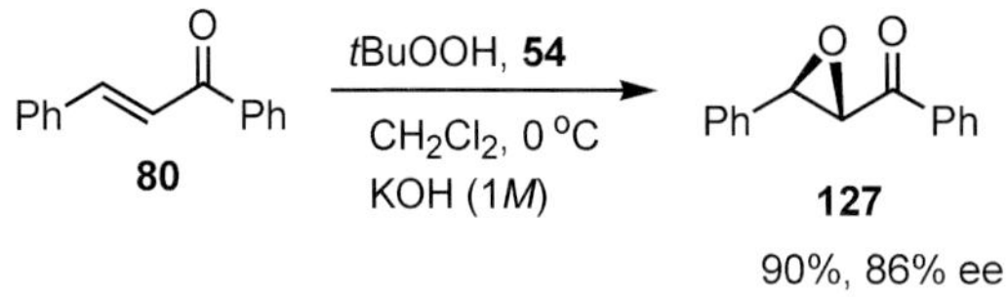

Scheme 3.35

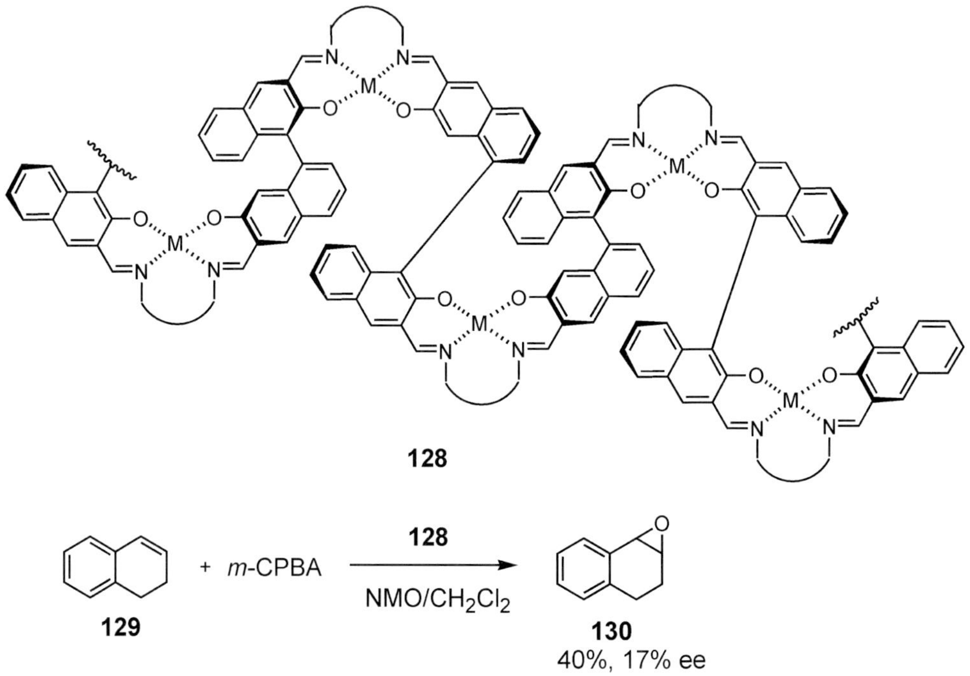

Scheme 3.36

epoxidation of unfunctionalized olefins [69–71]. Several types of immobilized versions of chiral complexes have been reported for the asymmetric epoxidation reaction. Although many types of inorganic support have been used for the immobilization of chiral salen ligands, the discussion here will center only on those asymmetric epoxidations catalyzed by the organic polymer-supported species.

A poly(binaphthyl metallosalen complex) **128** (Scheme 3.36) was prepared and used as a catalyst for the asymmetric epoxidation of alkene [72]. Although enantioselectivities obtained by using the polymeric catalyst were low, this represented a new type of polymeric chiral complex based on the main-chain helicity.

Polymer-supported chiral (salen)Mn complexes **131** were also used in other asymmetric epoxidation reactions (Scheme 3.37). For example, cis-β-methylstyrene **132** was efficiently epoxidized with m-CPBA/NMO in the presence of the polymeric catalyst [73]. For most of the tested substrates, the enantioselectivities

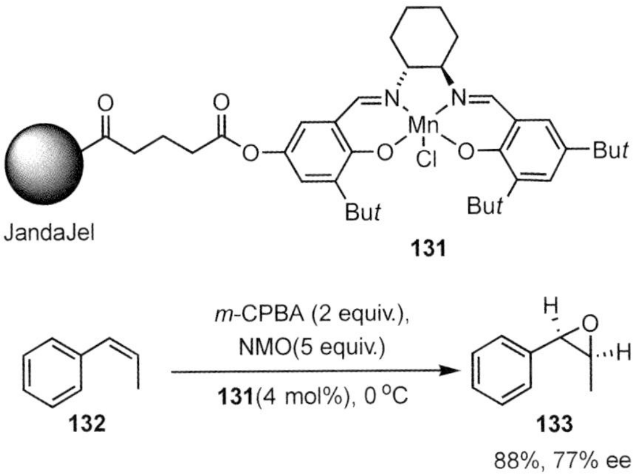

3.15 Asymmetric Epoxidation

Scheme 3.37

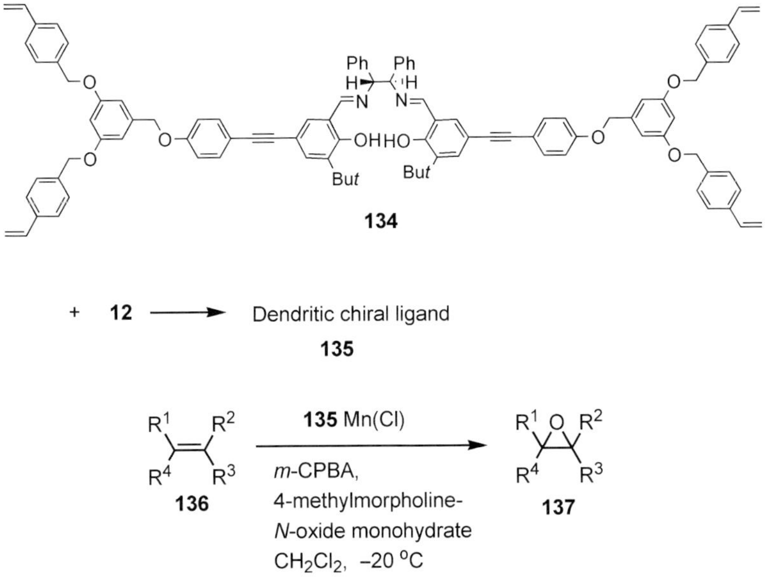

132 → (with m-CPBA (2 equiv.), NMO (5 equiv.), 131 (4 mol%), 0 °C) → 133
88%, 77% ee

Scheme 3.38

134 + 12 → Dendritic chiral ligand 135

136 → (with 135 Mn(Cl), m-CPBA, 4-methylmorpholine-N-oxide monohydrate, CH_2Cl_2, –20 °C) → 137

derived from JandaJel polymeric catalysts were almost equivalent to those for the low-molecular-weight catalyst.

Seebach introduced a novel concept for the immobilization of chiral ligands in PS. The ligand of choice was placed in the core of a styryl-substituted dendrimer **134**, which was copolymerized with styrene under suspension polymerization conditions to give the polymeric chiral ligand **135** [74]. The corresponding polymeric (salen)Mn complexes were used to catalyze the enantioselective epoxidation of alkene (Scheme 3.38), with the polymeric complexes being recycled ten times

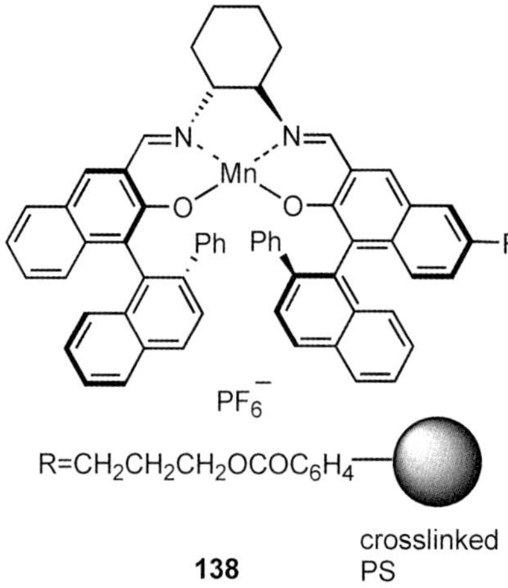

Scheme 3.39

R=CH₂CH₂CH₂OCOC₆H₄—crosslinked PS
138

in the reaction. The chiral epoxides were obtained without any reduction in enantioselectivity or conversion over the ten cycles. Seebach also showed that laser ablation inductively coupled plasma mass spectrometry (LA-ICP-MS) was well suited to assess the element distribution inside individual beads containing catalytically active transition metal sites and to detect spatial changes upon multiple use [75].

Katsuki-type (salen)Mn complexes **138** were attached to crosslinked PS through ester linkages (Scheme 3.39). The enantioselective epoxidation of 1,2-dihydronaphthalene using the polymeric chiral complex gave the chiral epoxide in somewhat lower yield compared to the result using the low-molecular-weight counterpart in a solution system. However, the enantioselectivities obtained with the polymeric catalyst were entirely analogous to those obtained with the homogeneous versions [76].

Polymer-supported salen catalysts were also developed by employing poly(norbornene)-immobilized salen complexes **139** of manganese and cobalt (Scheme 3.40) [77]. The poly(norbornene) complexes are highly active and selective catalysts for the epoxidation of olefins. The asymmetric epoxidation of *cis*-β-methylstyrene **132** occurred smoothly at −20 °C to give the chiral epoxide **133** in 100% conversion with 92% ee. Under the same reaction conditions, Jacobsen's catalyst (an unsupported salen complex) afforded the same product with 93% ee.

An alternative method for the immobilization of chiral Mn(salen) complexes is to use crosslinked PS having hydroxyphenyl groups or sulfonate groups. Chiral Mn(salen) complexes were immobilized axially onto these polymers by phenoxy or phenyl sulfonic groups, as shown in Scheme 3.41 [78]. In the presence of these polymeric complexes **140** the asymmetric epoxidation of **141** and **132** occurred smoothly to give the product with comparable or even higher enantioselectivities as compared to those obtained with homogeneous catalysts.

The polymer-supported chiral salen complex **146** was also prepared by the condensation reaction between (1S,2S)-1,2-cyclohexanediamine **144** and

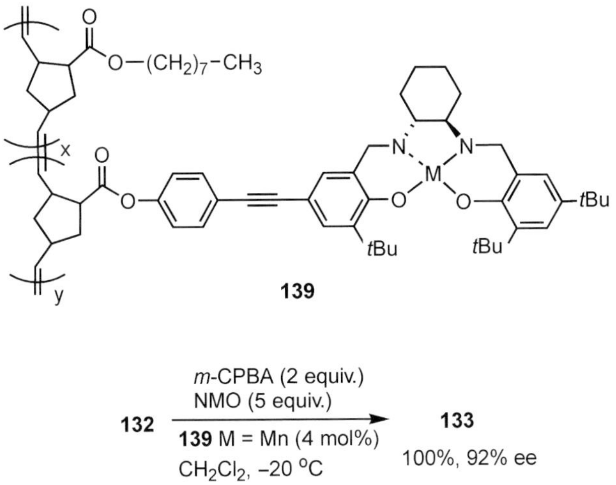
O—(CH2)7-CH3
x
y
N
M
N
O
O
tBu
tBu
tBu
139
m-CPBA (2 equiv.)
NMO (5 equiv.)
132
139 M = Mn (4 mol%)
CH2Cl2, -20 °C
133
100%, 92% ee

Scheme 3.40

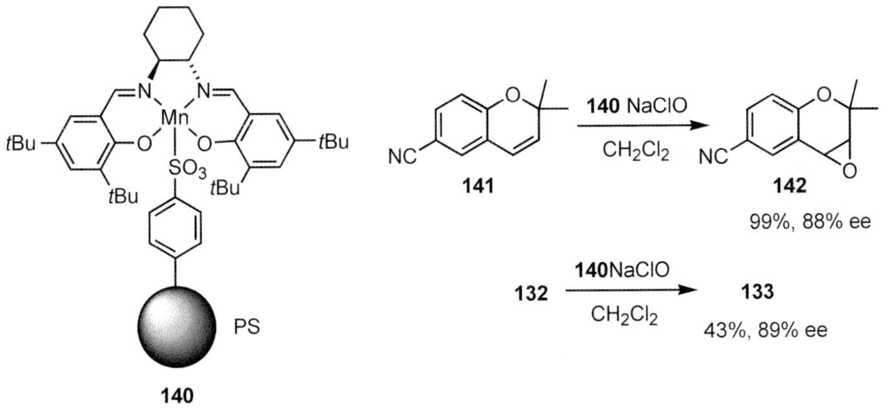
tBu
Mn
SO3
tBu
tBu
tBu
PS
140
NC
141
140 NaClO
CH2Cl2
NC
142
99%, 88% ee
132
140NaClO
CH2Cl2
133
43%, 89% ee

Scheme 3.41

5,5′-methylene-di-3-*tert*- butylsalicylaldehyde **143**, followed by complex formation (Scheme 3.42) [79]. The enantioselective epoxidation of chromenes, indene and styrene, mediated by the polymeric complexes **146** as catalyst using NaOCl as an oxidant, gave the chiral epoxides in >99% yield in all cases. The ee-values for the product epoxide were found to range from 75 to >99%, except for styrene (32–56%).

Sharpless asymmetric epoxidation was also conducted by using polymer-supported catalysts. Some very interesting phenomena were observed when methoxy PEG (MeO-PEG) -supported tartrate **147** was used as the polymeric chiral ligand (Scheme 3.43). In the epoxidation of **148** under Sharpless epoxidation conditions, (2*S*,3*S*)-*trans* **149** with 93% ee was obtained using **147** (MW = 750), while (2*R*,3*R*)-*trans* **149** with 93% ee was obtained using **147** (MW = 2000) [80]. More recently, Janda studied the precise effects of the molecular weight of the PEG chain on the

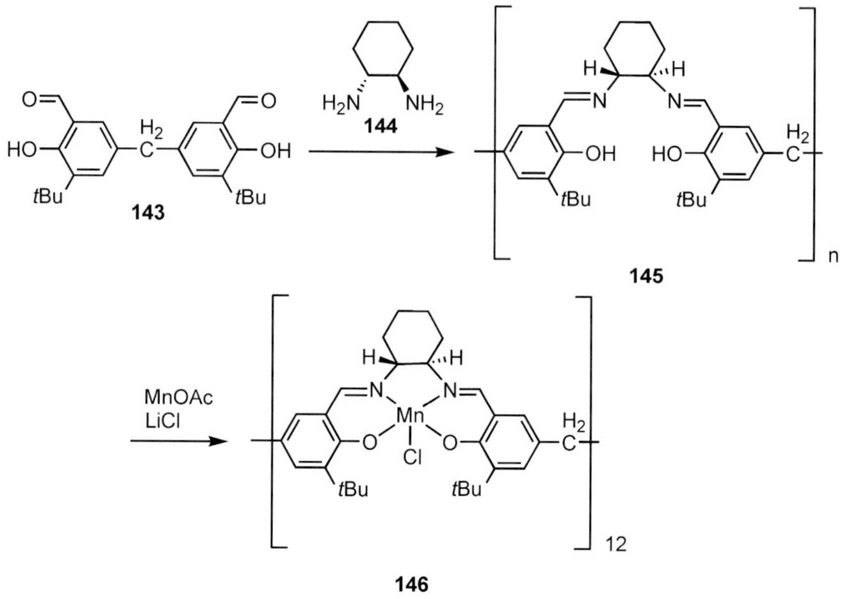

Scheme 3.42

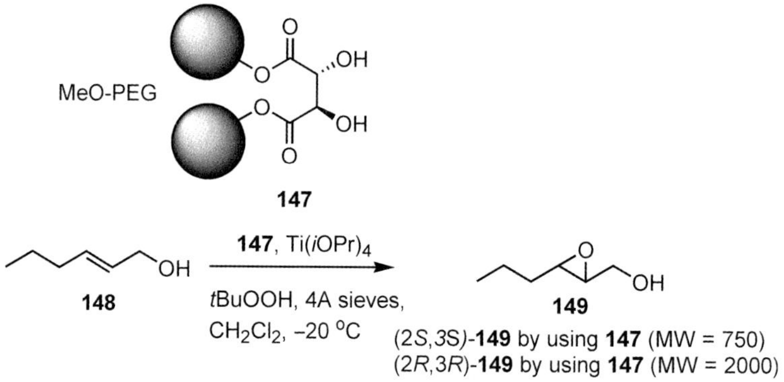

Scheme 3.43

same reaction, and revealed that the enantioselectivity of the reaction could be reproducibly reversed solely as a function of the appended PEG molecular weight [81].

3.16
Hydrolytic Kinetic Resolution of Terminal Epoxide

Jacobsen developed a powerful method for the preparation of enantiopure terminal epoxides by using the technique of hydrolytic kinetic resolution (HKR) [82, 83].

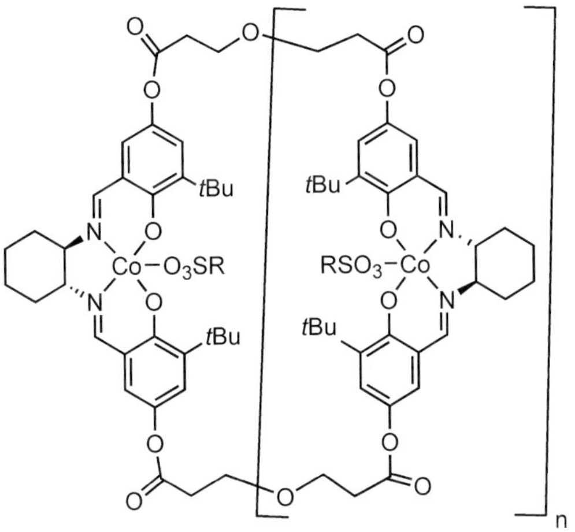

3.17 Enantioselective Borane Reduction of Ketone

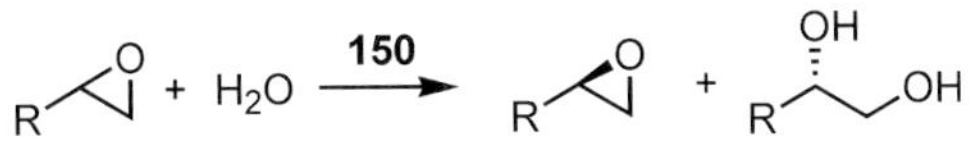

Scheme 3.44

Jacobsen also developed the solvent-free HKR of terminal epoxides catalyzed by an oligomeric (salen)–Co complex **150** (Scheme 3.44) [84–86]. Extremely low loadings (0.000 3–0.04 mol%) of catalyst were used to provide epoxides in good yields and >99% ee under ambient conditions, within 24 h.

Chiral Co(III)–salen complexes can also serve as efficient catalysts for HKR of terminal epoxides. Polymer-supported chiral salen complexes **156** were prepared from chiral Co complex **154** and ethylene glycol dimethacrylate **155**, as shown in Scheme 3.45. The chemical reduction of **156**, followed by treatment with acetic acid under aerobic conditions, produced the catalytically active polymer **157**, which was used in the HKR of propylene oxide [87]. Some other examples of polymeric salen–Co complexes have also been reported for the same reaction [88, 89].

A chiral cobalt–salen complex bearing BF_3 serves as an active catalyst for the HKR of terminal epoxides [90]. The polymeric salen–Co complex **158** (Scheme 3.46) also showed a high enantioselectivity in the same reaction [91].

3.17
Enantioselective Borane Reduction of Ketone

One of the most powerful asymmetric catalytic reductions of ketones is borane reduction with oxazaborolidine catalyst [92, 93]. Various types of polymer-supported chiral amino alcohols have been prepared and used for the formation

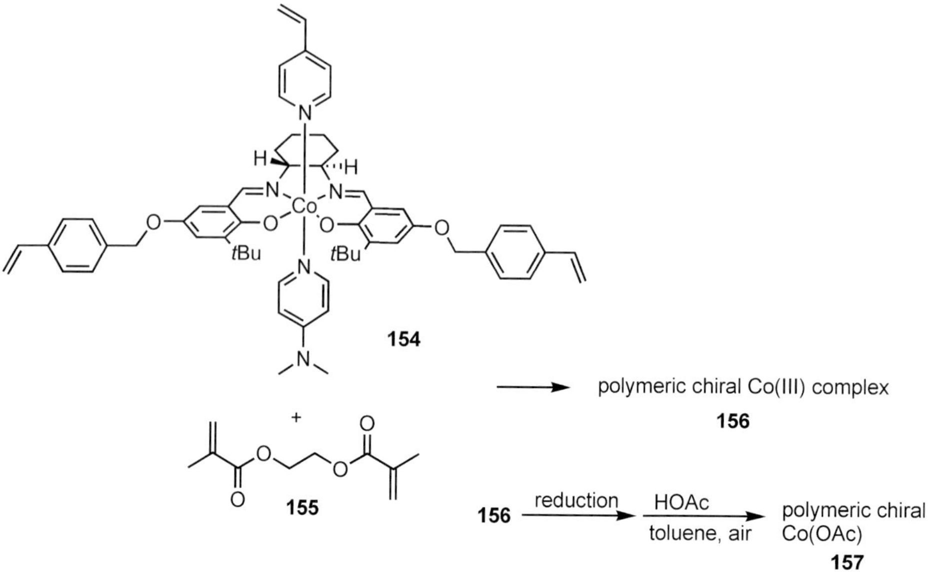

154

Scheme 3.45

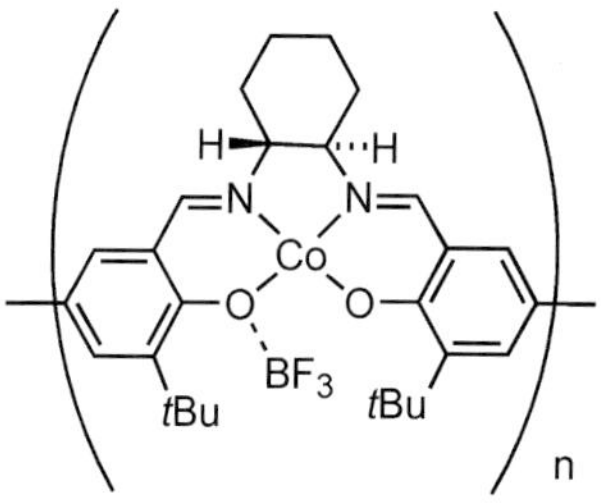

155

156 →(reduction, HOAc, toluene, air)→ polymeric chiral Co(OAc) **157**

154 → polymeric chiral Co(III) complex **156**

158

Scheme 3.46

of oxazaborolidines. One of the most recent examples involves the polymer-supported prolinol derivatives **159** and **160** (Scheme 3.47) [94]. Borane reduction of acetophenone in the presence of **160** gave 1-phenylethanol in 90% yield and 95% ee. The use of NaBH$_4$-TMSCl as reducing agent in the presence of **160** gave the alcohol with low enantioselectivity. Polymer-supported prolinol **161** having a sulfonamide linkage was also effective as a ligand for the borane reduction of acetophenone [95, 96]. Similar polymers **163** and **164** (Scheme 3.48) have been prepared and used for the enantioselective borane reduction [97]. Whilst the pendant-bound system (**163**) showed reduced stereoselectivity, the crosslinked version (**164**) afforded enantioselectivities which were almost identical to those of the nonsupported prolinol derivative.

The monolith-type prolinol **165** was prepared from the (trimethylsilyl)ethoxycarbonyl (Teoc)-protected monomers (Scheme 3.48). Although the asymmetric borane

3.17 Enantioselective Borane Reduction of Ketone

Scheme 3.47

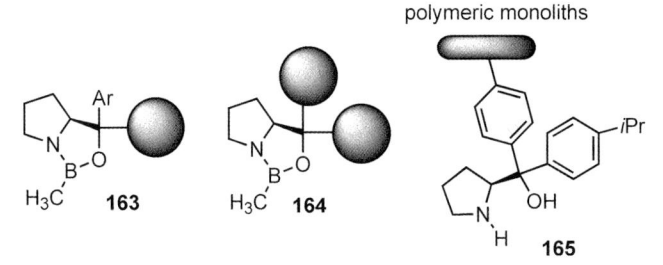

Polymer	Reducing agent	Yield (%)	ee (%)
159	BH$_3$·EMe$_2$S	99	44
160	BH$_3$·EMe$_2$S	90	95
159	NaBH$_4$, TMSCl	99	6
161	BH$_3$·EMe$_2$S	95	92.5

Scheme 3.48

reduction of acetophenone with **165** gave near-complete conversion within 1 h, the enantioselectivities obtained were very low (3% ee) [98].

Dendrimer-supported prolinol derivative **166** has been prepared and used as a chiral ligand in the asymmetric borane reduction of indanones **167** and tetralones (Scheme 3.49) [99]. From substituted indanones or tetralones, cis and trans isomers were obtained in a near 1:1 ratio. In the case of the reduction of substituted indanone, the cis isomers had an ee of about 80%, whereas the trans isomers had an ee of about 95%. In the case of tetralones, ee-values >90% ee were obtained for both the cis and trans isomers.

A large number of reducing agents and chiral catalysts have also been developed for the enantioselective reduction of ketones [100]. In the presence of polymer-

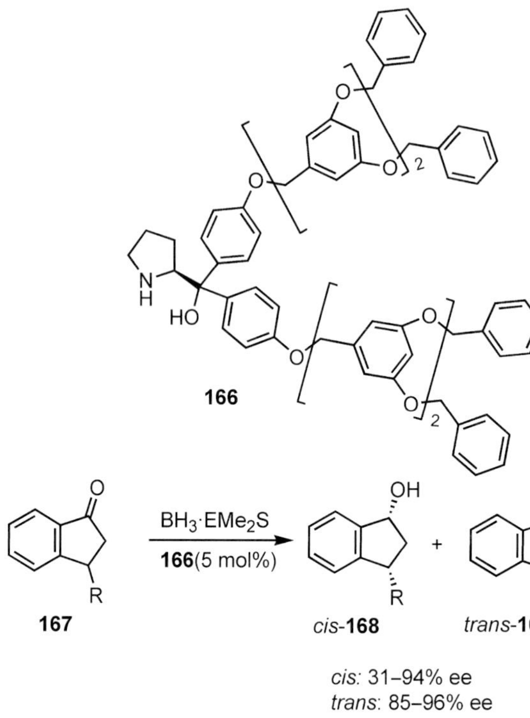

Scheme 3.49

cis: 31–94% ee
trans: 85–96% ee

supported chiral prolinol N-sulfonamide **161**, the aromatic ketones were asymmetrically reduced with $NaBH_4/ClSiMe_3$ or $NaBH_4/BF_3 \cdot OEt_2$ (Scheme 3.50).

3.18
Asymmetric Transfer Hydrogenation

A chiral diaminodiphosphine ligand was attached onto poly(acrylic acid) through an amide linkage. The ruthenium(II) complex of the resultant polymeric ligand (**169**) was then applied to the asymmetric transfer hydrogenation of acetophenone in 2-propanol (Scheme 3.51) [101]. *(S)*-1-phenylethanol was obtained in 95% yield with 96% ee by using **158**.

Norephedrine and its derivatives are efficient chiral ligands of the catalyst for asymmetric transfer hydrogenation. By using a 7 : 3 copolymer of a PEG ester and a hydroxyethyl ester, polymer-supported norephedrine **171** was prepared. The polymeric norephedrine/Ru(II) complex catalyzed the reduction of acetophenone in up to 95% yield and 81% ee (Scheme 3.52) [102]. Polymer-supported chiral amino alcohol **172** was also used for the same reaction [103].

A polyacetylene-type helical polymer having chiral amino alcohol pendant groups has also been prepared by the polymerization of chiral *(S)*-threonine-based

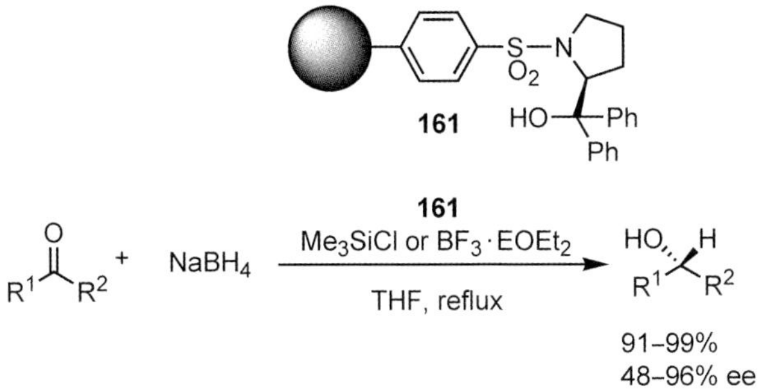

3.18 Asymmetric Transfer Hydrogenation

Scheme 3.50

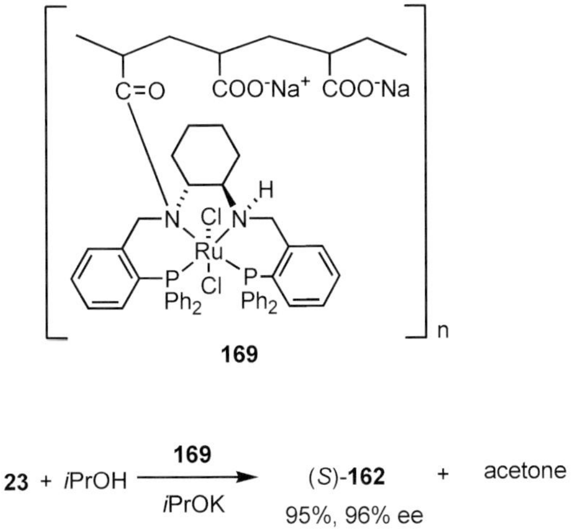

Scheme 3.51

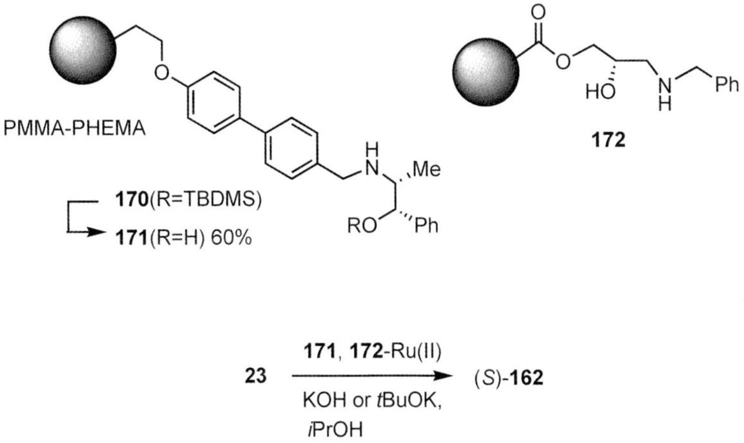

Scheme 3.52

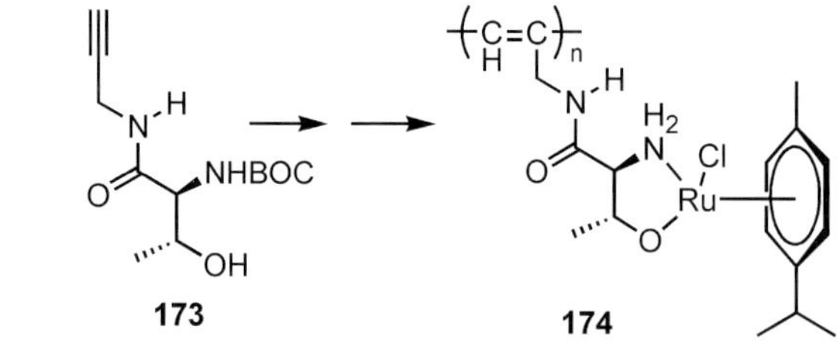

Scheme 3.53

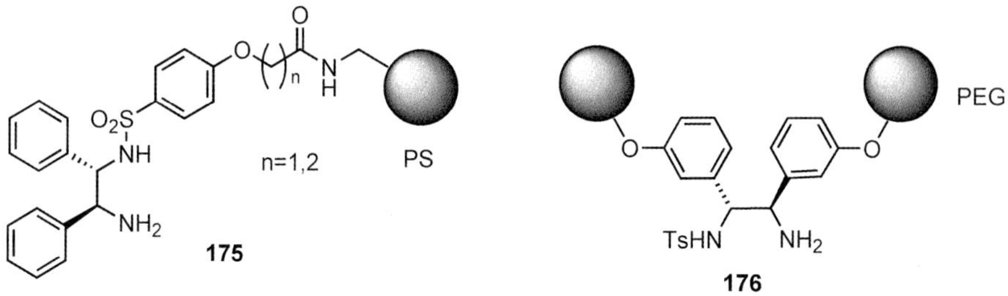

Scheme 3.54

N-propargylamide **173** (Scheme 3.53) [104]. The helical polymer–Ru complex **174** was used as a catalyst for the hydrogen-transfer reaction of ketones to give the alcohols with moderate enantioselectivities.

Among the various chiral catalysts reported for asymmetric transfer hydrogenation, the most notable has been the ruthenium complex with N-(p-toluenesulfonyl)-1,2-diphenylethylenediamine) (TsDPEN) developed by Noyori and Ikariya [105]. The first study on the polymer-supported version of this catalyst was reported by Lemaire [106]. Recently, a similar PS-supported TsDPEN **175** (Scheme 3.54) was utilized for the synthesis of (S)-fluoxetine [107].

As the transfer hydrogenation reactions using Ru–TsDPEN catalyst could take place in aqueous solution [108–111], the next stage was to develop a polymeric catalyst suitable for the aqueous conditions. One such example was the use of PEG as a polymer support, as reported by Xiao [112]. The PEG-supported TsDPEN **176** (Scheme 3.54) was highly effective in the Ru(II)-catalyzed transfer hydrogenation of simple ketones by sodium formate in water. The same polymeric catalyst was also effective for the same reaction by using a formic acid–triethylamine azeotrope [113].

An alternative approach is the use of a PS support bearing sulfonate pendant groups. For this, a quaternary ammonium salt of styrenesulfonic acid was copolymerized with a N-(p-styrenesulfonyl)-1,2-diphenylethylenediamine monomer. The polymeric chiral Ru complex was prepared from **177** and [RuCl$_2$(p-cymene)]$_2$ and applied to the asymmetric transfer hydrogenation of aromatic ketones in water (Scheme 3.55) [114]. The polymeric chiral complex was evenly suspended in water and the reaction proceeded smoothly to produce the alcohol in quantitative yield and with high enantioselectivity. For several of the aromatic ketones tested, higher

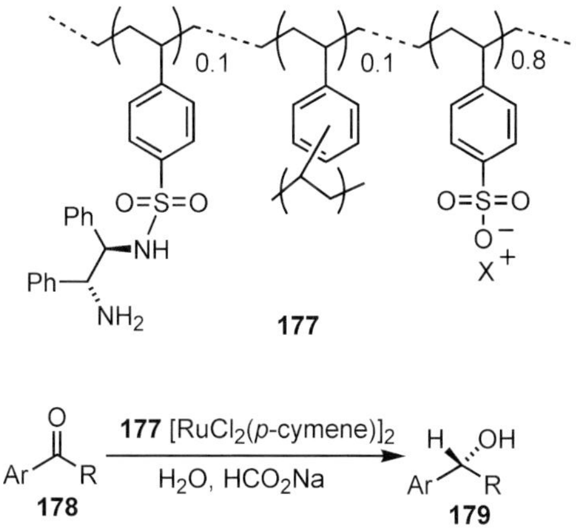

Scheme 3.55

Ar	R	ee (%)	
		177	TsDPEN
Ph	Me	98	94
Ph	Et	96	86
pClPh	Me	99	91
1-Naph	Me	97	87

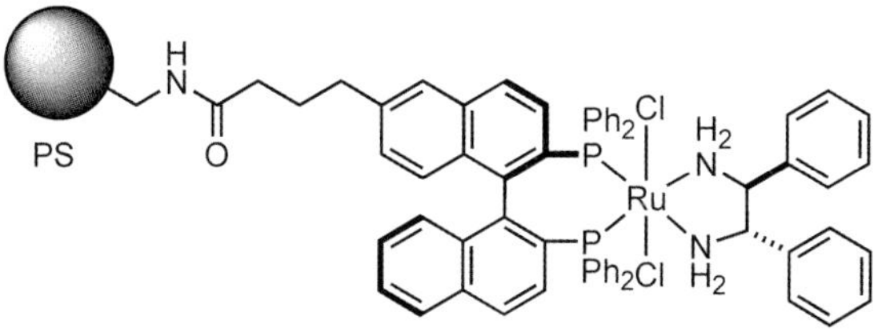

Scheme 3.56

enantioselectivities were obtained by using the polymeric catalyst than by using a low-molecular-weight catalyst in water.

3.19
Enantioselective Hydrogenation of Ketones

Among the many examples of asymmetric hydrogenation catalysts that have been developed, chiral complexes prepared from 1,2-diamines and $RuCl_2$/diphosphines provide one example of the most powerful catalysts for this reaction. Polymer-supported (R)-BINAP was treated with $RuCl_2$ and (R,R)-1,2-diphenylethylenediamine to give the polymeric chiral complex **180** (Scheme 3.56); this serves as an excellent precatalyst for the asymmetric hydrogenation of aromatic ketones to give the chiral secondary alcohols in quantitative yields with 84–97% ee-values [115]. For example, the asymmetric hydrogenation of 1′-acetonaphthone with (R,RR)-**180** occurred in quantitative conversion within 26 h with 98% ee. The enantioselectivity, turnover number (TON) and turnover frequency (TOF) in this reaction

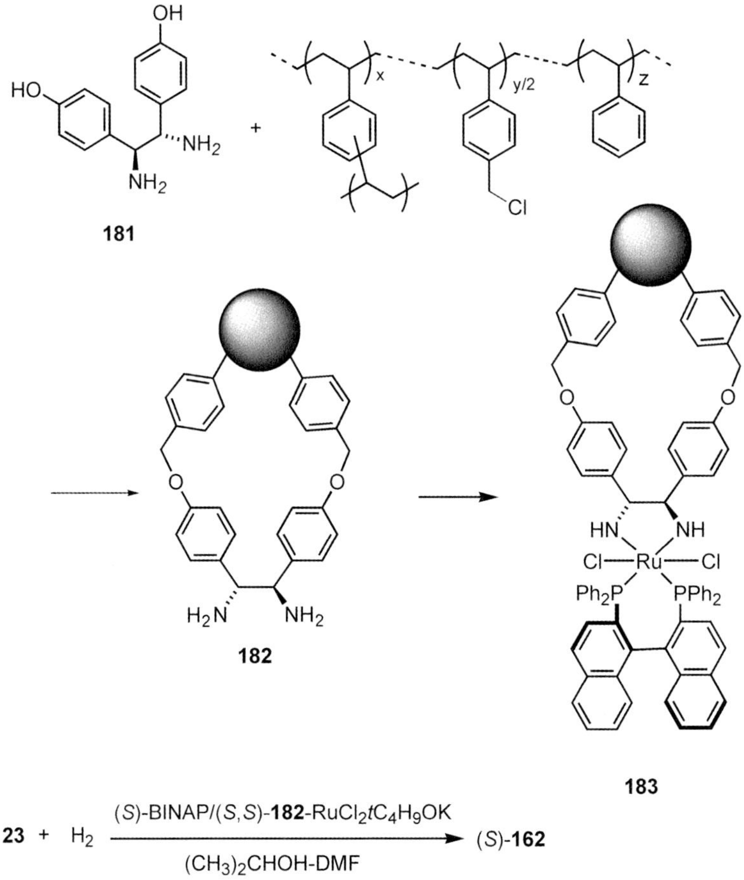

108 | *3 Heterogeneous Enantioselective Catalysis Using Organic Polymeric Supports*

Scheme 3.57

were comparable to those obtained under the otherwise identical conditions in the corresponding homogeneous catalyst solution system.

Polymer-supported Ru precatalysts were prepared from the polymeric chiral 1,2-diamines and the $RuCl_2$/BINAP complex [116]. Polymer-supported chiral 1,2-diamines were prepared by using either a polymer-reaction method or a polymerization method [117]. The polymer-reaction method involved the reaction of chiral 1,2-diamine **181** with chloromethylated PS (Scheme 3.57) [116]. Styrene-divinylbenzene (DVB)-based copolymers required the use of a good solvent for the PS chain in order to accelerate the hydrogenation reaction. A flexible crosslinking consisting of an oligooxyethylene chain allowed the use of 2-propanol as solvent [118].

In contrast, the chiral 1,2-diamine monomer was copolymerized with styrene to produce **184** (Scheme 3.58) [119]. As the chiral monomer possessed two polymerizable groups, the crosslinking polymers were obtained via a copolymerization. This flexible crosslinking structure, when compared to DVB crosslinkage, gave **184** with high reactivity and satisfactory enantioselectivities in the asymmetric hydrogena-

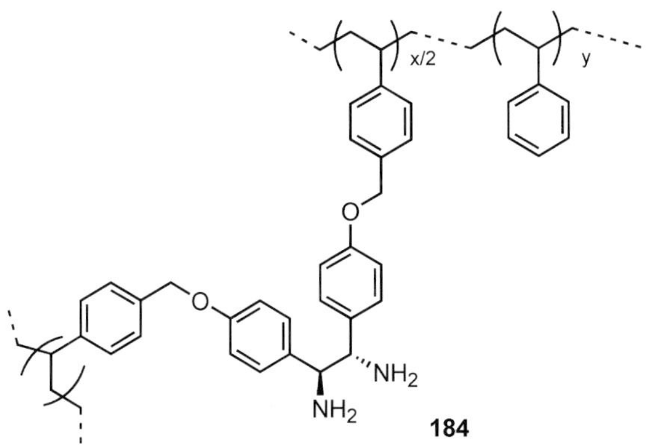

3.19 Enantioselective Hydrogenation of Ketones

Scheme 3.58

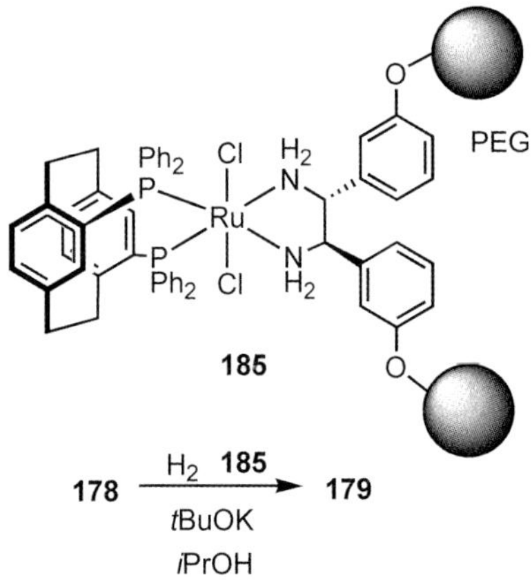

Scheme 3.59

tion of the ketone reaction; in fact, these parameters were better than those obtained when using **183** prepared with the polymer-reaction method. Other than the PS backbone, a variety of vinyl polymers have been also investigated as the polymer support. In this respect, poly(methacrylate)s possessing chiral 1,2-diamine moieties gave a higher reactivity and a somewhat higher enantioselectivity in the ketone hydrogenation than did **184** [120].

PEG 2000 monomethyl ether mesylate was attached to a 1,2-diphenylethylenediamine derivative, and the resultant soluble chiral diamine ligand was treated with the complex prepared from PhanePhos and [(benzene)RuCl$_2$] to give the polymeric chiral complex **185**. The latter compound was shown to be a highly effective soluble polymeric catalyst for the asymmetric hydrogenation of simple aromatic ketones (Scheme 3.59) [121]. A soluble, polymer-supported catalyst usually gave a higher yield compared to a crosslinked, insoluble polymeric catalyst. The PEG-supported catalyst was precipitated by Et$_2$O and could be recycled [122]. MeO-PEG-supported biphenylbisphosphine (BIPHEP) **186** (Scheme 3.60) was also prepared and complexed with Ru and DPEN. The polymeric complex was used for the same reaction [123].

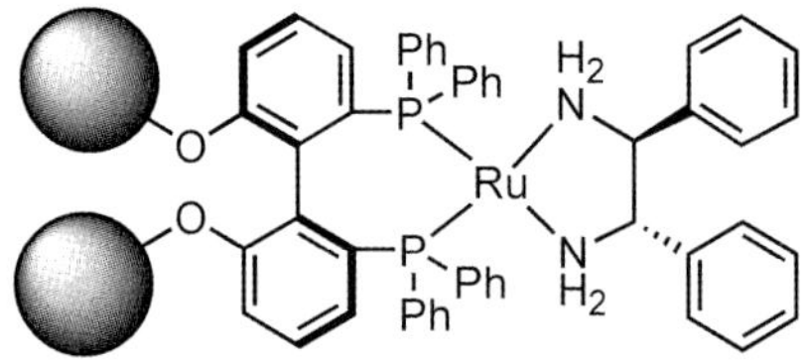

MeO-PEG   (S,SS)-**186**

Scheme 3.60

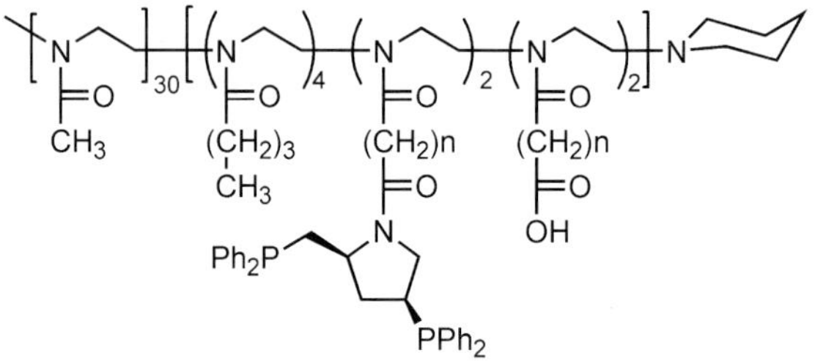

187

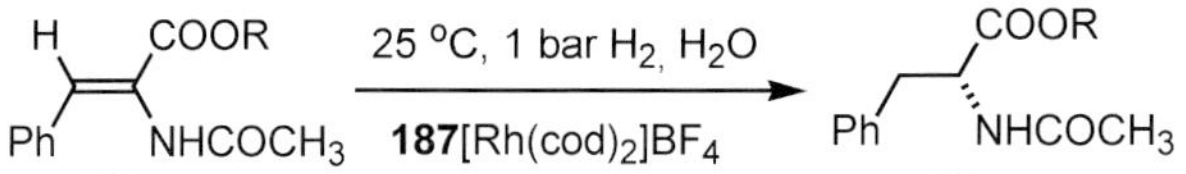

188 → **189**

25 °C, 1 bar H_2, H_2O
187[Rh(cod)$_2$]BF$_4$

Scheme 3.61

3.20
Asymmetric Hydrogenation of Enamine

Amphiphilic diblock copolymers based on 2-oxazoline derivatives with chiral diphosphine **187** were prepared (Scheme 3.61) and used in the asymmetric hydrogenation of methyl (Z)-α-acetamido cinnamate **188** in water to give the (R)-phenylalanine derivative **189** in 85% ee [124]. The polymeric catalyst could be recycled. This result illustrated the advantages of using amphiphilic copolymers for the efficient transformation of a hydrophobic substrate in water.

3.21
Enantioselective Hydrogenation of C=C Double Bonds

Chiral monophosphite ligands were attached to PEG and used as ligands for the Rh-catalyzed asymmetric hydrogenation of C=C double bonds [125]. The polymeric catalysts **190** and **191** performed well in the asymmetric hydrogenation of

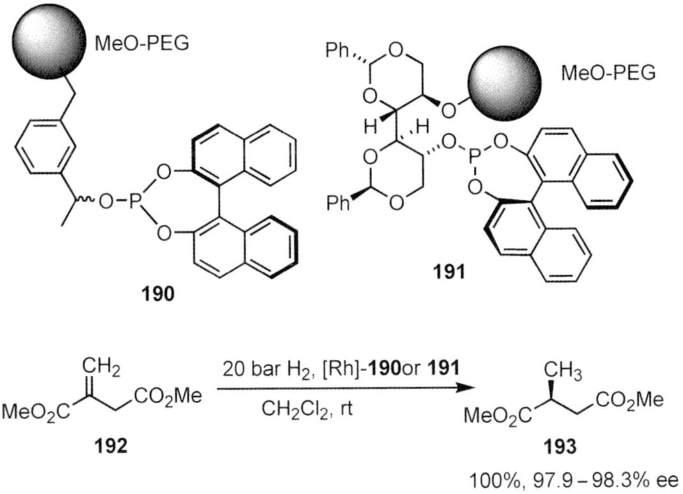

3.23 Asymmetric Allylic Alkylation

Scheme 3.62

Structures **190** and **191** (MeO-PEG supported phosphite/phosphoramidite ligands based on BINOL).

$$\text{MeO}_2\text{C}-\text{C}(=\text{CH}_2)-\text{CH}_2-\text{CO}_2\text{Me} \quad \xrightarrow[\text{CH}_2\text{Cl}_2,\ \text{rt}]{20\ \text{bar}\ \text{H}_2,\ [\text{Rh}]\text{-}\mathbf{190}\ \text{or}\ \mathbf{191}} \quad \text{MeO}_2\text{C}-\text{CH}(\text{CH}_3)-\text{CH}_2-\text{CO}_2\text{Me}$$

192 → **193** (100%, 97.9–98.3% ee)

itaconic ester in dichloromethane (DCM), as shown in Scheme 3.62, to give the corresponding saturated diester **193** with 97.9% ee and 98.3% ee, respectively. N-Acetylenamines were also hydrogenated using the same polymeric catalyst to give the optically active amine derivatives in quantitative conversion with 92–96% ee.

3.22
Enantioselective Hydrogenation of C=N Double Bonds

Chiral dendritic catalysts **194** derived from BINAP was prepared and used for the asymmetric hydrogenation of quinolines (Scheme 3.63) [126]. The corresponding cyclic amine products were obtained with high enantioselectivities up to 93% ee. The dendritic catalyst showed excellent catalytic activities (TOF up to 3450 h^{-1}) and productivities (TON up to 43 000). The dendritic catalyst was recovered by precipitation and filtration and reused at least six times, with similar enantioselectivity.

3.23
Asymmetric Allylic Alkylation

Cyclization of the bisurethane **198** to the oxazolidin-2-one **199** was examined by using the polymeric catalyst derived from **197** (Scheme 3.64), which was prepared by using ArgoGel-NH$_2$ as the polymer support. Although a somewhat lower enantioselectivity was obtained with the non-supported catalyst than with the supported catalyst, the former could be recycled several times without any significant differences in either yield or enantiomeric excess [127].

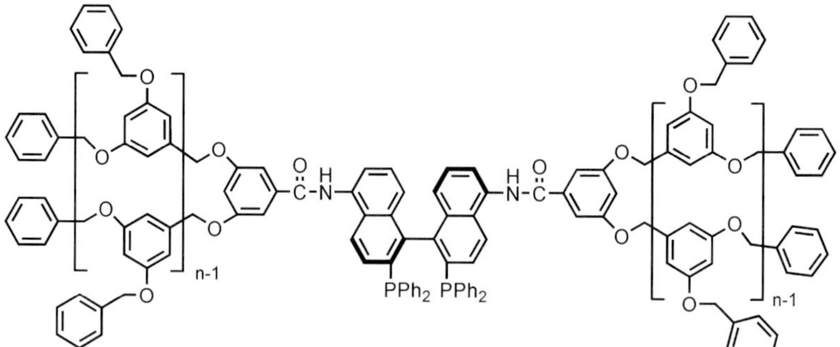

Scheme 3.63

(S)-**194** (n=1–4)

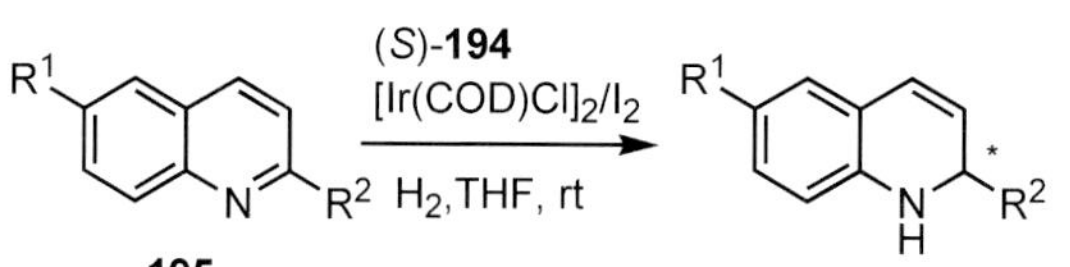

195 → **196** up to 93% ee

Reagents: (S)-**194**, [Ir(COD)Cl]$_2$/I$_2$, H$_2$, THF, rt

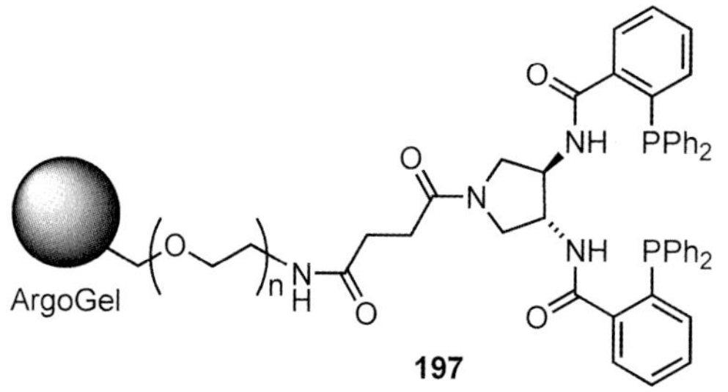

197 (ArgoGel-supported)

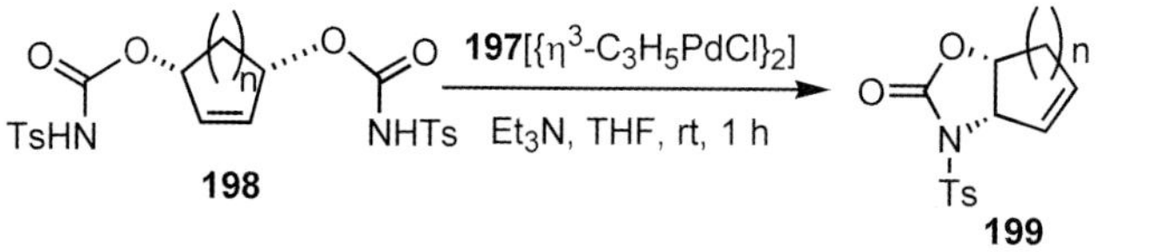

198 → **199**

Reagents: **197**[{η^3-C$_3$H$_5$PdCl}$_2$], Et$_3$N, THF, rt, 1 h

n = 1: 81%, 92% ee
n = 2: 78%, 90% ee

Scheme 3.64

Merrifield resin-supported chiral phosphinooxazolidine ligands **200** were prepared and used for the allylic alkylation of **201** and **202**. Near-complete conversion with 90% ee was obtained with the polymeric catalyst (Scheme 3.65) [128].

MeO-PEG-supported soluble polymeric chiral ligands **204** were synthesized and utilized in various asymmetric allylic substitution reactions (Scheme 3.66) [129].

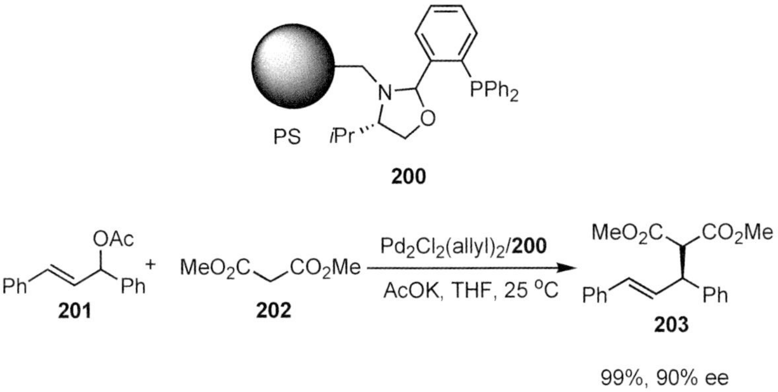

3.23 Asymmetric Allylic Alkylation | 113

Scheme 3.65

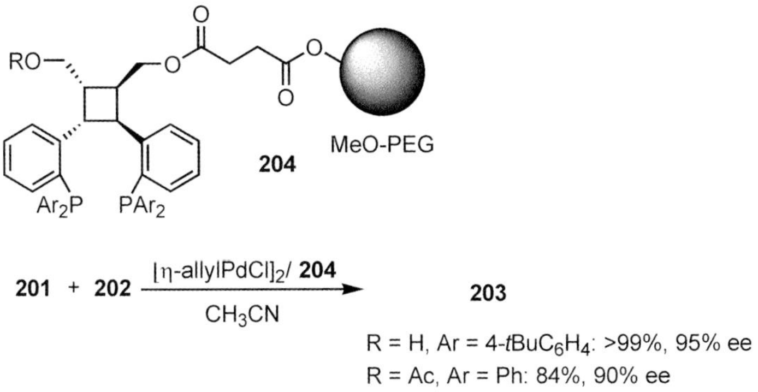

Scheme 3.66

In particular, the reaction of 1,3-diphenylprop-2-enyl acetate with dimethyl malonate by using **204** afforded the corresponding substituted product in high yield with excellent enantioselectivity.

An amphiphilic polystyrene-poly(ethylene glycol) (PS-PEG) resin-support has been used successfully as a suitable support for asymmetric catalysts in aqueous media. A polymeric *(R)*-2-(diphenylphosphino)binaphthyl (MOP) ligand **205** anchored onto the PS-PEG resin was shown to be an effective chiral ligand for the enantioselective *p*-allylic substitution under aqueous conditions (Scheme 3.67) [130].

An S—P-type chiral phosphinooxathiane was developed as an effective ligand for palladium-catalyzed allylic substitution reactions [131]. A polymer-supported chiral phosphinooxathiane **208** was also prepared and applied to asymmetric alkylations and aminations of acetate **201** [132]. Enantioselectivities of up to 99% ee were obtained in asymmetric Pd-catalyzed allylic amination of acetate **201** using the polymeric catalyst prepared from a PS-diethylsilyl support (Scheme 3.68).

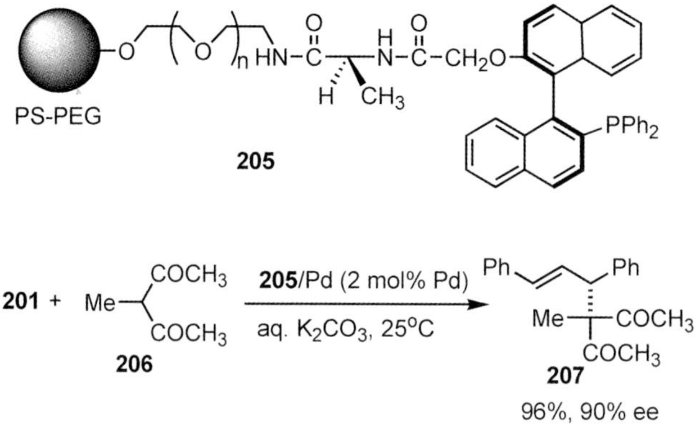

Scheme 3.67

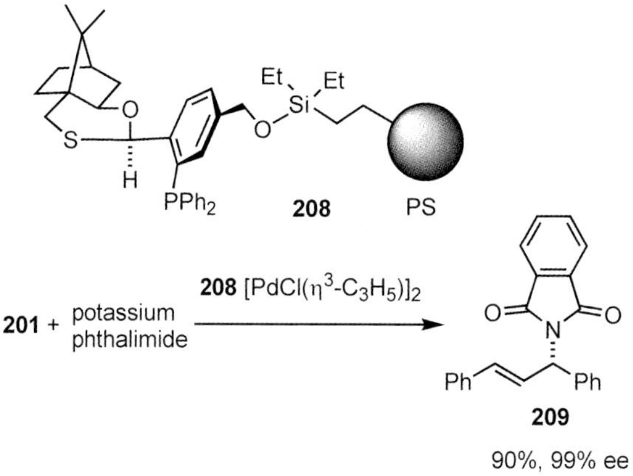

Scheme 3.68

The enantiopure phosphinooxazolinidines (POZ)s were a type of N–P chiral ligand developed by Nakano and used for asymmetric Pd-catalyzed allylic alkylations [133] and Diels–Alder reactions [59]. A PS-supported version of this type of chiral ligand **210** was prepared and applied to the asymmetric allylic alkylation of acetate **201** with dialkylmalonates [134], affording the product in excellent yield (99%) with 99% ee, as shown in Scheme 3.69.

An interesting approach to develop an asymmetric allylic alkylation catalyst by using a peptide-based phosphine ligand was examined by Gilbertson (Scheme 3.70) [135]. Phosphine-sulfide amino acid **211** was incorporated into a peptide sequence on a polymer support. After reduction of the phosphine sulfide to phosphine, the polymer-supported peptide sequence having phosphine-(support-Gly-Pps-D-Ala-Pro-Pps-D-Ala-Ac) was prepared. The complex of the peptide with Pd was utilized for the asymmetric addition of dimethyl malonate **202** to 3-acetoxycyclopentene **212** to give **213** in 59% yield and 66% ee.

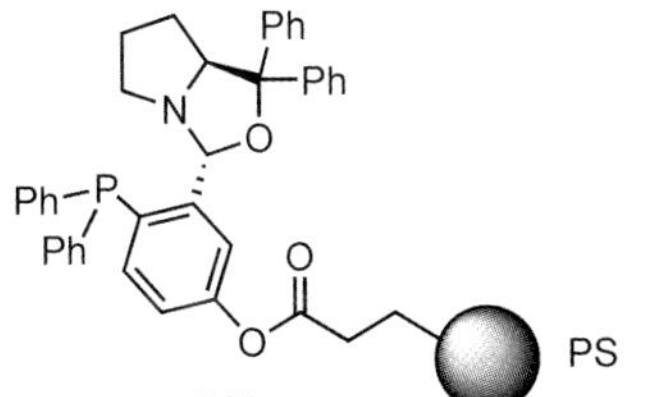

Scheme 3.69

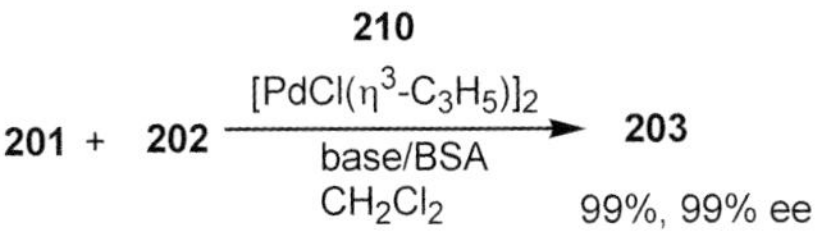

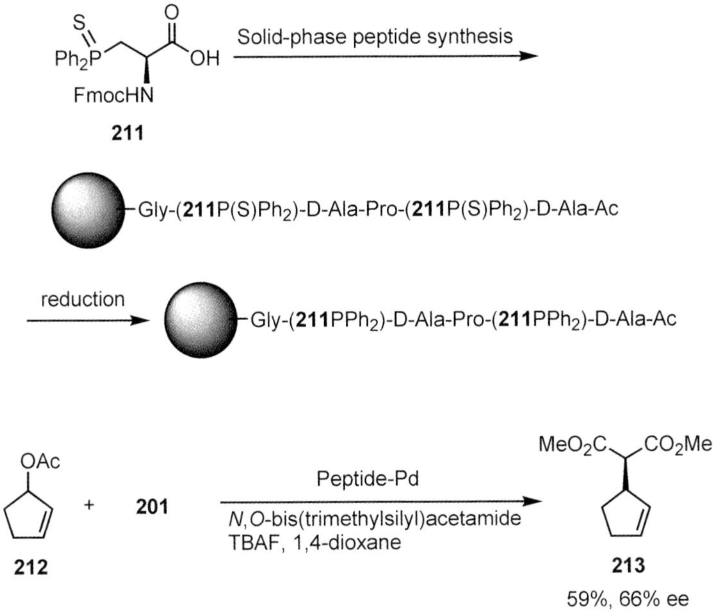

201 + 202 →[[PdCl(η³-C₃H₅)]₂, 210, base/BSA, CH₂Cl₂] 203

99%, 99% ee

Scheme 3.70

212 + 201 →[Peptide-Pd, N,O-bis(trimethylsilyl)acetamide, TBAF, 1,4-dioxane] 213

59%, 66% ee

3.24
Asymmetric Allylic Nitromethylation

Heterogeneous aquacatalytic palladium-catalyzed allylic substitution with nitromethane as the C1 nucleophile has been developed by Uozumi (Scheme 3.71). By using an amphiphilic PS-PEG polymer-supported chiral palladium complex, the asymmetric allylic nitromethylation of cycloalkenyl esters proceeded smoothly in water. For example, when polymer-supported palladium complex **214** was employed in the asymmetric nitromethylation of cycloheptenyl carbonate **215**

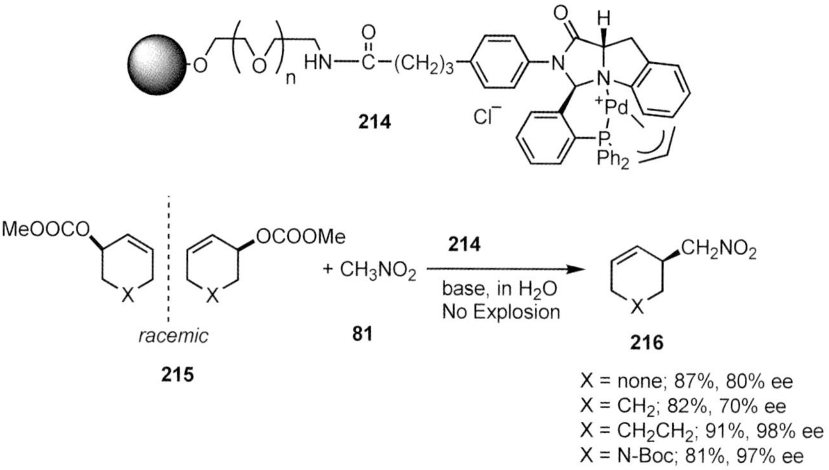

Scheme 3.71

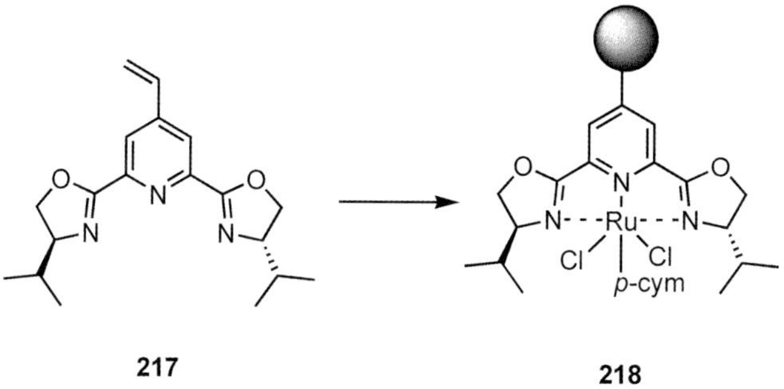

Scheme 3.72

(X = C$_2$H$_4$), the corresponding nitromethylated product **216** was obtained in 91% yield with 98% ee [136].

3.25
Asymmetric Cyclopropanation

Following Nishiyama's original discovery of an efficient chiral ligand (full name of Pybox) Pybox [137], many chiral complexes have been synthesized and utilized as catalysts in a variety of asymmetric transformations. Asymmetric cyclopropanation is one such application which uses the Pybox–Ru catalyst [138]. A polymer-supported version of the Pybox–Ru complex **218** was prepared by copolymerization of the chiral monomer **217** with styrene and DVB, followed by treatment of the resulting polymer with [RuCl$_2$(p-cymene)]$_2$ in CH$_2$Cl$_2$ (Scheme 3.72) [139]. The corresponding ruthenium complexes catalyzed the cyclopropanation reaction of

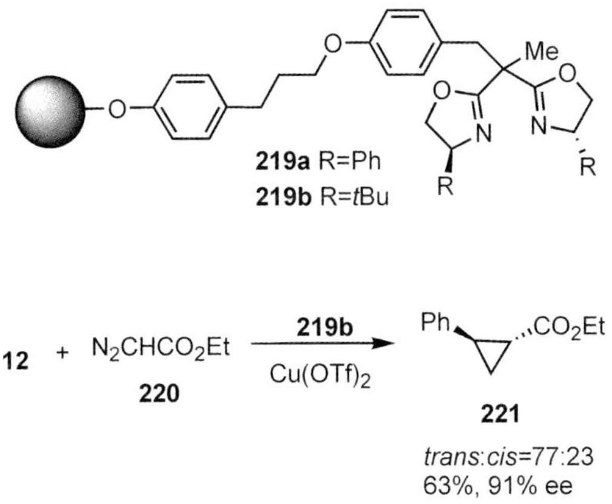

3.25 Asymmetric Cyclopropanation

Scheme 3.73

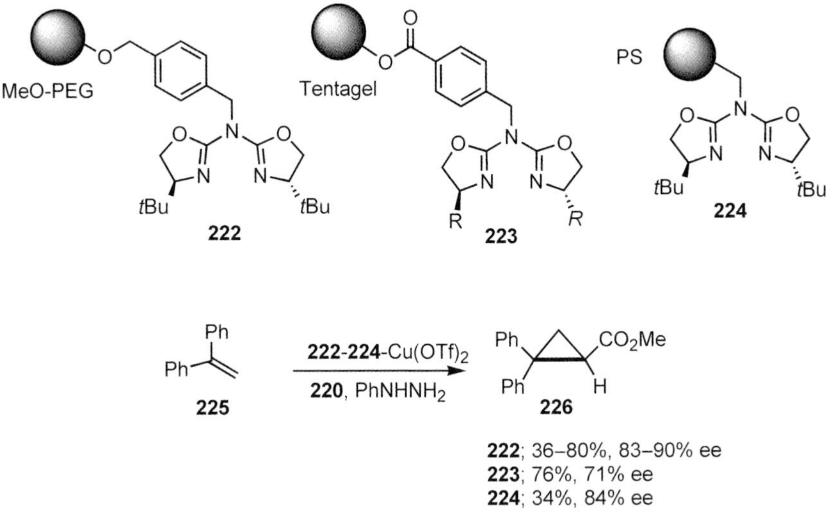

219a R=Ph
219b R=tBu

12 + N₂CHCO₂Et →(219b, Cu(OTf)₂) 221

220

trans:cis=77:23
63%, 91% ee

Scheme 3.74

MeO-PEG — 222 (tBu)
Tentagel — 223 (R)
PS — 224 (tBu)

225 (Ph₂C=CH₂ type, Ph/Ph) + 222–224-Cu(OTf)₂, 220, PhNHNH₂ → 226 (Ph, Ph, CO₂Me, H)

222: 36–80%, 83–90% ee
223: 76%, 71% ee
224: 34%, 84% ee

styrene with ethyl diazoacetate in 9–39% yield with up to 85% ee. When the polymeric catalyst was reused twice, a decrease in both selectivity and activity was observed in the second recycle.

Chiral bisoxazoline **219** supported on a modified PEG were prepared and employed in combination with CuOTf in the cyclopropanation of styrene and ethyl diazoacetate [140]. As shown in Scheme 3.73, the cyclopropane adduct **221** was obtained in 63% yield and 91% ee (*trans:cis* = 77:23). The same reaction with 1,1-diphenylethylene gave the adduct in 45% yield and 93% ee.

Azabis(oxazolines) [141] were attached to polymeric supports such as PEG, TentaGel or PS (Scheme 3.74) [142]. Copper(I)-catalyzed cyclopropanations were

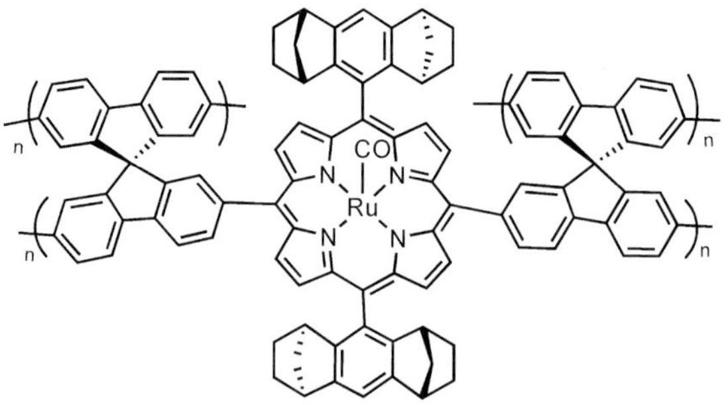

Scheme 3.75

performed with styrene and 1,1-diphenylethylene using 1–1.5 mol% of polymeric azabis(oxazoline) ligands **222–224**.

Another type of polymer-supported chiral catalyst for asymmetric cyclopropanation was obtained by electropolymerization of the tetraspirobifluorenylporphyrin ruthenium complex [143]. The cyclopropanation of styrene with diazoacetate, catalyzed by the polymeric catalyst **227**, proceeded efficiently at room temperature with good yields (80–90%) and moderate enantioselectivities (up to 53% at −40 °C) (Scheme 3.75). PS-supported versions of the chiral ruthenium–porphyrin complexes **231** (Scheme 3.76) were also prepared and used for the same reaction [144]. The cyclopropanation of styrene by ethyl diazoacetate proceeded well in the presence of the polymeric catalyst to give the product in good yields (60–88%) with high stereoselectivities (71–90% ee). The highest ee-value (90%) was obtained for the cyclopropanation of p-bromostyrene.

3.26
Enantioselective Olefin Metathesis

Hoveyda and Schrock developed chiral Mo complexes [145, 146] which have been used successfully as chiral catalysts for enantioselective olefin methathesis. A polymer-supported version of the chiral Mo complex **232** was also prepared. The supported chiral complex delivers appreciable levels of reactivity and excellent enantioselectivity [147] with, in most cases, the chiral catalyst providing as high levels of enantioselectivity as the corresponding homogeneous variant. For example, **236** was obtained in quantitative conversion in 30 min with 98% ee by using the polymeric catalyst **232** (Scheme 3.77).

The same research group also developed another type of polymer-supported chiral Mo-based complex **237** (Scheme 3.78) where, in many instances, the levels

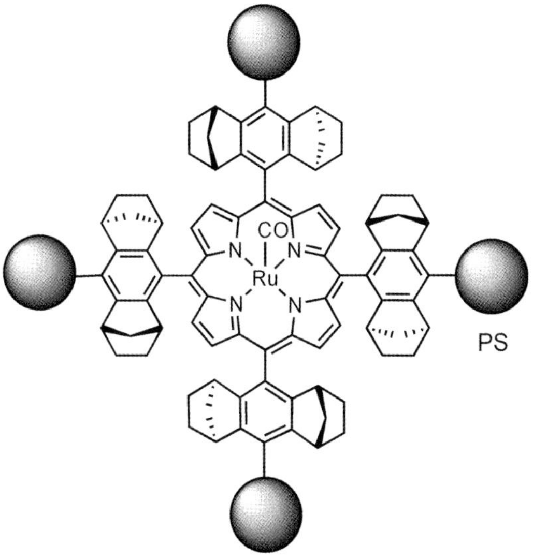

231

Scheme 3.76

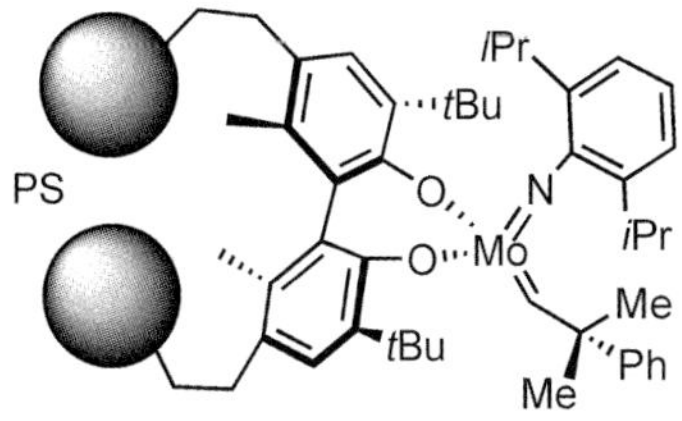

232

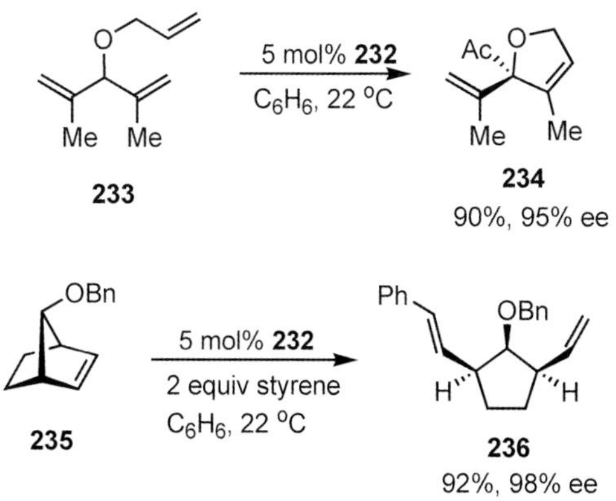

233 → **234** 90%, 95% ee

Conditions: 5 mol% **232**, C$_6$H$_6$, 22 °C

235 → **236** 92%, 98% ee

Conditions: 5 mol% **232**, 2 equiv styrene, C$_6$H$_6$, 22 °C

Scheme 3.77

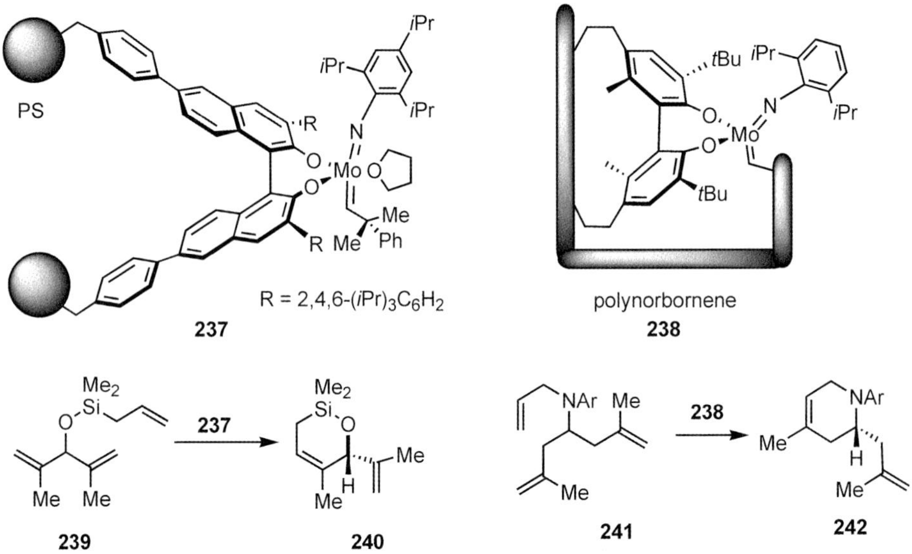

Scheme 3.78

of reactivity and enantioselectivity were competitive with those of the analogous homogeneous catalysts in a solution system [148].

3.27
Asymmetric Ring-Closing Metathesis

The immobilization of a chiral Schrock catalyst via the arylimido ligand was accomplished by Buchmeister and colleagues [149]. The linear PS was hydroxymethylated and reacted with 1,2,2-trifluoro-2-hydroxy-1-trifluoromehylethanesulfonic acid sultone and the resultant perfluoroalkanesulfonic acid converted into the corresponding silver salt. Using this support, the immobilized versions **243** and **246** were prepared and subjected to a variety of enantioselective ring-closing metathesis reactions (Scheme 3.79). Typically, the conversions were >90% using 5 mol% of the supported catalyst, while the enantiomeric excess obtained by means of the supported catalyst was basically identical to that achieved with the unsupported catalyst in a solution system.

3.28
Enantioselective Reissert-Type Reaction

The Reissert-type reaction is considered to proceed via two steps: (i) the generation of an acyl quinolinium intermediate by a nucleophilic attack of quinoline to an acid chloride; and (ii) the addition of cyanide [150]. Shibasaki has developed an

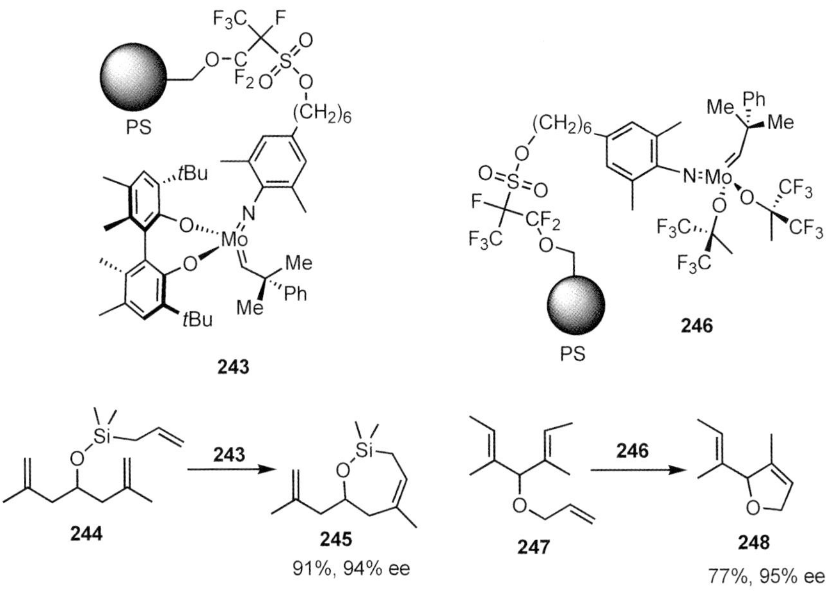

Scheme 3.79

efficient bifunctional catalyst for the reaction which consists of the Lewis acid (aluminum metal) and the Lewis base (oxygen atom of the phosphine oxide). This catalyst was attached to JandaJel. When using the polymeric chiral catalyst **249**, 2-furoyl chloride and TMSCN, the Reissert-type reaction of quinoline derivative **250** proceeded at a comparable rate as with the homogeneous catalyst, giving **251** in 92% yield with 86% ee (Scheme 3.80) [151].

3.29
Asymmetric Wacker-Type Cyclization

Uozumi developed polymer-supported 2,2′-bis(oxazolin-2-yl)-1,1′-binaphthyls **252** as chiral ligands for the asymmetric Wacker-type cyclization reaction (Scheme 3.81) [152]. In the presence of the polymeric catalyst derived from [Pd(MeCN)$_4$](BF$_4$)$_2$] and **252** o-allylphenol, **253** was asymmetrically cyclized to give **254** in 46% yield with 95% ee.

3.30
Enantioselective Hydrolysis

The *meso*-dicarboxylic diester **255** was enantioselectively hydrolyzed to the (4S,5S)-hemiester **256** (Scheme 3.82) with polymer-supported pig liver esterase (PLE) [153, 154]. The (4S,5S)-hemiester **256** was obtained in 90% yield with an enantiomeric

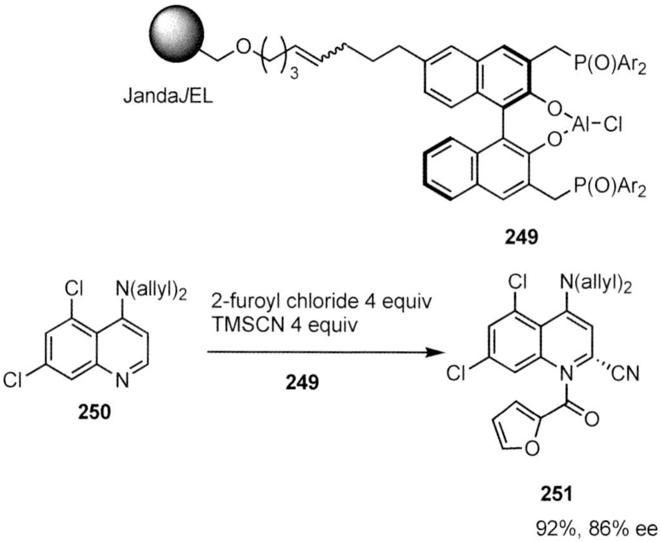
JandaJEL
O
3
P(O)Ar2
O
Al-Cl
O
P(O)Ar2
249
Cl
N(allyl)2
Cl
N
250
2-furoyl chloride 4 equiv
TMSCN 4 equiv
249
Cl
N(allyl)2
Cl
N
CN
O
O
251
92%, 86% ee

Scheme 3.80

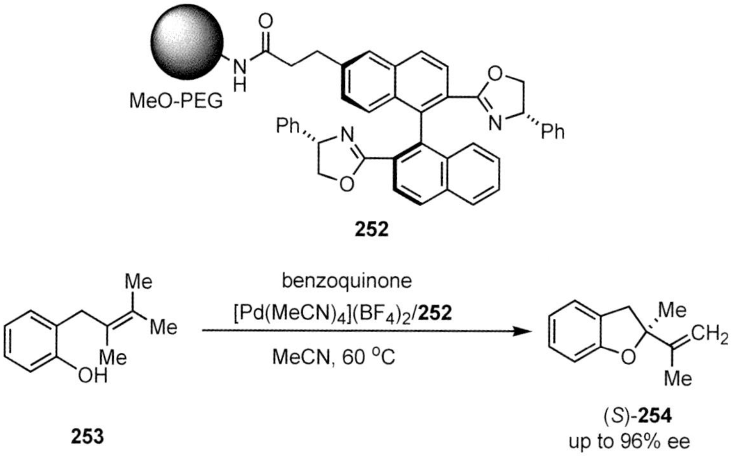
MeO-PEG
O
N
H
O
N
Ph
Ph
N
O
252
Me
Me
Me
OH
253
benzoquinone
[Pd(MeCN)4](BF4)2/252
MeCN, 60 °C
Me
CH2
O
Me
(S)-254
up to 96% ee

Scheme 3.81

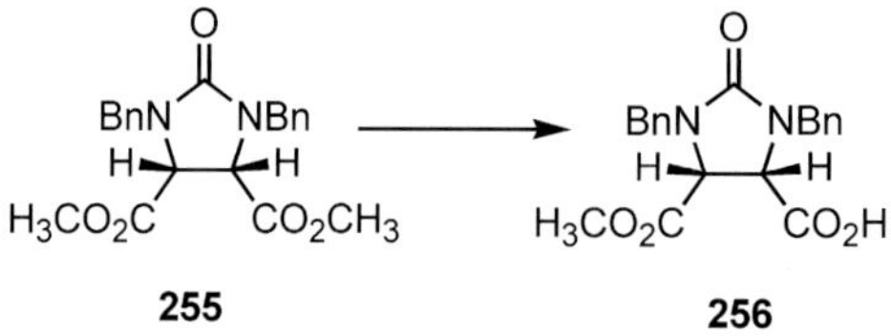
O
BnN
NBn
H
H
H3CO2C
CO2CH3
255
O
BnN
NBn
H
H
H3CO2C
CO2H
256

Scheme 3.82

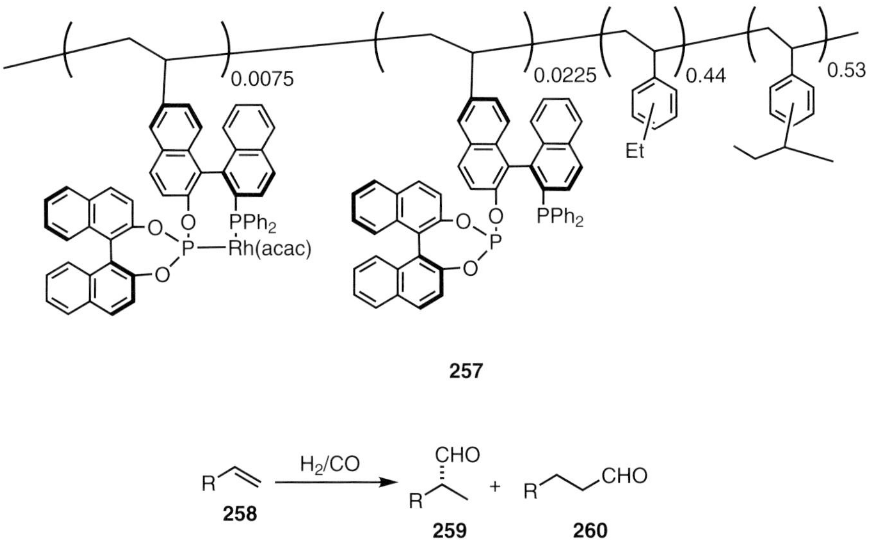

Scheme 3.83

excess of 91%, which was then upgraded to 98.5% ee by recrystallization from toluene.

3.31
Asymmetric Hydroformylation

Takaya and Nozaki invented an unsymmetrical phosphin-phosphite ligand, (R,S)-BINAPHOS, which was used in the Rh(I)-catalyzed asymmetric hydroformylation of a wide range of prochiral olefins, with excellent enantioselectivities [120, 155]. A highly crosslinked PS-supported (R,S)-BINAPHOS(257)-Rh(I) complex was prepared and applied to the same reaction (Scheme 3.83) [156]. Using the polymeric catalyst, the asymmetric hydroformylation of olefins was performed in the absence of organic solvents. The reaction of cis-2-butene, a gaseous substrate, provided (S)-methylbutanal with 100% regioselectivity and 82% ee upon treatment with H_2 and CO in a batchwise reactor equipped with a fixed bed.

3.32
Summary and Outlooks

As the efficacy of polymer-supported catalysts and reagents in various types of organic transformation has become increasingly recognized in synthetic chemistry, enantioselective catalysts have been immobilized in large numbers onto organic polymer supports. The development of polymer-supported catalysts for asymmetric synthesis is a rapidly growing field. Although both the activity and selectivity of the supported chiral catalyst were seen initially to be reduced compared to that of low-molecular-weight nonsupported catalysts in homogeneous solution systems, many of the recently developed supported chiral catalyst described in this chapter have demonstrated at least similar activities and enantioselectivities. Moreover, even higher enantioselectivities have been attained using polymeric chiral catalyst. Organic polymer-supported catalysts can be used successfully not only in organic solvents but also in aqueous media by using amphiphilic polymer supports. In addition, as the insoluble nature of the polymeric chiral catalyst means that they can be easily recovered, they can be recycled many times. The use of continuous-flow systems using polymeric catalysts has also demonstrated. Although most of the polymeric chiral catalysts described in this chapter consist of a crosslinked polymer based on PS, PEGs have also been also used in some instances. It follows that, in the near future, new polymeric chiral catalysts will be designed using a variety of polymers, each with a controlled architecture.

References

1 For recent reviews, see: (a) Itsuno, S., Haraguchi, N. and Arakawa, Y. (2005) *Recent Research Developments in Organic Chemistry*, **9**, 27–47.
(b) El-Shehawy, A.A. and Itsuno, S. (2005) *Current Topics in Polymer Research* (ed. R.K. Bregg), Nova Science Publishers, New York, Chapter 1, pp. 1–69.
(c) Clapham, B., Reger, T.S. and Janda, K.D. (2001) *Tetrahedron*, **57**, 4637–62.
(d) de Miguel, Y.R., Brule, E. and Margue, R.G. (2001) *Journal of the Chemical Society–Perkin Transactions 1*, 3085–94.
(e) Itsuno, S., Arakawa, Y. and Haraguchi, N. (2006) *Journal of the Society of the Rubber Industry of Japan*, **79**, 448–54.
(f) McNamara, C.A., Dixon, M.J. and Bradley, M. (2002) *Chemical Reviews*, **102**, 3275–300.
2 (a) Kell, R.J., Hodge, P., Nisar, M. and Watson, D. (2002) *Bioorganic and Medicinal Chemistry Letters*, **12**, 1803–7.
(b) Kell, R.J., Hodge, P., Snedden, P. and Watson, D. (2003) *Organic and Biomolecular Chemistry*, **1** (*18*), 3238–43.
3 Degni, S., Wilen, C.-E. and Leino, R. (2004) *Tetrahedron: Asymmetry*, **15**, 231–7.
4 Lesma, G., Daniele, B., Passarella, D., Sacchetti, A. and Silvani, A. (2003) *Tetrahedron: Asymmetry*, **14**, 2453–8.
5 Burguete, M.I., Verdugo, E.G., Vincent, M.J., Luis, S.V., Pennemann, H., Keyserling, N.G.v. and Martens, J. (2002) *Organic Letters*, **4**, 3947–50.
6 Svec, F. and Frechet, J.M.J. (1996) *Science*, **273**, 205–11.
7 Trnka, T.M. and Grubbs, R.H. (2001) *Accounts of Chemical Research*, **34**, 18–29.
8 Bilm, C., Tanyeli, C., Grenz, A. and Dinter, C.L. (2002) *Advanced Synthesis Catalysis*, **344**, 649–56.
9 Sellner, H., Faber, C., Rheiner, P.B. and Seebach, D. (2000) *Chemistry–A European Journal*, **6**, 3692–705.
10 Sellner, H., Rheiner, P.B. and Seebach, D. (2002) *Helvetica Chimica Acta*, **85**, 352–87.
11 Seebach, D., Beck, A.K. and Heckel, A. (2001) *Angewandte Chemie–International Edition*, **40**, 92–138.
12 Degni, S., Wilen, C.-E. and Leino, R. (2001) *Organic Letters*, **3**, 2551–4.
13 Altava, B., Burguete, M.I., Garcia-Verdugo, E., Luis, S.V. and Vicent, M.J. (2006) *Green Chemistry*, **8**, 717–26.
14 Anyanwu, U.K. and Venkataraman, D. (2003) *Tetrahedron Letters*, **44**, 6445–8.

15 Kamahori, K., Tada, S., Ito, K. and Itsuno, S. (1995) *Tetrahedron: Asymmetry*, **6**, 2547–55.
16 Itsuno, S., Matsumoto, T., Sato, D. and Inoue, T. (2000) *Journal of Organic Chemistry*, **65**, 5879–81.
17 Hui, X.-P., Chen, C.-A., Wu, K.-H. and Gau, H.-M. (2007) *Chirality*, **19**, 10–15.
18 Takizawa, S., Patil, M.L., Yonezawa, F., Mayubayashi, K., Tanaka, H., Kawai, T. and Sasai, H. (2005) *Tetrahedron Letters*, **46**, 1193–7.
19 Forrat, V.J., Prieto, O., Ramon, D.J. and Yus, M. (2006) *Chemistry – A European Journal*, **12**, 4431–45.
20 Forrat, V.J., Ramon, D.J. and Yus, M. (2006) *Tetrahedron: Asymmetry*, **17**, 2054–8.
21 Bolm, C., Hildebrand, J.P., Muniz, K. and Hermanns, N. (2001) *Angewandte Chemie – International Edition*, **40**, 3284.
22 Bolm, C. and Muniz, K. (1999) *Chemical Communications*, 1295–6.
23 Bolm, C., Hermanns, N., Clasen, A. and Muniz, K. (2002) *Bioorganic and Medicinal Chemistry Letters*, **12**, 1795–8.
24 Cozzi, P.G. (2003) *Angewandte Chemie – International Edition*, **42**, 2895–8.
25 Pathak, K., Bhatt, A.P., Abdi, S.H.R., Kureshy, R.I., Khan, N.-U.H., Ahmad, I. and Jasra, R.V. (2007) *Chirality*, **19**, 82–8.
26 Wei, C. and Li, C.-J. (2002) *Journal of the American Chemical Society*, **124**, 5638–9.
27 Kauffman, G.S., Harris, G.D., Dorow, R.L., Stone, B.P.R., Parsons, R.L., Pesti, J., Magnus, N.A., Fortunak, J.M., Confalone, P.N. and Nugent, W.A. (2000) *Organic Letters*, **2**, 3119–21.
28 Weissberg, A., Halak, B. and Portnoy, M. (2005) *Journal of Organic Chemistry*, **70**, 4556–9.
29 Itsuno, S., Watanabe, K. and El-Shehawy, A.A. (2001) *Advanced Synthesis Catalysis*, **343**, 89–94.
30 Lundgren, S., Lutsenko, S., Joensson, C. and Moberg, C. (2003) *Organic Letters*, **20**, 3663–5.
31 Ooi, T. and Maruoka, K. (2007) *Angewandte Chemie*, **46**, 4222–66.
32 (a) O'Donnell, M.J. and Polt, R.L. (1982) *Journal of Organic Chemistry*, **47**, 2663–6.
(b) O'Donnell, M.J., Bennett, W.D. and We, S. (1989) *Journal of the American Chemical Society*, **111**, 2353–5.
(c) O'Donnell, M.J. (2001) *Aldrichimica Acta*, **34**, 3–15.
33 Chinchilla, R., Mazon, P. and Najera, C. (2004) *Advanced Synthesis Catalysis*, **346**, 1186–94.
34 Wang, X., Yin, L., Yang, T. and Wang, Y. (2007) *Tetrahedron: Asymmetry*, **18**, 108–14.
35 Carreira, E.M. (1999) Aldol reactions, in *Comprehensive Asymmetric Catalysis III* (E.N. Jacobsen, A. Pfalts and H. Yamamoto), Springer, Berlin, Chapter 29.1, p. 997.
36 Ishihara, K., Kondo, S. and Yamamoto, H. (2000) *Journal of Organic Chemistry*, **65**, 9125–8.
37 Ishihara, K., Kondo, S. and Yamamoto, H. (1999) *Synlett*, 1283–5.
38 Itsuno, S., Arima, S. and Haraguchi, N. (2005) *Tetrahedron*, **61**, 12074–80.
39 (a) List, B., Lerner, R.A. and Barbas, C.F. III (2000) *Journal of the American Chemical Society*, **122**, 2395–6.
(b) Sakthivel, K., Notz, W., Bui, T. and Barbas, C.F. III (2001) *Journal of the American Chemical Society*, **123**, 5260–7.
(c) Hanessian, S. and Pham, V. (2000) *Organic Letters*, **2**, 2975–8. (d) List, B., Pojarliev, P. and Martin, H.J. (2001) *Organic Letters*, **3**, 2423–5.
40 Benaglia, M., Cinquini, M., Cozzi, F. and Puglisi, A. (2002) *Advanced Synthesis Catalysis*, **344**, 533–42.
41 Giacalone, F., Gruttadauria, M., Marculescu, A.M. and Noto, R. (2007) *Tetrahedron Letters*, **48**, 255–9.
42 Akagawa, K., Sakamoto, S. and Kudo, K. (2005) *Tetrahedron Letters*, **46**, 8185–7.
43 Akagawa, K., Sakamoto, S. and Kudo, K. (2007) *Tetrahedron Letters*, **48**, 985–7.
44 Krattiger, P., Revell, J.D., Kovasy, R., Ivan, S. and Wennemers, H. (2005) *Organic Letters*, **7**, 1101–3.
45 Revell, J.D., Gantenbein, D., Krattiger, P. and Wennemers, H. (2006) *Biopolymers (Peptide Science)*, **84**, 105–13.
46 Kitamoto, D., Imma, H. and Nakai, T. (1995) *Tetrahedron Letters*, **36**, 1861–4.

47 Yamada, Y.M.A., Ichinohe, M., Takahashi, H. and Ikegami, S. (2002) *Tetrahedron Letters*, **43**, 3431–4.

48 Sekiguti, T., Iizuka, Y., Takizawa, S., Jayaprakash, D., Arai, T. and Sasai, H. (2003) *Organic Letters*, **5**, 2647–50.

49 Manickam, G. and Sundaradajanm, G. (1999) *Tetrahedron*, **55**, 2721–36.

50 Sundararajan, G. and Prabagaran, N. (2001) *Organic Letters*, **3**, 389–92.

51 (a) Arai, T., Sasai, H., Aoe, K., Okamura, K., Date, T. and Shibasaki, M. (1997) *Angewandte Chemie – International Edition in English*, **35**, 104–6.
(b) Arai, T., Sasai, H., Yamaguchi, K. and Shibasaki, M. (1998) *Journal of the American Chemical Society*, **120**, 441–2.

52 Arai, T., Sekiguti, T., Otsuki, K., Takizawa, S. and Sasai, H. (2003) *Angewandte Chemie*, **115**, 2194–7.

53 Gu, L., Wu, Y., Zhang, Y. and Zhao, G. (2007) *Journal of Molecular Catalysis A – Chemical*, **263**, 186–94.

54 Bonfils, F., Cazaux, I., Hodge, P. and Caze, C. (2006) *Organic and Biomolecular Chemistry*, **4**, 493–7.

55 Yamada, H., Kawate, T., Nishida, A. and Nakagawa, M. (1999) *Journal of Organic Chemistry*, **64**, 8821–8.

56 Graf, C.-D., Malan, C., Harms, K. and Knochel, P. (1999) *Journal of Organic Chemistry*, **64**, 5581–8.

57 Ma, L. and Willard, P.G. (2006) *Tetrahedron: Asymmetry*, **17**, 3021–9.

58 Altava, B., Burguete, M.I., Verdugo, E.G., Luis, S.V. and Vicent, M.J. (2006) *Green Chemistry*, **8**, 717–26.

59 Nakano, H., Takahashi, K., Okuyama, Y., Senoo, C., Tsugawa, N., Suzuki, Y., Fujita, R., Sasaki, K. and Kabuto, C. (2004) *Journal of Organic Chemistry*, **69**, 7092–100.

60 Ahrendt, K.A., Borths, C.J. and MacMillan, D.W.C. (2000) *Journal of the American Chemical Society*, **122**, 4243–4.

61 Benaglia, M., Celentano, G., Cinquini, M., Puglisi, A. and Cozzi, F. (2002) *Advanced Synthesis Catalysis*, **344**, 149–52.

62 Puglisi, A., Benaglia, M., Cinquini, M., Cozzi, F. and Celentano, G. (2004) *European Journal of Organic Chemistry*, 567–73.

63 Lee, B.S., Mahajan, S. and Janda, K.D. (2005) *Tetrahedron Letters*, **46**, 4491–3.

64 Ishida, T., Akiyama, R. and Kobayashi, S. (2003) *Advanced Synthesis Catalysis*, **345**, 576–79.

65 Helder, R., Hummelen, J.C., Laane, R.W.P.M., Wiering, J.S. and Wynberg, H. (1976) *Tetrahedron Letters*, **17**, 1831–4.

66 Lygo, B. and Wainwright, P.G. (1999) *Tetrahedron Letters*, **55**, 6289–300.

67 Corey, E.J. and Zhang, F.-Y. (1999) *Organic Letters*, **1**, 1287–90.

68 Lu, J., Wang, X., Liu, J., Zhang, L. and Wang, Y. (2006) *Tetrahedron: Asymmetry*, **17**, 330–5.

69 Irie, R., Noda, K., Ito, Y., Matsumoto, N. and Katsuki, T. (1990) *Tetrahedron Letters*, **31**, 7345–8.

70 Zhang, W., Loebach, J.L., Wilson, S.R. and Jacobsen, E.N. (1990) *Journal of the American Chemical Society*, **112**, 2801–3.

71 Larrow, J.F. and Jacobsen, E.N. (2004) *Topics in Organometallic Chemistry*, **6**, 123–52.

72 Maeda, T., Furusho, Y. and Takata, T. (2002) *Chirality*, **14**, 587–90.

73 Reger, T.S. and Janda, K.D. (2000) *Journal of the American Chemical Society*, **122**, 6929–34.

74 Sellner, H., Karjalainen, J.K. and Seebach, D. (2001) *Chemistry – A European Journal*, **7**, 2873–87.

75 Sellner, H., Hametner, K., Gunther, D. and Seebach, D. (2003) *Journal of Catalysis*, **215**, 87–93.

76 Smith, K. and Liu, C.-H. (2002) *Chemical Communications*, 886–7.

77 Holbach, M. and Weck, M. (2006) *Journal of Organic Chemistry*, **71**, 1825–38.

78 Zhang, H., Zhang, Y. and Li, C. (2005) *Tetrahedron: Asymmetry*, **16**, 2417–23.

79 Kureshy, R.I., Khan, N.H., Abdi, S.H.R., Singh, S., Ahmed, I. and Jasra, R.V. (2004) *Journal of Molecular Catalysis A – Chemical*, **218**, 141–6.

80 Guo, H., Shi, X., Qiao, Z., Hou, S. and Wang, M. (2002) *Chemical Communications*, 118–19.

81 Reed, N.N., Dickerson, T.J., Boldt, G.E. and Janda, K. (2005) *Journal of Organic Chemistry*, **70**, 1728–31.

82 Tokunaga, M., Larrow, J.F., Kakiuchi, F. and Jacobsen, E.N. (1997) *Science*, **277**, 936–8.

83 Schaus, S.E., Brandes, B.D., Larrow, J.F., Tokunaga, M., Hansen, K.B., Gould, A.E., Furrow, M.E. and Jacobsen, E.N. (2002) *Journal of the American Chemical Society*, **124**, 1307–15.

84 White, D.E. and Jacobsen, E.N. (2003) *Tetrahedron: Asymmetry*, **14**, 3633–8.

85 Ready, J.M. and Jacobsen, E.N. (2002) *Angewandte Chemie – International Edition*, **41**, 1374–7.

86 Ready, J.M. and Jacobsen, E.N. (2001) *Journal of the American Chemical Society*, **123**, 2687–8.

87 Welbes, L.L., Scarrow, R.C. and Borovik, A.S. (2004) *Chemical Communications*, 2544–5.

88 Xiaoquan, Y.S., Chen, Y.H., Bai, C., Hu, X. and Zheng, Z. (2002) *Tetrahedron Letters*, **43**, 6625–7.

89 Song, Y., Chen, H., Hu, X., Bai, C. and Zheng, Z. (2003) *Tetrahedron Letters*, **44**, 7081–5.

90 Kim, G.-J., Lee, H. and Kim, S.-J. (2003) *Tetrahedron Letters*, **44**, 5005.

91 Shin, C.-K., Kim, S.-J. and Kim, G.-J. (2004) *Tetrahedron Letters*, **45**, 7429–33.

92 Itsuno, S. (1998) *Organic Reactions*, Vol. 52 (ed. L.A. Paquette), John Wiley & Sons, New York, pp. 395–576.

93 Itsuno, S. (1999) *Comprehensive Asymmetric Catalysis* (eds E.N. Jacobsen, A. Pfaltz and H. Yamamoto), Springer, pp. 289–315.

94 Degni, S., Wilen, C.-E. and Roslig, A. (2004) *Tetrahedron: Asymmetry*, **9**, 1495–9.

95 Hu, J.-B., Zhao, G., Yang, G.-S. and Ding, Z.-D. (2001) *Journal of Organic Chemistry*, **66**, 303–4.

96 Wang, G., Liu, X. and Zhao, G. (2005) *Tetrahedron: Asymmetry*, **16**, 1873–9.

97 Price, M.D., Sui, J.K., Kurth, M.J. and Schore, N.E. (2002) *Journal of Organic Chemistry*, **67**, 8086–9.

98 Varela, M.C., Dixon, S.M., Price, M.D., Merit, J.E., Berget, P.E., Shiraki, S., Kurth, M.J. and Schore, N.E. (2007) *Tetrahedron*, **63**, 3334–9.

99 Wang, G., Zheng, C. and Zhao, G. (2006) *Tetrahedron: Asymmetry*, **17**, 2074–81.

100 Hu, J.-B., Zhao, G. and Ding, Z.-D. (2001) *Angewandte Chemie – International Edition*, **40**, 1109.

101 Gao, J.-X., Yi, X.D., Tang, C.-L., Xu, P.-P. and Wan, H.-L. (2001) *Polymers for Advanced Technologies*, **12**, 716–19.

102 Bastin, S., Eaves, R.J., Edwards, C.W., Ichihara, O., Whittaker, M. and Wills, M. (2004) *Journal of Organic Chemistry*, **69**, 5405–12.

103 Herault, D., Saluzzo, C. and Lemaire, M. (2006) *Tetrahedron Letters*, **17**, 1944–51.

104 Sanda, F., Araki, H. and Masuda, T. (2005) *Chemistry Letters*, **34**, 1642–3.

105 Hashiguchi, S., Fujita, A., Takehara, J., Ikariya, T. and Noyori, R. (1995) *Journal of the American Chemical Society*, **117**, 7562–3.

106 Halle, R., Scults, E. and Lemaire, M. (1997) *Synlett*, 1257–8.

107 Li, Y., Li, Z., Wang, Q. and Tao, F. (2005) *Organic and Biomolecular Chemistry*, **3**, 2513–18.

108 Wu, X., Li, X., Hems, W., King, F. and Xiao, J. (2004) *Organic and Biomolecular Chemistry*, **2**, 1818–21.

109 Rhyoo, H.Y., Park, H.J. and Chung, Y.K. (2001) *Chemical Communications*, 2064–5.

110 Ma, Y., Chen, H.L., Cui, X., Zhu, J. and Deng, J. (2003) *Organic Letters*, **5**, 2103–6.

111 Bubert, C., Blacker, J., Brown, S.M., Crosby, J., Fitzjohn, S., Muxworthy, J.P., Thorpe, T. and Williams, J.M.J. (2001) *Tetrahedron Letters*, **42**, 4037–9.

112 Li, X., Wu, X., Chen, W., Hancock, F.E., King, F. and Xiao, J. (2004) *Organic Letters*, **6**, 3321–4.

113 Li, X., Chen, W., Hems, W., King, F. and Xiao, J. (2004) *Tetrahedron Letters*, **45**, 951–3.

114 Arakawa, Y., Haraguchi, N. and Itsuno, S. (2006) *Tetrahedron Letters*, **47**, 3239–43.

115 Ohkuma, T., Takeno, H., Honda, Y. and Noyori, R. (2001) *Advanced Synthesis Catalysis*, **343**, 369–75.

116 (a) Itsuno, S., Tsuji, A. and Takahashi, M. (2003) *Tetrahedron Letters*, **44**, 3825–8. (b) Itsuno, S., Takahashi, M. and Tsuji, A. (2004) *Macromolecular Symposia*, **217**, 191–202.

117 Itsuno, S., Haraguchi, N. and Takahashi, M. (2005) *Synthetic Organic Chemistry*, **63**, 1253–63.
118 Itsuno, S., Tsuji, A. and Takahashi, M. (2004) *Designed Monomers and Polymers*, **7**, 495–503.
119 Itsuno, S., Tsuji, A. and Takahashi, M. (2004) *Journal of Polymer Science Part A: Polymer Chemistry*, **42**, 4556–62.
120 Itsuno, S., Chiba, M., Takahashi, M., Arakawa, Y. and Haraguchi, N. (2007) *Journal of Organometallic Chemistry*, **692**, 487–94.
121 Sakai, N., Mano, S., Nozaki, K. and Takaya, H. (1993) *Journal of the American Chemical Society*, **115**, 7033–4.
122 Li, X., Chen, W., Hems, W., King, F. and Xiao, J. (2003) *Organic Letters*, **5**, 4559–61.
123 Chai, L.-T., Wang, W.-W., Wang, Q.-R. and Tao, F.-G. (2007) *Journal of Molecular Catalysis A–Chemical*, **270**, 83–8.
124 Zarka, M.T., Nuyken, O. and Weberskirch, R. (2003) *Chemistry–A European Journal*, **9**, 3228–34.
125 Chen, W., Roberts, S.M. and Whittall, J. (2006) *Tetrahedron Letters*, **47**, 4263–6.
126 Wang, Z.-J., Deng, G.-J., Li, Y., He, Y.-M., Tang, W.-J. and Fan, Q.-H. (2007) *Organic Letters*, **9**, 1243–6.
127 Trost, B.M., Pan, Z., Zambrano, J. and Kujat, C. (2002) *Angewandte Chemie–International Edition*, **41**, 4691–3.
128 Kim, Y.-M. and Jin, M.-J. (2005) *Bull. Korean Chem. Soc*, **26**, 373–4.
129 Zhao, D., Sun, J. and Ding, K. (2004) *Chemistry–A European Journal*, **10**, 5952–63.
130 Kobayashi, Y., Tanaka, D., Danjo, H. and Uozumi, Y. (2006) *Advanced Synthesis Catalysis*, **348**, 1561–6.
131 (a) Nakano, H., Okuyama, Y. and Hongo, H. (2000) *Tetrahedron Letters*, **41**, 4615.
(b) Nakano, H., Suzuki, Y., Kabuto, C., Fujita, R. and Hongo, H. (2002) *Journal of Organic Chemistry*, **67**, 5011–14.
132 Nakano, H., Takahashi, K., Suzuki, Y. and Fujita, R. (2005) *Tetrahedron: Asymmetry*, **16**, 609–14.
133 Okuyama, Y., Nakano, H. and Hongo, H. (2000) *Tetrahedron: Asymmetry*, **11**, 1193–8.
134 Nakano, H., Takahashi, K. and Fujita, R. (2005) *Tetrahedron: Asymmetry*, **16**, 2133–40.
135 Gilbertson, S.R. and Yamada, S. (2004) *Tetrahedron Letters*, **45**, 3917–20.
136 Uozumi, Y. and Suzuka, T. (2006) *Journal of Organic Chemistry*, **71**, 8644–6.
137 Nishiyama, H. (1995) *Journal of the Society of Synthetic Organic Chemistry of Japan*, **53**, 500–8.
138 Nishiyama, H., Itoh, Y., Matsumoto, H., Park, S.-B. and Itoh, K. (1994) *Journal of the American Chemical Society*, **116**, 2223–4.
139 Cornejo, A., Fraile, J.M., Garcia, J.I., Garcia-Verdugo, E., Gil, M.J., Legarreta, G., Luis, S.V., Martinez-Merino, V. and Mayoral, J.A. (2002) *Organic Letters*, **4**, 3927–30.
140 Annunziata, R., Benaglia, M., Cinquini, M., Cozzi, F. and Pitillo, M. (2001) *Journal of Organic Chemistry*, **66**, 3160–6.
141 (a) Geiger, C., Kreitmeier, P. and Reiser, O. (2005) *Advanced Synthesis Catalysis*, **347**, 249–54.
(b) Gissibl, A., Finn, M.G. and Reiser, O. (2005) *Organic Letters*, **7**, 2325.
142 Werner, H., Herrerias, C.I., Glos, M., Gissibl, A., Fraile, J.M., Perez, I., Mayoral, J.A. and Reiser, O. (2006) *Advanced Synthesis Catalysis*, **348**, 125–32.
143 Ferrand, Y., Poriel, C., Maux, P.L., Berthelot, J.R. and Simonneaux, G. (2005) *Tetrahedron: Asymmetry*, **16**, 1463–72.
144 Ferrand, Y., Maux, P.L. and Simonneaux, G. (2005) *Tetrahedron: Asymmetry*, **16**, 3829–36.
145 Alexander, J.B., La, D.S., Cefalo, D.R., Hoveyda, A.H. and Schrock, R.R. (1998) *Journal of the American Chemical Society*, **120**, 4041–2.
146 Hoveyda, A.H. and Schrock, R.R. (2001) *Chemistry–A European Journal*, **7**, 945–50.
147 Hultzsch, K.C., Jernelius, J.A., Hoveyda, A.M. and Schrock, R.R. (2002) *Angewandte Chemie–International Edition*, **41**, 589–93.
148 Dolman, S.J., Hultzsch, K.C., Pezet, F., Teng, X., Hoveyda, A.H. and Schrock, R.R. (2004) *Journal of the American Chemical Society*, **126**, 10945–53.

149 Wang, D., Kroll, R., Monika, M., Wurst, K. and Buchmeiser, M.R. (2006) *Advanced Synthesis Catalysis*, **348**, 1567–79.
150 Popp, F.D. (1973) *Heterocycles*, **1**, 165–80.
151 Takamura, M., Funabashi, K., Kanai, M. and Shibasaki, M. (2001) *Journal of the American Chemical Society*, **123**, 6801–8.
152 Hocke, H. and Uozumi, Y. (2002) *Synlett*, **12**, 2049–53.
153 Chen, F.-E., Chen, X.-X., Dai, H.-F., Kuang, Y.-Y., Xie, B. and Zhao, J.-F. (2005) *Advanced Synthesis Catalysis*, **347**, 549–54.
154 Laumen, K., Reimerdes, E.H., Schneider, M. and Gorisch, H. (1985) *Tetrahedron Letters*, **26**, 407–10.
155 Nozaki, K., Takaya, H. and Hiyama, T. (1997) *Topics in Catalysis*, **4**, 175.
156 Shibahara, F., Nozaki, K. and Hiyama, T. (2003) *Journal of the American Chemical Society*, **125**, 8555–60.

4
Enantioselective Catalysis Using Dendrimer Supports

Qing-Hua Fan, Guo-Jun Deng, Yu Feng, and Yan-Mei He

4.1
General Introduction

Dendrimers represent a new class of polymers which possess highly branched and well-defined molecular structures [1, 2]. These appealing macromolecules are composed of a central core, branching units and peripheral end groups, and obtained by an iterative sequence of reaction steps. In general, two synthetic strategies can be applied to the synthesis of dendrimers (Scheme 4.1). In the divergent approach the dendrimer is synthesized starting from the core and built up layer by layer (Scheme 4.1a) [3, 4], with each layer usually being referred to a 'generation' of growth. Relatively large dendrimers can be synthesized very quickly using this approach. However, in the past the high number of reactions which must be performed on a single molecule often led to problems with both purification and monodispersity. The alternative convergent approach developed by Hawker and Fréchet starts from the surface and ends up at the core, where the dendrons are coupled together in the final step (Scheme 4.1b) [5]. This method is characterized by reactions occurring at only one site, providing perfectly branched macromolecules which are free of defects. Unfortunately, this focal point soon becomes masked by the growing dendrons and therefore only relatively small dendrimers can be synthesized using this approach. Given the rapid development of synthetic procedures, a number of functionalized dendrimers, including chiral dendrimers [6], are readily available in sufficient quantities. Over the past decade, many potential applications for dendrimers, including catalysis [7–15], have been well documented in numerous reviews [16–19].

Since the pioneering studies reported by van Koten and coworkers in 1994 [20], dendrimers as catalyst supports have been attracting increasing attention. The metallodendrimers and their catalytic applications have been frequently reported and reviewed [7–15]. As a novel type of soluble macromolecular support, dendrimers feature homogeneous reaction conditions (faster kinetics, accessibility of the metal site, and so on) and enable the application of common analytical techniques such as thin-layer chromatography (TLC) and nuclear magnetic resonance

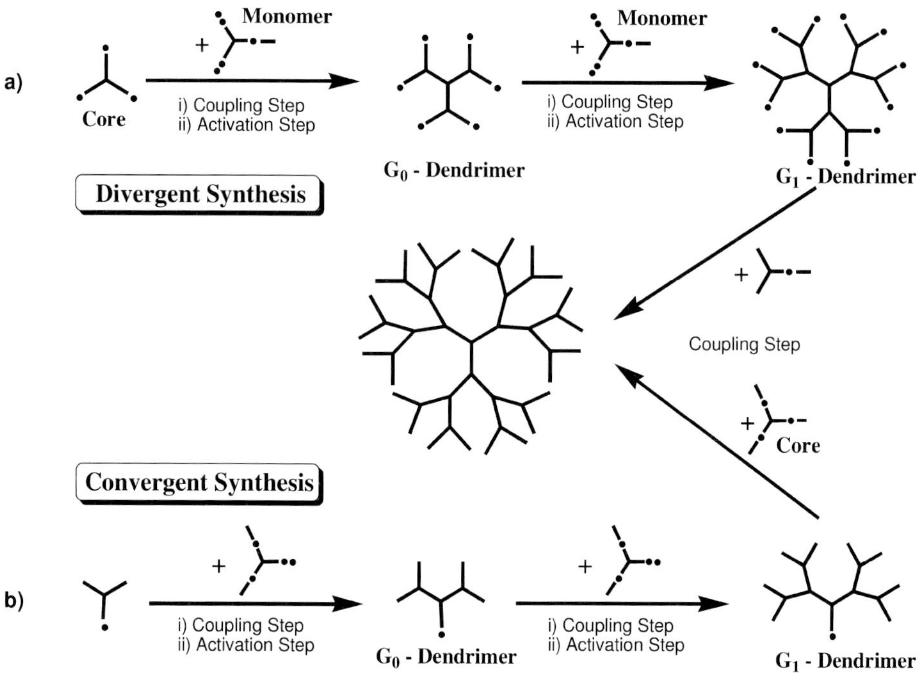

Scheme 4.1 Schematic representation of divergent and convergent approaches for dendrimer synthesis.

(NMR) and matrix-assisted laser desorption/ionization-time-of-flight (MALDI-TOF) spectroscopies. Compared to the linear, soluble, polymer-supported catalysts, the dendrimer architecture might offer a better control of the number and disposition of catalytic species on the support. Furthermore, the well-defined molecular architecture of dendrimers allows fine-tuning of the catalytic properties by precise dendritic ligand design. Immobilization of the catalyst onto a dendrimer is expected to give rise to different catalytic properties, which are not possible for the homogeneous analogues. Such special properties (the 'dendrimer effect') might include enhanced catalytic activity and/or enantioselectivity through steric isolation inside dendrimer, and/or cooperativity between two neighboring catalytic centers on the surface of dendrimer [13]. Therefore, such a novel class of catalysts may fill the gap between homogeneous and heterogeneous catalysis, and combine the advantages of homogeneous and heterogeneous catalysts. The very first attempt to carry out an asymmetric catalysis using a chiral dendrimer ligand was reported by Brunner's group [21]. In their first approach to dendrimer catalysis, these authors designed a 'dendrizyme' diphosphine ligand in which a dppe diphosphine core was surrounded by spacefilling dendritic chiral substituents. The rhodium complex of this ligand was studied in the asymmetric hydrogenation of acetamidocinnamic acid, which led to the product with poor asymmetric induction (2% enantiomeric excess (ee)). Since then, a number of efficient chiral dendrimer catalysts have been reported [14, 22].

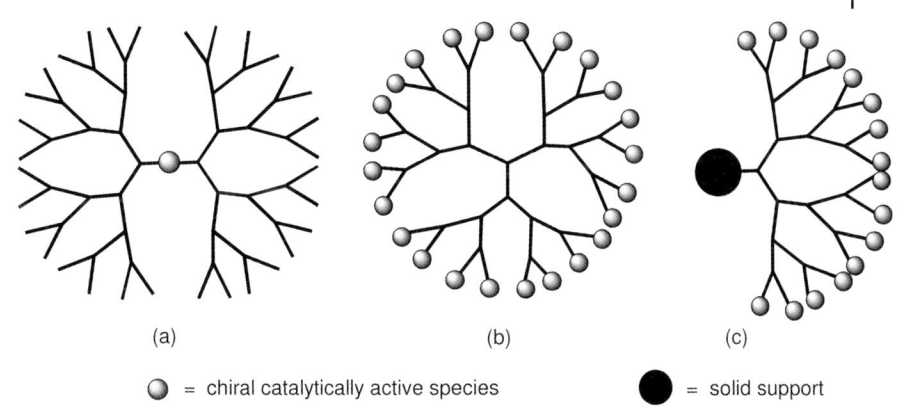

Figure 4.1 Commonly encountered chiral catalyst immobilization on dendritic polymer supports: (a) core-functionalized chiral dendrimers; (b) peripherally modified chiral dendrimers; (c) solid-supported dendritic chiral catalysts.

○ = chiral catalytically active species   ● = solid support

In general, two strategies can be applied for the construction of chiral dendrimer catalysts: (i) chiral metal complex (or organocatalyst) may be incorporated into the core of the dendrimer (Figure 4.1a); or (ii) multiple chiral metal complexes (or organocatalysts) may be located at the periphery of the dendrimer (Figure 4.1b). Recently, hybrids of dendrimer and crosslinked polymer as supports have been developed (Figure 4.1c) [12].

Catalyst recycling is an important practical objective in the development of supported catalysts, and in particular for dendrimer catalysts [10, 23]. Several methods have been developed and employed in the separation and recycling of chiral dendrimer catalysts (Figure 4.2). Taking advantage of the different solubility between dendrimers and small molecules, dendrimer catalysts can be recovered via solvent precipitation (Figure 4.2a). In addition, the large size and globular structure of the dendrimer, compared to the substrates and the products, can be used to facilitate catalyst–product separation by means of nanosized membrane filtration, which can be performed either batchwise or in a continuous-flow membrane reactor (Figure 4.2b). Recently, a novel catalytic system consisting of chiral dendrimer catalysts with a core/shell structure has been developed [24, 25]. By choosing a suitable solvent combination, the catalyst recycling was easily achieved via phase separation by the addition of a small amount of water (Figure 4.2c).

In this chapter, we attempt to summarize the recently developed chiral dendrimer catalysts with their chiral catalytically active species located either at the core or at the periphery of the dendritic macromolecular supports. The discussion will also be focused on dendrimer effects and the development of new methodologies for the recovery and reuse of chiral dendrimer catalysts, with special emphasis on their applications in enantioselective synthesis. The published data have been classified according to the type of reaction in each of the following three sections.

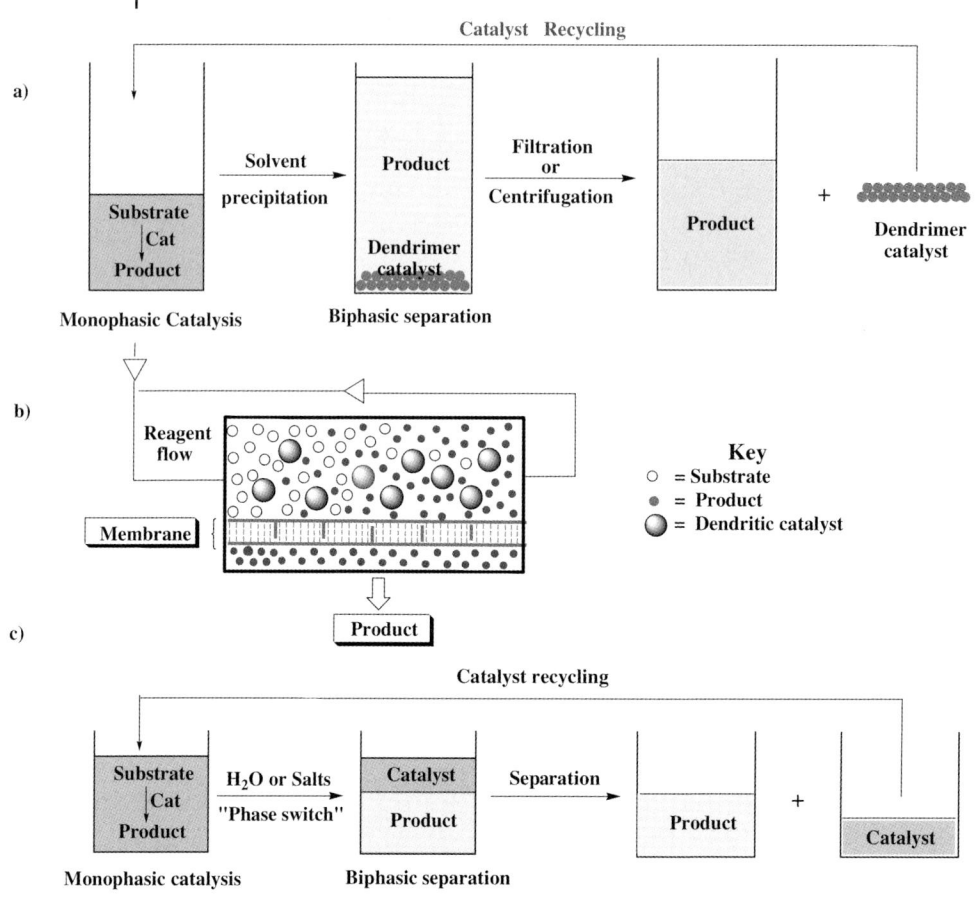

Figure 4.2 Schematic representation of dendrimer catalyst recycling via: (a) solvent precipitation; (b) membrane filtration; and (c) phase separation (latent biphasic system [24]).

4.2
Core-Functionalized Dendrimers in Asymmetric Catalysis

In the case of the core-functionalized dendrimers, it is expected that a steric shielding or blocking effect of the specific microenvironment created by the dendritic structure might modulate the catalytic behavior of the core [11, 26]. This site-isolation effects in dendrimer catalysts may be beneficial for some reactions, whereby the catalysts often suffer from deactivation caused by coordination saturation of the metal centers, or by the irreversible formation of an inactive metallic dimer under conventional homogenous reaction conditions. The encapsulation of such an organometallic catalyst into a dendrimer framework can specifically prevent the deactivation pathways and consequently enhance the stability and activity of the catalyst. The core-functionalized dendrimer catalysts may also benefit from the local environment (polarity difference) created by the dendritic wedges, which differs from the bulk solution. Also, there is a potential for shape- and size-selective catalysis when the catalytic site is at the core [27]. In addition, for the core-functionalized systems the solubility of the dendrimer catalysts can be tuned by changing the surface groups, an advantage which allows the recovery of dendrimer catalysts very easily via either solvent precipitation or phase separation [10, 25]. In contrast, the core-functionalized dendrimer catalysts may result in a low catalytic activity, particularly for the higher-generation catalysts. This negative dendrimer effect might be attributed to the steric shielding of the reactive center by the bulky dendritic wedges. Furthermore, the placing of a single catalytic group at the core of a large dendrimer results in a catalytic system with low catalyst loading.

Brunner's concept (dendrizyme) of attaching dendritic chiral wedges to a catalytically active achiral metal complex represents the first example of asymmetric catalysis using a core-functionalized dendrimer catalyst [21]. In view of the extremely poor asymmetric induction effected by the chiral dendritic structure, the bulk of the attention has been focused on the immobilization of the well-established chiral ligands and/or their metal complexes into an achiral dendrimer core. The important early examples included TADDOL-centered chiral dendrimers, which were reported by Seebach *et al.* in 1999 [28]. In this section, we attempt to summarize the recently reported chiral core-functionalized dendrimers with special emphasis on their applications in asymmetric synthesis.

4.2.1
Asymmetric Hydrogenation

Previously, transition metal-catalyzed asymmetric hydrogenation has been established as one of the most versatile and powerful tools for the preparation of a wide range of enantiomerically enriched compounds in organic synthesis. Thus, high activities and enantioselectivities have been observed using Rh, Ru and Ir complexes with chiral phosphine, phosphite or phosphoramidite ligands [29–31]. However, the high cost of both these chiral ligands and the noble metals, as well as the toxicity of trace metal contaminants in organic products, have often limited their use in industry. Therefore, the immobilization of these catalysts offers an attractive solution to these problems [23]. Among the reported supported catalysts, the chiral phosphorus-containing dendrimers, in particular, have attracted much recent attention due not only to their high catalytic activity and enantioselectivity but also their recyclability.

Recently, four generations of chiral diphosphine BINAP-centered dendrimers were synthesized by Fan and coworkers via the condensation of Fréchet-type dendrons with *(R)*-5,5′-diamino-BINAP (Figure 4.3) [32, 33]. The first- to third-generation BINAP dendrimers were tested in the Ru-catalyzed asymmetric hydrogenation of 2-[*p*-(2-methylpropyl)phenyl]-acrylic acid in methanol:toluene (1:1, v/v) at room temperature (Scheme 4.2) [32]. The size of the dendritic wedges was

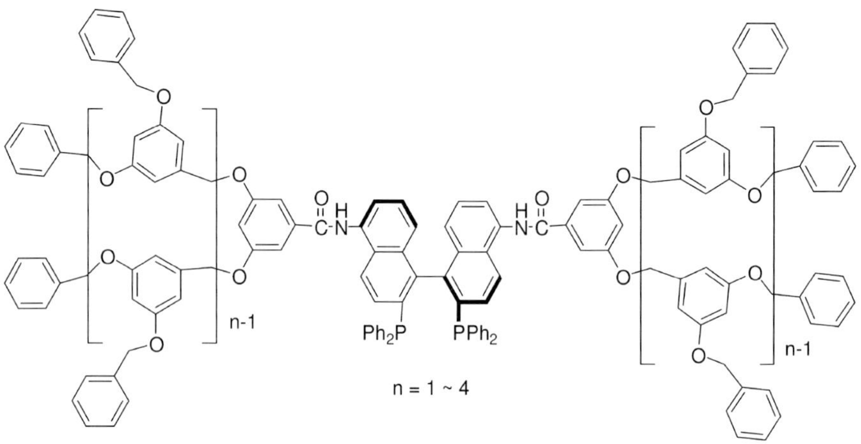
n-1
Ph2P
PPh2
n = 1 ~ 4
n-1

Figure 4.3 Chiral BINAP-cored dendrimers.

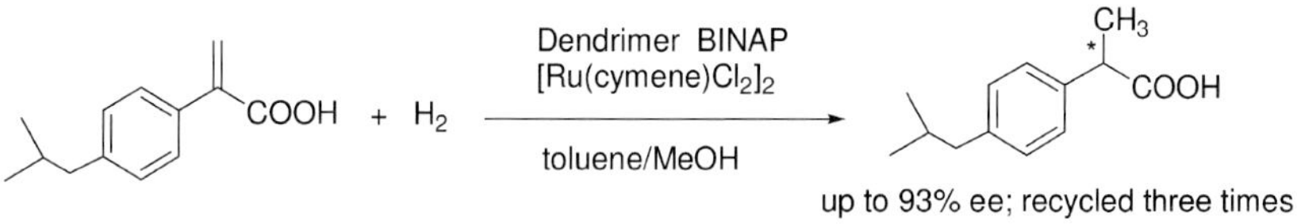
COOH
+ H2
Dendrimer BINAP
[Ru(cymene)Cl2]2
toluene/MeOH
CH3
COOH
up to 93% ee; recycled three times

Scheme 4.2

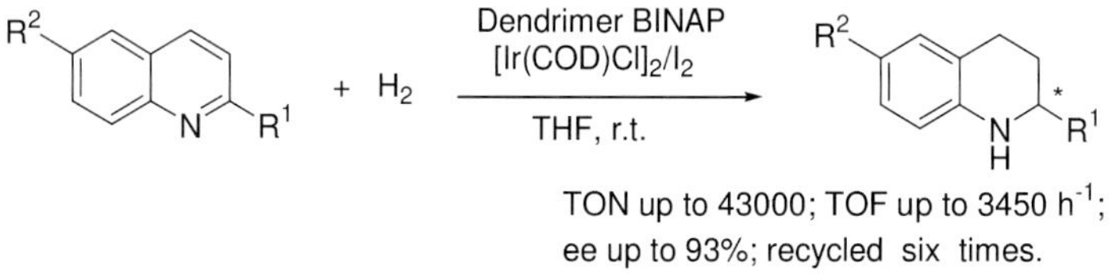
R2
R1
+ H2
Dendrimer BINAP
[Ir(COD)Cl]2/I2
THF, r.t.
R2
R1
TON up to 43000; TOF up to 3450 h-1;
ee up to 93%; recycled six times.

Scheme 4.3

found to clearly influence the reactivity of the ruthenium catalysts but, unexpectedly, the rate of the reaction increased with the higher-generation catalysts; this effect was most pronounced when moving from second- to third-generation catalysts. The catalyst containing the third-generation dendrimer ligand was recovered quantitatively by precipitation with methanol, and the catalyst was subsequently reused at least three times with similar activities and enantioselectivities.

Most recently, these BINAP-cored dendrimers were further employed in the Ir-catalyzed asymmetric hydrogenation of quinolines by Fan *et al.* (Scheme 4.3) [33]. Unlike the asymmetric hydrogenation of prochiral olefins, ketones and imines, the hydrogenation of heteroaromatic compounds proved to be rather difficult [34–37]. All four generations of dendrimer catalysts generated *in situ* from BINAP-cored dendrimers and $[Ir(COD)Cl]_2$ were found to be effective, even at an extremely high substrate : catalyst ratio in the asymmetric hydrogenation of quinaldine with

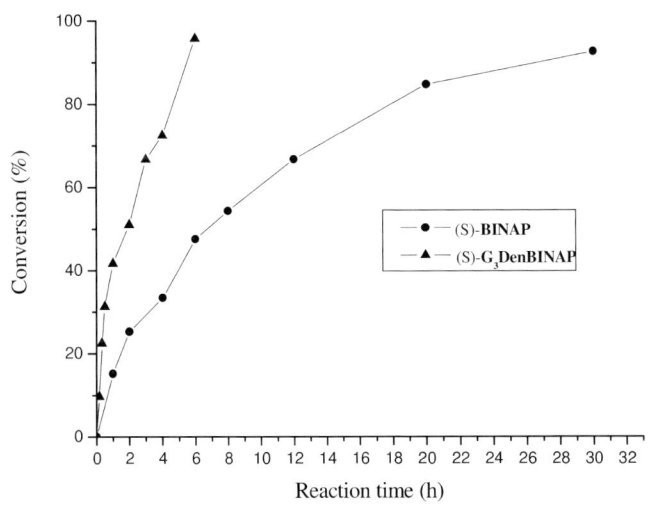

Figure 4.4 Time courses of the Ir-catalyzed hydrogenation of quinaldine using the third-generation dendrimer and BINAP.

I_2 as an additive. For example, in the presence of 0.01 mol% of the second-generation catalyst, the reaction proceeded smoothly giving 88% ee and more than 95% conversion in 20 h. Notably, the reaction performed well under a rather low catalyst loading in a large-scale reaction (~18 g substrate was used), giving a turn-over number (TON) of 43 000, which is the highest TON reported to date for such reaction [37]. Compared to the *(S)*-BINAP system, which demonstrated a turnover frequency (TOF) of 430 h^{-1}, with 71% ee, higher activities and enantioselectivities (89–90% ee) were achieved for the four generations of dendrimer catalysts. Interestingly, the catalytic activity was seen gradually to increase in line with the dendrimer generation number, with the TOF values for the first- to fourth-generation catalysts being 1000, 1500, 1580 and >1900 h^{-1}, respectively. On occasion, the maximum initial TOF from the third-generation catalyst reached 3450 h^{-1}, which represented the highest TOF obtained to date for the asymmetric hydrogenation of quinolines [37]. This rate enhancement of dendrimer catalysts was further demonstrated by the time–conversion curves (Figure 4.4). The isolation effect of the steric dendritic shell might reduce the formation of an inactive iridium dimer during the catalytic reaction [38] and thus enhance the stability and activity of the catalyst. In addition, the third-generation dendrimer catalyst could be quantitatively recovered by precipitation with methanol, and reused at least six times with similar enantioselectivities, but at the expense of relatively low catalytic activities.

The ternary catalyst system of Ru–chiral diphosphine–chiral diamine–KOH developed by Noyori and coworkers has proved to be effective in the asymmetric hydrogenation of simple prochiral ketones without a secondary coordinating functional group [39]. Recently, chiral BINAP- and DPEN-cored dendrimer ligands were used in the above-described Noyori's catalyst system by the groups of Fan [40] and Deng [41], respectively (Figure 4.5). Both catalytic systems displayed high

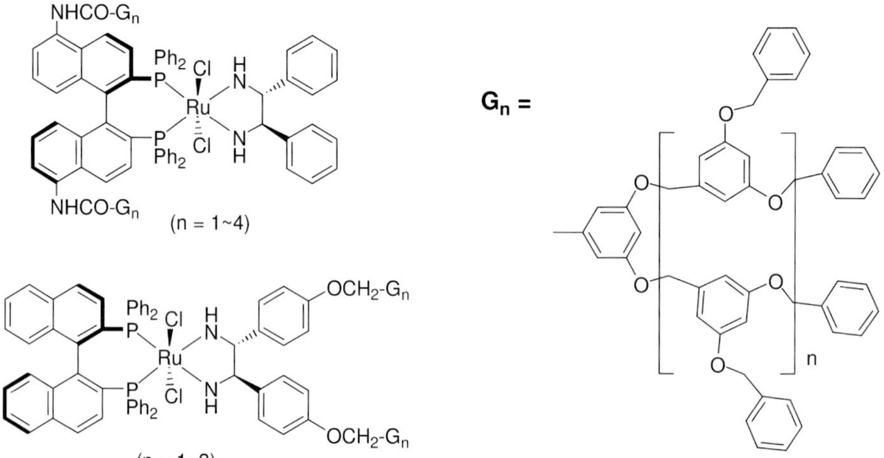

Figure 4.5 Chiral dendritic ruthenium catalysts containing BINAP- and DPEN-cored dendrimer ligands.

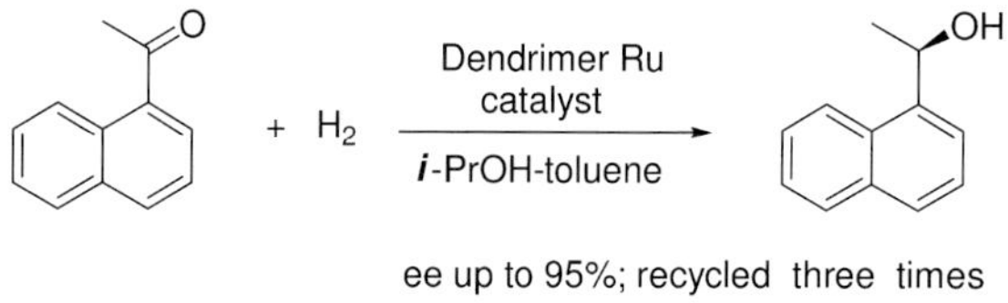

Scheme 4.4

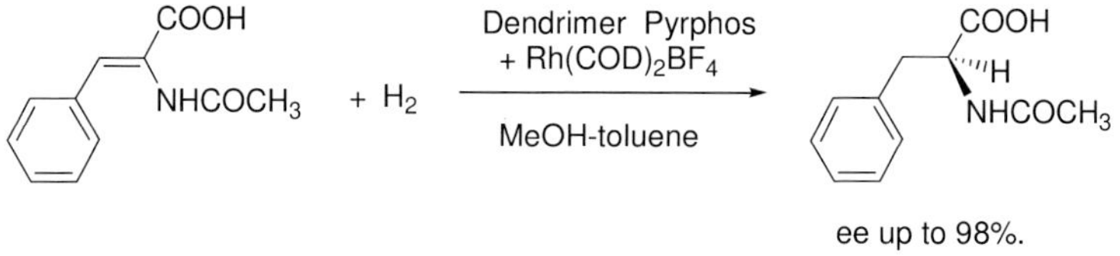

ee up to 95%; recycled three times

Scheme 4.5

ee up to 98%.

catalytic activity and enantioselectivity and allowed facile catalyst recycling. In the case of 1-acetonaphthone (Scheme 4.4), high enantioselectivities (up to 95% ee) were observed for both systems, which was comparable to that obtained with Noyori's homogeneous catalyst under the similar conditions, and higher than that for the heterogeneous poly(BINAP)–Ru catalyst reported by Pu et al. [42].

In 2004, Fan and coworkers synthesized a series of chiral dendrimer ligands bearing a chiral diphosphine Pyrphos at the focal point of the Fréchet-type polyether dendrons (Figure 4.6) [43]. The relationship between the primary structure of the dendrimer and its catalytic properties was established in the Rh-catalyzed hydrogenation of α-acetamidocinnamic acid (Scheme 4.5). High enantioselectivi-

Figure 4.6 Chiral dendrimer diphosphine ligands bearing Pyrphos at the focal point of the Fréchet-type dendrimer.

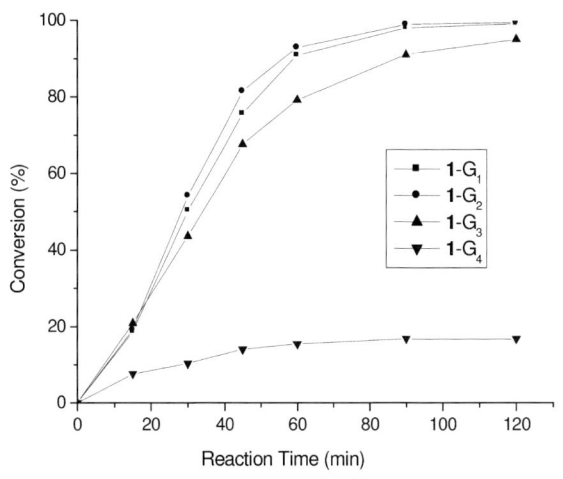

Figure 4.7 Time courses of the Rh-catalyzed hydrogenation of α-acetamidocinnamic acid for different generation dendrimer catalysts.

ties (up to 98%) were achieved for the first- to fourth-generation dendrimer catalysts, these being higher than those obtained when using the soluble, polymer-supported catalyst (96% ee) reported previously by the same group [44]. In contrast to the dendrimer Ir(BINAP) catalysts described above, the rate of reaction decreased when the higher-generation catalysts were used (Figure 4.7). In fact, while moving from the third to the fourth generation the dendrimer had almost totally lost its catalytic activity. In general, a core-functionalized single-site dendrimer catalyst often shows a gradual decrease in reactivity in line with increasing dendrimer generation, due to the increasing steric shielding. The sudden loss of activity for the fourth-generation dendrimer catalyst might be attributed to the change in dendrimer conformation from a loose, extended structure to a more rigid globular

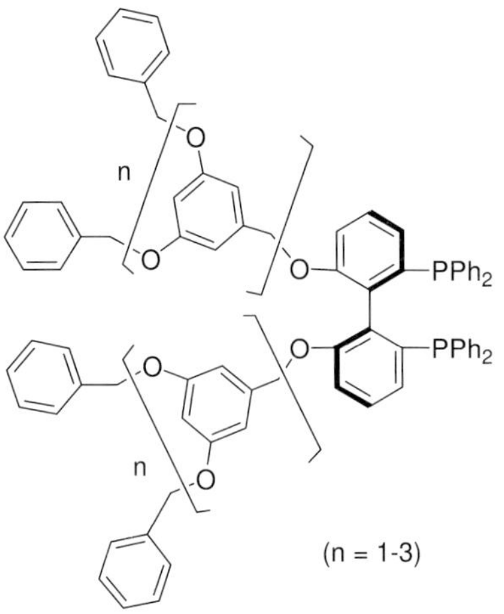

Figure 4.8 Chiral BIPHEP-cored dendrimers.

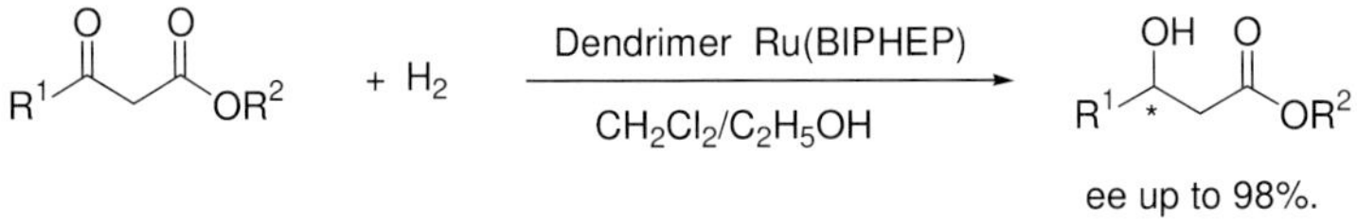

Scheme 4.6

structure. This may in turn have resulted in an encapsulation of the active center by the dendrimer, and influenced the diffusion of substrates into the catalytically active core of the dendrimer. Hence, the asymmetric hydrogenation reaction may also be used for probing the dendrimer microstructure and overall shape [45].

The group of Fan also reported the synthesis of chiral BIPHEP-cored dendrimer ligands by replacing the methoxyl substituents at the 6,6'-positions of the biphenyl backbone with different generations of Fréchet-type dendrons (Figure 4.8) [46]. These dendrimer ligands, which have tunable dihedral angles, were employed in the Ru-catalyzed asymmetric hydrogenation of β-ketoesters (Scheme 4.6). All of the Ru catalysts exhibited excellent catalytic activities, but the enantioselectivities changed dramatically. For example, methyl 3-oxo-3-phenylpropanoate was reduced with 93% ee by using the zeroth-generation dendrimer catalyst, while enantioselectivity was decreased to 92% with the first-generation catalyst and reached a minimum of 87% ee with the second-generation catalyst. Upon moving from the second to third generation, the enantioselectivity was increased slightly to 91% ee. This profound size effect is most likely due to the steric bulk of the dendritic wedges, which were expected to increase the dihedral angle of the two phenyl rings in the Ru–BIPHEP complex and thereby influence the selectivity of the catalysts.

After being neglected for about 30 years, monodentate phosphorus ligands [47] have recently begun to attract increasing amounts of attention. Chiral monoden-

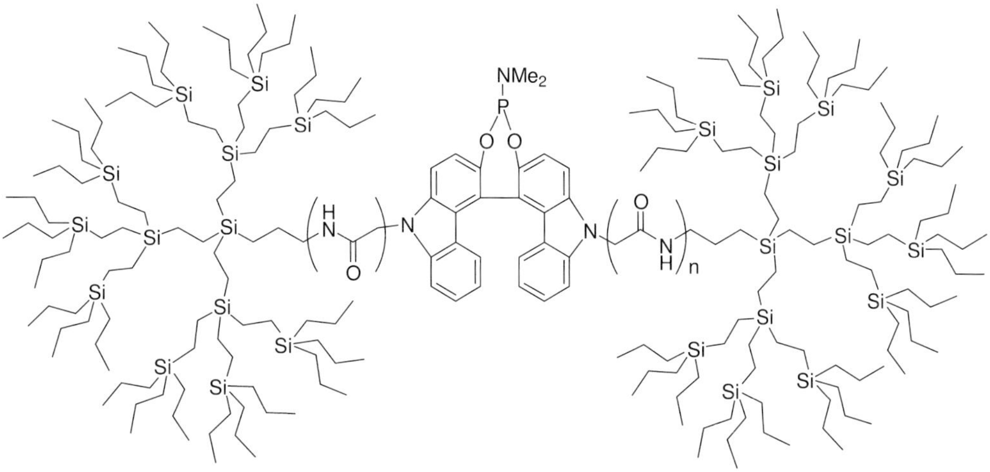

Figure 4.9 Chiral BICOL-type phosphoramide-cored dendrimers.

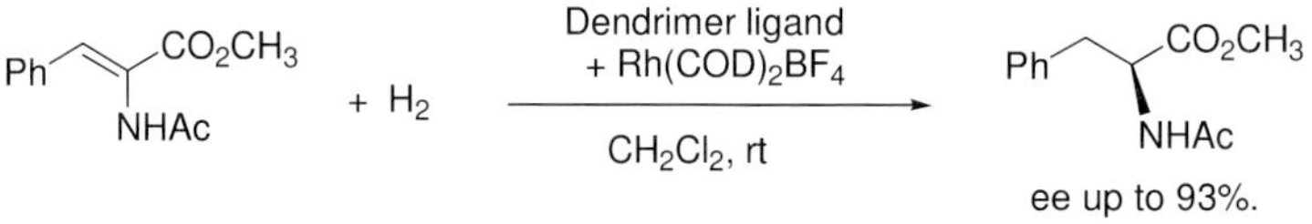

Scheme 4.7

tate phosphonite, phosphoramidite or phosphite have proven capable of inducing excellent enantioselectivity in rhodium-catalyzed asymmetric hydrogenation reactions, which was comparable to or better than those obtained with diphosphine ligands [48–51]. Recently, Reek et al. reported the first example of dendrimer monodentate phosphoramide ligands. The axially chiral BICOL (9H, 9H′-[4, 4′]bicarbazole-3,3′-diol) backbone was functionalized with two third-generation carbosilane dendritic wedges (Figure 4.9) [52]. These dendrimer catalysts showed high enantioselectivities in the rhodium-catalyzed asymmetric hydrogenation of methyl 2-acetamidocinnamate (Scheme 4.7). When a ligand:rhodium ratio of 2.2 was used, the enantiomeric excess was 93% with complete conversion, which was comparable to results obtained with the corresponding parent catalyst. Most importantly, when the ligand:Rh ratio was increased to 3.2, no loss in either reactivity or enantioselectivity was observed, which was in direct contrast to the results obtained with the parent MonoPhos [53]. The authors suggested that the steric dendritic bulk of the ligand had suppressed the formation of other unwanted rhodium species during the hydrogenation.

The easy synthesis and modular structure of monodentate phosphoramidites in general allow for fine-tuning of the ligand to produce highly effective catalysts. Initially, replacement of the alkyl group(s) on the amidite was the most frequently used strategy to tune the catalytic efficiency [50], but very recently Fan et al. developed a new class of dendritic monodentate phosphoramide ligands through

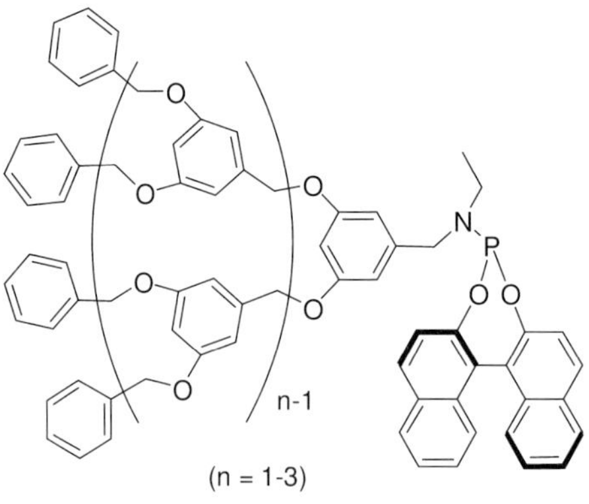

Figure 4.10 Chiral MonoPhos-cored dendrimers.

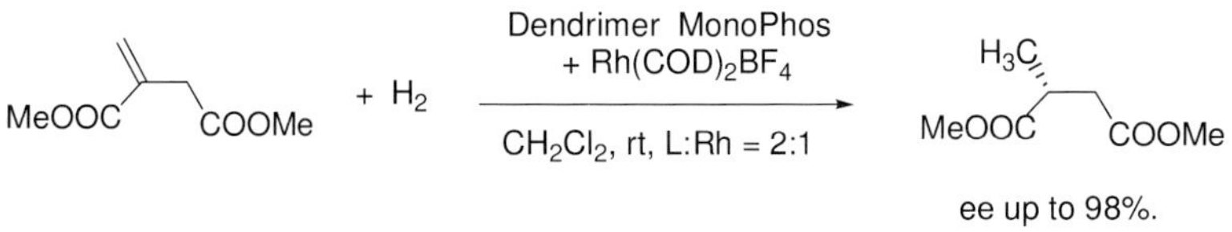

Scheme 4.8

substitution on the dimethylamino moiety in MonoPhos by the Fréchet-type dendritic wedges (Figure 4.10) [54]. In the Rh-catalyzed asymmetric hydrogenation of α-dehydroamino acid esters and dimethyl itaconate (Scheme 4.8), high enantioselectivities (up to 98%) and catalytic activities (TOF up to 4850 h^{-1}) were obtained, which were better than (or comparable with) those obtained when using the Rh–MonoPhos catalyst. Notably, the steric shielding of the dendritic wedges tended to stabilize the rhodium complex against decomposition caused by its hydrolysis in protic solvents. For example, when CH$_2$Cl$_2$/methanol (1:1, v/v) was used as the solvent, the second-generation dendrimer catalyst gave the reduced product with 96% ee, whereas the Rh–MonoPhos only gave 88% ee under similar conditions. High enantioselectivities and activities were also achieved under a higher ligand:rhodium ratio.

Dendronized polymers are also commonly used for the construction of dendrimer-supported catalysts, which are macromolecules with dendritic side chains attached to the polymeric cores [55]. Unlike dendrimers, these rod-shaped and nanosized polymers can be more easily synthesized. The immobilization of achiral catalytically active centers onto the polymeric core was originally demonstrated by Fréchet et al. [56]. The first example of the Ru-catalyzed asymmetric hydrogenation of prochiral olefins and ketones using dendronized polymeric chiral BINAP ligands (Figure 4.11) was recently reported by Fan et al. [57], whose Ru complexes were used as catalysts for the asymmetric hydrogenation of simple aryl ketones and 2-arylacrylic acids. These dendronized poly(Ru–BINAP) catalysts exhibited high catalytic activities and enantioselectivities, which were similar to

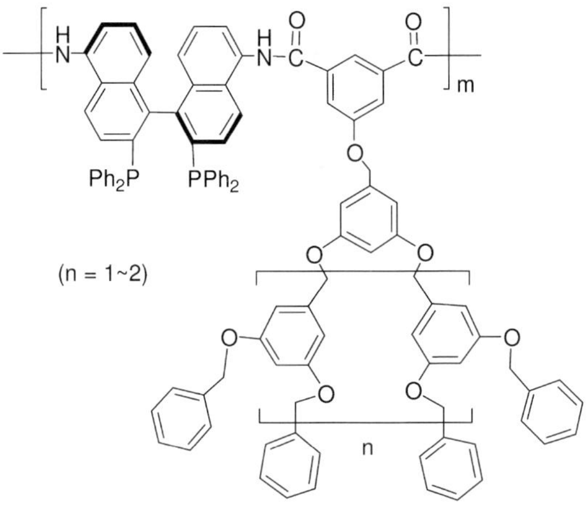

Figure 4.11 Chiral dendronized poly(BINAP)s.

those obtained with the corresponding parent Ru(BINAP) and the Ru(BINAP)-cored dendrimers [40]. An important feature of the dendronized polymeric catalysts was the easy and reliable separation of the chiral catalyst based on its rather large molecular size and different solubilities in various organic solvents. The third-generation catalyst could be quantitatively recovered by solvent precipitation and the recovered catalyst was reused at least three times in the asymmetric hydrogenation of 2-methylactophenone. The enantioselectivities not only remained similar under repeat use but were also better than those obtained with the Ru(BINAP)-cored dendrimers under similar conditions [40].

Although it has been well established that the core-functionalized dendrimer catalysts may be easily recovered by using solvent precipitation, this method requires the use of a large amount of organic solvent for catalyst precipitation. Recently, Fan and coworkers developed a novel system consisting of BINAP-cored dendrimers functionalized with multiple alkyl chains at the periphery (Figure 4.12) for asymmetric hydrogenation using a mixture of ethanol/hexane as solvent [25]. This binary solvent system provided complete miscibility over a broad range of reaction temperatures, and this enabled the catalytic reaction to be carried out in a homogeneous manner. Upon completion of the reaction, the addition of a small amount of water induced complete phase separation and provided facile catalyst recycling (the so-called 'latent biphasic system' [24], as shown in Figure 4.13). These highly apolar first- and second-generation dendrimer BINAP ligands were synthesized by the condensation of 5,5′-diamino-BINAP with Fréchet-type wedges containing long-chain alkyl tails on the periphery. Their ruthenium complexes were employed in the homogeneous hydrogenation of relatively polar substrates (2-phenylacrylic acid and 2-[p-(2-methylpropyl) phenyl]acrylic acid) in a 1:1 (v/v) ethanol:hexane mixture (Scheme 4.9). These dendrimer catalysts exhibited excellent reactivities and enantioselectivities which were identical to those obtained with BINAP as the ligand.

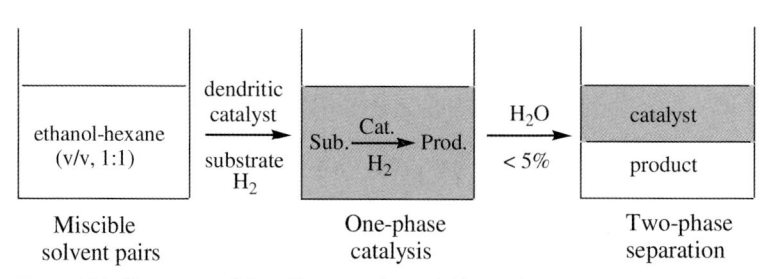

Figure 4.12 Chiral BINAP-cored dendrimers with core/shell structures.

Figure 4.13 Illustration of the effective and recyclable catalyst system.

Scheme 4.9

R = *t*Bu; R = H

Dendrimer BINAP [Ru(cymene)Cl$_2$]$_2$
ethanol/hexane (1:1, v/v)

up to 91% ee; recycled 4 times.

Recycling experiments were performed using 2-[p-(2-methylpropyl)phenyl]acrylic acid as a substrate. Upon the completion of the reaction, a small amount of water (2.5%) was added to the homogeneous reaction solution to accomplish complete phase separation. More than 99% of the second-generation catalyst was separated and extracted to the nonpolar hexane phase, whereas the polar product mainly stayed in the ethanol phase. The catalyst was reused at least four times, and showed similar reactivities and enantioselectivities on each occasion.

4.2.2
Asymmetric Transfer Hydrogenation

The asymmetric reduction of ketones via hydrogen transfer can be regarded as an alternative for the hydrogenation of ketones in the preparation of chiral secondary alcohols. Asymmetric transfer hydrogenation is a very attractive procedure, particularly in small-scale reactions, for avoiding the inconvenient handling of hydrogen, which sometimes is under high pressure. In 1996, Noyori et al. discovered an excellent ligand, (S,S)-TsDPEN [(S,S)-N-(p-tolylsulfonyl)-1,2-diphenylethylenediamine], for the ruthenium-catalyzed transfer hydrogenation reactions [58]. In order to improve the handling and separation of the expensive catalysts from the reduced products, a number of approaches for the immobilization of the catalyst have been reported [59].

Recently, Deng and coworkers designed and synthesized a series of chiral dendritic TsDPEN-type ligands by attaching the monomeric TsDPEN onto the focal point of the Fréchet-type dendrons (Figure 4.14) [60]. The ruthenium complexes up to the fourth generation were prepared *in situ* and employed in the asymmetric transfer hydrogenation of acetophenone (Scheme 4.10). Compared to the monomer catalyst, slightly enhanced reactivities were observed for the dendrimer catalysts with high enantioselectivities (up to 97% ee). The fourth-generation catalyst was quantitatively recovered by solvent precipitation and reused at least four times with very similar enantioselectivities and activities.

Very recently, Chen and Deng reported the synthesis of the chiral dendrimer 1,2-diaminocyclohexane (DACH) by using a very similar synthetic approach

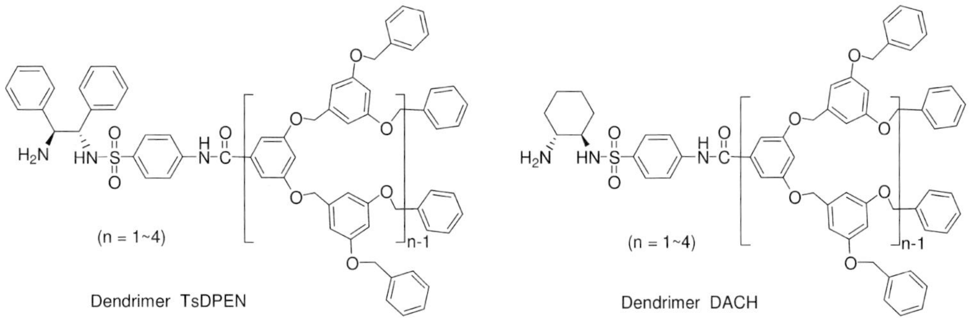

Figure 4.14 Chiral diamine ligands attached to Fréchet-type dendrons.

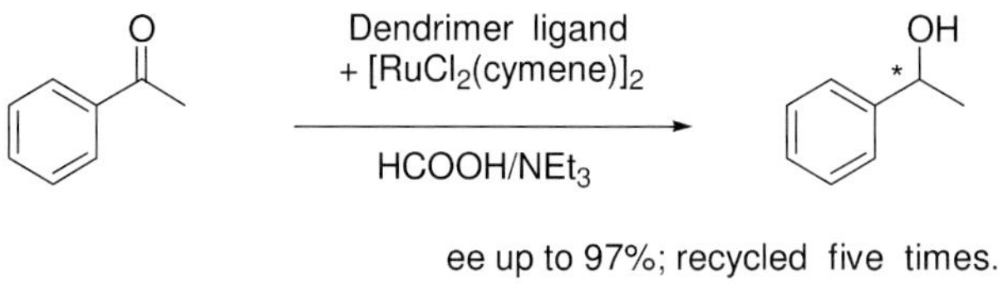

Scheme 4.10

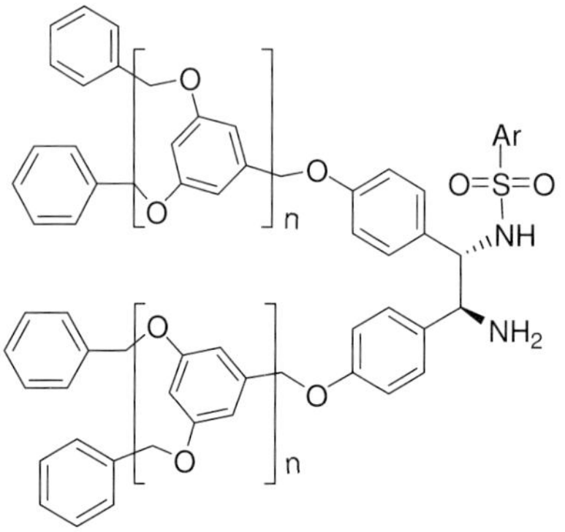

ee up to 97%; recycled five times.

Figure 4.15 Chiral tunable dendritic diamine ligands.

Ar = 4-CH$_3$C$_6$H$_4$, n = 1~3

Ar = 2, 4, 6-Et$_3$-C$_6$H$_2$, n = 2

Ar = 2, 4, 6-^{i}Pr$_3$-C$_6$H$_2$, n = 2

Ar = 1-naphthyl, n = 2

(Figure 4.14) [61]. The ruthenium and rhodium complexes produced were tested in the transfer hydrogenation of ketones in organic or aqueous solutions, and a much better catalytic efficacy was achieved in an aqueous system using HCOONa as the hydrogen source. These highly hydrophobic dendrimer catalysts were well dissolved in the liquid substrates of the reaction mixture and exhibited high catalytic activities and enantioselectivities for a range of ketones. The catalyst loading could be decreased to 0.01 mol%, and good conversion still obtained with excellent enantioselectivity (95% ee). The second-generation dendrimer catalyst could be easily precipitated from the reaction mixture by adding hexane, and then reused five times with very similar enantioselectivities.

Another approach towards immobilizing the DPEN ligand was functionalization of the phenyl rings on the chiral diamine backbone, as exemplified in Figure 4.15 [62]. The resultant chiral DPEN-cored dendrimer diamines were expected to be fine-tuned by modifying one of the amino groups with different arylsulfonyls, leading to highly enantioselective catalysts. The ruthenium complexes containing the first- and second-generation dendrimers showed high catalytic and recyclable activities, with enantioselectivities comparable to those of Noyori's prototype catalyst in the asymmetric transfer hydrogenation of an extended range of substrates, such as ketones, keto esters and olefins. However, the third-generation catalyst gave a significantly decreased activity due to encapsulation of the catalytically active center by the dendritic wedges.

In 2003, two other series of chiral dendrimer ligands based on (1S,2R)-norephedrine were reported by Deng and Tu, which were used in ruthenium-

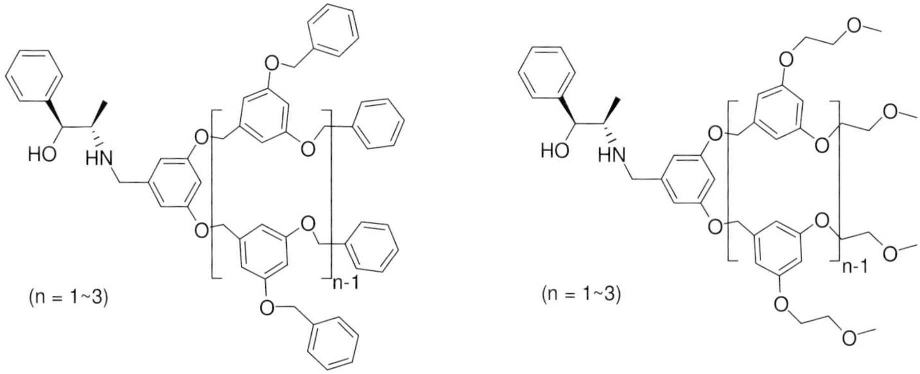

Figure 4.16 Chiral (1S,2R)-norephedrine ligands attached to Fréchet-type dendrons.

catalyzed asymmetric transfer hydrogenations (Figure 4.16) [63]. These dendrimer catalysts were found to be efficient for the catalytic reactions and, compared to the monomer catalyst, enhanced reactivities and near-identical enantioselectivities were observed for the low-generation dendrimer catalysts. However, the third generation of both types of dendrimer catalyst gave decreased reactivities and enantioselectivities due to their poor solubility in isopropanol.

4.2.3
Asymmetric Borane Reduction of Ketones

The enantioselective reduction of ketones also represents an important method of preparing chiral secondary alcohols, and a large number of reducing reagents and/or chiral catalysts have been developed in this respect. One of the most successful reactions is the oxazaborolidine-mediated borane reduction, in which chiral ligands such as chiral prolinol or ephedrine-type ligands have proved to be very effective [64, 65]. The first asymmetric borane reduction using chiral dendrimer catalyst was reported by Bolm et al. in 1999 [66], since then, several dendrimers bearing optically active amino alcohol located at the focal point have been synthesized for the asymmetric borane reduction of prochiral ketones, affording the corresponding secondary alcohols with good yields and excellent enantioselectivities.

Recently, Zhao et al. described the use of chiral dendritic amino alcohol ligands for the asymmetric borane reduction of ketones [67]. An optically active prolinol was attached to the focal point of the Fréchet-type polyether dendrons (Figure 4.17). Compared to the monomer catalyst (94% ee), a slightly higher enantioselectivity (97% ee) and a good yield were obtained using the third-generation catalyst with p-nitroacetophenone as the substrate (Scheme 4.11). The highest enantioselectivity (98% ee) in the reduction of 2-(phenylsulfonyl)acetophenone was observed when the third-generation catalyst was used.

Following up these investigations, the same group further employed the third-generation dendrimer prolinol as a catalyst in the asymmetric reduction of

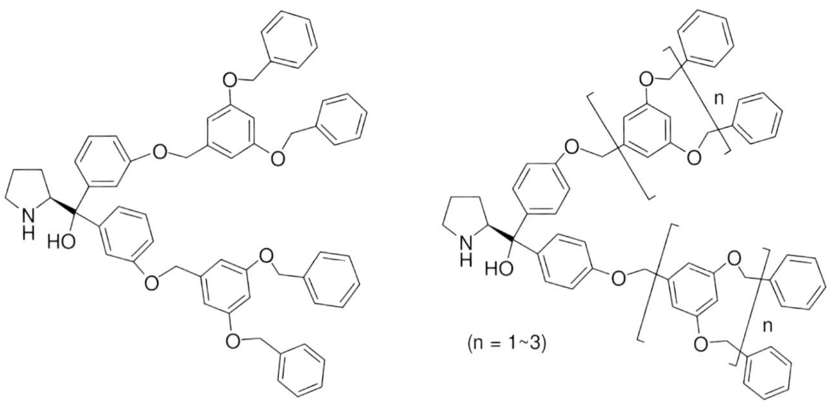

Figure 4.17 Chiral prolinol-cored dendrimers.

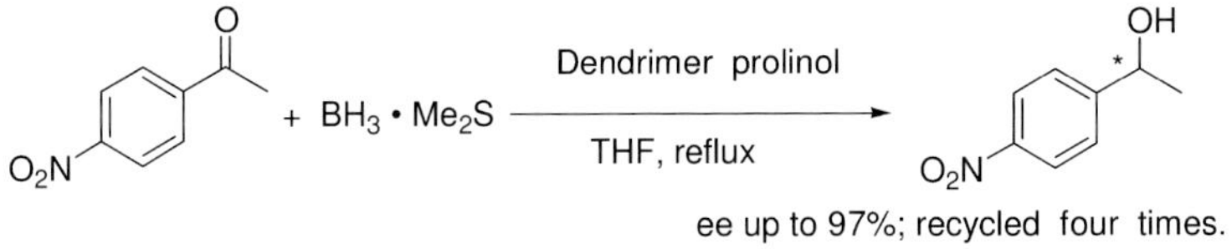

Scheme 4.11

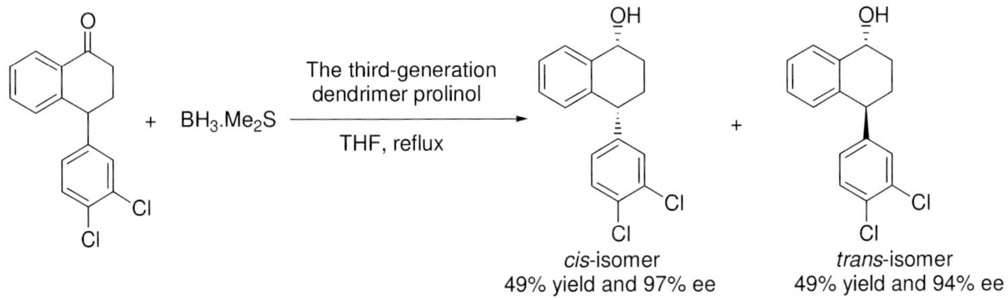

ee up to 97%; recycled four times.

Scheme 4.12

cis-isomer
49% yield and 97% ee

trans-isomer
49% yield and 94% ee

indanones and tetralones [68]. The obtained products are of current interest in the pharmaceutical industry and, therefore, their derivatives have been extensively evaluated for their biological activities. The dendrimer catalysts were found to be effective in the reduction of 2- and 3-substituted indanones, and 4-substituted tetralones with borane-dimethyl sulfide complex as the reducing reagent. In the case of the reduction of 3-substituted indanones, the *cis*- and *trans*-isomers were obtained in a 1:1 molar ratio, although the enantioselectivities for the *trans*-isomers (up to 96% ee) were higher than those of the *cis*-isomers (up to 82% ee). When the substrate was changed to substituted indanone, both isomers were obtained in excellent enantioselectivities (Scheme 4.12) and the dendrimer catalyst

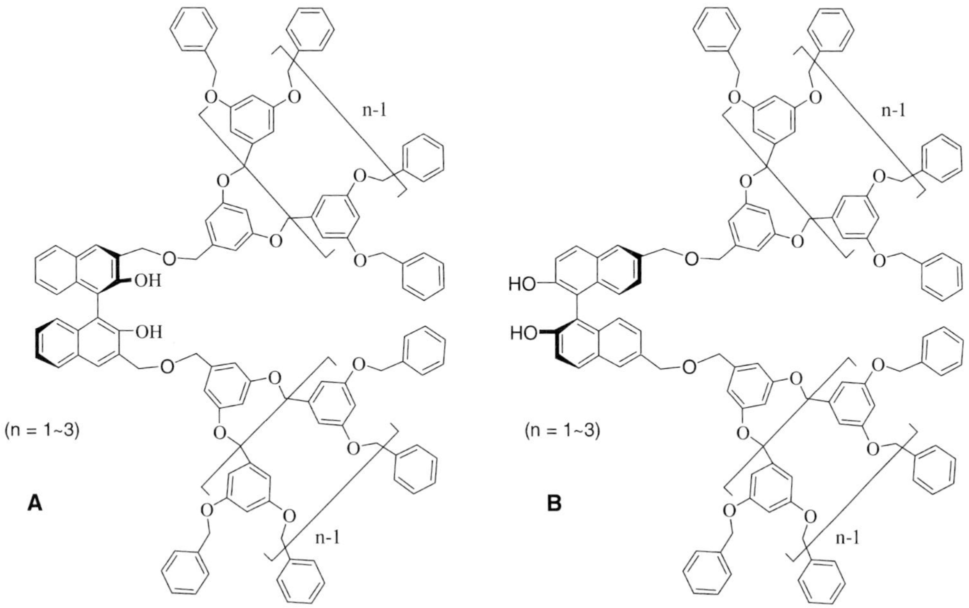

Figure 4.18 Chiral BINOL-cored dendrimers.

was quantitatively recovered through precipitation by the addition of methanol. The recovered catalyst could be reused at least five times, without any loss of catalytic efficiency.

4.2.4
Asymmetric Addition of Organometallic Compounds to Aldehydes

During the past two decades the homogeneous and heterogeneous catalytic enantioselective addition of organozinc compounds to aldehydes has attracted much attention because of its potential in the preparation of optically active secondary alcohols [69]. Chiral amino alcohols (such as prolinol) and titanium complexes of chiral diols (such as TADDOL and BINOL) have proved to be very effective chiral catalysts for such reactions. The important early examples included Bolm's flexible chiral pyridyl alcohol-cored dendrimers [70], Seebach's chiral TADDOL-cored Fréchet-type dendrimers [28], Yoshida's BINOL-cored Fréchet-type dendrimers [71] and Pu's structurally rigid and optically active BINOL-functionalized dendrimers [72]. All of these dendrimers were used successfully in the asymmetric addition of diethylzinc (or allyltributylstannane) to aldehydes.

Recently, Fan *et al.* reported a series of chiral BINOL-cored dendrimers via substitution at the 3,3′-positions of the binaphthyl backbone by different generations of Fréchet-type dendrons (Figure 4.18) [73]. The proximity of the dendritic wedges to the catalytic center is expected to induce catalytic properties different from BINOL. In the absence of Ti(OiPr)$_4$, the chiral dendrimer ligands showed much

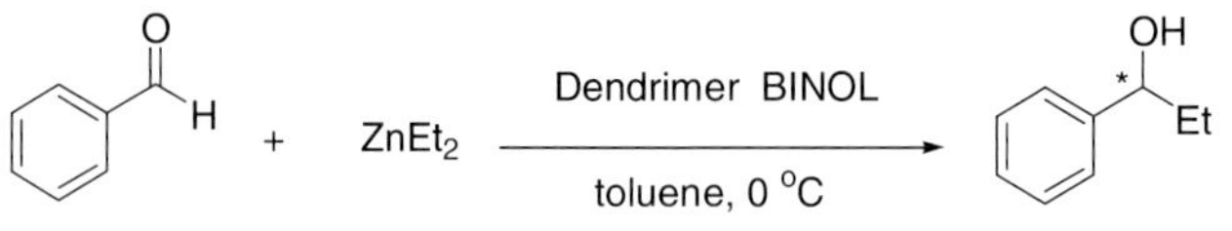

Dendrimer A without Ti(O*i*Pr)$_4$: up to 97.5% conversion and 61.9% ee;
Dendrimer B with Ti(O*i*Pr)$_4$: up to 99% conversion and 87% ee.

Scheme 4.13

higher catalytic activities and enantioselectivities than BINOL in the asymmetric addition of diethylzinc to benzaldehyde (Scheme 4.13). For example, in the presence of 20 mol% of the dendrimer catalyst in toluene, 97.5% conversion and 61.9% ee were observed in 7 h at 0 °C, while only 19.0% conversion and 4.6% ee were observed under similar conditions when BINOL was used. This dendrimer effect was most likely due to the existence of an oxygen atom on the linkages between the dendritic wedges and the binaphthyl backbone. The coordination of such oxygen atoms to zinc species might suppress the formation of intermolecular Zn–O–Zn bonds, which is considered to be catalytically inactive. Another advantage of the dendrimer over BINOL was its easy recovery and reuse in the subsequent catalytic cycles, with the same catalytic efficiency.

In order to further study the effect of the linking position and the generation of dendrimer supports on the catalyst properties, the same group synthesized another type of chiral BINOL derivatives bearing Fréchet-type polyether dendritic wedges at the 6,6′-positions of the binaphthyl backbone [74]. These dendrimer ligands were tested in the enantioselective addition of diethylzinc to benzaldehyde. High enantioselectivities (up to 87% ee) were observed in the presence of Ti(O*i*Pr)$_4$, which were higher than those obtained from the 3,3′-derived dendrimer BINOLs with or without Ti(O*i*Pr)$_4$. The effect of the position and size of the dendritic wedges on enantioselectivity was thought to depend on the following two factors: (i) the potentially coordinating ether linkage at the 3,3′-position on the binaphthyl backbone played an important role on the activity and enantioselectivity; and (ii) the bulk dendritic wedges attached to the 3,3′- or 6,6′-positions may affect the dihedral angle of the two naphthalene rings of the dendrimer BINOL and thus affect the activity and enantioselectivity.

The asymmetric aryl transfer onto aromatic aldehydes represents yet another method of obtaining chiral diarylmethanols, which are versatile and important chiral building blocks for the preparation of biologically important agents. Recently, Zhao and coworkers reported, for the first time, the use of chiral dendrimer amino alcohols as catalysts for the enantioselective aryl transfer reactions of aldehydes [75]. The most effective pyrrolidinylmethanol ligand was chosen as the model ligand and attached onto the Fréchet-type polyether dendrimer supports (Figure 4.19) via a convergent approach. When these dendrimer catalysts were tested in the phenyl transfer reactions of *p*-chlorobenzaldehyde using phenyl boronic acid as the phenyl source (Scheme 4.14), high reactivities (up to 98% yield) and enantioselectivities (up to 93% ee) were observed. The size/generation of the dendrimer had a slight influence on the enantioselectivity. When using triphenylcyclotri-

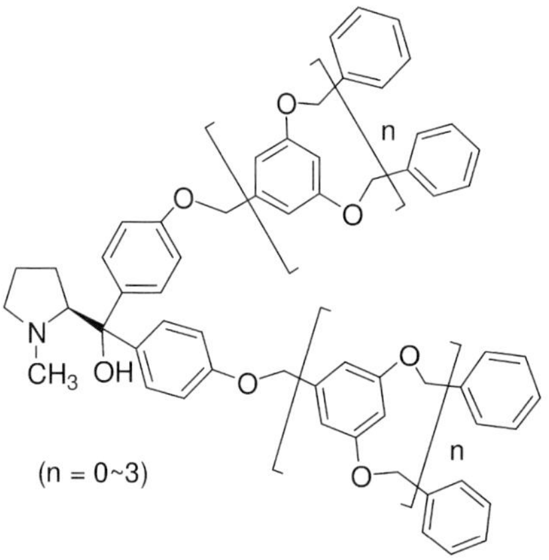

Figure 4.19 Chiral pyrrolidinylmethanol-cored dendrimers.

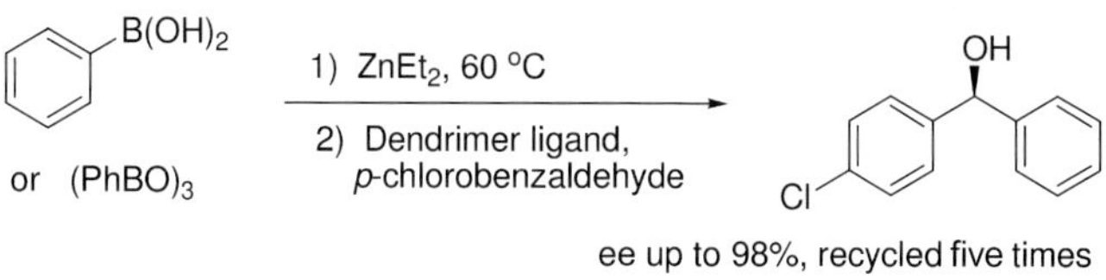

Scheme 4.14

boroxane (PhBO)$_3$ in place of phenyl boronic acid as the phenyl source, the dendrimer catalysts showed excellent enantioselectivities (up to 98% ee), which were higher than those obtained from the monomer catalyst. In addition, the second-generation dendrimer catalyst was easily separated by precipitation with methanol, recovered via filtration and reused at least five times, with very similar catalytic activities and enantioselectivities.

4.2.5
Asymmetric Michael Addition

The Michael addition reaction is commonly recognized as one of the most important carbon–carbon bond-forming reactions in organic synthesis, and major efforts have been made to develop efficient catalytic systems for this type of transformation. In particular, the Michael addition of a carbon nucleophile to nitroalkenes is a useful synthetic method for the preparation of nitroalkanes [76], which are versatile synthetic intermediates owing to the various possible easy transformations of the nitro group into other useful functional groups, such as amino groups and nitrile oxides.

Recently, Zhao et al. reported the use of 2-trimethylsilanyloxy-methyl-pyrrolidine functionalized chiral dendrimer catalysts (Figure 4.20) for catalytic enantio- and diastereoselective Michael addition of aldehydes to nitrostyrenes (Scheme 4.15)

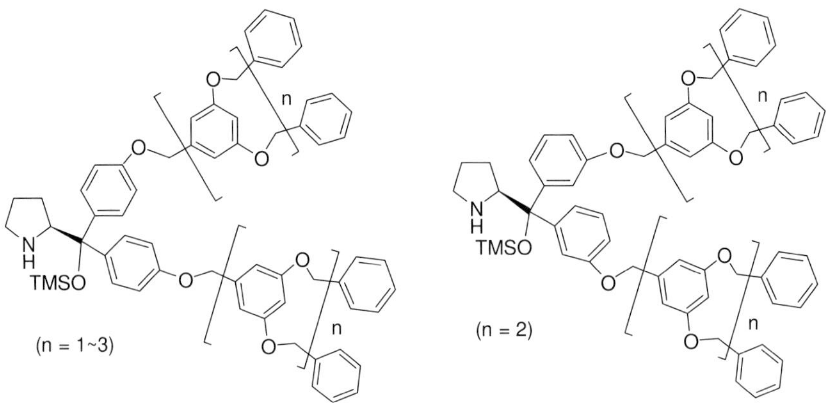

Figure 4.20 Chiral 2-trimethylsilanyloxy-methyl-pyrrolidine-based dendrimers.

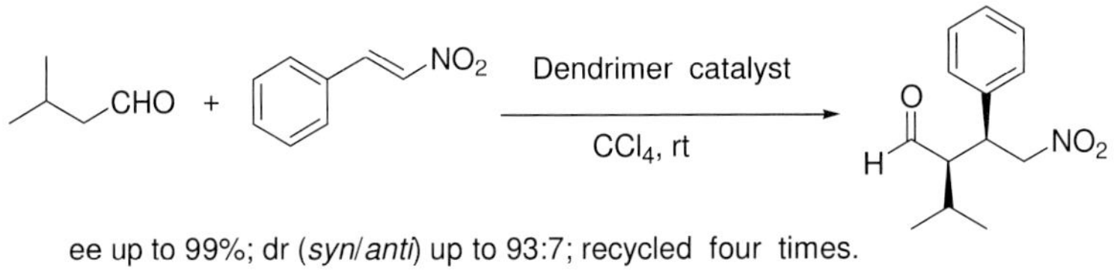

ee up to 99%; dr (*syn/anti*) up to 93:7; recycled four times.
Scheme 4.15

[77]. Excellent enantioselectivities for all of these dendrimers were observed in the Michael addition reactions of isovaleraldehyde with nitrostyrene. For example, in the presence of 10 mol% of the first-generation catalyst, the addition product 2-isopropyl-4-nitro-3-phenyl-butyraldehyde was isolated in 86% yield with 99% ee and moderate diastereoselectivity (diastereomeric ratio (dr) 80:20). Higher diastereoselectivities were observed when the *meta*-substituted dendrimer catalysts were used, but the yields and enantioselectivities were relatively low. In addition, the second-generation dendrimer catalyst could be easily recovered via precipitation with methanol and reused at least five times, with only a slight loss of catalytic activity.

4.2.6
Asymmetric Allylic Substitution

Allylic substitutions are among the most important carbon–carbon bond-forming reactions in organic synthesis. Palladium-catalyzed allylic substitutions and their asymmetric version have been extensively studied and widely used in a variety of total syntheses [78]. The palladium catalysis mostly requires soft nucleophiles such as malonate carbanions to achieve high stereo- and regioselectivity.

In 2002, Malmström and coworkers attached the first- to fourth-generation dendritic substituents based on 2,2-bis(hydroxymethyl)propinic acid and

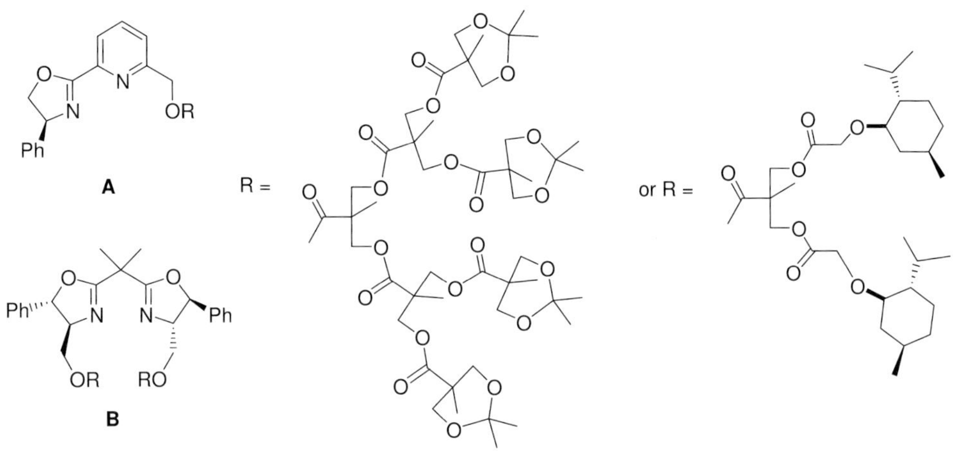

Figure 4.21 Chiral dendrimer oxazoline ligands.

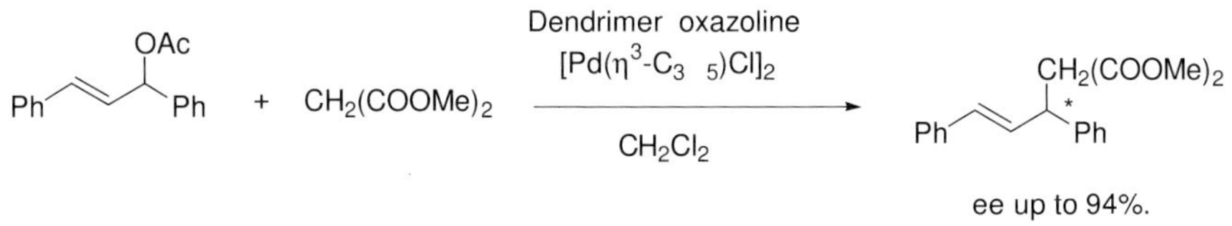

Scheme 4.16

(1R,2S,5R)-menthoxyacetic acid to 2-(hydroxymethyl)-pyridinooxazoline and bis-[4-(hydroxymethyl)oxazoline] compounds (Figure 4.21) [79]. These chiral dendrimers were tested in the enantioselective palladium-catalyzed allylic substitutions of rac-1,3-diphenyl-2-propenyl acetate with dimethyl malonate (Scheme 4.16). The first type of dendritic ligand (A in Figure 4.21) exhibited enantioselectivity similar to that of the corresponding benzoyl ester ligand, whereas the latter type of ligand (B in Figure 4.21) afforded products with higher selectivity than the corresponding monomeric benzoyl ester ligand. Although, the activity of the dendrimer catalysts decreased with increasing dendrimer generation, the introduction of chiral dendritic substituents to the chiral alcohol center had no effect on enantioselectivity for either type of ligand.

4.2.7
Asymmetric Aldol Reaction

The aldol reaction is one of the most efficient methods for extending the carbon framework of an organic synthon. Since the discovery of the Lewis acid-catalyzed asymmetric aldol reaction of silyl enol ethers by Mukaiyama, numerous variations of this type of reaction have been reported [80]. Recently, more attention has been focused on the development of new organocatalysts for the asymmetric direct aldol

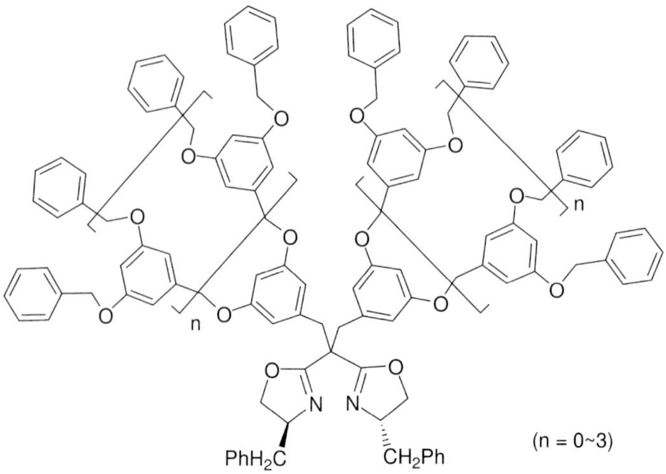

Figure 4.22 Chiral bis(oxazoline)-cored dendrimers.

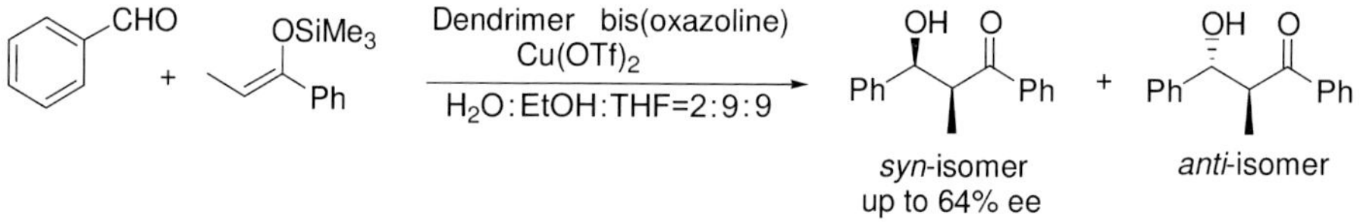

Scheme 4.17

reactions [81], while some supported chiral ligands and catalysts have also been developed.

In 2003, Fan et al. reported a series of chiral bis(oxazoline)-cored dendrimer ligands for enantioselective aldol reaction in aqueous media [82]. The dendritic substituents were introduced by double alkylation of the methylene bridge of chiral bis(oxazoline) with Fréchet-type dendritic benzyl bromides (Figure 4.22). The copper(II)-catalyzed asymmetric aldol reaction of a silyl enol ether with benzaldehyde was chosen as the model reaction to evaluate these chiral dendrimers in aqueous media (Scheme 4.17). Considering the insolubility of the dendrimer catalysts in alcohol and water, a mixed solvent system containing water, tetrahydrofuran (THF) and ethanol was used. These dendrimer catalysts exhibited good activities and moderate enantioselectivities, which were comparable to those obtained from the monomer catalyst. The second- and third-generation catalysts even gave slightly higher enantioselectivities and/or yields. The second-generation catalyst was easily separated by adding cold methanol, with the recovered catalyst showing relatively low yield and enantioselectivity.

In 2006, Zhao and coworkers reported the details of direct aldol reactions catalyzed by chiral dendrimer catalysts derived from N-prolylsulfonamide [83]. The first- to third-generation dendrimer catalysts were synthesized by the condensation of N-4-aminobenzenesulfonyl-L-Boc-proline with the Fréchet-type polyether dendritic wedges bearing a carboxyl group at the focal point (Figure 4.23), and their catalytic activities investigated by performing a model reaction of 4-nitrobenzaldehyde

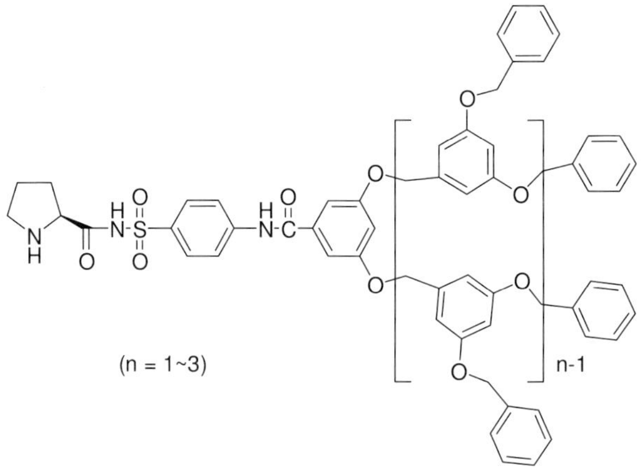

Figure 4.23 Chiral N-prolylsulfonamide-functionalized dendrimer organocatalysts.

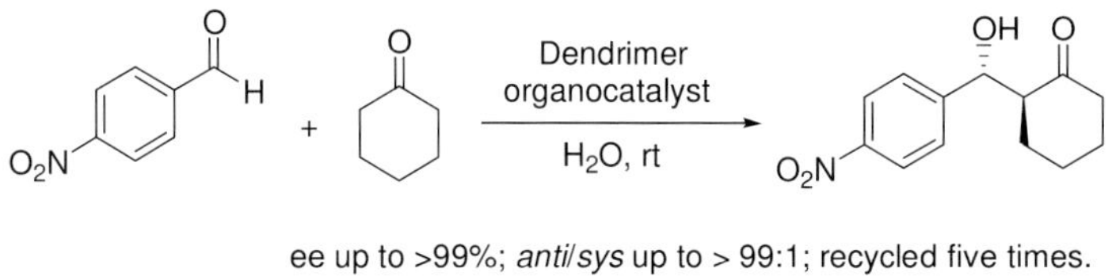

Scheme 4.18

ee up to >99%; anti/sys up to > 99:1; recycled five times.

with cyclohexanone (Scheme 4.18). The aldol reactions were found to proceed efficiently in water in the presence of these dendrimer catalysts, with the *anti* product being obtained as the major isomer in good to excellent yields (58–99%) with high diastereoselectivities and enantioselectivities. The highly hydrophobic dendrimer catalysts showed better reactivities and higher selectivities than the small model catalyst, and the second-generation catalyst exhibited the best performance (99% yield, *anti/syn* = 99:1 and 99% ee). In order to recover and reuse the dendrimer catalyst, several different solvent systems were tested and a mixed solvent of *n*-hexane/EtOAc (1:1, v/v) was found to be the best choice. The second-generation catalyst was quantitatively recovered by solvent precipitation, and reused five times without any loss of reactivity and enantioselectivity.

4.2.8
Asymmetric Hetero-Diels–Alder Reaction

During recent years, the homogeneous Lewis acid-catalyzed asymmetric Diels–Alder reactions and hetero-Diels–Alder (HDA) reactions have each undergone extensive study. Various chiral Lewis acids including aluminum, titanium or boron, and chiral ligands such as chiral amino alcohols, diols, salen, bisoxazoline or *N*-sulfonylamino acids have been used as the catalysts [84]. Much efforts have also been made in the investigation of heterogeneous diastereoselective Diels–Alder reactions.

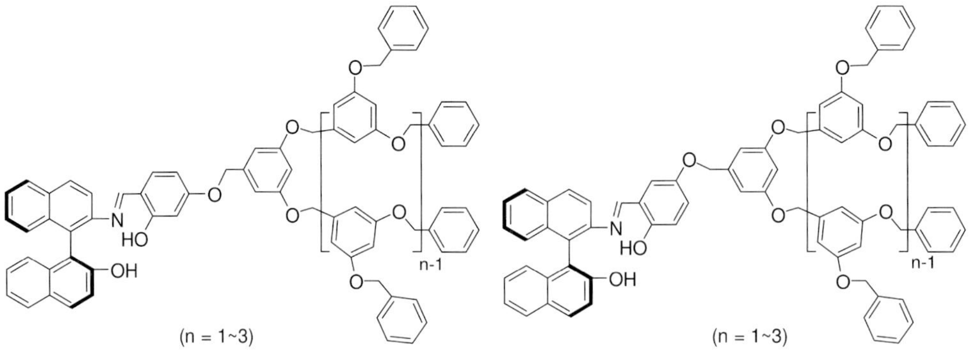

Figure 4.24 Chiral NOBIN-cored dendrimers.

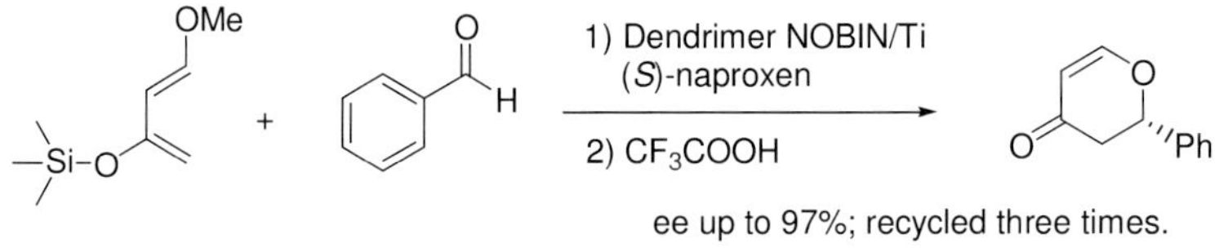

Scheme 4.19

Recently, a series of dendritic 2-amino-2′-hydroxy-1,1′-binaphthyl (NOBIN) ligands (Figure 4.24) were synthesized by Ding et al. and applied to the titanium-catalyzed HDA reactions of 1-methoxy-3-(trimethylsilyloxy) buta-1,3-diene (Danishefsky's diene) with aldehydes (Scheme 4.19) [85]. The products, 2-substituted 2,3-dihydro-4H-pyran-4-ones, are a class of heterocycles with extensive synthetic applications in both natural and non-natural products. Excellent enantioselectivities (up to 97% ee) and catalytic activities (>99% yield) were found for both types of the dendrimer catalysts. An enhanced enantioselectivity could also be achieved by optimizing the disposition and size of the dendritic wedges in the ligands. In addition, the dendrimer catalyst could be quantitatively recovered by solvent precipitation, and reused without any further addition of the Ti source or a carboxylic acid additive for at least three cycles, with similar activities and enantioselectivities. This good catalyst reusability was most likely due to the stabilization of the titanium active species by steric hindrance of the dendritic wedges. A higher degree of asymmetric amplification in the catalytic HDA reactions was also observed with the chiral dendrimer catalysts as compared to the monomer catalyst.

4.3
Peripherally Modified Dendrimers in Asymmetric Catalysis

In contrast to the core-functionalized systems, periphery-functionalized dendrimers have their catalytically active groups located at the surface of the dendrimer supports, and these active sites are therefore directly available to substrate. Furthermore, the 'proximity effect' between the peripheral catalytic sites can enhance the catalytic activity by multiple complexation or even cooperation ('positive' dendrimer effect). In contrast, in some cases a 'negative' dendrimer effect was observed which was caused by an undesired interaction between the neighboring peripheral catalytic centers (a bimetallic deactivation mechanism).

The first example of a catalytically active metallodendrimer, having catalytic groups at the periphery, was reported by van Koten, van Leeuwen and coworkers [20]. These authors prepared the nickel(II) complexes containing carbosilane dendrimers, which were successfully employed in the homogeneous regioselective Kharasch addition of polyhalogenoalkanes to the terminal C=C double bonds. Since these early studies there has been a steadily increasing number of dendrimer catalysts which have been synthesized and studied [15]. In this section, the details of peripherally modified chiral dendrimer catalysts for different asymmetric catalytic reactions will be summarized.

4.3.1
Asymmetric Hydrogenation

The first example of peripherally modified chiral diphosphine dendrimers for asymmetric hydrogenation were reported by Togni et al. in 1998 [86, 87]. Chiral dendrimers with six, eight, 12 and 16 peripheral Josiphos units were synthesized and employed in the Rh-catalyzed hydrogenation of dimethyl itaconate (Scheme 4.20). The enantioselectivities afforded by the dendrimer catalyst bearing eight Josiphos units (Figure 4.25) were very similar to those obtained with the mononuclear Josiphos catalyst (98% versus 99% ee). On completion of the reaction, the dendrimer catalysts could be separated from the reaction mixture by using a nanofiltration membrane.

By using commercially available zero- to fourth-generation poly(propyleneimine) (PPI) as the supports, Gade et al. recently synthesized a series of chiral dendrimer diphosphine ligands with two, four, eight, 16 and 32 Pyrphos units at the periphery [88]. Metallation of the multi-site diphosphines with [Rh(COD)$_2$]BF$_4$ cleanly yielded the cationic rhododendrimers containing up to 32 metal centers (Figure 4.26). The relationship between the size/generation of the dendrimers and their catalytic properties was established in the asymmetric hydrogenation of Z-methyl-α-acetamidocinammate (Scheme 4.21) and dimethyl itaconate. The activity and enantioselectivity of the dendrimer catalysts decreased as the generation of dendrimers increased. This 'negative' dendrimer effect was most likely due to the

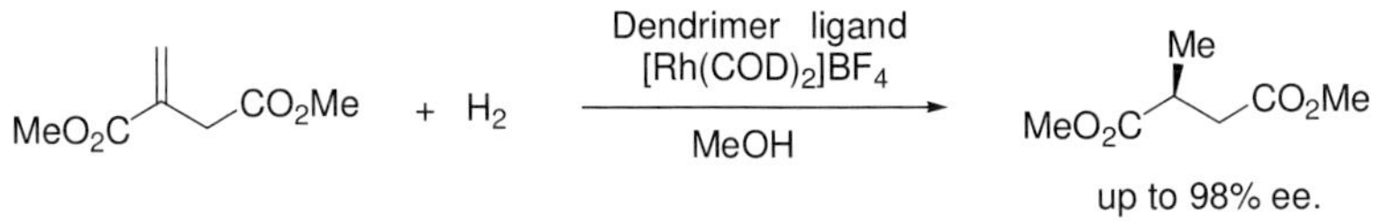

Scheme 4.20

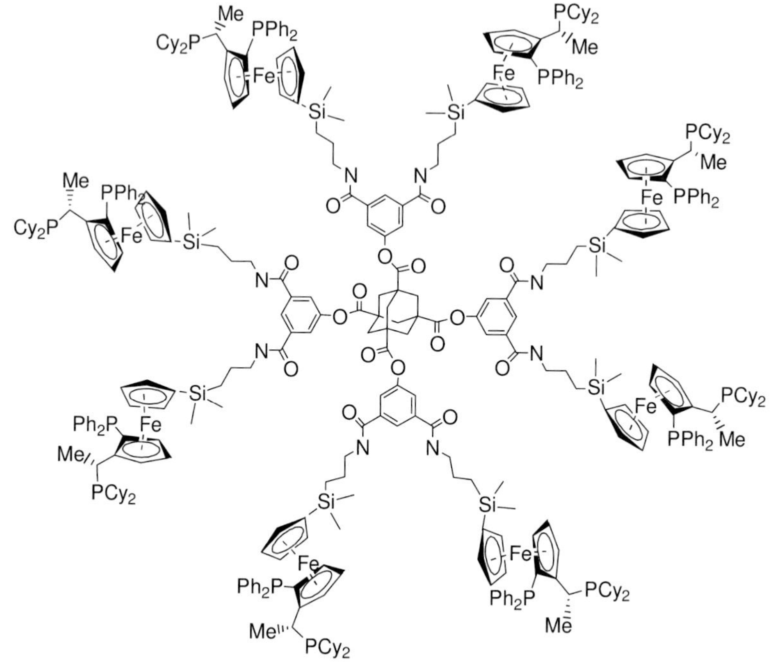

Figure 4.25 Chiral dendrimer bearing Josiphos units located at the periphery.

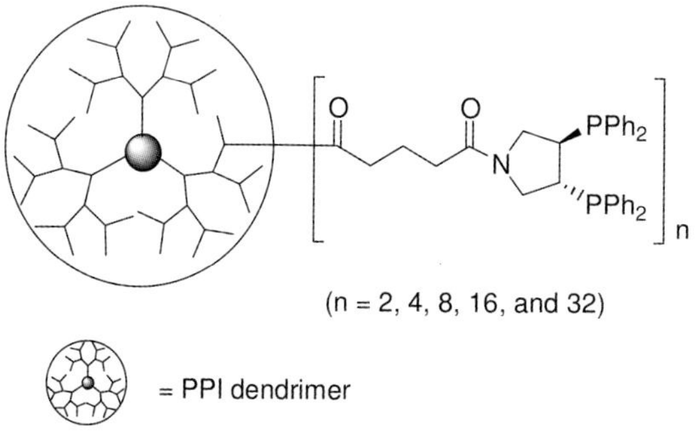
PPh2
PPh2
N
n
(n = 2, 4, 8, 16, and 32)
= PPI dendrimer

Figure 4.26 PPI dendrimers bearing chiral diphosphine Pyrphos units located at the periphery.

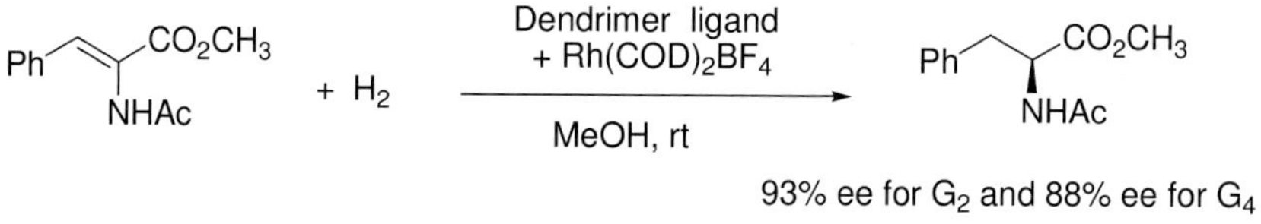
Ph
CO2CH3
NHAc
+ H2
Dendrimer ligand
+ Rh(COD)2BF4
MeOH, rt
Ph
CO2CH3
NHAc
93% ee for G2 and 88% ee for G4

Scheme 4.21

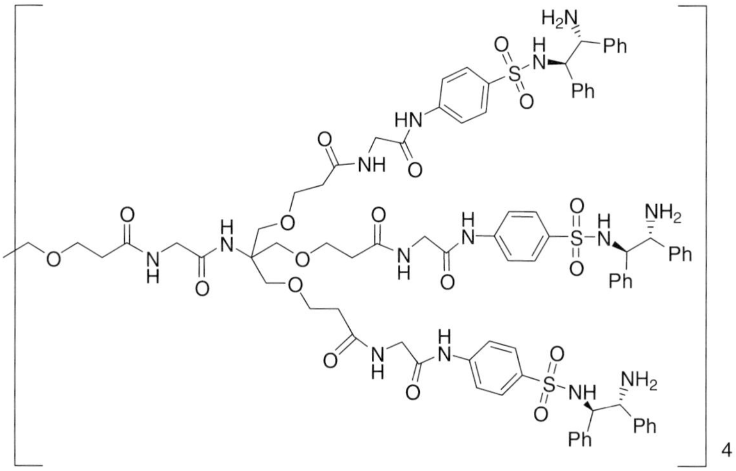

Figure 4.27 Chiral dendrimer bearing 12 TsDPENs located at the periphery.

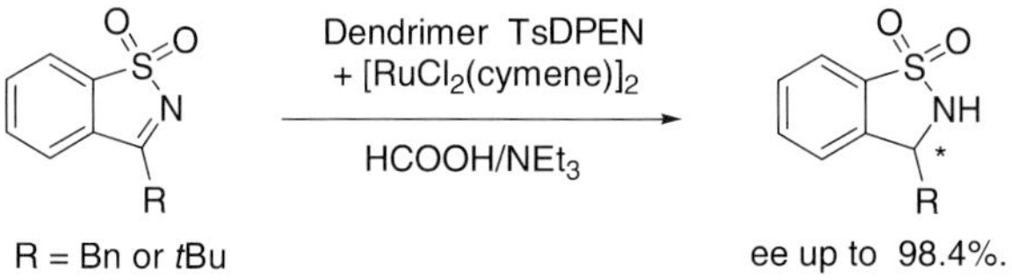

Scheme 4.22

R = Bn or tBu

ee up to 98.4%.

flexibility of the dendrimer core, which allowed the attached rhodium complexes to bend back inside the globular molecule.

4.3.2
Asymmetric Transfer Hydrogenation

In 2002, Deng and coworkers synthesized the first- and second-generation dendrimers containing up to 12 chiral diamines (TsDPEN) at the periphery (Figure 4.27) via a convergent approach [89]. Their ruthenium complexes exhibited high catalytic activities and enantioselectivities in the asymmetric transfer hydrogenation of ketones and imines (Scheme 4.22). Quantitative yields and, in some cases slightly higher enantioselectivities, were observed compared to those obtained with the monomeric catalyst.

In order to facilitate recycling of the multiple TsDPEN-functionalized dendrimer catalysts, the same group recently reported the synthesis of a novel form of 'hybrid' dendrimer ligands by coupling polyether dendrons with peripherally TsDPEN-functionalized Newkome-type poly(ether-amide) dendrimer (Figure 4.28) [90]. The solubility of these hybrid dendrimers was found to be affected by the generation of the polyether dendron. The ruthenium complexes produced were applied in the asymmetric transfer hydrogenation of ketones, enones, imines and activated

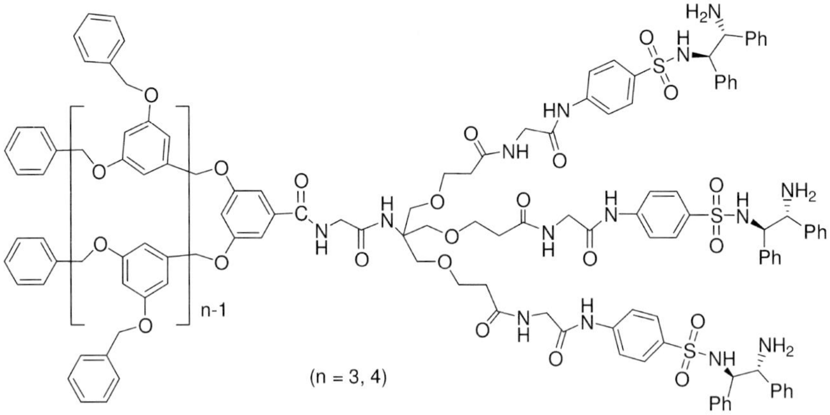

Figure 4.28 Chiral 'hybrid' dendrimers bearing TsDPENs located at the periphery.

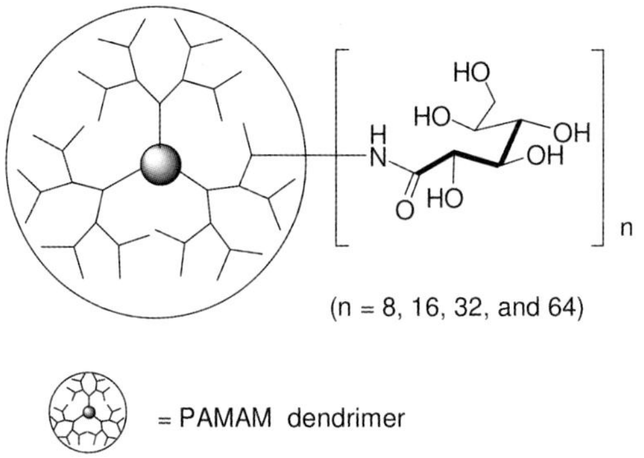

Figure 4.29 Chiral amphiphilic dendrimers bearing D-gluconolactones located at the periphery.

olefins. As a result, moderate to excellent enantioselectivities were achieved which were comparable to those obtained with the monomeric catalyst. Unlike the multiple TsDPEN-functionalized dendrimer catalysts displayed in Figure 4.27, the hybrid dendrimer catalysts were easily separated by solvent precipitation. However, the hybrid catalyst showed a reduced recyclability compared to the high generation of TsDPEN-cored dendrimer catalyst (as shown in Figure 4.15) [62].

4.3.3
Asymmetric Borane Reduction of Ketones

Rico-Lattes et al. synthesized four generations of peripherally D-gluconolactone-functionalized polyamidoamine (PAMAM) dendrimers (Figure 4.29) [91, 92]. These amphiphilic dendrimers were used as chiral ligands for the reduction of ketones with sodium borohydride, either homogeneously in water or heterogeneously in THF (Scheme 4.23). Although only the third-generation amphiphilic

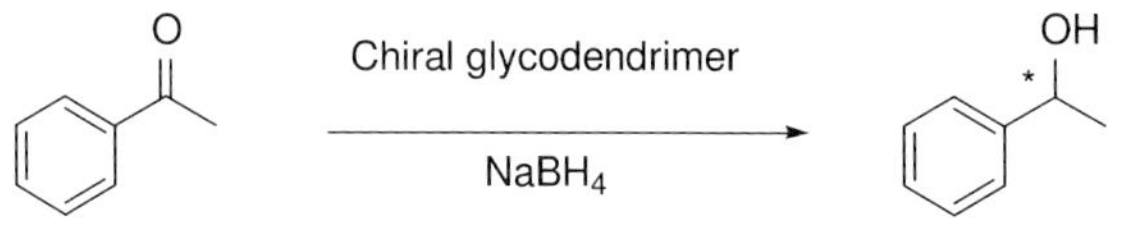

4.3 Peripherally Modified Dendrimers in Asymmetric Catalysis | 161

[Scheme 4.23: Reduction of acetophenone to 1-phenylethanol using chiral glycodendrimer and NaBH₄]

ee (H_2O): 0% for G_1; 0% for G_2; 50% for G_3; 98% for G_4.
ee (THF): 0% for G_1; 0% for G_2; 99% for G_3; 3% for G_4.

Scheme 4.23

dendrimer was found to achieve high enantioselectivity in the reduction of acetophenone under heterogeneous conditions (in THF), in water the highest enantioselectivity was obtained by using the fourth-generation amphiphilic dendrimer. In addition, the third-generation dendrimer was easily recovered from the heterogeneous reaction mixture by nanofiltration; moreover, when regenerated in HCl/MeOH the dendrimer was reused 10 times, each with similar catalytic properties.

In order to elucidate the nature of the observed 'positive' dendrimer effect, the mechanistic insights of such asymmetric induction were studied by using molecular modeling [13], carbon-NMR (^{13}C NMR) spectroscopy, induced circular dichroism, a systematic variation of the reaction parameters and the molecular structure of the chiral sugar moieties [93]. These studies demonstrated that the main factor affecting the enantioselectivity was the ordering and specific orientation of the ketones at the chiral interface when the reaction was carried out homogeneously in water. Under heterogeneous conditions (in THF), the situation appeared to be more complex.

4.3.4
Asymmetric Ring Opening of Epoxides

Epoxides are valuable intermediates for the stereocontrolled synthesis of complex organic compounds, and the asymmetric ring opening of chiral epoxides is one of their extended utilities [94].

In 2000, Breinbauer and Jacobsen reported the use of dendrimer-supported Co(salen) complexes for the asymmetric ring opening of epoxides and demonstrated the first example of a 'positive' dendrimer effect in asymmetric catalysis [95]. A series of dendrimers with up to 16 catalytic sites at the periphery were synthesized by covalently attaching Co–salen to the commercially available PAMAM dendrimers with NH_2-terminals (Figure 4.30).

When using (rac)-vinylcyclohexane epoxide as a standard substrate (Scheme 4.24), the dendrimer catalysts exhibited significantly higher catalytic activities in the hydrolytic kinetic resolution of terminal epoxides as compared to the monomer or the dimer catalysts. Among the dendrimer catalysts, the first-generation dendrimer catalyst gave the best results and the efficiency of the catalyst on a per-metal basis was in the following order: 4-Co(salen)–PAMAM > 8-Co(salen)–PAMAM > 16-Co(salen)–PAMAM. This dendrimer effect was considered to arise from the restricted conformation imposed by the dendrimer structure, which enhanced the

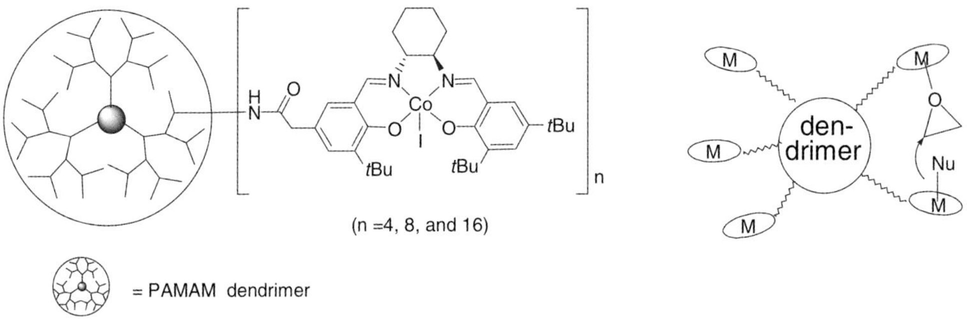

Figure 4.30 The PAMAM dendrimer-bound chiral [Co(salen)] catalysts for asymmetric ring opening of epoxides.

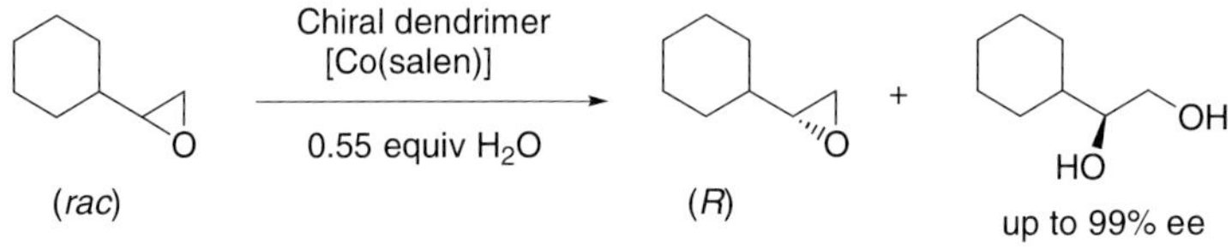

Scheme 4.24

cooperative interactions between neighboring Co–salen units on the surface of the PAMAM dendrimer.

4.3.5
Asymmetric Addition of Dialkylzincs to Aldehydes and Imine Derivatives

The early examples of peripherally functionalized dendrimers, PPI-supported amino alcohol and dendrimer-bound Ti-TADOLates, in the asymmetric addition of diethylzinc to benzaldehyde were reported by the groups of Meijer [96, 97] and Seebach [98], respectively.

Recently, Soai et al. reported the synthesis of series of chiral dendrimer amino alcohol ligands based on PAMAM, hydrocarbon and carbosilane dendritic backbones (Figure 4.31) [99–102]. These chiral dendrimers were used as catalysts for the enantioselective addition of dialkylzincs to aldehydes and N-diphenylphosphinylimines (Scheme 4.25). The molecular structures of the dendrimer supports were shown to have a significant influence on the catalytic properties. The 'negative' dendrimer effect for the PAMAM-bound catalysts was considered due to the fact that the nitrogen and oxygen atoms on the dendrimer skeleton could coordinate to zinc.

In contrast to the PAMAM-bound catalysts, chiral dendrimers based on rigid poly(phenylethyne) and flexible carbosilane backbones without heteroatoms proved to be highly efficient catalysts in such alkylation reactions. Excellent enantioselectivities in the asymmetric addition of dialkylzincs to aldehydes (up to 93% ee) and N-diphenylphosphinylimines (up to 94% ee) were achieved.

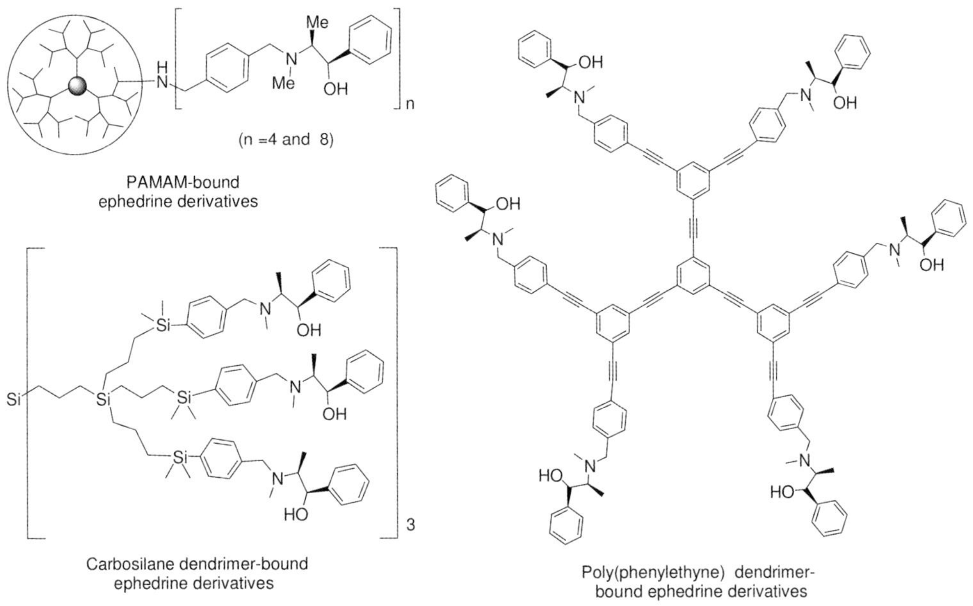

Figure 4.31 The dendrimer-bound ephedrine derivatives.

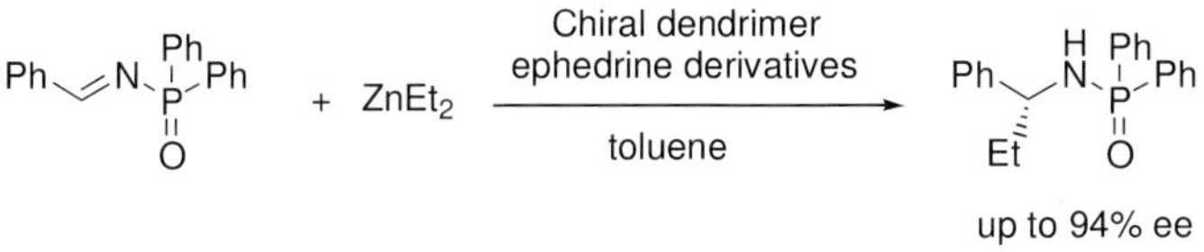

Scheme 4.25

Hu *et al.* reported the synthesis of two optically active ephedrine-bearing dendronized polymers (Figure 4.32) by using Suzuki coupling polymerization [103]. Both polymers were soluble in common organic solvents such as THF, toluene and chloroform, but were insoluble in methanol and hexane. These dendronized polymer catalysts were tested in the asymmetric addition of diethylzinc to benzaldehyde, and good enantioselectivities and high conversions were observed. The combined features of more catalytic sites, better solubility and nanoscopic dimensions of these optically active dendronized polymers made them more efficient than the corresponding linear polymeric and dendritic chiral catalysts. In addition, this rod-like polymer catalyst could be easily recovered by solvent precipitation, with the recovered catalysts showing similar reactivities and enantioselectivities.

4.3.6
Asymmetric Allylic Amination and Alkylation

In an effort to extend the use of the peripherally Pyrphos-functionalized dendrimers to asymmetric allylic aminations, Gade *et al.* recently prepared two types

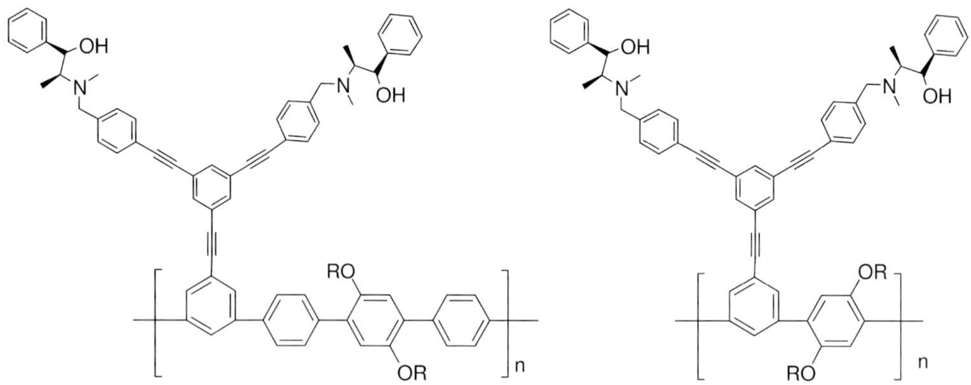

Figure 4.32 Chiral ephedrine-bearing dendronized polymers.

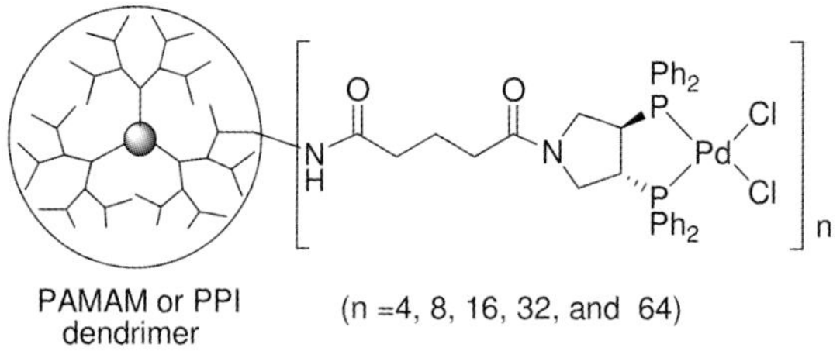

$R = n\text{-}C_6H_{13}$

Figure 4.33 The peripherally Pd(Pyrphos)-functionalized PPI and PAMAM dendrimers.

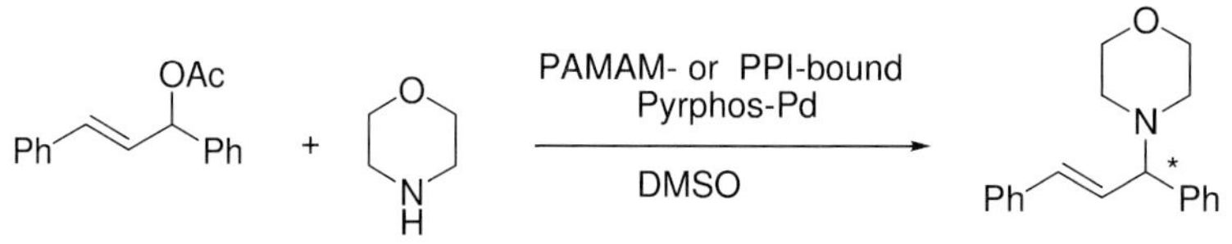

PAMAM or PPI dendrimer (n = 4, 8, 16, 32, and 64)

Scheme 4.26

of palladium catalyst containing the PPI dendrimers mentioned above (Figure 4.26) or the analogous PAMAM dendrimers with up to 64 Pyphos units (Figure 4.33) [104]. The amination of 1,3-diphenyl-2-propenyl acetate with morpholine was chosen as the model reaction (Scheme 4.26). In contrast to the 'negative' dendrimer effect observed in the Rh-catalyzed asymmetric hydrogenation, a remarkable and unprecedented enhancement in catalyst selectivity was observed as a function of the dendrimer generation. This steady increase in ee-values for the allylic amination was less pronounced for the PPI-bound catalysts (40% ee for the

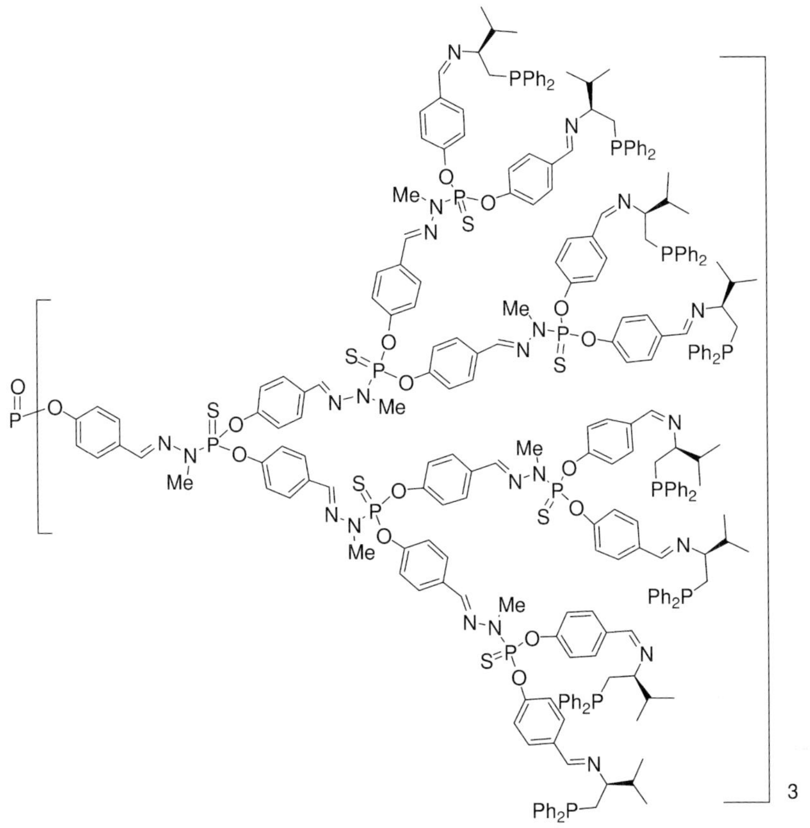

Figure 4.34 The third-generation dendrimers bearing 24 chiral iminophosphine end groups.

fifth-generation PPI catalyst) than the palladium–PAMAM dendrimer catalysts, for which an increase in selectivity from 9% ee for a mononuclear complex to 69% ee for the fifth-generation dendrimer catalyst was found. In addition, the same general trend was observed in the asymmetric allylic alkylation of 1,3-diphenyl-2-propenyl acetate with sodium dimethylmalonate.

Caminade and coworkers reported the synthesis of a third-generation phosphorus-containing dendrimer possessing 24 chiral iminophosphine end groups (Figure 4.34) [105], which was derived from (2S)-2-amino-1-(diphenylphosphinyl)-3-methylbutane. The palladium catalyst, generated *in situ* by mixing the dendrimer ligand with [Pd(η^3-C$_3$H$_5$)Cl]$_2$, was used in the asymmetric allylic alkylation of *rac-(E)*-diphenyl-2-propenyl acetate and pivalate (Scheme 4.27). Good to excellent enantioselectivities were obtained (up to 95% ee) for both cases. In addition, the dendrimer catalyst was easily recovered by precipitation with diethyl ether, and could be reused at least twice with almost the same efficiency.

In 2007, using a similar type of phosphorus dendrimer support, the same group synthesized a series of functionalized dendrimers bearing chiral ferrocenyl

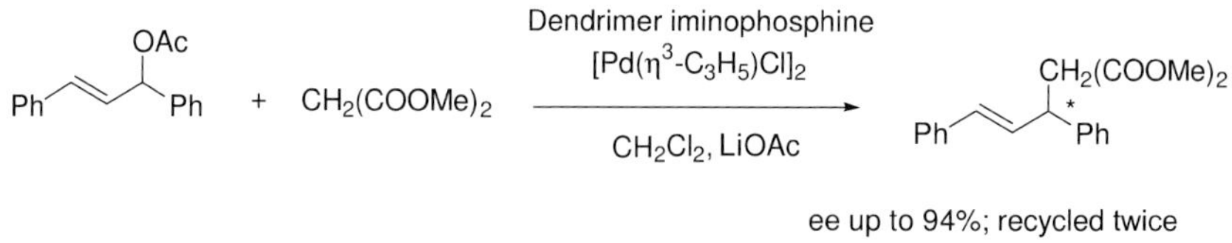

Scheme 4.27

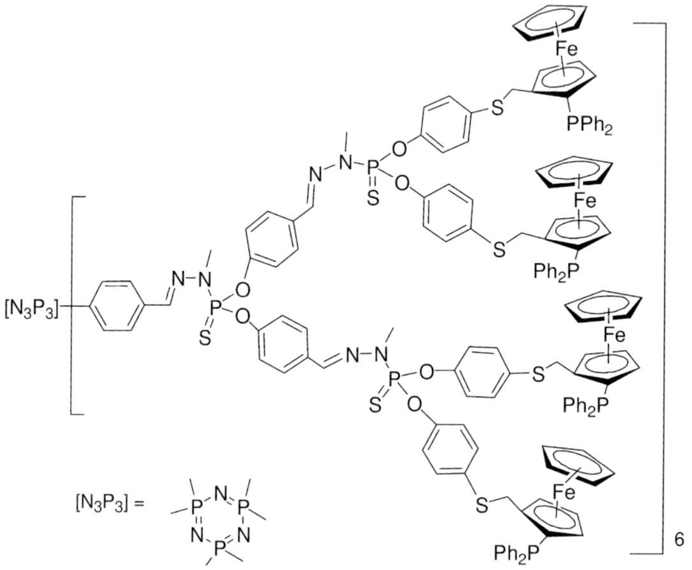

Figure 4.35 The third-generation dendrimers bearing 24 chiral ferrocenyl phosphine-thioether end groups.

phosphine-thioether ligands located at the periphery (Figure 4.35) [106]. These dendrimer ligands from the first to the fourth generation (up to 96 ferrocenyl P, S ligands) were successfully used in the palladium-catalyzed asymmetric allylic alkylation of *rac-(E)*-1,3-diphenyl-2-propenyl acetate with dimethylmalonate. All dendrimer catalysts exhibited very high activities and enantioselectivities (up to 92% ee), which were similar to those obtained with the corresponding monomer catalyst.

4.3.7
Asymmetric Michael Addition

In 2002, Sasai *et al.* reported the synthesis of dendrimer heterobimetallic multifunctional chiral catalysts, containing up to 12 chiral BINOL units at the periphery (Figure 4.36) [107]. The insoluble dendritic heterobimetallic multifunctional chiral AlLibis(binaphthoxide) (ALB) complexes were obtained by treating these dendrimer ligands with AlMe$_3$ and *n*-BuLi. The resulting dendrimer-supported ALB

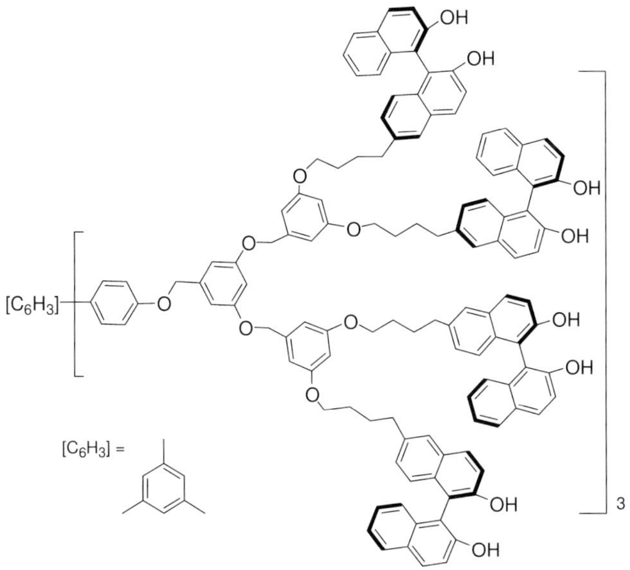

Figure 4.36 The peripherally BINOL-functionalized chiral dendrimers.

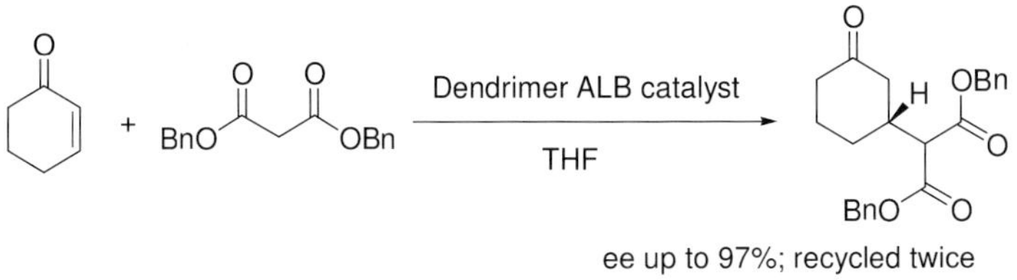

Scheme 4.28

promoted the asymmetric Michael reaction of 2-cyclohexenone with dibenzyl malonate to give the corresponding adduct with up to 97% ee (Scheme 4.28). For example, the first-generation dendrimer ALB containing six BINOL units gave the Michael adduct in 63% yield with 91% ee after 48 h. Furthermore, the insoluble dendrimer catalyst was easily recovered and reused twice without any loss of activity, but with even higher enantioselectivities (93% ee and 94% ee, respectively). In contrast, the ALB derived from BINOLs which were randomly attached on a polystyrene resin only gave an essentially racemic product.

4.3.8
Asymmetric Diels–Alder Reaction

Recently, Wan and Chow reported the synthesis of the zeroth- and first-generation peripherally BINOL-functionalized chiral dendrimer ligands using an oligo(arylene) framework as the rigid support (Figure 4.37) [108]. Hexyl groups were grafted onto the surface of the dendrimers to improve their solubility in organic solvents. Their

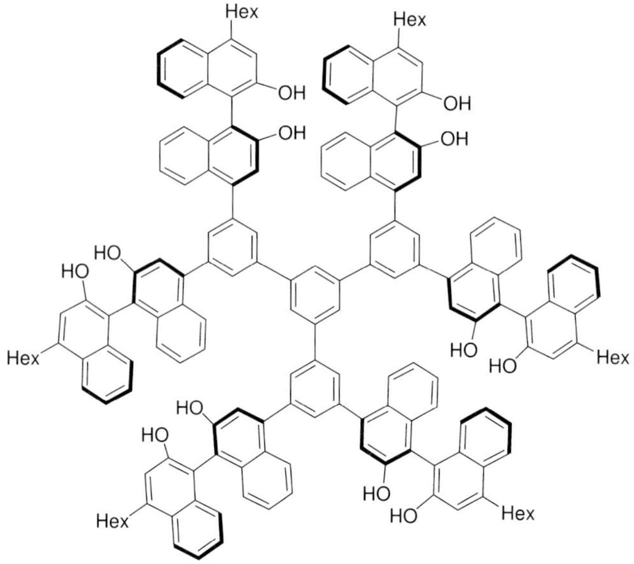

Figure 4.37 The peripherally BINOL-functionalized chiral dendrimers.

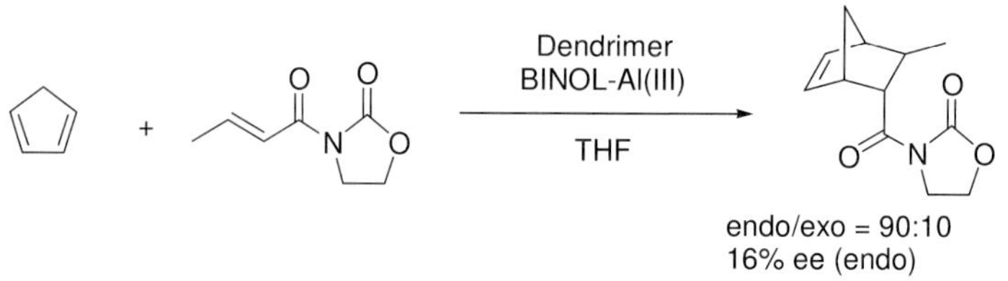

Scheme 4.29

endo/exo = 90:10
16% ee (endo)

aluminum complexes, generated *in situ* by mixing the dendrimer ligands with Me$_2$AlCl, exhibited slightly better reactivities and enantioselectivities than that of a monomer catalyst in the Diels–Alder reaction between cyclopenta-1,3-diene and 3-[(E)-but-2-enoyl]oxazolidin-2-one (Scheme 4.29). The reactivity and enantioselectivity were found to be independent of dendrimer generation, most likely due to the absence of intramolecular interactions among the catalytic centers.

4.3.9
Asymmetric Aldol Reaction

In 2005, Bellis and Kokotos synthesized a series of proline-based PPI chiral dendrimers possessing up to 126 proline end groups (Figure 4.38) [109]. These dendrimers were evaluated as the catalyst for asymmetric aldol reactions (Scheme 4.30). Using 6.5 mol% of the second-generation dendrimer catalyst, the products of the aldol reactions were obtained in moderate yields and enantioselectivities (up

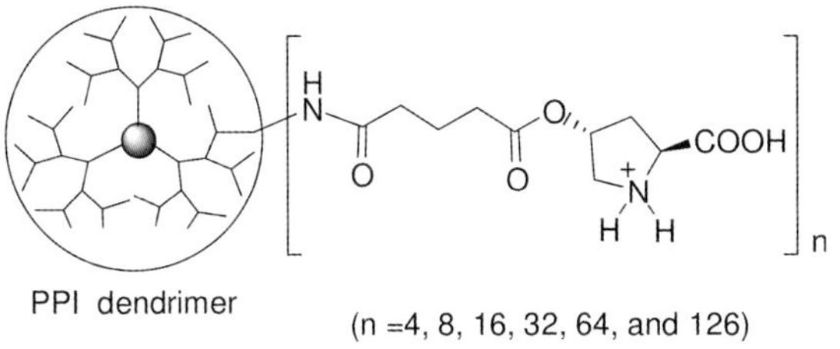

Figure 4.38 The peripherally proline-functionalized chiral dendrimers.

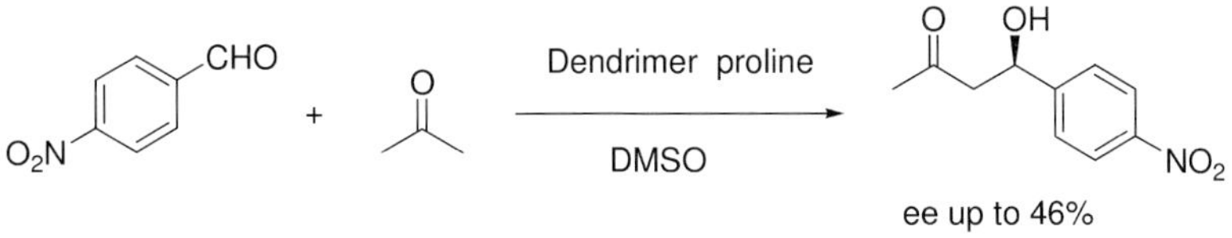

Scheme 4.30

to 65%), which were comparable to those obtained from proline itself. In addition, the improved solubility of the dendrimer catalysts in organic solvents provided a completely homogeneous reaction system, and thus enhanced the catalytic activity.

4.3.10
Asymmetric Hydrovinylation

Transition metal-catalyzed hydrovinylation is one of a few practically useful carbon–carbon bond-forming reactions utilizing feedstock carbon sources for the synthesis of high-value fine chemicals. Asymmetric hydrovinylation has many potential applications in the synthesis of pharmacologically important compounds, such as ibuprofen and naproxen, and has attracted much attention [110]. Recently, chiral monodentate phosphines have proven to be highly efficient ligands for the asymmetric hydrovinylation of α-alkyl vinylarenes [111].

Very recently, Rossell and coworkers reported the synthesis of carbosilane dendrimers containing P-stereogenic monophosphines at the periphery (Figure 4.39) [112]. The palladodendrimers, generated by mixing the chiral dendrimers with the dinuclear allyl complex [Pd(μ-Cl)(η^3-2-MeC$_3$H$_4$)]$_2$, were tested in the asymmetric hydrovinylation of styrene (Scheme 4.31). The catalytic performance was found to depend heavily on the nature of the phosphine and the halide abstractor. The best result was obtained when using first-generation dendrimer containing the 2-biphenyl-substituted phosphine and NaBARF (79% ee). Unexpectedly, the zeroth-generation Pd catalyst containing 9-phenanthryl-substituted phosphine gave *(R)*-3-phenyl-1-butene as the predominant isomer, in contrast to its analogous model system which yielded *(S)*-3-phenyl-1-butene.

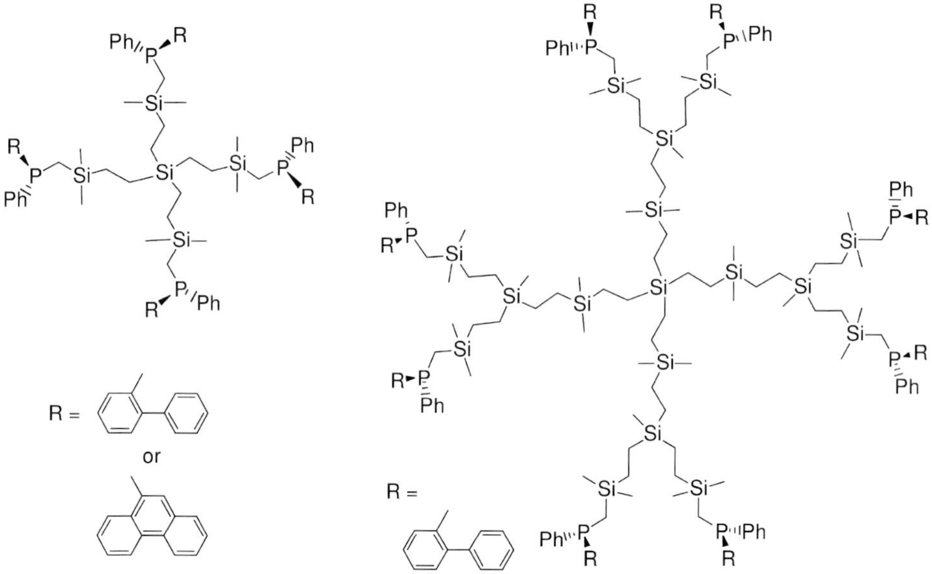

Figure 4.39 The peripherally chiral monophosphine-functionalized carbosilane dendrimers.

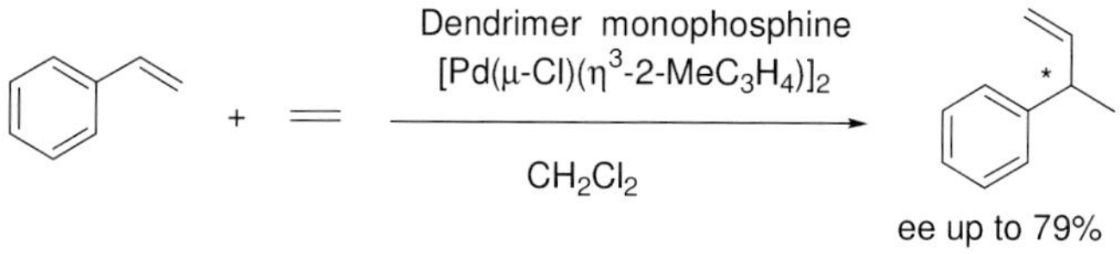

Scheme 4.31

4.4
Solid-Supported Chiral Dendrimer Catalysts for Asymmetric Catalysis

Despite the advantage of their easy separation, the use of conventional insoluble polymer-supported catalysts often suffered from a reduced catalytic activity and stereoselectivity, due either to diffusion problems or to a change of the preferred conformations within the chiral pocket created by the ligand around the metal center. In order to circumvent these problems, a new class of crosslinked macromolecule – namely dendronized polymers – has been developed and employed as catalyst supports. In general, two types of such solid-supported dendrimer have been reported: (i) with the dendrimer as a linker of the polymer support; and (ii) with dendrons attached to the polymer support [12, 113].

4.4.1
Solid-Supported Internally Functionalized Chiral Dendrimer Catalysts

In 1997, Seebach et al. introduced a novel concept for the immobilization of chiral ligands in PS [114–116]. In this approach, the chiral ligand to be immobilized was

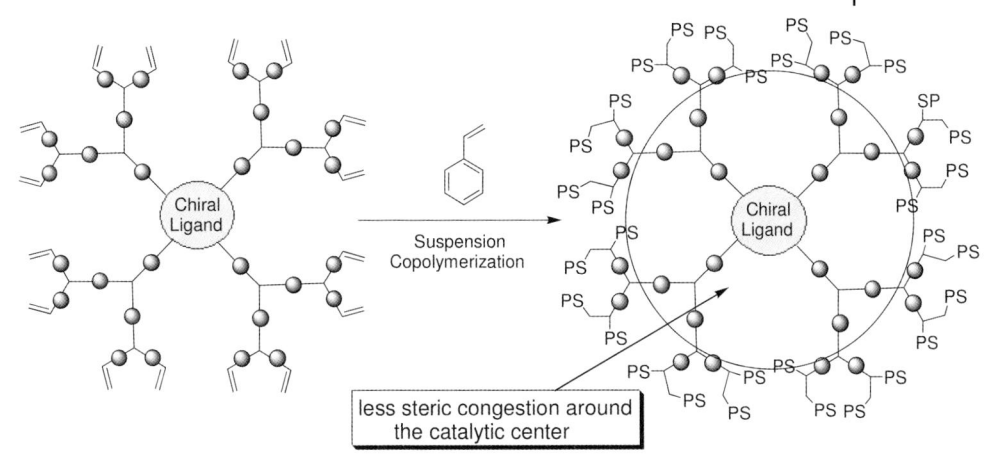

Figure 4.40 Copolymerization of dendritically surrounded chiral ligands with styrene.

placed at the core of a styryl-substituted dendrimer, followed by a suspension copolymerization with styrene (Figure 4.40). The dendritic branches were thought to act as spacer units keeping the 'obstructing' PS backbone away from the catalytic centers; this led to a better accessibility and thus enhanced the catalytic activity.

Based on this concept, Seebach et al. developed the first example of TADDOL-cored dendrimers (Figure 4.41) immobilized in a PS matrix [116]. The resultant internally dendrimer-functionalized polymer beads were loaded with Ti(OiPr)$_4$, leading to a new class of supported Ti-TADDOLate catalysts for the enantioselective addition of diethylzinc to benzaldehyde. Compared to the conventional insoluble polymer-supported Ti-TADDOLate catalysts, these heterogeneous dendrimer catalysts gave much higher catalytic activities, with turnover rates close to those of the soluble analogues. The polymer-supported dendrimer TADDOLs were recovered by simple phase separation and reused for at least 20 runs, with similar catalytic efficiency.

The same group also developed a series of homogeneous and heterogeneous chiral dendritic BINOLs using an approach similar to that employed in the polymer-supported dendrimer TADDOLs [116]. Chiral BINOL-cored Fréchet-type dendrimers bearing styryl groups located at the periphery (Figure 4.42) were synthesized and used as crosslinkers for the subsequent copolymerization with styrene, giving the insoluble polymer-supported dendrimer BINOL ligands [117]. The resultant nonpolymerized and polymer-supported dendrimer BINOL ligands were loaded with Ti(OiPr)$_4$ to afford catalytically active BINOLate-type catalysts for the enantioselective addition of diethylzinc to benzaldehyde. The enantioselectivities (up to 93%) and conversions obtained with the polymer-supported dendrimer catalysts were, in most cases, identical to those obtained with the soluble analogues under homogeneous conditions. Moreover, the polymer-supported dendrimer catalysts could be recovered by simple phase separation and reused in up to 20 consecutive catalytic runs with only a minor loss of enantioselectivity.

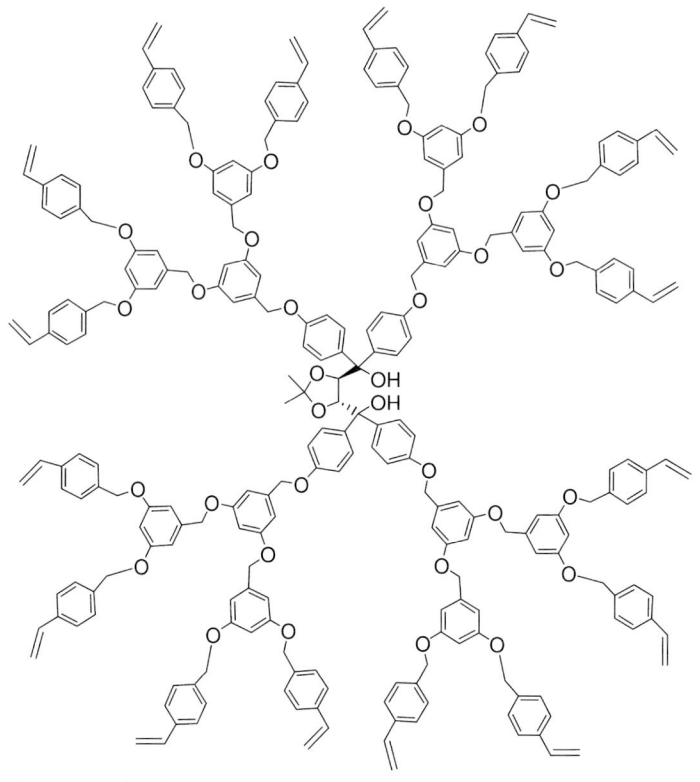

Figure 4.41 Chiral TADDOL-cored dendrimer with styryl groups located at the periphery.

Figure 4.42 Chiral BINOL- and salen-cored dendrimers with styryl groups located at the periphery.

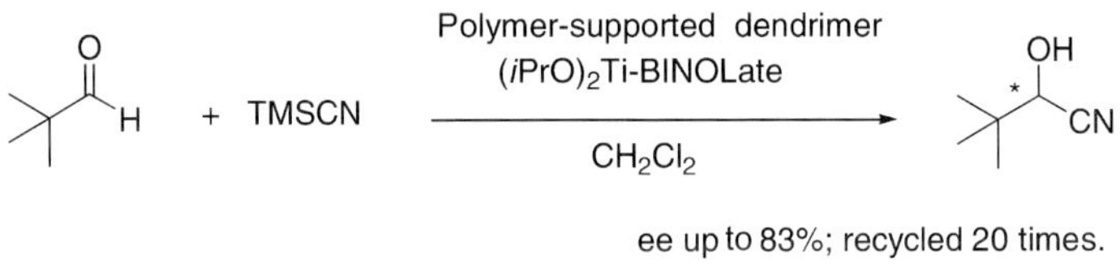

Scheme 4.32

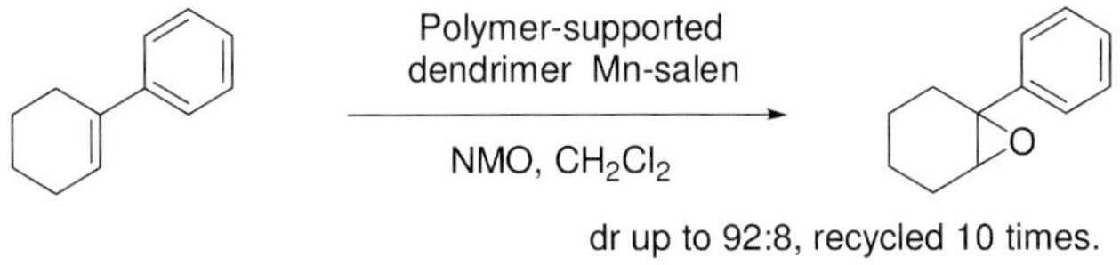

Pivaldehyde + TMSCN → cyanohydrin (OH, CN)
Reagents: Polymer-supported dendrimer (iPrO)₂Ti-BINOLate, CH₂Cl₂
ee up to 83%; recycled 20 times.

Scheme 4.33

Phenylcyclohexene → epoxide
Reagents: Polymer-supported dendrimer Mn-salen, NMO, CH₂Cl₂
dr up to 92:8, recycled 10 times.

The polymer-supported second-generation dendrimer BINOL ligand was also used in the Ti-catalyzed asymmetric cyanosilylation of pivalaldehyde (Scheme 4.32). The enantioselectivity of the cyanohydrine obtained in the first run was as high as that of the homogeneous reaction (72% ee). Unexpectedly, the enantioselectivity increased gradually during the following catalytic runs, and finally reached a value of 83% after five runs.

Epoxides are versatile synthetic intermediates that can be readily converted into a large variety of useful compounds by means of regioselective ring opening. Among several catalytic procedures to optically active epoxides, the asymmetric epoxidation of unfunctionalized alkenes catalyzed by chiral Mn-salen complexes initially developed by Jacobsen *et al.* is considered one of the most effective methods to be discovered during the past 20 years [118]. In following their initial studies with TADDOLs and BINOLs, Seebach *et al.* synthesized salen-cored dendrimers bearing two to eight styryl end groups, which were subsequently immobilized in a PS matrix according to a similar approach described above [119]. The resulting internally dendrimer-functionalized polymer salens were employed for the stereoselective Mn-catalyzed epoxidation of phenyl-substituted olefins (Scheme 4.33), as well as for Cr-catalyzed HDA reactions of Danishefsky's diene with aldehydes. The polymer-supported manganese catalysts were stable in air, and showed high selectivities in the epoxidation reactions, which were comparable to the results obtained in homogeneous reactions. Furthermore, the immobilized Mn–salen catalysts could be reused up to ten times, without loss of activity. The high catalytic activity of the polymer-supported dendritic salens was also shown in the Cr–salen-catalyzed HDA reactions of Danishefsky's diene with aldehydes, affording the corresponding cycloadducts in similar selectivities, as under homogeneous conditions. In addition, multiple reuse of the supported Cr–salen catalysts in consecutive catalytic runs were achieved.

4.4.2
Solid-Supported Peripherally Functionalized Chiral Dendrimer Catalysts

Soluble dendrimers bearing catalytic centers located at the periphery can be covalently attached onto the surface of conventional solid supports (such as polymer beads or silica gels), leading to another type of solid-supported dendrimer catalyst. It is expected that this type of immobilized catalysts would combine the advantages of both the traditional supported catalysts and the dendrimer catalysts. First, the catalytically active species at the dendrimer surface are more easily solvated, which makes the catalytic sites more available in the reaction solutions (relative to cross-linked polymers). Second, the insoluble supported dendrimers are easily removed from the reaction mixtures as precipitates or via filtration (relative to soluble dendrimers). These solid-supported peripherally functionalized chiral dendrimer catalysts have attracted much attention over the past few years [12, 113], but their number of applications in asymmetric catalysis is very limited.

In 2002, Rhee et al. synthesized a series of silica-supported chiral dendrimer catalysts for the enantioselective addition of diethylzinc to benzaldehyde (Figure 4.43) [120–122]. The preparation of the peripherally ephedrine-functionalized dendrimers on silica was accomplished via two different approaches: (i) a stepwise propagation of dendrimers from the solid support; or (ii) direct immobilization of the 'ready-made' dendrimers onto the solid support. In the first catalyst preparation, symmetric hyperbranching was found to be a prerequisite in order to suppress the unfavorable racemic reaction which occurred on the naked surface. Moreover, the control of hyperbranching was important not only to render the accessibility of reagents to the active sites, but also to relieve the multiple interactions between the chiral active sites. In the alternative catalyst preparation, the participation of surface silanol groups in the racemic reaction could be effectively suppressed by appropriate functionalization. In addition, the substitution of terminal end groups with a long alkyl chain spacer was found to be effective in relieving the multiple interactions between the active sites. Moderate enantioselectivities (up to 60% ee) were observed with the supported dendrimer catalyst, while the catalyst was easily recovered and reused three times, with similar activities and enantioselectivities.

Very recently, Portnoy et al. described the design and synthesis of insoluble polymer-supported dendrimers bearing proline end groups for asymmetric aldol reactions [123]. The zeroth- to third-generation supported dendrimer catalysts were prepared by attaching (2S,4R)-4-hydroxyproline onto the insoluble PS-bound

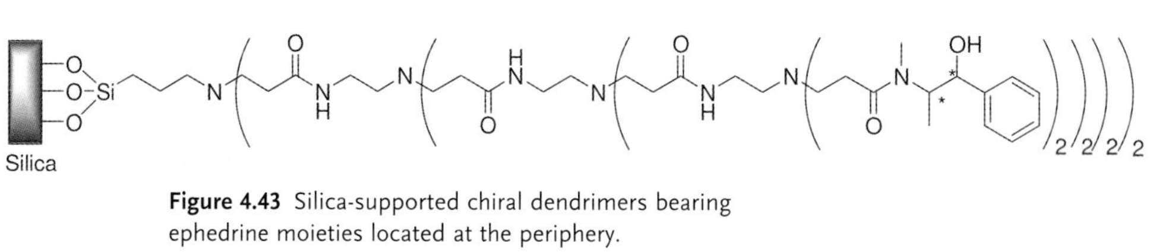

Figure 4.43 Silica-supported chiral dendrimers bearing ephedrine moieties located at the periphery.

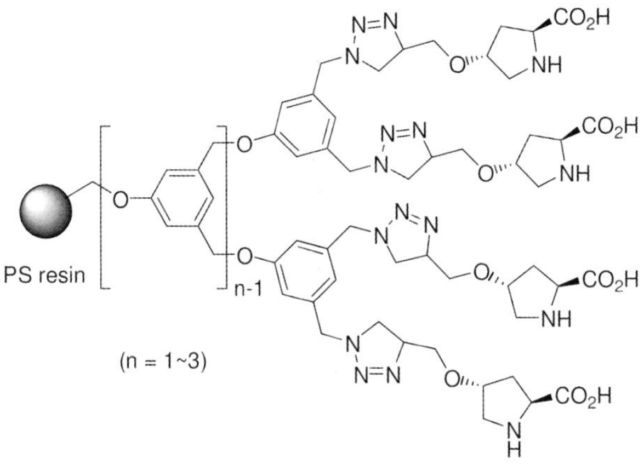

Figure 4.44 Insoluble polystyrene-supported dendrimers bearing proline end groups.

poly(aryl benzyl ether) dendrons via a 'click strategy' (Figure 4.44). When carried out in dimethylsulfoxide (DMSO) at room temperature, the aldol reactions revealed a remarkable influence of the dendronization on conversion, yield and enantioselectivity. For benzaldehyde, quantitative conversion and good yield (58%) were achieved in 4 days with the second-generation catalyst, which were higher than those of the nondendritic analogue (34% yield and 42% conversion). Moreover, the enantioselectivity for the dendrimer catalyst was also higher than that of the nondendritic catalyst (68 versus 27% ee). Notably, the enantioselectivities with the first- and second-generation catalysts were even higher than those achieved in homogeneous reaction with L-proline. This 'positive' dendrimer effect was demonstrated to be caused by the dendritic branched architecture instead of the lengthening of the tethers between prolines and the polymer support.

4.5
Conclusion and Perspectives

Since the first reports described by Brunner [21], Meijer [97] and Seebach [98], a number of examples of chiral dendrimer catalysts based on different dendritic backbones functionalized at the core or at the periphery have been reported. Most of these have been successfully employed as recyclable chiral catalysts for asymmetric syntheses, such as asymmetric hydrogenation, transfer hydrogenation, borane reduction of ketones, alkylation of carbonyl compounds and the addition of organozinc compounds to aldehydes. In most cases, the chiral dendrimer catalysts behaved like homogeneous catalysts during the reaction and could be easily separated upon completion of the reaction. Consequently, the attachment of chiral catalysts to dendrimer supports offers a potential combination of the advantages of homogeneous and heterogeneous asymmetric catalysis. In addition, it is possible to systematically adjust the structure, size, shape and solubility of the dendrimer support at will, and thus to fine-tune the catalytic activity and enantioselectivity of the supported chiral catalysts. In some particular cases, interesting dendrimer effects have been observed, including increased/decreased catalytic activity [32, 33, 43, 95], stereoselectivity [92, 104, 123], and stability [54, 85]. However, the nature of the observed dendrimer effects remains to be elucidated.

Efforts have also been concentrated on the recovery and reuse of the often expensive chiral dendrimer catalysts. Several separation approaches are applicable to these functionalized dendrimers, including solvent precipitation, nanofiltration and immobilization of the dendrimer to an insoluble support (e.g. PS, silica). Among these methods, nanofiltration – which can be performed batchwise and in a continuous-flow membrane reactor (CFMR) [124] – is expected to have great potential for large-scale applications. The common problems involved in catalyst recycling may include dendrimer catalyst decomposition and metal leaching, as well as catalyst deactivation (such as the formation of inactive dimer). Although the attachment of dendrimer catalysts to conventional solid supports has facilitated the separation and recycling of catalyst [12, 113], the heterogeneous reaction nature of this strategy may also cause some problems, such as extended reaction times and difficulties in characterizing the dendritic hybrids. The use of dendrimer catalysts with core/shell structure in a biphasic solvent system [25] – that is, a 'latent biphasic system' [24] – has provided an alternative means of facilitating catalyst separation and recycling, and therefore will be actively studied in the future.

To date, no example of the successful industrial application of a chiral dendrimer catalyst has been reported. The main problem here is the difficulty and high cost in synthesizing these perfect, yet complicated, macromolecules. Divergent and convergent approaches represent two general routes for the synthesis of dendrimers, but both methods involve multiple steps and time-consuming purifications, which often result in low yields, thereby limiting their potential practical applications. In order to overcome these drawbacks, hyperbranched polymers [125], dendronized polymers and Janus dendrimers [90, 126] have recently been used as catalyst supports. Among these, the use of hyperbranched polymers rather than perfect dendrimers as catalyst supports has attracted much attention, due to their similar homogeneous reaction properties, better accessibility and lower costs [125]. However, so far, the synthesis and application of hyperbranched polymer-supported chiral catalyst has not been reported. Alternatively, the noncovalent functionalization of dendrimer (or hyperbranched polymer) support with catalyst has also been reported [127]. This novel strategy offers several advantages over the traditional covalent approaches, such as simple catalyst preparation and easy recycling of the expensive dendrimer support. Moreover, the supramolecular dendritic catalysis may provide the possibility of attaching different catalytic units onto the dendritic supports at several well-defined locations [128], which can potentially be used for tandem and/or multicompoent catalytic asymmetric reactions [129].

In conclusion, chiral dendrimer catalysts combining the advantages of both homogeneous and traditional immobilized catalysts bridge the gap between homogeneous and heterogeneous asymmetric catalysis, and are expected to have a promising future. Finding more general and simple methods to prepare easily recyclable enantioselective dendrimer (or hyperbranched polymer) catalysts with high stability, activity and productivity remains an important challenge.

References

1 Newkome, G.R., Moorefield, C.N. and Vögtle, F. (2001) *Dendrimers and Dendrons: Concepts, Synthesis, Applications*, Wiley-VCH Verlag GmbH, Weinheim.
2 Fréchet, J.M.J. and Tomalia, D.A. (2002) *Dendrimers and Other Dendritic Polymers*, John Wiley & Sons, Ltd, Chichester, England.
3 Tomalia, D.A., Baker, H., Dewald, J., Hall, M., Kallos, G., Martin, S., Roeck, J., Ryder, J. and Smith, P. (1985) *Polymer Journal*, **17**, 117–32.
4 Newkome, G.R., Yao, Z.Q., Baker, G.R. and Gupta, V.K. (1985) *Journal of Organic Chemistry*, **50**, 2003–4.
5 Hawker, C.J. and Fréchet, J.M.J. (1990) *Journal of the American Chemical Society*, **112**, 7638–47.
6 Romagnoli, B. and Hayes, W. (2002) *Journal of Materials Chemistry*, **12**, 767–99.
7 Oosterom, G.E., Reek, J.N.H., Kamer, P.C.J. and van Leeuwen, P.W.N.M. (2001) *Angewandte Chemie – International Edition*, **40**, 1828–49.
8 Astruc, D. and Chardac, F. (2001) *Chemical Reviews*, **101**, 2991–3023.
9 Crooks, R.M., Zhao, M.Q., Sun, L., Chechik, V. and Yeung, L.K. (2001) *Accounts of Chemical Research*, **34**, 181–90.
10 van Heerbeek, R., Kamer, P.C.J., van Leeuwen, P.W.N.M. and Reek, J.N.H. (2002) *Chemical Reviews*, **102**, 3717–56.
11 Twyman, L.J., King, A.S.H. and Martin, I.K. (2002) *Chemical Society Reviews*, **31**, 69–82.
12 Dahan, A. and Portnoy, M. (2005) *Journal of Polymer Science – Part A Polymer Chemistry*, **43**, 235–62.
13 Helms, B. and Fréchet, J.M.J. (2006) *Advanced Synthesis Catalysis*, **348**, 1125–48.
14 Gade, L.H. (2006) *Topics in Organometallic Chemistry*, **20**, 61–96.
15 Reek, J.N.H., Arévalo, S., van Heerbeek, R., Kamer, P.C.J. and van Leeuwen, P.W.N.M. (2006) *Advanced in Catalysis*, **49**, 71–151.
16 Zimmerman, S.C. and Lawless, L.J. (2001) *Topics in Current Chemistry*, **217**, 95–120.
17 Smith, D.K. (2006) *Chemical Communications*, 34–44.
18 Jiang, D.L. and Aida, T. (2005) *Progress in Polymer Science*, **30**, 403–22.
19 Gillies, E.R. and Fréchet, J.M.J. (2005) *Drug Discovery Today*, **10**, 35–43.
20 Knapen, J.W.J., van der Made, A.W., de Wilde, J.C., van Leeuwen, P.W.N.M., Wijkens, P., Grove, D.M. and van Koten, G. (1994) *Nature*, **372**, 659–63.
21 Brunner, H. (1995) *Journal of Organometallic Chemistry*, **500**, 39–46.
22 Ribourdouille, Y., Engel, G.D. and Gade, L.H. (2003) *Comptes Rendus Chimie*, **6**, 1087–96.
23 Fan, Q.H., Li, Y.M. and Chan, A.S.C. (2002) *Chemical Reviews*, **102**, 3385–466.
24 Bergbreiter, D.E., Osburn, P.L., Smith, T., Li, C.M. and Frels, J.D. (2003) *Journal of the American Chemical Society*, **125**, 6254–60.
25 Deng, G.J., Fan, Q.H., Chen, X.M., Liu, D.S. and Chan, A.S.C. (2002) *Chemical Communications*, 1570–1.
26 Hecht, S. and Fréchet, J.M.J. (2001) *Angewandte Chemie – International Edition*, **40**, 74–91.
27 Niu, Y., Yeung, L.K. and Crook, R.M. (2001) *Journal of the American Chemical Society*, **123**, 6840–6.
28 Rheiner, P. B. and Seebach, D. (1999) *Chemistry – A European Journal*, **5**, 3221–36.
29 Noyori, R. (1994) *Asymmetric Catalysis in Organic Synthesis*, John Wiley & Sons, Ltd, New York.
30 Lin, G.Q., Li, Y.M. and Chan, A.S.C. (2001) *Principles and Applications of*

Asymmetric Synthesis, Wiley-Interscience, New York.
31 Tang, W. and Zhang, X. (2003) *Chemical Reviews*, **103**, 3029–69.
32 Fan, Q.H., Chen, Y.M., Chen, X.M., Jiang, D.Z., Xi, F. and Chan, A.S.C. (2000) *Chemical Communications*, 789–90.
33 Wang, Z.J., Deng, G.J., Li, Y., He, Y.M., Tang, W.J. and Fan, Q.-H. (2007) *Organic Letters*, **9**, 1243–6.
34 Glorius, F. (2005) *Organic and Biomolecular Chemistry*, **3**, 4171–5.
35 Wang, W.B., Lu, S.M., Yang, P.Y., Han, X.W. and Zhou, Y.G. (2003) *Journal of the American Chemical Society*, **125**, 10536–7.
36 Xu, L.J., Lam, K.H., Ji, J.X., Wu, J., Fan, Q.H., Lo, W.H. and Chan, A.S.C. (2005) *Chemical Communications*, 1390–2.
37 Tang, W.J., Zhu, S.F., Xu, L.J., Zhou, Q.L., Fan, Q.H., Zhou, H.F., Lam, K.H. and Chan, A.S.C. (2007) *Chemical Communications*, 613–15.
38 Blaser, H.U., Pugin, B., Spindler, F. and Togni, A. (2002) *Comptes Rendus Chimie*, **5**, 379–85.
39 Noyori, R. and Ohkima, T. (2001) *Angewandte Chemie – International Edition*, **40**, 40–73.
40 Deng, G.J., Fan, Q.H., Chen, X.M. and Liu, G.H. (2003) *Journal of Molecular Catalysis A – Chemical*, **193**, 21–5.
41 Liu, W.G., Cui, X., Cun, L.F., Wu, J., Zhu, J., Deng, J.G. and Fan, Q.H. (2005) *Synlett*, 1591–5.
42 Yu, H.B., Hu, Q.S. and Pu, L. (2000) *Tetrahedron Letters*, **41**, 1681.
43 Yi, B., Fan, Q.H., Deng, G.J., Li, Y.M., Qiu, L.Q. and Chan, A.S.C. (2004) *Organic Letters*, **6**, 1361–4.
44 Fan, Q.H., Deng, G.J., Lin, C.C. and Chan, A.S.C. (2001) *Tetrahedron: Asymmetry*, **12**, 1241–7.
45 Hawker, C.J., Wooley, K.L. and Fréchet, J.M.J. (1993) *Journal of the American Chemical Society*, **115**, 4375–6.
46 Deng, G.J., Li, G.R., Zhu, L.Y., Zhou, H.F., He, Y.M., Fan, Q.H. and Shuai, Z.G. (2006) *Journal of Molecular Catalysis A – Chemical*, **244**, 118–23.
47 Knowles, W.S. and Sabacky, M.J. (1968) *Journal of the Chemical Society D – Chemical Communications*, 1445–6.
48 Jerphagnon, T., Renaud, J.L. and Bruneau, C. (2004) *Tetrahedron: Asymmetry*, **15**, 2101–11.
49 van den Berg, M., Minnaard, A.J., Schudde, E.P., van Esch, J., de Vries, A.H.M., de Vries, J.G. and Feringa, B.L. (2000) *Journal of the American Chemical Society*, **122**, 11539–40.
50 Hu, A.G., Fu, Y., Xie, J.H., Zhou, H., Wang, L.X. and Zhou, Q.L. (2002) *Angewandte Chemie – International Edition*, **41**, 2348–50.
51 Liu, Y. and Ding, K.L. (2005) *Journal of the American Chemical Society*, **127**, 10488–9.
52 Botman, P.N.M., Amore, A., van Heerbeek, R., Back, J.W., Hiemstra, H., Reek, J.N.H. and van Maarseveen, J.H. (2004) *Tetrahedron Letters*, **45**, 5999–6002.
53 van den Berg, M., Minnaard, A.J., Haak, R.M., Leeman, M., Schudde, E.P., Meetsma, A., Feringa, B.L., de Vries, A.H.M., Maljaars, C.E.P., Willans, C.E., Hyett, D., Boogers, J.A.F., Henderickx, H.J.W. and de Vries, J.G. (2003) *Advanced Synthesis Catalysis*, **345**, 308–23.
54 Tang, W.J., Huang, Y.Y., He, Y.M. and Fan, Q.H. (2006) *Tetrahedron: Asymmetry*, **17**, 536–43.
55 Schlüter, A.D. and Rave, J.P. (2000) *Angewandte Chemie – International Edition*, **39**, 864–83.
56 Liang, C.O., Helms, B., Hawker, C.J. and Fréchet, J.M.J. (2003) *Chemical Communications*, 2524–5.
57 Deng, G.J., Yi, B., Huang, Y.Y., Tang, W.J., He, Y.M. and Fan, Q.H. (2004) *Advanced Synthesis Catalysis*, **346**, 1440–4.
58 Noyori, R. and Hashiguchi, S. (1997) *Accounts of Chemical Research*, **30**, 97–102.
59 Gladiali, S. and Alberico, E. (2006) *Chemical Society Reviews*, **35**, 226–36.
60 Chen, Y.C., Wu, T.F., Deng, J.G., Liu, H., Jiang, Y.Z., Choi, M.C.K. and Chan, A.S.C. (2001) *Chemical Communications*, 1488–9.
61 Jiang, L., Wu, T.F., Chen, Y.C., Zhu, J. and Deng, J.G. (2006) *Organic and Biomolecular Chemistry*, **4**, 3319–24.
62 Liu, W.G., Cui, X., Cun, L.F., Zhu, J. and Deng, J.G. (2005) *Tetrahedron: Asymmetry*, **16**, 2525–30.

63 Liu, P.N., Chen, Y.C., Li, X.Q., Tu, Y.Q. and Deng, J.G. (2003) *Tetrahedron: Asymmetry*, **14**, 2481–5.

64 Corey, E.J., Bakshi, R.K. and Shibata, S. (1987) *Journal of the American Chemical Society*, **109**, 5551–3.

65 Deloux, L. and Srebnik, M. (1993) *Chemical Reviews*, **93**, 763–84.

66 Bolm, C., Derrien, N. and Seger, A. (1999) *Chemical Communications*, 2087–8.

67 Wang, G.Y., Liu, X.Y. and Zhao, G. (2006) *Synlett*, 1150–4.

68 Wang, G.Y., Zheng, C.W. and Zhao, G. (2006) *Tetrahedron: Asymmetry*, **17**, 2074–81.

69 Pu, L. and Yu, H.B. (2001) *Chemical Reviews*, **101**, 757–824.

70 Bolm, C., Derrien, N. and Seger, A. (1996) *Synlett*, 387–8.

71 Yamago, S., Furukawa, M., Azuma, A. and Yoshita, J.I. (1998) *Tetrahedron Letters*, **39**, 3783–6.

72 Hu, Q.S., Pugh, V., Sabat, M. and Pu, L. (1999) *Journal of Organic Chemistry*, **64**, 7528–36.

73 Fan, Q.H., Liu, G.H., Chen, X.M., Deng, G.J. and Chan, A.S.C. (2001) *Tetrahedron: Asymmetry*, **12**, 1559–65.

74 Liu, G.H., Tang, W.J. and Fan, Q.H. (2003) *Tetrahedron*, **59**, 8603–11.

75 Liu, X.Y., Wu, X.Y., Chai, Z., Wu, Y.Y., Zhao, G. and Zhu, S.Z. (2005) *Journal of Organic Chemistry*, **70**, 7432–5.

76 Christoffers, J. and Baro, A. (2003) *Angewandte Chemie–International Edition*, **42**, 1688–90.

77 Li, Y.W., Liu, X.Y. and Zhao, G. (2006) *Tetrahedron: Asymmetry*, **17**, 2034–9.

78 Trost, B.M. and Crawley, M.L. (2003) *Chemical Reviews*, **103**, 2921–43.

79 Malkoch, M., Hallman, K., Lutsenko, S., Hult, A., Malmström, E. and Moberg, C. (2002) *Journal of Organic Chemistry*, **67**, 8197–202.

80 Mahrwald, R. (1999) *Chemical Reviews*, **99**, 1095–120.

81 Dalko, P.I. and Moisan, L. (2004) *Angewandte Chemie–International Edition*, **43**, 5138–75.

82 Yang, B.Y., Chen, X.M., Deng, G.J., Zhang, Y.L. and Fan, Q.H. (2003) *Tetrahedron Letters*, **44**, 3535–8.

83 Wu, Y.Y., Zhang, Y.Z., Yu, M.L., Zhao, G. and Wang, S.W. (2006) *Organic Letters*, **8**, 4417–20.

84 Jørgensen, K.A. (2000) *Angewandte Chemie–International Edition*, **39**, 3558–88.

85 Ji, B.M., Yuan, Y., Ding, K.L. and Meng, J.B. (2003) *Chemistry–A European Journal*, **9**, 5989–96.

86 Köllner, C., Pugin, B. and Togni, A. (1998) *Journal of the American Chemical Society*, **120**, 10274–5.

87 Schneider, R., Köllner, C., Weber, I. and Togni, A. (1999) *Chemical Communications*, 2415–16.

88 Engel, G.D. and Gade, L.H. (2002) *Chemistry–A European Journal*, **8**, 4319–29.

89 Chen, Y.C., Wu, T.F., Deng, J.G., Liu, H., Cui, X., Zhu, J., Jiang, Y.Z., Choi, M.C.K. and Chan, A.S.C. (2002) *Journal of Organic Chemistry*, **67**, 5301–6.

90 Chen, Y.C., Wu, T.F., Jiang, L., Deng, J.G., Liu, H., Zhu, J. and Jiang, Y.Z. (2005) *Journal of Organic Chemistry*, **70**, 1006–10.

91 Schmitzer, A., Perez, E., Rico-Lattes, I. and Lattes, A. (1999) *Tetrahedron Letters*, **40**, 2947–50.

92 Schmitzer, A., Franceschi, S., Perez, E., Rico-Lattes, I., Lattes, A., Thion, L., Erard, M. and Vidal, C. (2001) *Journal of the American Chemical Society*, **123**, 5956–61.

93 Schmitzer, A., Perez, E., Rico-Lattes, I. and Lattes, A. (2003) *Tetrahedron: Asymmetry*, **14**, 3719–30.

94 Pastor, I.M. and Yus, M. (2005) *Current Organic Chemistry*, **9**, 1–29.

95 Breinbauer, R. and Jacobsen, E.N. (2000) *Angewandte Chemie–International Edition*, **39**, 3604–7.

96 Sanders-Hovens, M.S.T.H., Jansen, J.F.G.A., Vekemans, J.A.J.M. and Meijer, E.W. (1995) *Polymeric Materials Science and Engineering*, **73**, 338.

97 Peerlings, H.W.I. and Meijer, E.W. (1997) *Chemistry–A European Journal*, **3**, 1563–70.

98 Seebach, D., Marti, R.E. and Hintermann, T. (1996) *Helvetica Chimica Acta*, **79**, 1710–40.

99 Suzuki, T., Hirokawa, Y., Ohtake, K., Shibata, T. and Soai, K. (1997) *Tetrahedron: Asymmetry*, **8**, 4033–40.

100 Sato, I., Shibata, T., Ohtake, K., Kodaka, R., Hirokawa, Y., Shirai, N. and Soai, K. (2000) *Tetrahedron Letters*, **41**, 3123–6.

101 Sato, I., Kodaka, R., Shibata, T., Hirokawa, Y., Shirai, N., Ohtake, K. and Soai, K. (2000) *Tetrahedron: Asymmetry*, **11**, 2271–5.

102 Sato, I., Hosoi, K., Kodaka, R. and Soai, K. (2002) *European Journal of Organic Chemistry*, 3115–18.

103 Hu, Q.S., Sun, C.D. and Monaghan, C.E. (2002) *Tetrahedron Letters*, **43**, 927–30.

104 Ribourdouille, Y., Engel, G.D., Richard-Plouet, M. and Gade, L.H. (2003) *Chemical Communications*, 1228–9.

105 Laurent, R., Caminade, A.M. and Majoral, J.P. (2005) *Tetrahedron Letters*, **46**, 6503–6.

106 Routaboul, L., Vincendeau, S., Turrin, C.O. Caminade, A.M., Majoral, J.P., Daran, J.C. and Manoury, E. (2007) *Journal of Organometallic Chemistry*, **692**, 1064–73.

107 Arai, T., Sekiguti, T., Lizuka, Y., Takizawa, S., Sakamoto, S., Yamaguchi, K. and Sasai, H. (2002) *Tetrahedron: Asymmetry*, **13**, 2083–7.

108 Chow, H.F. and Wang, C.W. (2002) *Helvetica Chimica Acta*, **85**, 3444–53.

109 Bellis, E. and Kokotos, G. (2005) *Journal of Molecular Catalysis A–Chemical*, **241**, 166–74.

110 RajanBabu, T.V. (2003) *Chemical Reviews*, **103**, 2845–60.

111 Shi, W.J., Zhang, Q., Xie, J.H., Zhu, S.F., Hou, G.H. and Zhou, Q.L. (2006) *Journal of the American Chemical Society*, **128**, 2780–1.

112 Rodríguez, L.I., Rossell, O., Seco, M., Grabulosa, A., Muller, G. and Rocamora, M. (2006) *Organometallics*, **25**, 1368–76.

113 King, A.S.H. and Twyman, L.J. (2002) *Journal of the Chemical Society–Perkin Transactions 1*, 2209–18.

114 Rheiner, P. B., Sellner, H. and Seebach, D. (1997) *Helvetica Chimica Acta*, **80**, 2027–31.

115 Sellner, H. and Seebach, D. (1999) *Angewandte Chemie–International Edition*, **38**, 1918–20.

116 Sellner, H., Rheiner, P.B. and Seebach, D. (2002) *Helvetica Chimica Acta*, **85**, 352–87.

117 Sellner, H., Faber, C., Rheiner, P.B. and Seebach, D. (2000) *Chemistry–A European Journal*, **6**, 3692–705.

118 Katsuki, T. (1995) *Coordination Chemistry Reviews*, **140**, 189–214.

119 Sellner, H., Karjalainen, J.K. and Seebach, D. (2001) *Chemistry–A European Journal*, **7**, 2873–87.

120 Chung, Y.M. and Rhee, H.K. (2002) *Chemical Communications*, 238–9.

121 Chung, Y.M. and Rhee, H.K. (2002) *Catalysis Letters*, **82**, 249–53.

122 Chung, Y.M. and Rhee, H.K. (2003) *Comptes Rendus Chimie*, **6**, 695–705.

123 Kehat, T. and Portnoy, M. (2007) *Chemical Communications*, 2823–5.

124 Vankelecom, I.F.J. (2002) *Chemical Reviews*, **102**, 3779–810.

125 Hajji, C. and Haag, R. (2006) *Topics in Organometallic Chemistry*, **20**, 149–76.

126 Feng, Y., He, Y.M., Zhao, L.W., Huang, Y.Y. and Fan, Q.H. (2007) *Organic Letters*, **9**, 2261–4.

127 Ribaudo, F., van Leeuwen, P.W.N.M. and Reek, J.N.H. (2006) *Topics in Organometallic Chemistry*, **20**, 39–59.

128 Chen, R., Bronger, R.P.J., Kamer, P.C.J., van Leeuwen, P.W.N.M. and Reek, J.N.H. (2004) *Journal of the American Chemical Society*, **126**, 14557–66.

129 Voit, B. (2006) *Angewandte Chemie–International Edition*, **45**, 4238–40.

5
Enantioselective Fluorous Catalysis
Gianluca Pozzi

5.1
Introduction

Homogeneous catalysis in multiphase liquid systems holds considerable scientific and practical interest [1]. Throughout the 1980s and the early 1990s, much of the research conducted in this field was devoted to the development of aqueous–organic biphase systems where the catalyst is confined in the aqueous phase and the organic phase contains reagents and products [2]. The complementary 'Fluorous Biphase System' (FBS) concept was introduced in 1994 by Horváth and Rábai, who demonstrated that the Rh-catalyzed hydroformylation of C_8–C_{12} olefins could be conveniently run in a perfluorocarbon–organic system endowed with temperature- and pressure-dependent miscibility behavior [3]. The perfluorocarbon-rich phase of the FBS was referred to as the 'fluorous phase', while the catalyst designed to dissolve in it became a 'fluorous soluble' catalyst or, more simply, a 'fluorous' catalyst. The low miscibility of perfluorocarbons (that is saturated fluids such as alkanes, alkenes, ethers or amines in which all hydrogen atoms have been replaced by fluorine atoms) with most organic solvents (and water) at room temperature allowed an efficient separation of the fluorous phase containing the catalyst from reaction products showing the usual affinity for organic solvents.

Besides the use of fluorous catalysts, the characteristics of which will be briefly outlined in the next section, two features of the original FBS concept are also worth mentioning:

- Although both CO/H_2 and, to a much lesser extent, C_8–C_{12} olefins are soluble in perfluorocarbons, the formation of a single liquid phase by increasing the temperature was crucial for achieving high reaction rates. However, this is neither a strict requirement nor a general behavior of perfluorocarbon–organic solvent mixtures, and examples of FBS where the fluorous and the organic layers remained immiscible during the reaction step were reported soon after the publication of the seminal hydroformylation paper [4–6]. Indeed, even reagents poorly soluble in the fluorous phase can react at a reasonable rate at the interface of the two immiscible layers.

Handbook of Asymmetric Heterogeneous Catalysis. Edited by K. Ding and Y. Uozumi
Copyright © 2008 WILEY-VCH Verlag GmbH & Co. KGaA, Weinheim
ISBN: 978-3-527-31913-8

- Carbonyl compounds are more polar than olefins and therefore much less soluble in perfluorocarbons, so the aldehyde products were expelled from the fluorous phase and efficiently taken up in the organic phase. Most catalytic FBS reactions developed to date likewise have involved substrates of low polarity that are converted to more polar products. Some exceptions to this rule of thumb are known, however, as exemplified by an efficient FBS process for the esterification of equimolar amounts of an alcohol and a carboxylic acid [7].

During the past decade the development of the FBS approach with its advantageous separation and recycling properties was reported across a broad spectrum of metal-catalyzed processes [8, 9], among which enantioselective reactions where the effective recovery of precious homogeneous chiral catalysts is a sought-after goal [10]. The meaning of the adjective 'fluorous' also became broader, and today it indicates perfluoroalkyl-labeled species or highly fluorinated saturated organic materials, molecules or molecular fragments in general [11]. At the same time, the research field has expanded to include complementary strategies that do not specifically depend on the use of perfluorocarbons as reaction media, but that are nonetheless based on the ease of separation of fluorous from nonfluorous compounds [12–14]. The present chapter provides an overview of both FBS and homogeneous enantioselective reactions promoted by fluorous (in the broad sense) catalysts.

5.2
Designing Fluorous Catalysts

The presence of one or more fluorocarbon domains increases the affinity of a given molecule for perfluorocarbons, reflecting a 'like dissolves like' effect. [3] Accordingly, conventional homogeneous catalysts can be rendered fluorous-soluble by incorporating fluorocarbon moieties (called 'fluorous ponytails') to their structures in appropriate size and number. The same applies to organic, inorganic and organometallic reagents. In the case of organometallic catalysts, the attachment of fluorous ponytails to the ligand is the preferred strategy, whereas the active metal site is not modified. However, highly fluorinated counteranions (e.g. $C_nF_{2n+1}COO^-$ and BF_4^-) can be used in addition to fluorous ponytails to increase the fluorous phase affinity of cationic metal complexes [15]. The most useful fluorous ponytails are linear or branched perfluoroalkyl C_6–C_{12} chains (often indicated as R_F) that may contain other heteroatoms.

Another crucial element in the design of fluorous catalysts is the minimization of the electronic consequences of incorporating many extremely electron-withdrawing fluorine atoms into the ligand. Indeed, the attachment of fluorous ponytails could significantly change the electronic properties and consequently the reactivity of fluorous catalysts. Therefore, a certain number of insulating groups must be inserted between the fluorous ponytails and the reactive site of a molecule made fluorous, in order to maintain its original properties. These concepts are

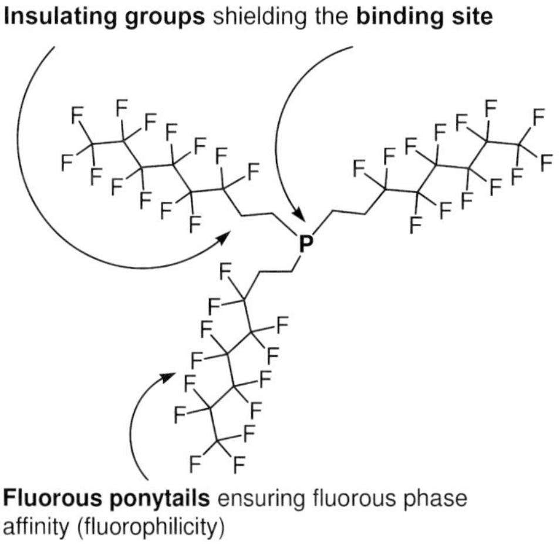

Insulating groups shielding the **binding site**

Fluorous ponytails ensuring fluorous phase affinity (fluorophilicity)

Figure 5.1 Phosphine 1 and the key features of a fluorous ligand.

embodied in the archetype fluorous phosphine 1 (Figure 5.1), where $-(CH_2)_2-$ units were used to minimize the electronic influence of the perfluoroalkyl substituents on the donor atom [3]. Theoretical calculations [16] and experimental spectroscopic data [17] later revealed that even very long hydrocarbon inserts do not completely insulate the phosphorus atom from the electronic influence of the fluorinated units, and that beyond a $-(CH_2)_3-$ group the variation of the shielding effect with additional $-CH_2-$ units is minimal. Besides methylene spacers, insulating units containing aryl rings, heteroatoms – or a combination of all of these elements – have been successfully employed. In particular, spectroscopic and catalytic studies indicate that aryl rings are better insulators than the $-(CH_2)_2-$ spacer, although they do not completely shield the donor atom(s) from the electronic influence of the fluorinated substituent(s) [18].

Fluorous catalysts can be partially, preferentially or even exclusively soluble in a fluorous liquid phase, depending on the quantity and nature of the ponytails they bear. In any case, fluorous catalysts are 'fluorophilic' molecules; that is, they show an affinity for fluorous media under a given set of conditions while standard organic molecules do not. It should be noted that, besides perfluorocarbons, fluorous media can include fluorinated greases and fluorous-coated solid phases. Fluorophilicity may be used interchangeably with fluorous phase affinity, and it is often quantified by mean of a perfluorocarbon/organic liquid/liquid partition coefficient P defined as $[Catalyst]_{Perfluorocarbon}/[Catalyst]_{Organic\ Solvent}$. Rábai, Gladysz and coworkers have collected many published data regarding the measured partition coefficients of various types of fluorous compound [19]. On careful examination of these data the following empirical rules were established, which are also supported by computational models predicting the fluorophilicity of a wide range of molecules [20–23]:

- At least 60 wt% fluorine is required to obtain fluorous molecules that are preferentially soluble in the fluorous phase (partition coefficient $P > 1$).

- Under the same fluorine loading, an increased number of fluorous ponytails usually results in higher P values.
- Molecules bearing longer fluorous ponytails show increased P values, together with decreased absolute solubilities in both phases.
- Fluorophilicity decreases in the presence of functional groups capable of intermolecular attractive interactions (through orientation forces, hydrogen bonds or induction forces). This effect can be counterbalanced by introducing more fluorous ponytails in the molecular structure.
- Branching and flexibility of the fluorous ponytails could have influence on the phase behavior of a fluorous molecule.

According to a loose convention, 'heavy fluorous' (F > 60%) and 'light fluorous' (F < 60%) molecules can be distinguished on the bases of their fluorine content. This should reflect their higher affinity for the fluorous- and the organic phase, respectively. As other factors have an influence on fluorous phase affinities, Gladysz and Curran suggested reserving the term 'heavy fluorous' for cases where two or more ponytails emerge from a common atom or molecular fragment [11]. In any case, heavy fluorous catalysts are better suited for FBS applications, whereas light fluorous catalysts showing limited solubility in perfluorocarbons can be conveniently employed under standard homogeneous conditions and then quickly recovered by means of separation methods based primarily on the peculiar phase-behavior ensured by the presence of fluorous domains in the catalyst structure [12–14]. The extraction of crude reaction mixtures with perfluorocarbons (fluorous liquid-liquid extraction; F-LLE) [24] and solid-phase extraction on silica gel modified by perfluoroalkyl chains (fluorous solid-phase extraction; F-SPE) [25] are the most popular among these methods and they are extensively applied in fluorous enantioselective catalysis.

5.3
C—O Bond Formation

5.3.1
Epoxidation

Catalytic oxidation reactions were among the earliest explored applications of the FBS concept because of the chemical and thermal stability of perfluorocarbons, the convenient partition of the final polar products into the organic phase, and the possible increased lifetime of the catalysts confined in the fluorous phase [4–6]. The first example of enantioselective catalysis under FBS conditions, reported in 1998, also dealt with an oxidation process, namely the epoxidation of prochiral alkenes [26]. Mn(III)-complexes of salen ligands **2** and **3** were found to catalyze the asymmetric epoxidation of indene in a two-phase system CH_2Cl_2/perfluorooctane at 20 °C under an atmospheric pressure of oxygen in the presence of pivalaldehyde (Scheme 5.1).

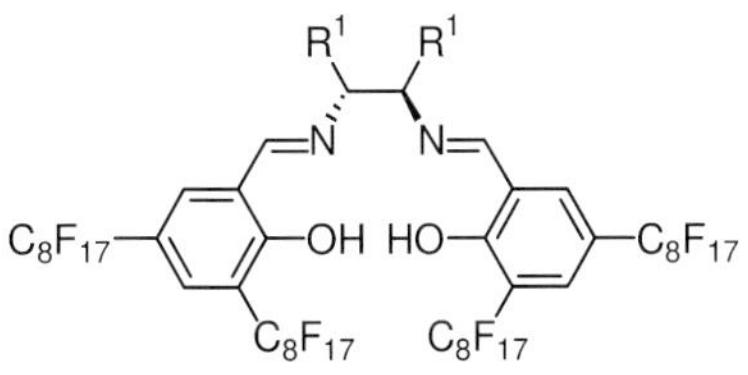

5.3 C—O Bond Formation

2 $R^1, R^1 = -(CH_2)_4-$
3 $R^1 = Ph$

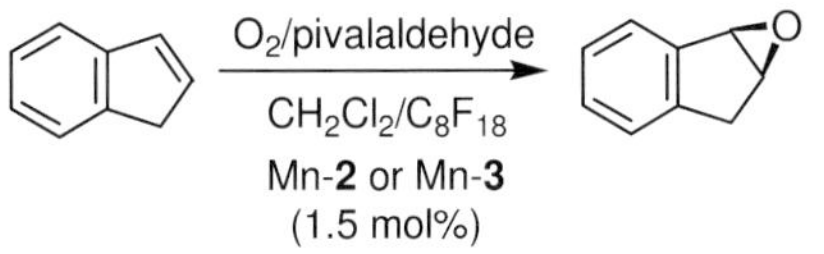

Scheme 5.1 Enantioselective epoxidation of indene under FBS conditions.

Catalysts Mn-**2** and Mn-**3** gave comparable epoxide yields (83 and 77%, respectively) and enantiomeric excess (ee) values of 92 and 90%, respectively. Recycling of the fluorous phase without any loss of catalytic activity and enantioselectivity in a second run was also possible. However, the same catalytic system afforded very low enantioselectivities in the epoxidation of other cyclic alkenes, such as dihydronaphthalene, and this occurred whichever oxidant and solvent system was used [27]. Despite the limited success in catalysis, these studies showed that enantiopure fluorous ligands could be prepared by attaching fluorous building blocks to a pre-existing chiral scaffold. At the same time, it was made clear that the insertion of fluorous ponytails could strongly influence the level of enantioselection attainable with chiral catalysts. Such a notion was further supported by the results obtained with the Mn(III)-complexes of fluorous salen ligands **4–7** bearing bulky t-butyl groups and differently spaced fluorous ponytails in the in the 3,3′- and 5,5′ positions of the aryl rings, respectively (Scheme 5.2) [28, 29]. These compounds were synthesized with the idea of ensuring high steric hindrance and better shielding the metal site from the electronic-withdrawing effects of the R_F substituents. They were first evaluated as catalysts in the epoxidation of dihydronaphthalene with a combination of meta-chloroperbenzoic acid and N-methylmorpholine N-oxide. Reactions performed at −50 °C in a homogeneous solvent mixture CH_2Cl_2/benzotrifluoride (BTF)[1] afforded the epoxide with enantioselectivities ranging from 33% ee (catalyst = Mn-**5**) to 63% ee (catalysts = Mn-**7**), and much higher than those obtained with Mn-**2** and Mn-**3** (16 and 12% ee, respectively). Mn(III)-complexes of ligands **6** and **7** featuring the fluorinated $C_7F_{15}COO^-$ counteranion were next used for the epoxidation of dihydronaphthalene and other cyclic and

---

[1] Benzotrifluoride ($C_6H_5CF_3$) or BTF is a partially fluorinated solvent that dissolves both fluorous- and purely organic molecules.

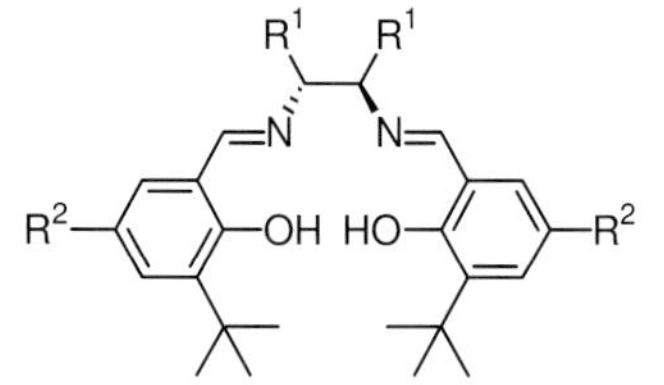

4 $R^1, R^1 = -(CH_2)_4-$, $R^2 = C_8F_{17}$
5 $R^1 = Ph$, $R^2 = C_8F_{17}$
6 $R^1, R^1 = -(CH_2)_4-$, $R^2 = 3,5-(C_8F_{17})_2C_6H_3-$
7 $R^1, R^1 = -(CH_2)_4-$, $R^2 = 2,3,4-(C_8F_{17}C_2H_4O)_3C_6H_2-$

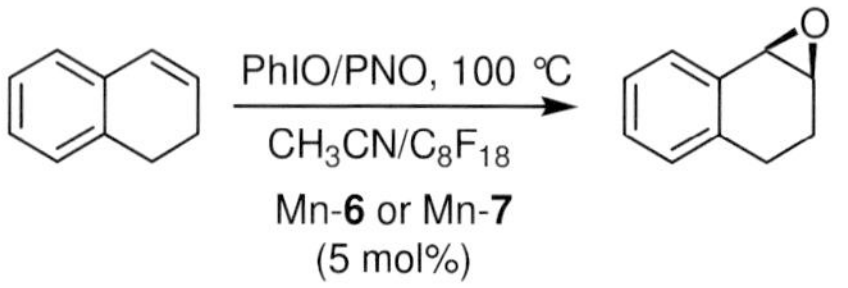

Scheme 5.2 Enantioselective epoxidation of dihydronaphthalene under FBS conditions.

linear alkenes with iodosylbenzene and pyridine N-oxide (PNO) in a FBS of CH_3CN/perfluorooctane. The two catalysts behaved quite similarly, affording epoxides in 68–98% yields and 50–92% ee, very close to the values obtained using standard Mn(III)-salen complexes in CH_3CN. For the FBS epoxidations both reaction yields and enantioselectivities rose with the temperature, the best results being obtained at 100 °C, which corresponded to the boiling point of n-perfluorooctane. The unforeseen beneficial effect on enantioselectivity was ascribed to the increased miscibility of the two liquid layers, which however did not form a single phase during the reaction step [29]. Finally, the fluorous layer containing the catalyst could be efficiently recycled and the same activities and enantioselectivities were maintained for three consecutive runs. The lower activity generally observed for the fourth run was due mainly to the oxidative decomposition of the catalyst; this was analogous to the situation observed in the case of nonfluorous Mn(salen) complexes immobilized onto soluble or insoluble polymers [30].

5.3.2
Hydrolytic Kinetic Resolution

Cationic Co(III)-complexes of selected chiral salen ligands efficiently catalyze the hydrolytic kinetic resolution (HKR) of terminal epoxides to provide enantioenriched terminal epoxides and 1,2-diols in almost theoretical yields (50% each) and excellent enantioselectivities (up to 99% ee) [31]. Similar results were obtained using Co(III)-complexes of ligands **2**, **4** and **5** as catalysts for the HKR of terminal epoxides in homogeneous systems where the epoxide acted as the solvent (Scheme 5.3) [32]. The catalytically active Co(III) species were obtained prior to reaction by aerobic oxidation of the corresponding Co(II)-complexes in the presence of

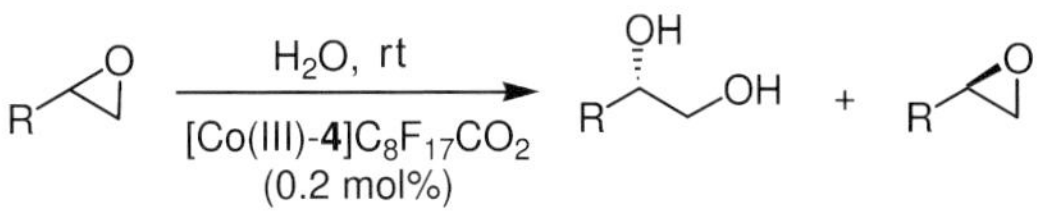

Scheme 5.3 Hydrolytic kinetic resolution of terminal epoxides.

$C_8F_{17}COOH$. The perfluorocarboxylate counteranion enhanced the activity of the cationic fluorous catalysts, among which the complex prepared from ligand **4** gave the best results. This light fluorous catalyst could be conveniently recovered by distilling off the products, and reactivated upon treatment with $C_8F_{17}COOH$ in air. After four subsequent runs the activity of the catalyst was somewhat decreased, although the chemical yields and enantioselectivities were almost unaffected. In comparison, specific fluorous separation techniques such as F-LLE in a continuous apparatus or F-SPE on fluorous reverse-phase silica allowed the efficient separation of the products, but led to incomplete recovery of the catalyst which could not entirely removed from the fluorous phase.

Heavy fluorous $[Co(III)\text{-}6]^+X^-$ complexes were next tested in the HKR of terminal epoxides under epoxide/perfluorooctane FBS conditions [33]. The nature of X^- strongly influenced the activity of the Co(III) catalysts, and once again optimal results were obtained with $X^- = C_8F_{17}CO_2^-$. As an example, HKR of racemic 1-hexene oxide in the presence of 0.2 mol% of $[Co(III)\text{-}6]^+C_8F_{17}CO_2^-$ dissolved in perfluorooctane afforded (S)-1,2-hexane diol and (R)-1-hexene oxide in 49 and 46% yields, respectively, with ee-values of 99%. The reaction products were readily isolated by fractional distillation of the upper organic layer, but the bottom fluorous layer could not be used as such for a subsequent run as it contained variable amounts of Co(II)-**6**. Regeneration of the active Co(III)-complex was attempted, but a significant drop in catalytic activity was observed, regardless of the oxidation procedure.

5.3.3
Allylic Oxidation

Although several enantioselective versions of the Cu-catalyzed oxidation of olefins with peresters to give allyl esters (often referred to as the Kharasch–Sosnovsky reaction) have been reported [34], recycling of the chiral catalysts employed remains an issue. The first attempt to solve this problem by means of a fluorous approach was reported by Fache and Piva, who studied the allylic oxidation of cyclohexene with t-butyl perbenzoate (Scheme 5.4) in the presence of Cu_2O (5 mol%) and fluorous proline **7** (13 mol%) in various fluorinated and nonfluorinated solvents [35]. The catalytic system was completely inactive under FBS conditions, and the highest allyl ester yield (77%) and enantioselectivity (20% ee) were achieved using pure hexafluoroisopropanol as a solvent. It should be noted that the analogous nonfluorous system proline/Cu_2O in benzene afforded the same product in 59% yield and 45% ee [36]. Nonetheless, the fluorous catalyst recovered after evaporation of the

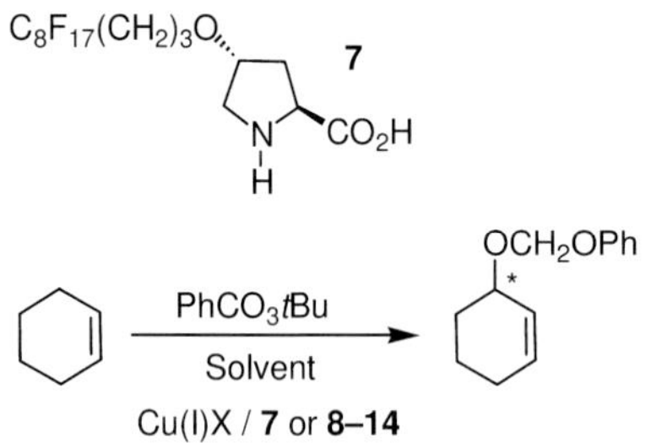

Scheme 5.4 Allylic oxidation of cyclohexene.

Reagents/conditions shown in scheme:
- Substrate: cyclohexene + PhCO$_3$tBu, Solvent, Cu(I)X / **7** or **8–14** → 3-(PhCO$_2$)-cyclohexene (OCH$_2$OPh shown as OC(O)Ph product)

Ligand **7**: C$_8$F$_{17}$(CH$_2$)$_3$O-substituted pyrrolidine-2-carboxylic acid (N–H, CO$_2$H)

Bis(oxazoline) ligands:

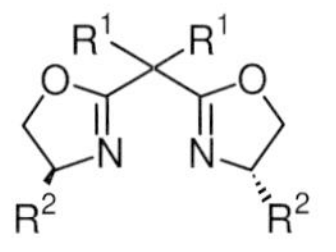

8 R^1 = C$_8$F$_{17}$(CH$_2$)$_3$-, R^2 = Ph
9 R^1 = C$_8$F$_{17}$(CH$_2$)$_3$-, R^2 = iPr
10 R^1 = C$_{10}$F$_{21}$(CH$_2$)$_3$-, R^2 = Ph
11 R^1 = C$_{10}$F$_{21}$(CH$_2$)$_3$-, R^2 = iPr
12 R^1 = C$_8$F$_{17}$(CH$_2$)$_3$-, R^2 = [4-(C$_7$F$_{15}$CH$_2$O)C$_6$H$_4$]CH$_2$-
13 R^1 = H, R^2 = C$_8$F$_{17}$(CH$_2$)$_3$OCH$_2$-
14 R^1 = C$_{11}$H$_{23}$, R^2 = iPr

solvent and extraction of the organic compounds with petroleum ether was reused in a subsequent run affording the allyl ester in 54% yield and 13% ee.

Later, Bayardon and Sinou prepared a series of enantiopure fluorous bis(oxazoline)s **8–13** with fluorine contents ranging between 53 and 59% (Scheme 5.4), and demonstrated their use in representative catalytic asymmetric processes, including the Kharasch–Sosnovsky reaction [37, 38]. The allylic oxidation of cyclohexene was performed using a monophasic organic system CHCl$_3$/CH$_3$CN or a FBS CH$_3$CN/FC-72,[2] in the presence of a combination of fluorous bis(oxazoline) (8 mol%) and Cu(I)X (X = OTf or PF$_6$, 5 mol%). Enantioselectivities ranging from 43% (ligand = **13**) to 73% ee (ligand = **8**) were obtained, depending on the fluorous bis(oxazoline) used. The nature of the reaction medium and that of the Cu precursor had little or no influence on these results, neither did the presence of properly spaced R$_F$ substituents significantly modify the activity and enantioselectivity of the ligands with respect to their nonfluorous analogues, as demonstrated by comparison between ligands **9** (yield = 67%, ee = 61%, FBS conditions) and **14** (yield = 64%, ee = 61%, homogeneous CHCl$_3$/CH$_3$CN conditions). Attempts at recycling the fluorous layer as such were foiled by the limited affinity of the postulated catalytic species for perfluorocarbons. Nevertheless, evaporation of the volatiles followed by addition of hexane allowed the precipitation of the Cu(I)-complex derived from **9** that was reused in a next run, affording the product in 53% yield and 37% ee.

---

2) Commercially available perfluorocarbons used as reaction media are often mixtures of compounds and are indicated with their trade names. In particular, FC-72 is mainly a mixture of perfluorohexanes.

5.4
C—H Bond Formation

5.4.1
Reduction of Ketones

The catalytic asymmetric transfer hydrogenation of ketones either with 2-propanol or with a HCO_2H/Et_3N mixture as a hydride source represents a valuable alternative to asymmetric reductions with H_2, in particular for small- to medium-scale reactions. Numerous catalytic systems based on homogeneous transition-metal complexes of bidentate chiral ligands such as diamines, diimines and β-amino alcohols have been developed to this purpose, and enantioselectivities up to 99% ee have been obtained by using a Ru catalyst bearing mono-N-tosylated diphenylethylenediamine as the ligand [39].

Ir-complexes (5 mol%) associated with fluorous chiral salen **2–6** and diimine **15** (10 mol%) have also been shown to be effective catalysts in the hydrogen transfer reduction of aryl alkyl ketones to give the corresponding secondary alcohols (Scheme 5.5) [40, 41]. The reactions were carried out at 70 °C in a FBS where the hydride source isopropanol acted also as the organic solvent, in the presence of KOH as a promoter. Enantioselectivities of up to 60% ee were achieved using salen **5** in the reduction of ethyl phenyl ketone; these were slightly higher than those obtained using comparable nonfluorous chiral aldimines under homogeneous conditions [40]. However, recycling of the fluorous layer gave poor results, as diimine-type ligands partly decomposed under basic reaction conditions and the

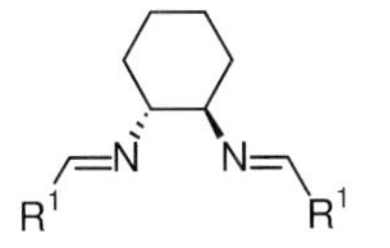

15 $R^1 = 3,5-(C_8F_{17})_2C_6H_3$

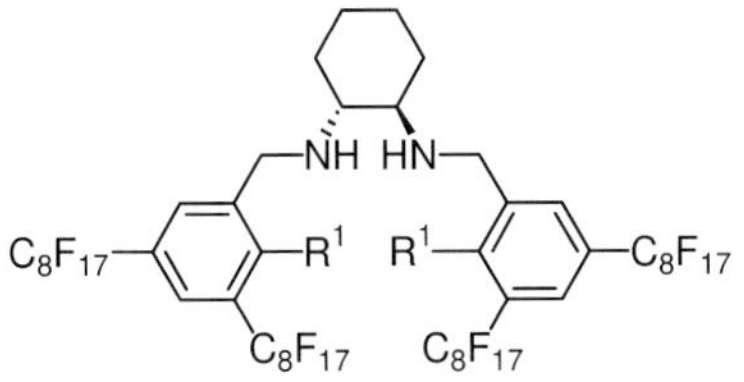

16 $R^1 = OH$

17 $R^1 = H$

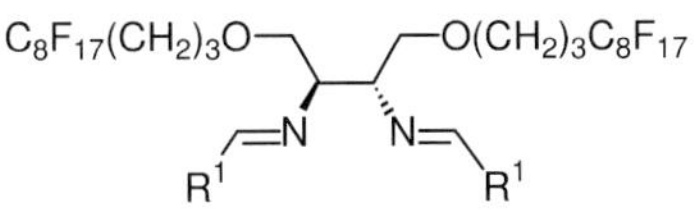

18 $R^1 = 3,5-(C_8F_{17})_2C_6H_3$

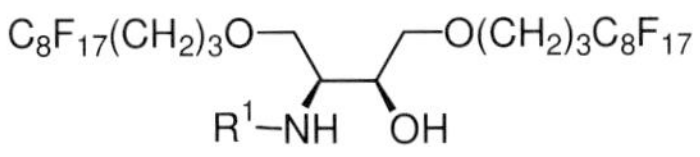

19 $R^1 = [3,5-(C_8F_{17})_2C_6H_3]CH_2-$
20 $R^1 = H$

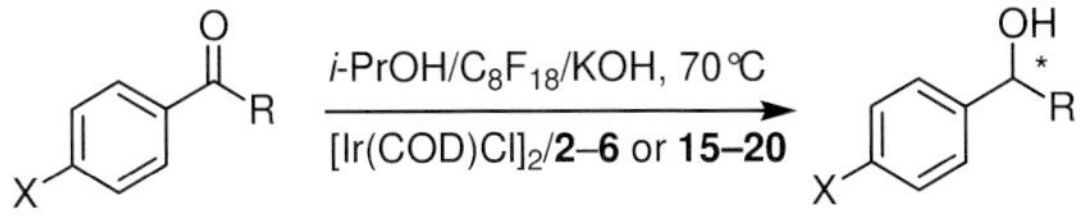

Scheme 5.5 Hydrogen-transfer reduction of ketones under FBS conditions.

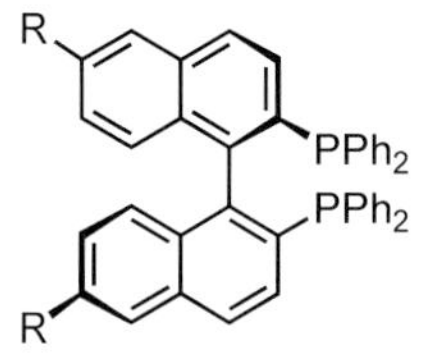

21 R = C$_6$F$_{13}$ **22** R = C$_6$F$_{13}$CH=CH-   **23** R = C$_6$F$_{13}$(CH$_2$)$_2$-

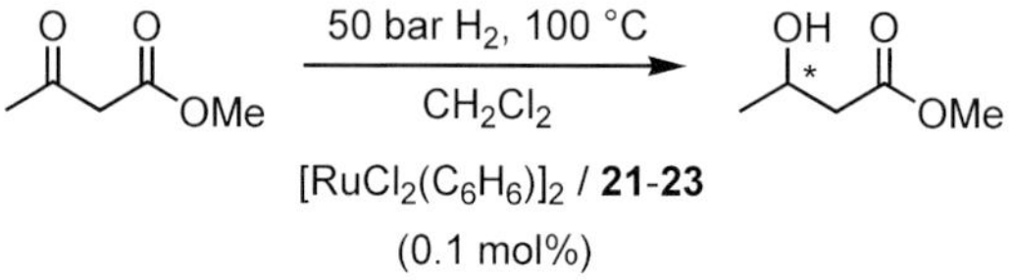

Scheme 5.6 Hydrogenation of methyl acetoacetate: (S)-enantiomers of **21–23** were also tested.

metal was massively lost into the organic phase. In order to circumvent this problem, fluorous diamines such as **16** and **17** obtained by reduction of the C=N groups of **2** and **15**, respectively, were used as ligands of [Ir(COD)Cl]$_2$ [41]. The best results were obtained for diamines without chelating hydroxy groups in the 1,1′ positions, as shown by the behavior of **17** which gave similar activities to salen **5** and enantioselectivities up to 79% ee in the hydrogen-transfer reduction of acetophenone. The catalytically active fluorous layer could be recycled up to four times with moderate loss of Ir (4% in the first run, then ≤1%).

The hydrogen-transfer reduction of acetophenone under FBS conditions was also readily achieved in the presence of fluorous chiral diamines, diimines and β-amino alcohols derived from tartaric acid (e.g. **18–20**; Scheme 5.5) in combination with [Ir(COD)Cl]$_2$ or [Ru(p-cymene)Cl$_2$]$_2$ [42], but much lower enantioselectivities (up to 31% ee in the case of **18**/[Ir(COD)Cl]$_2$) were obtained.

Hope and coworkers successfully employed a series of enantiopure fluorous analogues of 2,2′-bis(diphenylphosphino)-1,1′-binaphthyl (BINAP) in the Ru-catalyzed asymmetric hydrogenation of methyl acetoacetate (Scheme 5.6) [43]. As expected for bulky compounds with only two R$_F$ substituents, ligands **21–23** partitioned almost exclusively into the organic phase of a toluene/perfluoro-1,3-dimethylcyclohexane biphase system. The reactions were thus run in CH$_2$Cl$_2$ and the results compared directly to those obtained for the parent catalyst formed with enantiopure BINAP. The presence of R$_F$ substituents has no apparent effect on the activity or enantioselectivity of the fluorous catalysts in this system. Indeed, the substrate was quantitatively consumed in 1 h and the product readily isolated using F-SPE with ee-values ranging between 72% (ligand = **21**) and 80% (ligand = **23**), as compared with 76% ee achieved with (S)-BINAP. Recovery of the catalytically active species using F-SPE was unsuccessful, because the metal bound irreversibly to the surface of the fluorous reverse-phase silica. However, the fluorous BINAP ligands contaminated with small amounts of the corresponding phosphine

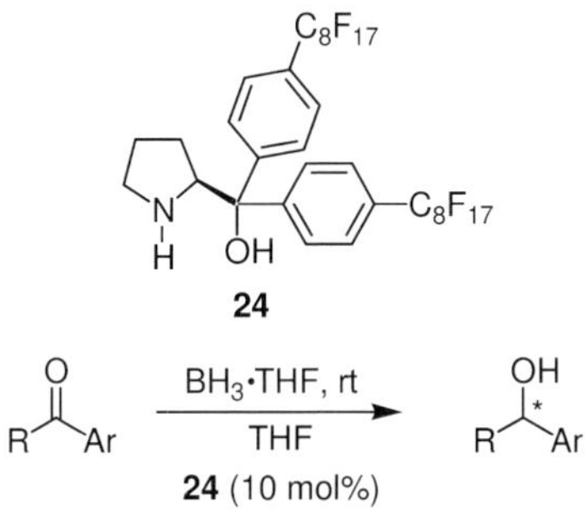

Scheme 5.7 Fluorous Corey–Bakshi–Shibata reduction of ketones.

oxides could be recovered by elution with degassed CH_2Cl_2, and reused in a second run after the addition of fresh aliquots of $[RuCl_2(C_6H_6)]_2$, without compromising activity or enantioselectivity.

The hydrolysis of fluorous chiral oxaborolidine catalysts under F-SPE recycling conditions was exploited by Soós and coworkers for the development of an efficient and operationally simple enantioselective borane reduction of ketones [44]. The fluorous prolinol **24** treated *in situ* with boron sources such as $B(OMe)_3$ or $BH_3 \cdot THF$ smoothly generated the corresponding oxazaborolidines, fluorous analogues of the Corey–Bakshi–Shibata (CBS) catalyst [45], which promoted the reduction of prochiral aryl alkyl, biaryl and cycloalkyl ketones (Scheme 5.7). The reaction mixtures were quenched with MeOH and H_2O and loaded onto fluorous, reverse-phase silica cartridges. The secondary alcohols were quickly removed from the fluorous support by washing with aqueous CH_3CN and isolated in excellent yields (up to 93%) and enantioselectivities (up to 95% ee). The fluorous precatalyst **24** was recovered quantitatively in a second-pass elution using THF and recycled twice more, without loss of catalytic activity and enantioselectivity.

5.4.2
Reduction of C=N and C=C Bonds

Despite the high degree of efficiency reached with traditional homogeneous chiral catalysts, the enantioselective reduction of imines and olefins with various substitution patterns in the presence of fluorous chiral catalysts has been only minimally explored. Early examples concerned reactions in supercritical carbon dioxide (scCO$_2$), the solvent properties of which are roughly comparable to those of hexane. In fact, the insertion of R_F substituents, as well as the use of highly lipophilic fluorinated counteranions, is known to enhance the affinity of polar organometallic complexes for scCO$_2$. Leitner, Pfaltz and coworkers thus prepared the cationic Ir(I)-complexes **25a–27a** featuring an enantiopure fluorous phosphinodihydrooxazole ligand and the lipophilic counteranions tetrakis(3,5-bis(trifluoromethyl)phenyl)borate (BARF), PF_6^- and Ph_4B^-, respectively [46]. These complexes were tested as catalysts for the hydrogenation of prochiral imines in CH_2Cl_2 or scCO$_2$ (Scheme 5.8). The nature of the counteranion had little influence on the hydrogenation of

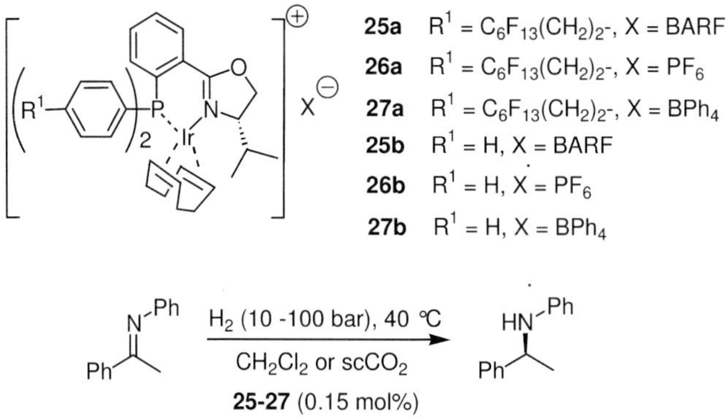

Scheme 5.8 Enantioselective hydrogenation of N-(1-phenylethylidene)aniline.

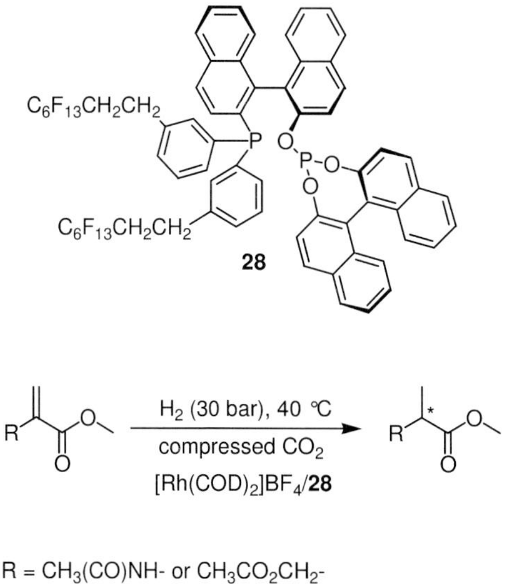

Scheme 5.9 Enantioselective Rh-catalyzed hydrogenation reactions.

the model substrate N-(1-phenylethylidene)aniline in CH_2Cl_2, which occurred with quantitative conversions and enantioselectivities ranging between 80 and 86% ee, as observed in reactions catalyzed by the corresponding nonfluorous complexes **25b–27b**. On the other hand, the level of enantioselection depended heavily on the counteranion for reactions carried out in $scCO_2$, the highest enantioselectivity (80% ee) being obtained using either catalyst **25a** or its nonfluorous analogue **25b**.

Ligand **28**, developed by Leitner's group, was applied (among others) in the Rh-catalyzed hydrogenation of 2-acetamido methylacrylate and dimethyl itaconate in compressed CO_2 (Scheme 5.9). Both substrates were reduced quantitatively in the presence of 0.01 mol% of catalyst, with enantioselectivities up to 97% ee [47].

The asymmetric hydrogenation of dimethyl itaconate catalyzed by Ru-complexes of **21** and **23** in MeOH as a solvent was reported by Stuart and coworkers [48]. As

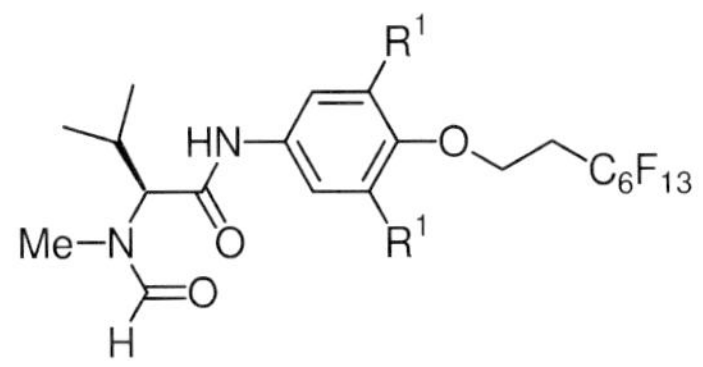

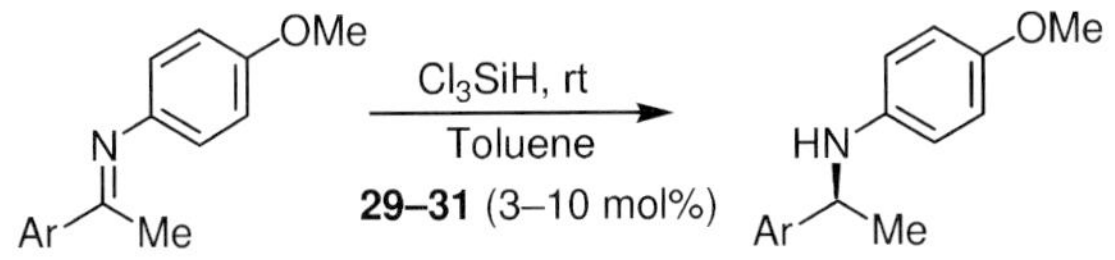

Scheme 5.10 Enantioselective reduction of imines.

found in the hydrogenation of methyl acetoacetate (5.4.1), the level of enantioselectivity (>95% ee) was similar to that achieved using enantiopure BINAP, but conversion of dimethyl itaconate after 15 min was 42% (ligand = **21**), 83% (**23**) and 88% (BINAP), which indicated an adverse influence of the electron-withdrawing R_F substituents on the activity of the hydrogenation catalyst. This deactivating effect could be mostly avoided using an ethylene spacer between the ponytail and the binaphthyl framework.

Fluorous valine-based *N*-methylformamides **29–31** have been recently introduced as efficient and recoverable organocatalysts for the enantioselective reduction of imines derived from methyl aryl ketones with Cl_3SiH (Scheme 5.10) [49]. The reactions were carried out in toluene, and the conversions and enantioselectivities matched those obtained with related catalysts bearing H or OR groups (instead of $O(CH_2)_2R_F$) in the 4-position of the aryl ring. As observed with the latter, the addition of bulky 3,5-substituents had a beneficial effect on the performance of the fluorous catalysts. For instance, the *p*-methoxyanilide of acetophenone was reduced to the corresponding amine in 80% yield and 84% ee in the presence of **29**, whereas **30** and **31** afforded the amine in 90 and 98% yield, with 91 and 89% ee, respectively. The fluorous catalysts and the product were readily separated following a F-SPE protocol, and the isolated catalysts were repeatedly used for the reduction of the model substrate; thus, up to five subsequent runs were carried out in the case of **31**. Although the catalyst recovery was not quantitative, this had only a marginal effect on both the reactivity and selectivity.

5.5
C–C Bond Formation

5.5.1
Addition of Organometallic Reagents to Aldehydes

A wealth of catalytic systems have been developed which promote the nucleophilic 1,2-addition of diorganozinc reagents to aldehydes to give secondary alcohols with excellent enantioselectivities [50]. This reaction has also been studied extensively in the fluorous field, starting with an example reported by van Koten and coworkers [51]. Fluorous chiral ethylzinc arenethiolates **32–34** (Scheme 5.11) were found to be active as catalysts in the asymmetric addition of diethylzinc to benzaldehyde in hexane, their enantioselectivity (up to 94% ee using **34**) being even superior to that of their nonfluorous equivalents (up to 72% ee). The same reaction was then performed with comparable results in a FBS hexane/perfluoromethylcyclohexane. In this case, the fluorous layer was separated and reused four more times, although a considerable drop in enantioselectivity was generally observed after two runs. For instance, 92% ee was achieved with fresh catalyst **34**, whereas 92 and 76% ee were achieved in the second and third runs, respectively.

Amino alcohols are well known to react *in situ* with dialkylzincs to generate Zn-based chiral Lewis acid complexes which can further coordinate with both the aldehyde substrates and the dialkylzinc reagents to conduct the catalytic addition [50]. Thus, Nakamura, Takeuchi and coworkers examined the addition of Et_2Zn to benzaldehyde in the presence of 10 mol% of ephedrine-based fluorous β-amino alcohols **35–37** (Figure 5.2) as catalysts, in toluene or toluene/hexane as solvents [52]. Ligands **35** and **36** gave similar results (yield = 90%, ee = 83%) in a mixture of toluene/hexane as the solvent at room temperature. The more sterically hindered, heavy fluorous ligand **37** was insoluble in toluene and had to be tested in a homogeneous mixture of BTF/hexane, affording the secondary alcohol in lower yield (54%) and poor enantioselectivity (ee = 25%). The enantioselective addition

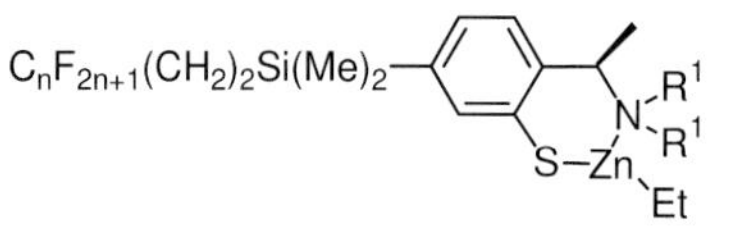

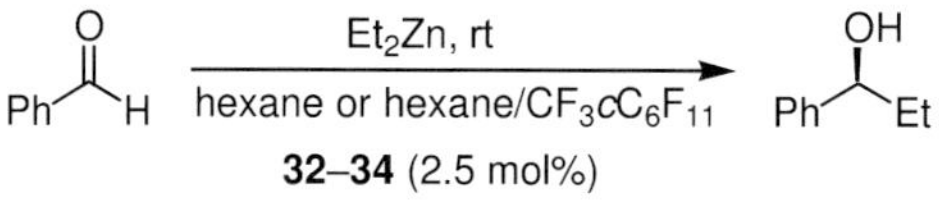

Scheme 5.11 Nucleophilic 1,2 addition of Et_2Zn to benzaldehyde.

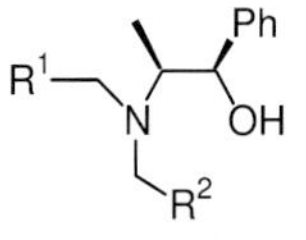

Figure 5.2 Fluorous β-amino alcohols.

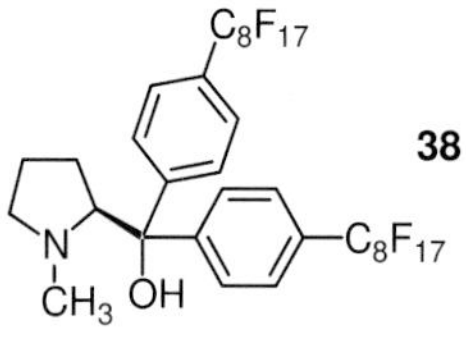

of Et$_2$Zn to representative aldehydes in the presence of **35** was then demonstrated, with enantioselectivities ranging from 84% ee (4-chlorobenzaldehyde) to 70% ee (*trans*-cinnamaldehyde and 3-phenylpropanal). Finally, ligand **35** could be recovered by F-SPE and recycled up to ten times with no apparent loss in catalytic activity and enantioselectivity.

The addition of Et$_2$Zn to benzaldehyde catalyzed by fluorous prolinol **38** (Figure 5.2) in the presence of *n*BuLi was reported by Bolm, Kim and coworkers [53]. The reactions were carried out in either pure hexane or in a FBS hexane/FC-72 using a catalyst loading of 3 mol% with respect to the aldehyde. Using FBS conditions, the ee-value of the product increased from 81%, 86% to 92% when the reactions were performed at 0, 20 and 40 °C, respectively. A further increase of the reaction temperature to 60 °C resulted in a small drop in enantioselectivity (ee = 88%). This temperature effect, which resembled that previously observed in enantioselective FBS epoxidation [28], was tentatively ascribed to the increased solubility of **38** at elevated temperature. Separation of the fluorous layer, followed by the addition of fresh solutions of *n*BuLi and Et$_2$Zn in hexane, restored the catalytic system. Up to six consecutive runs were thus performed at 40 °C without any significant loss of enantioselectivity and reactivity. However, the ee-value of the product fell to 86% in the seventh run and was as low as 81% in the ninth run, a value comparable to that obtained using 1 mol% of fresh **38** at the same temperature. The performance of **38** (10 mol%) in the FBS addition of Ph$_2$Zn to 4-chlorobenzaldehyde was also examined. A mixture of Ph$_2$Zn and Et$_2$Zn was used as the aryl source, and the reactions were better carried out at 40 °C in the absence of *n*BuLi. The highest enantioselectivity (88% ee) was achieved in toluene/FC-72.

Several examples of enantioselective addition of organometallic reagents to aromatic aldehydes catalyzed by Ti-complexes of enantiopure fluorous 1,1′-bi(2-naphthol) (BINOL) derivatives (Figure 5.3) have been disclosed [54–58].

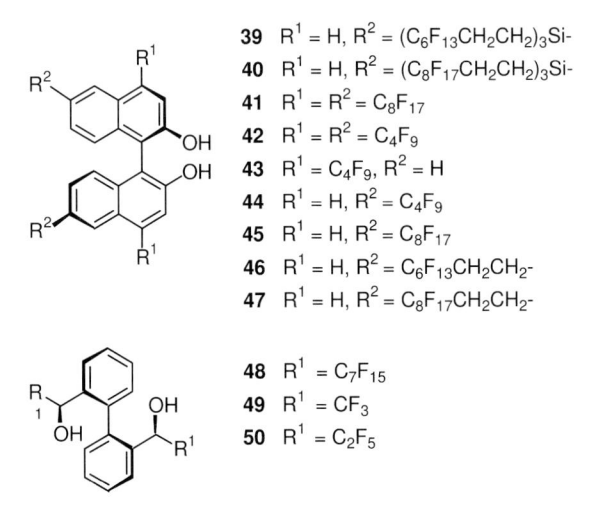

39 R^1 = H, R^2 = (C$_6$F$_{13}$CH$_2$CH$_2$)$_3$Si-
40 R^1 = H, R^2 = (C$_8$F$_{17}$CH$_2$CH$_2$)$_3$Si-
41 R^1 = R^2 = C$_8$F$_{17}$
42 R^1 = R^2 = C$_4$F$_9$
43 R^1 = C$_4$F$_9$, R^2 = H
44 R^1 = H, R^2 = C$_4$F$_9$
45 R^1 = H, R^2 = C$_8$F$_{17}$
46 R^1 = H, R^2 = C$_6$F$_{13}$CH$_2$CH$_2$-
47 R^1 = H, R^2 = C$_8$F$_{17}$CH$_2$CH$_2$-

48 R^1 = C$_7$F$_{15}$
49 R^1 = CF$_3$
50 R^1 = C$_2$F$_5$

Figure 5.3 Fluorous BINOLs and biphenyl alcohols.

Ligands **39** and **40** were developed by the group of Nakamura and Takeuchi, who studied their behavior in the Ti-catalyzed addition of Et$_2$Zn to benzaldehyde [54, 55]. The different size of the fluorous ponytails influenced the partition coefficients of **39** and **40** between toluene and FC-72 only to a limited extent, and both ligands seemed to be suitable for FBS applications. However, when the reaction was performed in the presence of **39** (20 mol%) and Ti(*i*OPr)$_4$ in a FBS toluene-hexane/FC-72, about 10% of the fluorous ligand was recovered from the organic phase after the hydrolytic work-up procedure required to liberate the alcohol. For experiments carried out in toluene/FC-72 the amount of **39** leached into the organic phase was less than 1%, but the enantioselectivity (ee = 78%) was lowered with respect to that achieved in toluene-hexane/FC-72 (ee = 83%). On the other hand, hexane did not enhance the leaching of **40** into the organic phase (consistently <1%) [55]. However, the enantioselectivity observed using this bulkier ligand (ee = 79%) was slightly inferior to that obtained with **39**. The addition of Et$_2$Zn to other aromatic aldehydes catalyzed by Ti(*i*OPr)$_4$ in combination with **39** was studied under homogeneous conditions using BTF as the solvent [55]. The organic products and fluorous ligand were easily separated by F-SPE, and the secondary alcohols were isolated in excellent yields (>90%) with enantioselectivities ranging from 78% ee in the case of 2-naphthylbenzaldehyde to 91% ee in the case of 1-naphthylbenzaldehyde.

Chan's group demonstrated the Ti-catalyzed addition of Et$_2$Zn to benzaldehyde using the BINOL ligand **41** bearing four C$_8$F$_{17}$ ponytails linked directly to the binaphthyl moiety [56]. The fluorous ligand (20 mol%) dissolved in perfluoro(methyldecalin) was added to a solution of Ti(*i*OPr)$_4$ in hexane and the resulting biphase system was heated to 45 °C. A single phase was thus formed to which Et$_2$Zn in hexane and benzaldehyde were added sequentially. On completion of the reaction the homogeneous system was cooled to 0 °C and demixed into two layers which were readily separated. The upper organic phase was subjected to the usual aqueous work-up and purification to give enantiomerically enriched 1-phenylpropanol in 80% yield and 55–60% ee, whereas the lower, fluorous phase in which **41** was immobilized could be reused (up to nine times) after the addition of fresh Ti(*i*PrO)$_4$. Ethylation of aromatic aldehydes with triethylaluminum catalyzed by **41** in combination with Ti(*i*PrO)$_4$ was analogously carried out at 53 °C in hexane/perfluoro(methyldecalin) [57]. Higher enantioselectivities (up to 82% ee) were achieved in the case of benzaldehyde in comparison with Et$_2$Zn addition, but only minor differences were observed with other substrates. Chan's group also reported the synthesis of BINOL derivatives **42–45** (as well as their enantiomers) with varying numbers of fluorous ponytails of different lengths [57]. These light fluorous ligands were completely soluble in CH$_2$Cl$_2$ and could be compared directly to (*R*)-BINOL in the Ti-catalyzed addition of Et$_2$Zn to benzaldehyde, where they gave lower enantioselectivities (70–77% ee) than their parent compound (88% ee).

Zhao and coworkers performed the enantioselective allylation of benzaldehyde with allyltributyltin in the presence of 20 mol% fluorous BINOL **46** or **47** and Ti(*i*PrO)$_4$ [58]. The highest enantioselectivities (up to 90% ee) were obtained using a FBS hexane/FC-72 at 0 °C. The two ligands gave similar results and, because of their relatively low fluorine loading, they were recovered by phase separation followed by repeated extraction of the organic layer with FC-72 and evaporation of the combined fluorous layers.

Kumadaki and coworkers have recently synthesized the enantiopure biphenyl alcohol **48** (Figure 5.3) characterized by the hindered rotation around the aryl–aryl bond and by the presence of two stereogenic quaternary carbons bearing both perfluoroheptyl- and hydroxy groups [59]. Hence, the acidity of the latter was strongly enhanced and **48** could efficiently interact with Ti(iPrO)$_4$ to generate a chiral complex which was found to catalyze the addition of Et$_2$Zn to benzaldehyde in toluene/hexane at −30 °C. Enantioselectivities of up to 97% ee were observed in the presence of 5 mol% **48**, the yield being 96%. Under the same conditions, 85% ee and 91% ee were achieved using ligands **49** and **50**, respectively, thus showing a significant dependence of the level of enantioselection on the size of the perfluoroalkyl substituents. Ligand **48**, which was recovered almost quantitatively from the reaction mixture by repeated extraction with FC-72 and evaporation of the combined fluorous layers, was reused after the addition of fresh Ti(iPrO)$_4$. Seven subsequent runs were carried out with no apparent loss of catalytic activity and enantioselectivity [60]. The catalytic system based on **48** and Ti(iPrO)$_4$ was also active in the methylation of aromatic and aliphatic aldehydes using Me$_2$Zn generated *in situ* from ZnCl$_2$ and MeMgBr [61]. 1-Phenylethanol was obtained from benzaldehyde in 89% yield with 94% ee using a ligand loading of 20 mol%. Enantioselectivities ranging from 12% ee (*p*-anisaldehyde) to 92% ee (*p*-iPr-benzaldehyde) were observed in the methylation of other aromatic aldehydes, whereas extremely high enantioselective methylation was attained with aliphatic aldehydes such as octanal (99% ee). As in the case of Et$_2$Zn addition, the ligand was recovered and recycled by F-LLE.

5.5.2
Pd-Catalyzed Reactions

Enantiopure 2-(diphenylphosphino)-2'-alkoxy-1,1'-binaphthyls (chiral monodentate phosphine ligands; MOPs) are useful ligands for several Pd-catalyzed asymmetric transformations [62]. Sinou, Pozzi and coworkers reported the preparation and use of fluorous MOP derivative **51** for the Pd(0)-catalyzed asymmetric allylic alkylation of 1,3-diphenyl-2-propenyl acetate with various carbon nucleophiles (Scheme 5.12) [63]. Fluorous MOP ligands **52** and **53** independently developed by Sinou and coworkers were also tested in the same C—C bond-forming process [64]. In the case of **51**, the reactions were conveniently carried out in toluene or BTF at room temperature or at 50 °C in the presence of a mild base such as BSA (bis(trimethylsilyl)acetamide)/KOAc. The substitution products were obtained in moderate to excellent yields (67–99%), with enantioselectivities up to 87% ee being attained using dimethyl malonate as a nucleophile in toluene as a solvent. In comparison, catalytic systems based on ligands **52** and **53** showed reduced enantioselectivities (37 and 24% ee, respectively) and a strong base such as NaH was required to activate dimethyl malonate. This was possibly due to steric effects arising from mutual interaction of the perfluoroalkyl chains placed onto the binaphthyl moiety of ligands **52** and **53**. Indeed, electron-withdrawing effects could not be the determining factor as the interposition of $-CH_2CH_2Si-$insulating spacers did not improve the enantioselectivities obtained with ligand **53** in comparison to those obtained with **52**. Pd-complexes generated from ligands **51–53** did not contain enough fluorine to be used under FBS conditions. However, they were easily removed from homogeneous reaction mixtures by liquid–liquid extraction with perfluorooctane. Unfortunately, the recovered fluorous material did not show any catalytic activity when tested in a second run [63].

The chiral diaminophosphine **54** prepared from *(S)*-prolinol (Scheme 5.12) was analogously applied as a Pd(0)-ligand by Mino and coworkers [65]. Various organic solvents were tested as homogeneous reaction media, among which Et_2O gave the best results. The solubility of the Pd-complex prepared from $[Pd(\eta^3-C_3H_5)Cl]_2$ and **54** in Et_2O increased about 15-fold between $-20\,°C$ and $30\,°C$, and this was matched by a considerable increase in the reaction rate, whereas the enantioselectivity of the catalytic system was only slightly reduced. Indeed, ee-values of 94% and 90% were obtained in the asymmetric allylic alkylation of 1,3-diphenyl-2-propenyl acetate with dimethyl malonate at $-20\,°C$ and $30\,°C$, respectively, but at $-20\,°C$ the yield was only 5% after 48 h against 96% after 5 h at $30\,°C$. The fluorous Pd-complex was easily recovered from the reaction mixture by cooling to $0\,°C$, followed by the removal of Et_2O and extraction of the organic products with cold hexane. When diethyl malonate was used as a nucleophile, this protocol allowed the fluorous complex to be recycled up to five times, without any deterioration of its activity and enantioselectivity.

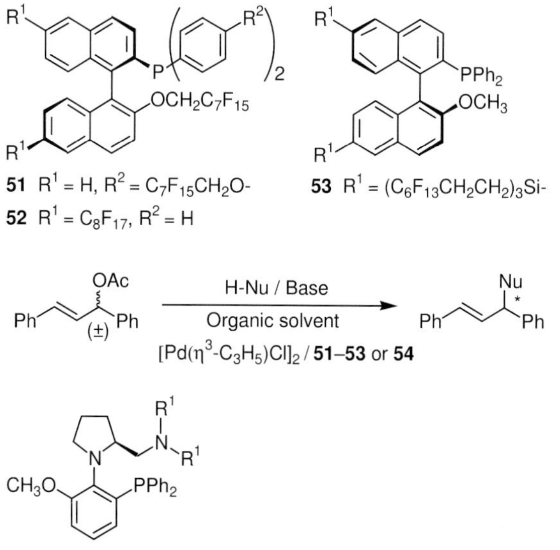

51 $R^1 = H$, $R^2 = C_7F_{15}CH_2O-$
52 $R^1 = C_8F_{17}$, $R^2 = H$

53 $R^1 = (C_6F_{13}CH_2CH_2)_3Si-$

54 $R^1 = C_{11}F_{23}COOCH_2CH_2-$

Scheme 5.12 Pd(0)-catalyzed asymmetric allylic alkylation of 1,3-diphenyl-2-propenyl acetate.

The allylic alkylation of 1,3-diphenyl-2-propenyl acetate catalyzed by Pd(0)-complexes of fluorous bis(oxazolines) **8–13** (see Scheme 5.4) was thoroughly investigated by the group of Sinou [37, 38, 66]. The best results were achieved when the nucleophile was dimethyl malonate, in the presence of BSA/KOAc as base and CH_2Cl_2 as solvent. Under such conditions, the influence of the ligand substitution pattern on catalytic behavior was irrelevant, and a quantitative conversion of the starting material was achieved with enantioselectivities ranging between 92% ee and 94% ee; these were as high as those reported in the literature using nonfluorous bis(oxazolines). The low partition coefficients of the formed Pd-catalysts between perfluorocarbons and CH_2Cl_2 hampered their direct recovery by F-LLE. In addition, the formation of Pd black occurred readily during the reaction step; hence, recovery of the ligands and their subsequent reuse following the addition of a fresh Pd source was attempted. Ligands **8, 10** and **11**, when used in allylic substitutions with dimethyl malonate, were recovered in 76%, 68% and 84% yields, respectively, by evaporation of the organic solvent followed by extraction of the residue with FC-72 [37]. The same procedure allowed the quantitative recovery of **12** which was reused in a second run, affording the substitution product with 98% conversion and 92% ee [38].

Nakamura and coworkers synthesized the fluorous BINAP analogue **55** (Scheme 5.13) with six fluorous ponytails attached via Si atoms to the binaphthyl moiety, and applied it to the asymmetric Heck reaction between 2,3-dihydrofuran and 4-chlorophenyltriflate [67]. Catalytic tests were conducted in either BTF or benzene, in the presence of **55** (6 mol%) and Pd(OAc)$_2$ (3 mol%). The major product, 2-(4-chlorophenyl)-2,3-dihydrofuran, was obtained in similar chemical yields (59%) and enantioselectivities (up to 92% ee with benzene), but the isomer ratio was 88:12 using BTF and 72:28 using benzene. The same reaction, when catalyzed by Pd(OAc)$_2$ in the presence of BINAP, proceeded about threefold faster and afforded the major isomer in 67% yield and 76% ee in BTF and 71% yield and 91% ee in benzene, with increased regioselectivity (isomer ratio = 92:8 in BTF) with respect to **55**. The fluorous ligand was next used in a FBS benzene/FC-72, affording the

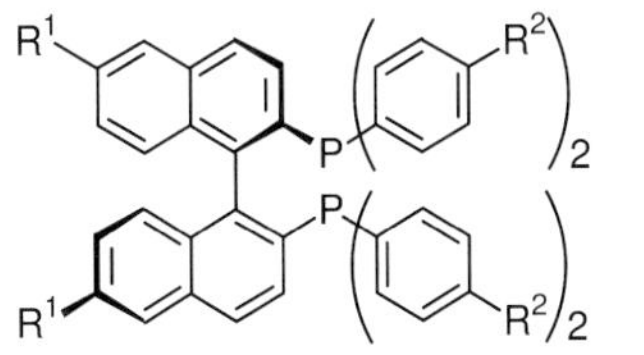

55 R^1 = (C$_6$F$_{13}$CH$_2$CH$_2$)$_3$Si-, R^2 = H
56 R^1 = H, R^2 = C$_7$F$_{15}$CH$_2$O-

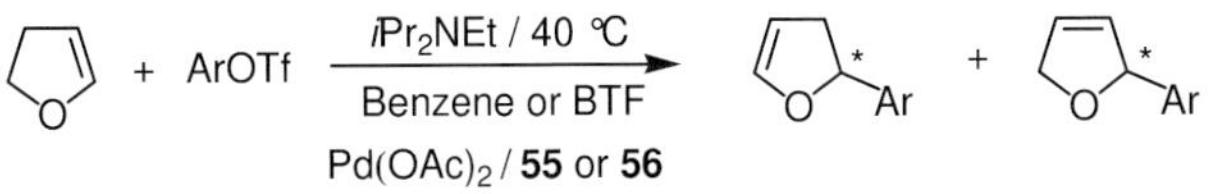

Scheme 5.13 Asymmetric Heck arylation of 2,3-dihydrofuran.

major isomer with an excellent 93% ee, but in reduced yield (39%) and regioselectivity (isomer ratio = 69:31). Attempts to reuse the fluorous phase in further reactions failed due to oxidation of the ligand to the corresponding phosphine oxide.

Pd-catalyzed arylation of 2,3-dihydrofuran with various triflates was later investigated using fluorous BINAP **56** as a ligand, under conditions otherwise equivalent to those reported by the group of Nakamura [68]. The best results were obtained with 4-chlorophenyl triflate as arylating agent in BTF as a solvent. In this case too, 2-(4-chlorophenyl)-2,3-dihydrofuran was obtained as the major product, with moderate enantioselectivity (68% ee), but very high regioselectivity (isomer ratio = 97:3). Triflates such as phenyl triflate and naphthyl triflate gave lower enantioselectivities and/or regioselectivities, while aryl triflates bearing electron-donating substituents were found to be unreactive. Ligand **56** did not contain enough fluorine to be used with success under FBS conditions, but it could be quickly separated from the product by F-LLE, as already shown in the case of MOP **51**.

5.5.3
Cyclopropanation of Styrene

Fluorous chiral bis(oxazolines) featuring a different number of R_F substituents attached to the bridging carbon atom through long oxygenated spacers were designed by Benaglia and coworkers and employed as ligands in various Cu(I)- and Cu(II)-catalyzed reactions [69, 70]. In particular, complexes generated from ligands **57** and **58** in combination with an equimolar amount of CuOTf were evaluated as catalysts in the cyclopropanation of styrene involving ethyl diazoacetate (Scheme 5.14) and compared to related catalysts based on soluble, polymer-supported bis(oxazolines) [70]. The use of the *gem*-disubstituted ligand **57** (10 mol%) in a FBS CH_2Cl_2/perfluorooctane at 20 °C led to a *trans/cis* mixture of cyclopropane adducts (73:27) in 55% overall yield, the major *trans* isomer being obtained with 60% ee. The complex based on the less fluorinated (and less cumbersome) ligand **58** was found to be completely soluble in CH_2Cl_2 and when the cyclopropanation was run in such a solvent a 65:35 mixture of *trans/cis* cyclopropanes was obtained in 68% yield, with enantioselectivity up to 78% ee for the major *trans* isomer. Under

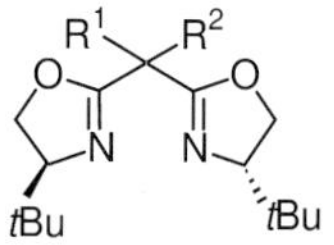

57 $R^1 = R^2 = [3,5$-bis$(C_8F_{17})C_6H_3]CH_2OC_6H_4CH_2$-
58 $R^1 = [3,5$-bis$(C_8F_{17})C_6H_3]CH_2OC_6H_4CH_2$-, R^2 = Me
59 $R^1 = R^2 = C_8F_{17}(CH_2)_3$-

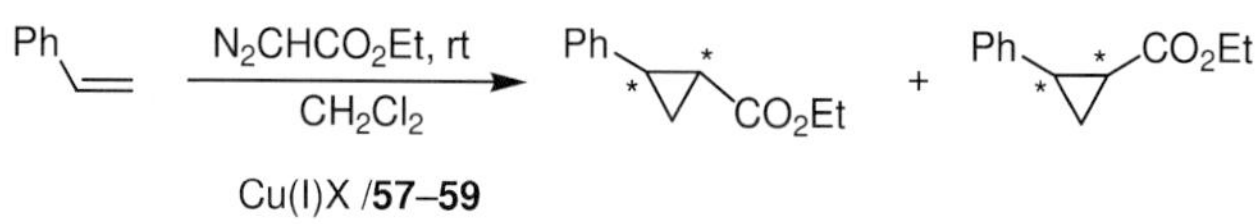

Scheme 5.14 Cu(I)-catalyzed asymmetric cyclopropanation of styrene.

the same conditions, a poly(ethyleneglycol)-supported bis(oxazoline) structurally similar to **58** gave the major *trans* isomer with up to 91% ee, whereas up to 99% ee was reported in the literature for *gem*-dimethyl-substituted bis(oxazoline)s [71]. This trend was tentatively explained as a result of the reduced bite angle for the box moiety arising from repulsion – both steric and electronic in origin – experienced by the bulky fluorinated substituents on the bridging carbon atom. This should result in a more difficult complexation of the Cu cation leading to the presence in solution of ligand-free catalytically active species that promote a poorly enantioselective transformation.

Results obtained in the same reaction operated in the presence of bis(oxazoline) **59** (Scheme 5.14) were in agreement with such a view [72]. Indeed, at 20 °C this ligand bearing two uncomplicated fluorous ponytails at the bridging carbon atom afforded a 68:32 mixture of *trans/cis* cyclopropanes in 22% yield, with 84% ee and 81% ee for the major and minor isomers, respectively. The catalyst loading was only 2 mol%, and the isolated yield could be increased up to 63% with virtually identical diastereo- and enantioselectivity by simply running the reaction at 40 °C. Bis(oxazolines) **8** and **9** (see Scheme 5.4) with the same substitution pattern at the bridging atom as **59** were then evaluated under optimized conditions in the cyclopropanation of styrene involving different alkyl diazoacetates, such as ethyl diazoacetate, *t*-butyl diazoacetate and methyl phenyldiazoacetate. As expected, the use of more bulky diazoacetates resulted in higher *trans/cis* cyclopropane ratios, whatever the ligand used. Mixtures containing up to 98% *trans* and only 2% *cis* isomer were obtained with methyl phenyldiazoacetate in the presence of ligands **9**. In general, ligand **59** afforded slightly higher enantioselectivities than **8** and **9**, with the noteworthy exception of the cyclopropanation of styrene with methyl phenyldiazoacetate, where it gave the major *trans* isomer with only 27% ee.

Finally, the fluorous Cu-catalyst obtained from ligand **59** was recovered by simple evaporation of CH_2Cl_2 followed by addition of cold hexane. Decantation of the liquid layer containing the products afforded a solid that could be reused without further addition of Cu source or ligand. The diastereoselectivities and enantioselectivities were maintained over five runs, although a slight decrease in chemical yield was observed.

The scope and limitations of a variety of enantiopure fluorous C_2-symmetric diimines and diamines as Cu(I) ligands have been established in the cyclopropanation of styrene under both homogeneous and FBS conditions [73, 74]. In contrast to that found for cyclopropanations carried out in the presence of fluorous bis(oxazolines), neither the introduction of perfluoroalkyl substituents in the ligand structure nor the application of FBS conditions had a detrimental effect on chemical yields and selectivities obtained in the presence of fluorous diamine ligands, which remained as high as those obtained using their nonfluorous analogues in halogenated solvents. Moreover, the nature of the Cu(I) source strongly affected the outcome of the reactions and the use of $Cu(CH_3CN)_4BF_4$ or $Cu(CH_3CN)_4PF_6$ was found to be highly beneficial. The best results were achieved with relatively rigid diamines derived from *trans*-1,2-diaminocyclohexane. As an example, when styrene was reacted with ethyl diazoacetate in a FBS CH_2Cl_2/

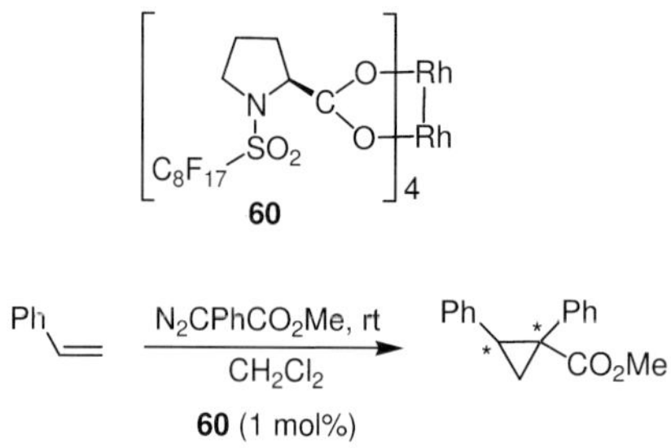

Scheme 5.15 Rh(II)-catalyzed asymmetric cyclopropanation of styrene.

perfluoroctane at 20 °C in the presence of the complex generated from $Cu(CH_3CN)_4PF_6$ (10 mol%) and diamine **17** (see Scheme 5.5; 20 mol%), a 67/33 mixture of *trans/cis* cyclopropane adducts was obtained in 77% overall yield, the ee-value of the major *trans* isomer being 62% [73]. The catalyst was easily removed from the products by simply decanting the fluorous phase that was reused as such in a second run maintaining the same yield and diastereoselectivity. However, the enantioselectivity fell to 46% ee due to partial decomposition of the ligand [73].

Biffis and coworkers prepared a fluorous tetrakis-dirhodium(II)-prolinate catalyst **60** (Scheme 5.15) which was applied in the cyclopropanation of styrene with methyl phenyl diazoacetate [75]. Although complex **60** had a quite low weight percentage of fluorine (about 50%), it showed high affinity for both perfluorocarbons and fluorous reverse-phase silica, and this allowed an assessment to be made of the efficiency of different strategies for catalyst recovery and recycling. Catalytic cyclopropanation tests were first performed under homogeneous conditions in CH_2Cl_2, affording the *trans* product in 82% yield and almost complete diastereoselectivity, but with lower enantioselectivity (ee = 48%) with respect to a related nonfluorous, tetrakis-dirhodium(II)-prolinate (ee = 60%). The catalyst was readily recovered and recycled twice upon extraction from the reaction mixture into perfluoro(methylcyclohexane), followed by solvent evaporation and redissolution in a fresh styrene solution in CH_2Cl_2. Next, the cyclopropanation of styrene was performed in the presence of **60** adsorbed onto fluorous, reversed-phase silica. The enantioselectivity of this reaction reached 60% ee using pentane a solvent, but the cyclopropane yield fell to 66%. The best results were obtained by mixing a solution of **60** in perfluoro(methylcyclohexane) with styrene in excess and no other organic solvent. The products formed upon addition of methyl phenyl diazoacetate were soluble in styrene, whereas the catalyst remained confined in the fluorous phase that was removed from the organic phase by simple decantation on completion of the reaction. Only 2 ppm of rhodium were found in the organic phase, corresponding to 0.1% metal leaching. Under these conditions, the reaction yield attained 79% and the enantioselectivity 62% ee. The enantioselectivity rose to 72% ee in a successive run carried out with the recovered fluorous phase, with no substantial variation of chemical yield, while in the third run the yield fell to 65%.

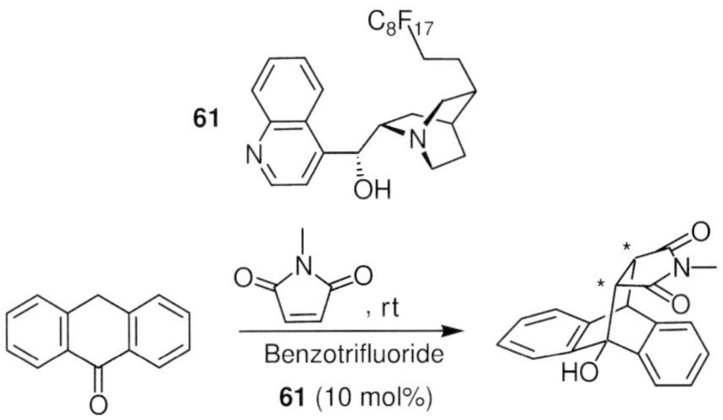

Scheme 5.16 Diels–Alder reaction between anthrone and N-methyl maleidimide.

5.5.4
Metal-Free Catalytic Processes

As outlined in the preceding sections, most fluorous chiral catalysts developed so far have consisted of transition-metal complexes of fluorous ligands. However, a few examples of enantioselective C–C bond-forming reactions catalyzed by metal-free fluorous chiral compounds have also been reported. Fache and Piva made a start in this promising field by preparing a family of fluorous cinchona derivatives that were tested in the base-catalyzed Diels–Alder reaction between anthrone and N-methyl maleidimide (Scheme 5.16) in various solvent systems [76]. Enantioselectivities up to 40% ee and quantitative conversions were achieved using cinchonidine **61** in BTF at room temperature. Both, the catalyst and the reagents were freely soluble in such a solvent, but not the Diels–Alder adduct, about 75% of which precipitated out and could be recovered by simple filtration. Although the liquid phase maintained its catalytic activity in a second run, the enantioselectivity diminished to less than 20% ee.

In the same way, the fluorous proline-derivative **7** (see Scheme 5.4) was applied as a catalyst (30 mol%) in the intermolecular aldol reaction between acetone and p-nitrobenzaldehyde in BTF as a solvent. The chemical yield (72%) and enantioselectivity (73% ee) were fully comparable to those obtained with proline in dimethyl sulfoxide (DMSO), but attempts to recover and reuse the fluorous catalyst failed [35].

The first example of a fully recyclable fluorous chiral metal-free catalyst was reported by Maruoka and coworkers, who described the enantioselective alkylation of a protected glycine derivative (Scheme 5.17) with various benzyl- and alkyl bromides, in the presence of the quaternary ammonium bromide **62** as a phase-transfer catalyst [77]. Reactions were performed in a 50% aqueous KOH/toluene biphasic system in which **62** was poorly soluble. Nevertheless, the alkylated products were obtained in good yields (from 81 to 93%), with enantioselectivity ranging from 87 to 93% ee. Catalyst **62** was recovered by extraction with FC-72, followed by evaporation of the solvent, and could be used at least three times without any loss of activity and selectivity.

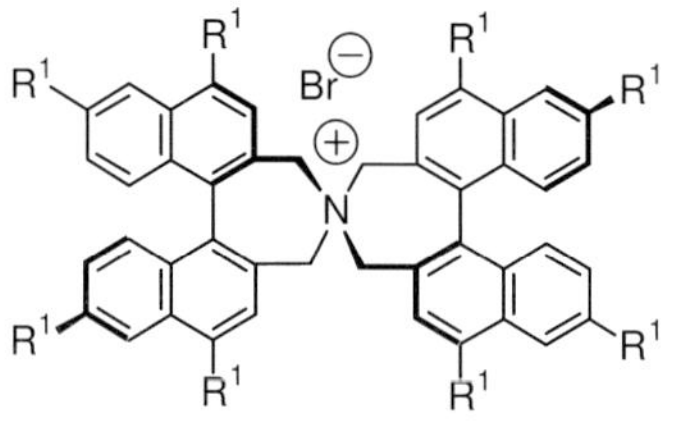

62 R^1 = C$_8$F$_{17}$(CH$_2$)$_2$Si(Me)$_2$-

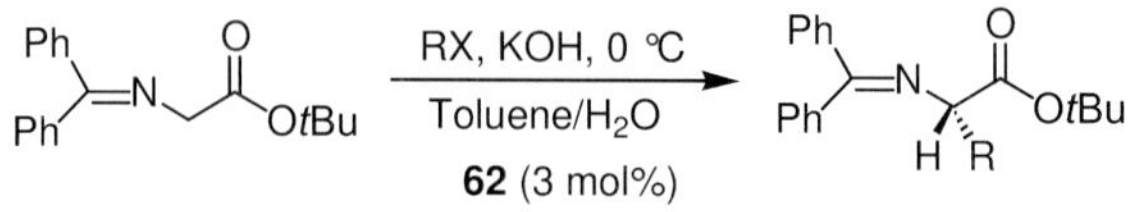

Scheme 5.17 Enantioselective alkylation of a protected glycine derivative.

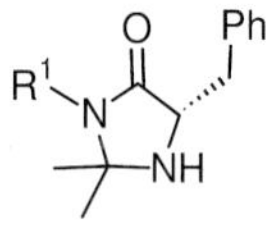

63 R^1 = 4-[C$_8$F$_{17}$(CH$_2$)$_2$]C$_6$H$_4$CH$_2$-
64 R^1 = Me

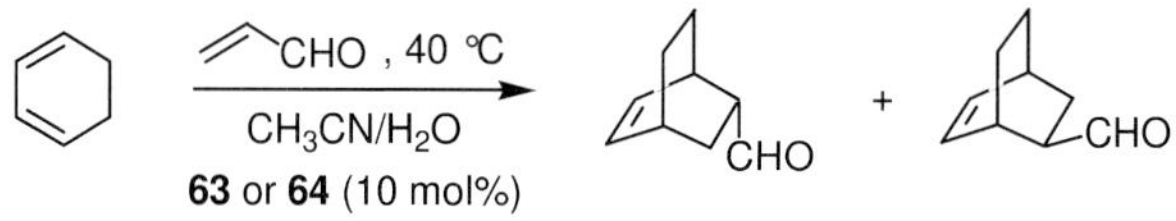

Scheme 5.18 Diels–Alder reaction between acrolein and cyclohexadiene.

A typical Diels–Alder reaction of acrolein and cyclohexadiene was conducted by Zhang and coworkers using either fluorous imidazolidinone **63** or its standard counterpart **64** as the catalyst (Scheme 5.18) [78]. The product yield (86%), *endo/exo* ratio (93.4:6.6) and enantioselectivity (93.4% ee for the *endo* isomer) afforded by **63** were comparable to those of the control experiment with **64**. The fluorous catalyst was examined in Diels–Alder reactions of other dienes and α,β-unsaturated aldehydes, providing consistently high enantioselectivities. In addition, **63** could be readily separated from the reaction products by F-SPE and recovered in 80–84% yields with excellent purity.

Wang and coworkers concentrated their efforts on developing recoverable chiral catalysts for Michael addition reactions of carbonyl compounds with nitroolefins. Fluorous prolinol trimethylsilyl ether **65** and pyrrolidine sulfonamides **6–68** (Scheme 5.19) were fit for the purpose [79, 80]. The former catalyst was found to promote Michael addition reactions between aliphatic aldehydes and *trans*-β-nitrostyrene derivatives at room temperature in BTF as a solvent, affording the corresponding γ-nitro carbonyl compounds in 81–91% yields, with enantioselectivities generally higher than 99% ee and a high degree of diastereoselectivity (*syn/anti* ratio up to 29:1) [79]. The hydrophobic sulfonamide **67** served as an effective catalyst for analogous reactions in an aqueous environment [80]. In that case, cyclic ketones also underwent the catalytic process efficiently to give products with good

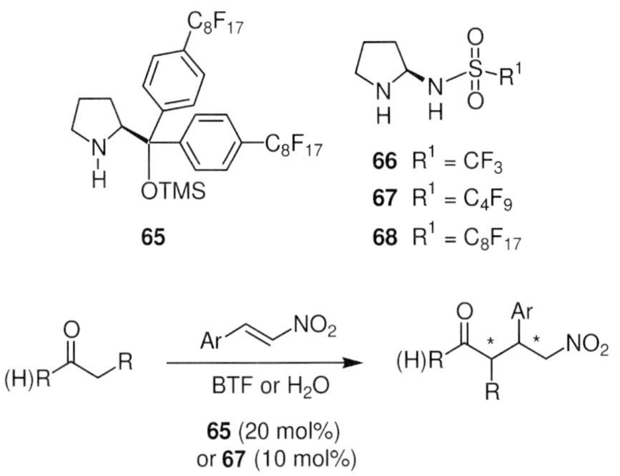

Scheme 5.19 Michael addition reactions of carbonyl compounds with nitroolefins.

to excellent enantioselectivities (68–95% ee) and diastereomeric ratios of up to 50:1. Both catalysts could be recovered from the reaction mixtures in about 90% yield by F-SPE. Recycling experiments performed with an initial catalyst loading of 20 mol% revealed that, despite a steady decrease in activity possibly due to material loss upon F-SPE, both **65** and **67** retained their stereoselectivities even after over six consecutive runs.

5.6
Conclusions

During the past decade significant examples of enantioselective reactions catalyzed by fluorous chiral catalysts have been reported, and slowly the effects of the incorporation of medium-sized perfluoroalkyl groups on both their phase-properties and stereoselectivity are being understood. Accordingly, the location and number of perfluoroalkyl substituents must be carefully chosen and, with a few interesting exceptions, the efficient shielding of the catalytically active site from the electron-withdrawing effect of these substituents must be ensured in order to attain reasonably good catalytic activities and selectivities. For charged fluorous transition-metal complexes, the nature of the counterion can also affect the outcome of the catalytic processes.

The most appealing feature of fluorous chiral catalysts is their easy separation from the reaction products, and possible recycling. To this end, emphasis was initially placed on the FBS approach, which is only effective for heavy fluorous catalysts with high or at least moderate partition coefficients in perfluorocarbons. Besides being synthetically demanding, heavy fluorous chiral catalysts can exhibit unpredictably low activities and stereoselectivities, even after careful optimization of the FBS reaction conditions. These seem to be less-compelling issues for relatively simple light fluorous chiral catalysts that can be quickly evaluated under the established conditions for their nonfluorous analogues, with little or no need for optimization. The separation stage remains straightforward due to the application of techniques, such as F-SPE, which perhaps are less elegant but are more easily put into practice than the original FBS approach. Finally, although the reuse of several heavy and light chiral fluorous catalysts has been demonstrated, their activity is lost after a limited number of recyclings. These findings are similar to results obtained with other supported homogeneous systems, and must be improved if fluorous chiral catalysts are to have large-scale applications in the future. Consequently, many avenues are still to be investigated in this rapidly expanding field in order to meet the demanding requirements of recyclable enantioselective catalysts.

References

1 Cornils, B., Herrmann, W.A., Horváth, I.T., Leitner, W., Mecking, S., Olivier-Bourbigou, H. and Vogt, D. (2005) *Multiphase Homogeneous Catalysis*, Wiley-VCH Verlag GmbH, Weinheim.
2 Cornils, B. and Herrmann, W.A. (2004) *Aqueous-Phase Organometallic Catalysis. Concept and Applications*, 2nd edn, Wiley-VCH Verlag GmbH, Weinheim.
3 Horváth, I.T. and Rábai, J. (1994) *Science*, **266**, 72–5.
4 DiMagno, S.G., Dussault, P.H. and Schultz, J.A. (1996) *Journal of the American Chemical Society*, **118**, 5312–13.
5 Pozzi, G., Montanari, F. and Quici, S. (1997) *Chemical Communications*, 69–70.
6 Vincent, J.-M., Rabion, A., Yachandra, V.K. and Fish, R.H. (1997) *Angewandte Chemie – International Edition in English*, **36**, 2346–9.
7 Xiang, J., Orita, A. and Otera, J. (2002) *Angewandte Chemie – International Edition*, **41**, 4117–19.
8 Rábai, J., Szlavik, Z., Horváth, I.T. in Clark, J. and Macquarrie, D. (2002) *Handbook of Green Chemistry and Technology*, Blackwell Science, Oxford, Chap. 22, pp. 502–23.
9 Dobbs, A.P. and Kimberley, M.R. (2002) *Journal of Fluorine Chemistry*, **118**, 3–17.
10 Sinou, D. in Gladysz, J.A., Curran, D.P. and Horváth, I.T. (2004) *Handbook of Fluorous Chemistry*, Wiley-VCH Verlag GmbH, Weinheim, Chap 10, pp. 306–15.
11 Gladysz, J.A. and Curran, D.P. (2002) *Tetrahedron*, **58**, 3823–5.
12 Curran, D.P. (2001) *Green Chemistry*, **3**, G3–7.
13 Zhang, W. (2004) *Chemical Reviews*, **104**, 2531–56.
14 Curran, D.P. (2006) *Aldrichimica Acta*, **39**, 3–9.
15 van den Broeke, J., de Wolf, E., Deelman, B.-J. and van Koten, G. (2003) *Advanced Synthesis Catalysis*, **345**, 625–34.
16 Horváth, I.T., Kiss, G., Cook, R.A., Bond, J.E., Stevens, P.A., Rábai, J. and Mozeleski, E.J. (1998) *Journal of the American Chemical Society*, **120**, 3133–43.
17 Jiao, H., Le Stang, S., Soós, T., Meier, R., Kowski, K., Rademacher, P., Jafarpour, L., Hamard, J-B., Nolan, S.P. and Gladysz, J.A. (2002) *Journal of the American Chemical Society*, **124**, 1516–23.
18 Fawcett, J., Hope, E.G., Kemmitt, R.D.W., Paige, D.R., Russell, D.R. and Stuart, A.M. (1998) *Journal of the Chemical Society – Dalton Transactions*, 3751–63.
19 Gladysz, J.A., Emnet, C., Rábai, J. in Gladysz, J.A., Curran, D.P. and Horváth, I.T. (2004) *Handbook of Fluorous Chemistry*, Wiley-VCH Verlag GmbH, Weinheim, Chap. 6, pp. 56–90.
20 Kiss, L.E., Kövesdi, I. and Rábai, J. (2001) *Journal of Fluorine Chemistry*, **108**, 95–109.
21 Huque, F.T.T., Jones, K., Saunders, R.A. and Platts, J.A. (2002) *Journal of Fluorine Chemistry*, **115**, 119–28.
22 de Wolf, E., Ruelle, P., van den Broeke, J., Deelman, B.-J. and van Koten, G. (2004) *The Journal of Physical Chemistry B*, **108**, 1458–66.

23 Mercader, A.G., Duchowicz, P.R., Sanservino, M.A., Fernández, F.M. and Castro, E.A. (2007) *Journal of Fluorine Chemistry*, **128**, 484–92.
24 Curran, D.P. and Hadida, S. (1996) *Journal of the American Chemical Society*, **118**, 2531–2.
25 Curran, D.P., Hadida, S. and He, M. (1997) *Journal of Organic Chemistry*, **62**, 6714–15.
26 Pozzi, G., Cinato, F., Montanari, F. and Quici, S. (1998) *Chemical Communications*, 877–8.
27 Pozzi, G., Cavazzini, M., Cinato, F., Montanari, F. and Quici, S. (1999) *European Journal of Organic Chemistry*, 1947–55.
28 Cavazzini, M., Manfredi, A., Montanari, F., Quici, S. and Pozzi, G. (2000) *Chemical Communications*, 2171–2.
29 Cavazzini, M., Manfredi, A., Montanari, F., Quici, S. and Pozzi, G. (2001) *European Journal of Organic Chemistry*, 4639–49.
30 Reger, T.S. and Janda, K.D. (2000) *Journal of the American Chemical Society*, **122**, 6929–34.
31 Jacobsen, E.N. (2000) *Accounts of Chemical Research*, **33**, 421–31.
32 Cavazzini, M., Quici, S. and Pozzi, G. (2002) *Tetrahedron*, **58**, 3943–9.
33 Shepperson, I., Cavazzini, M., Pozzi, G. and Quici, S. (2004) *Journal of Fluorine Chemistry*, **125**, 175–80.
34 Andrus, M.B. and Lashley, J.C. (2002) *Tetrahedron*, **58**, 845–66.
35 Fache, F. and Piva, O. (2003) *Tetrahedron: Asymmetry*, **14**, 139–43.
36 Levina, A. and Muzart, J. (1995) *Tetrahedron: Asymmetry*, **6**, 147–56.
37 Bayardon, J. and Sinou, D. (2004) *Journal of Organic Chemistry*, **69**, 3121–8.
38 Bayardon, J. and Sinou, D. (2005) *Tetrahedron: Asymmetry*, **16**, 2965–72.
39 Noyori, R. and Hashiguchi, S. (1997) *Accounts of Chemical Research*, **30**, 40–73.
40 Maillard, D., Nguefack, C., Pozzi, G., Quici, S., Valadé, B. and Sinou, D. (2000) *Tetrahedron: Asymmetry*, **11**, 2881–4.
41 Maillard, D., Pozzi, G., Quici, S. and Sinou, D. (2002) *Tetrahedron*, **58**, 3971–6.
42 Bayardon, J., Maillard, D., Pozzi, G. and Sinou, D. (2004) *Tetrahedron: Asymmetry*, **15**, 2633–40.
43 Hope, E.G., Stuart, A.M. and West, A.J. (2004) *Green Chemistry*, **6**, 345–50.
44 Dalicsek, Z., Pollreisz, F., Gömöry, A. and Soós, T. (2005) *Organic Letters*, **7**, 3243–6.
45 Corey, E.J. and Helal, C.J. (1998) *Angewandte Chemie – International Edition*, **37**, 1986–2012.
46 Kainz, S., Brinkmann, A., Leitner, W. and Pfaltz, A. (1999) *Journal of the American Chemical Society*, **121**, 6421–9.
47 Franciò, G., Wittmann, K. and Leitner, W. (2001) *Journal of Organometallic Chemistry*, **621**, 130–42.
48 Birdsall, D.J., Hope, E.G., Stuart, A.M., Chen, W., Hu, Y. and Xiao, J. (2001) *Tetrahedron Letters*, **42**, 8551–3.
49 Malkov, A.V., Figlus, M., Ston ius, S. and Ko ovský, P. (2007) *Journal of Organic Chemistry*, **72**, 1315–25.
50 Pu, L. and Yu, H.-B. (2001) *Chemical Reviews*, **101**, 757–824.
51 Kleijn, H., Rijnberg, E., Jastrzebski, J.T.B.H. and van Koten, G. (1999) *Organic Letters*, **1**, 853–5.
52 Nakamura, Y., Takeuchi, S., Okumura, K. and Ohgo, Y. (2001) *Tetrahedron*, **57**, 5565–71.
53 Park, J.K., Lee, H.G., Bolm, C. and Kim, B.M. (2005) *Chemistry – A European Journal*, **11**, 945–50.
54 Nakamura, Y., Takeuchi, S., Ohgo, Y. and Curran, D.P. (2000) *Tetrahedron Letters*, **41**, 57–60.
55 Nakamura, Y., Takeuchi, S., Okumura, K., Ohgo, Y. and Curran, D.P. (2002) *Tetrahedron*, **58**, 3963–9.
56 Tian, Y. and Chan, K.S. (2000) *Tetrahedron Letters*, **41**, 8813–16.
57 Tian, Y., Yang, Q.C., Mak, T.C.W. and Chan, K.S. (2002) *Tetrahedron*, **58**, 3951–61.
58 Yin, Y.-Y., Zhao, G., Qian, Z.-S. and Yin, W.-X. (2003) *Journal of Fluorine Chemistry*, **120**, 117–20.
59 Omote, M., Nishimura, Y., Sato, K., Ando, A. and Kumadaki, I. (2005) *Tetrahedron Letters*, **46**, 319–22.
60 Omote, M., Nishimura, Y., Sato, K., Ando, A. and Kumadaki, I. (2006) *Tetrahedron*, **62**, 1886–94.
61 Omote, M., Tanaka, N., Tarui, A., Sato, K., Kumadaki, I. and Ando, A. (2007) *Tetrahedron Letters*, **48**, 2989–91.

62 Hayashi, T. (2000) *Accounts of Chemical Research*, **33**, 354–62.
63 Cavazzini, M., Quici, S., Pozzi, G., Maillard, D. and Sinou, D. (2001) *Chemical Communications*, 1220–1.
64 Maillard, D., Bayardon, J., Kurichiparambil, J.D., Nguefack-Fournicr, C. and Sinou, D. (2002) *Tetrahedron: Asymmetry*, **13**, 1449–56.
65 Mino, T., Sato, Y., Saito, A., Tanaka, Y., Saotome, H., Sakamoto, M. and Fujita, T. (2005) *Journal of Organic Chemistry*, **70**, 7979–84.
66 Bayardon, J. and Sinou, D. (2003) *Tetrahedron Letters*, **44**, 1449–51.
67 Nakamura, Y., Takeuchi, S., Zhang, S., Okumura, K. and Ohgo, Y. (2002) *Tetrahedron Letters*, **43**, 3053–6.
68 Bayardon, J., Cavazzini, M., Maillard, D., Pozzi, G., Quici, S. and Sinou, D. (2003) *Tetrahedron: Asymmetry*, **14**, 2215–24.
69 Annunziata, R., Benaglia, M., Cinquini, M., Cozzi, F. and Pozzi, G. (2003) *European Journal of Organic Chemistry*, 1191–7.
70 Simonelli, B., Orlandi, S., Benaglia, M. and Pozzi, G. (2004) *European Journal of Organic Chemistry*, 2669–73.
71 Evans, D.A., Woerpel, K.A., Hinman, M.M. and Faul, M.M. (1991) *Journal of the American Chemical Society*, **113**, 726–8.
72 Bayardon, J., Holczknecht, O., Pozzi, G. and Sinou, D. (2006) *Tetrahedron: Asymmetry*, **17**, 1568–72.
73 Shepperson, I., Quici, S., Pozzi, G., Nicoletti, M. and O'Hagan, D. (2004) *European Journal of Organic Chemistry*, 4545–51.
74 Bayardon, J., Sinou, D., Holczknecht, O., Mercs, L. and Pozzi, G. (2005) *Tetrahedron: Asymmetry*, **16**, 2319–27.
75 Biffis, A., Braga, M., Cadamuro, S., Tubaro, C. and Basato, M. (2005) *Organic Letters*, **7**, 1841–4.
76 Fache, F. and Piva, O. (2001) *Tetrahedron Letters*, **42**, 5655–7.
77 Shirakawa, S., Tanaka, Y. and Maruoka, K. (2004) *Organic Letters*, **6**, 1429–31.
78 Chu, Q., Zhang, W. and Curran, D.P. (2006) *Tetrahedron Letters*, **47**, 9287–90.
79 Zu, L., Li, H., Wang, J., Yu, X. and Wang, W. (2006) *Tetrahedron Letters*, **47**, 5131–4.
80 Zu, L., Wang, J., Li, H. and Wang, W. (2006) *Organic Letters*, **814**, 3077–9.

6
Heterogeneous Asymmetric Catalysis in Aqueous Media
Yasuhiro Uozumi

6.1
Introduction

The development of catalytic asymmetric organic transformations has emerged as one of the most exciting and challenging areas in modern synthetic chemistry. Today, homogeneous chiral catalysts are widely used for a variety of organic transformations, and transition-metal complexes of homochiral ligands, in particular, have been recognized as very powerful tools in the arsenal of the synthetic organic chemist. The heterogeneous-switching of homogeneous catalytic processes has become a very useful means for high-throughput synthesis as well as for the industrial production of fine chemicals, where the catalyst residue can be readily removed by simple manipulation before being subjected to the next reaction (reused and/or recycled) [1, 2]. Indeed, the efficient removal of chiral metal complexes from the reaction mixture of a catalytic asymmetric process would allow not only the recovery of costly noble metal species and the chiral auxiliary, but also the production of chiral compounds uncontaminated by metal species to provide compounds with improved biological utility.

Recently, organic reactions conducted in water have begun to attract considerable attention due to the advantages of water as a readily available, safe and environmentally benign solvent [3, 4]. Nevertheless, with the catalysts designed for use under conventional organic conditions, it is more difficult to achieve high catalytic performance in water than in organic media because of the low water-compatibility of both the catalysts and the organic substrates. To date, several approaches to achieve asymmetric catalysis in water have been reported, the most representative of these being: (i) Hydrophilic modification of the chiral catalysts, for example, sulfonation, quaternary salt formation, and so on; and (ii) the use of amphiphilic additives, for example, detergent-like additives, phase-transfer catalysts, and so on. Although these approaches make water-based catalytic reactions possible, the catalytic species is often difficult to recover from the aqueous phase. The asymmetric catalytic protocol performed in water with a solid-supported chiral catalyst under heterogeneous conditions would come close to realizing an ideal asymmetric

Handbook of Asymmetric Heterogeneous Catalysis. Edited by K. Ding and Y. Uozumi
Copyright © 2008 WILEY-VCH Verlag GmbH & Co. KGaA, Weinheim
ISBN: 978-3-527-31913-8

reaction, where the advantages of both aqueous- and heterogeneous-switching are combined in one system. This chapter describes the development of asymmetric processes in water with solid-supported chiral catalysts. Asymmetric catalysis performed *in the presence of* water will not be included here [5].

6.2
Chiral-Switching of Heterogeneous Aquacatalytic Process

Recently, Uozumi and coworkers have achieved a variety of catalytic transformations in water under heterogeneous conditions by the use of amphiphilic polystyrene-poly(ethylene glycol) copolymer (PS-PEG) resin-supported catalysts [6]. The amphiphilic property of the PS-PEG resin [7] is essential to promote organic transformations in water where both the hydrophobic reaction matrix (PS) and the hydrophilic PEG region interacting with the water-soluble reactants are combined in one system. Thus, typically, a palladium complex anchored to the PS-PEG resin via coordination to the phosphine ligand group covalently tethered to the resin (Figure 6.1) efficiently catalyzed π-allylic substitution, carbonylation, the Heck reaction, cycloisomerization, Suzuki–Miyaura coupling, Sonogashira coupling, and so on, in water without a need for any organic cosolvents (Scheme 6.1). Details of the chiral-switching of heterogeneous aquacatalytic systems (Figure 6.2) using PS-PEG resin-based polymeric palladium complexes will be introduced in the following section.

6.2.1
Combinatorial Approach

A diversity-based approach was examined to identify a polymeric palladium–phosphine complex exhibiting high stereoselectivity in water [8]. Thus, a library of

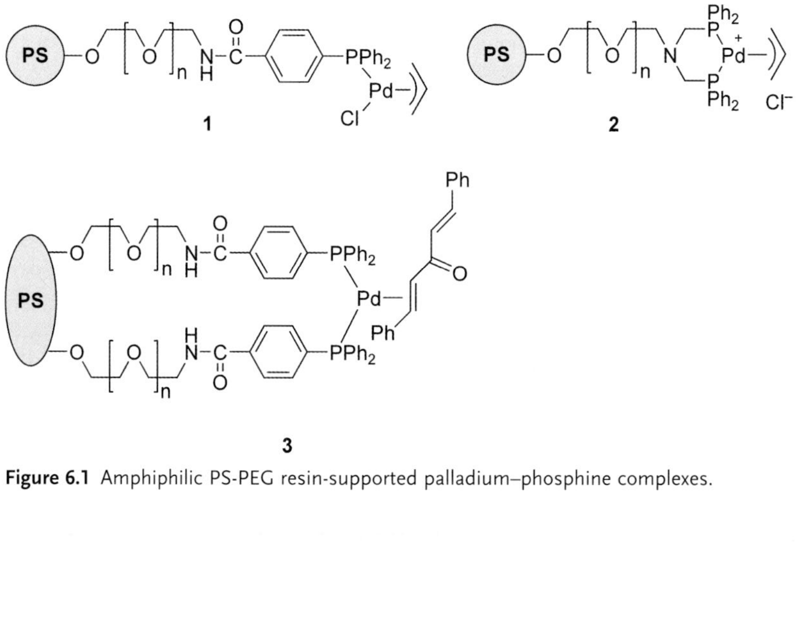

Figure 6.1 Amphiphilic PS-PEG resin-supported palladium–phosphine complexes.

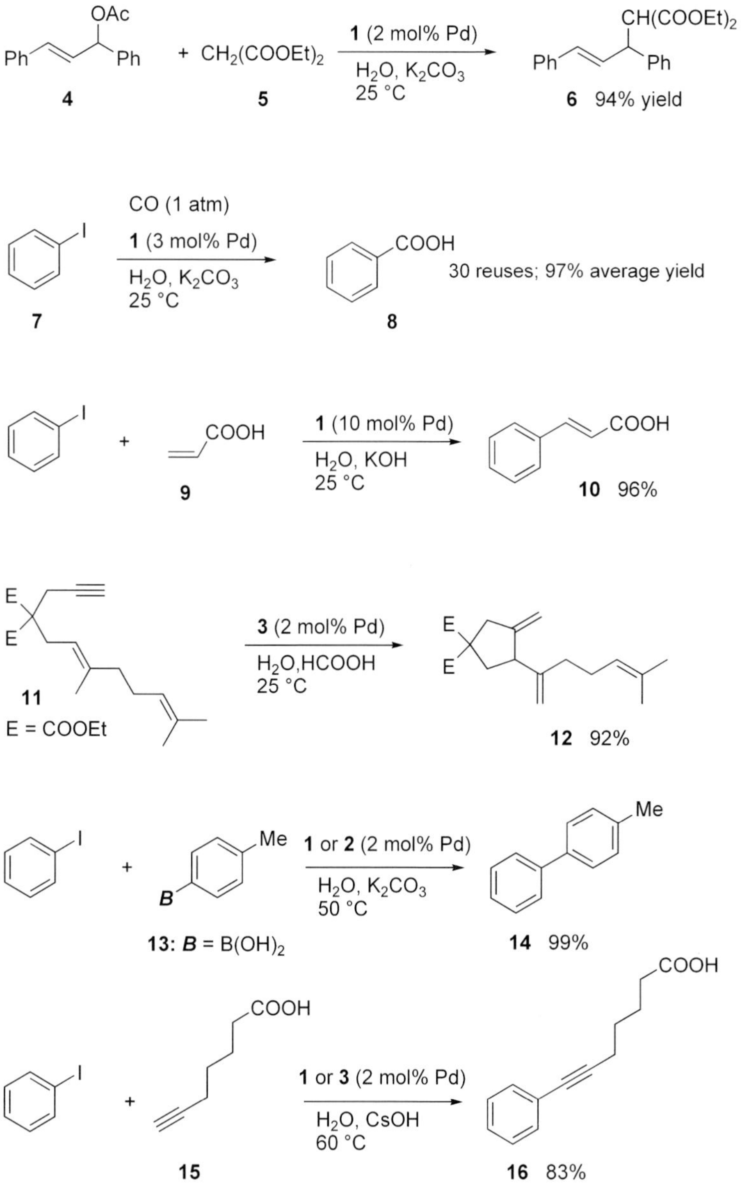

6.2 Chiral-Switching of Heterogeneous Aquacatalytic Process

Scheme 6.1 Representatives of heterogeneous aquacatalysis with PS-PEG resin-supported palladium catalysts.

84 chiral phosphine ligands **17** was prepared on amphiphilic PS-PEG resin beads from the achiral and chiral amino acids **A–L** and the phosphines **a–g**, including the axially chiral MOP ligand derivatives **d–g** (Figure 6.3). With this library of the amphiphilic PS-PEG resin-supported chiral phosphine ligands **17{A–L, a–g}**, the enantiocontrolling ability and catalytic potency of the polymeric

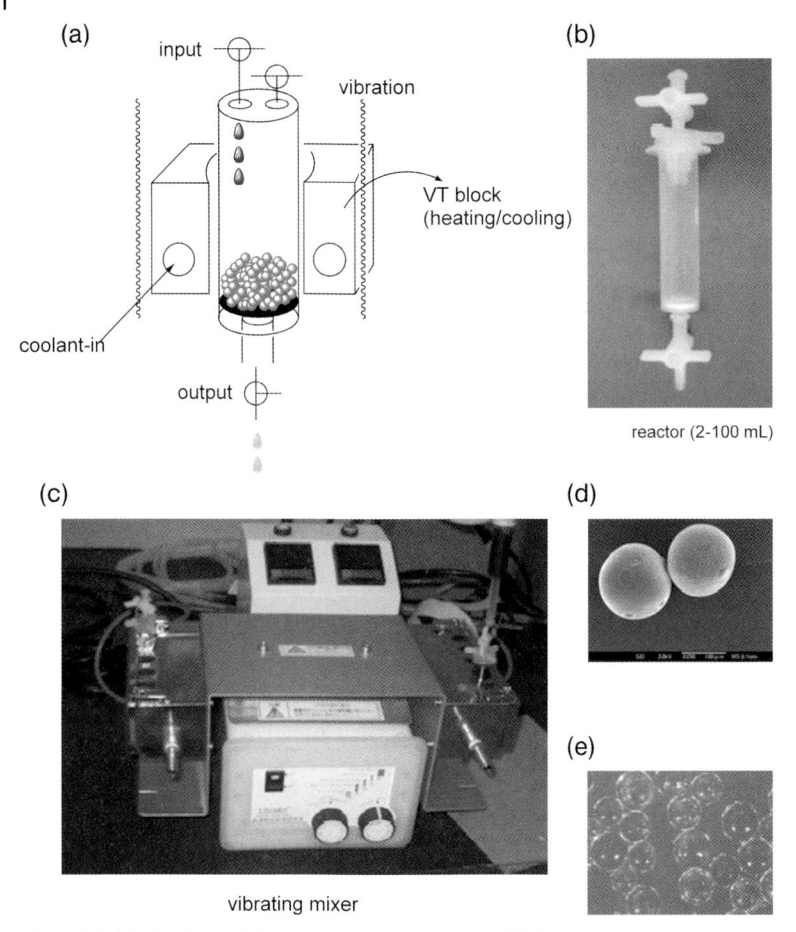

Figure 6.2 (a) A schematic image of a reaction system; (b) A photographic image of a reactor; (c) A photographic image of a vibrating mixer; (d) A scanning electron microscopy image of the polymeric catalyst; (e) A microscopic photo image of the polymeric catalyst.

ligands were examined in water for palladium-catalyzed π-allylic alkylation of 1,3-diphenylpropenyl acetate with 3-methyl-2,4-pentanedione (**18**). Through preliminary screening, the highest enantioselectivity was obtained with the palladium complex of the PS-PEG resin-supported phosphine **17Cg**, which was prepared from *(S)*-alanine and *(R)*-C(O)CH$_2$O-MOP **(g)** (Scheme 6.2). The **17Cg–Pd** was resynthesized to confirm its asymmetric aquacatalytic efficiency. Thus, the allylic ester, 1,3-diphenylpropenyl acetate, reacted with the 1,3-diketone **18** in aqueous potassium carbonate at 25 °C in the presence of 2 mol% palladium of the polymeric palladium complex **17Cg–Pd** to afford 96% yield of 1,3-diphenyl-4-acetyl-4-methyl-1-hexen-5-one (**19**) with 90% enantiomeric excess (ee) *(S)*.

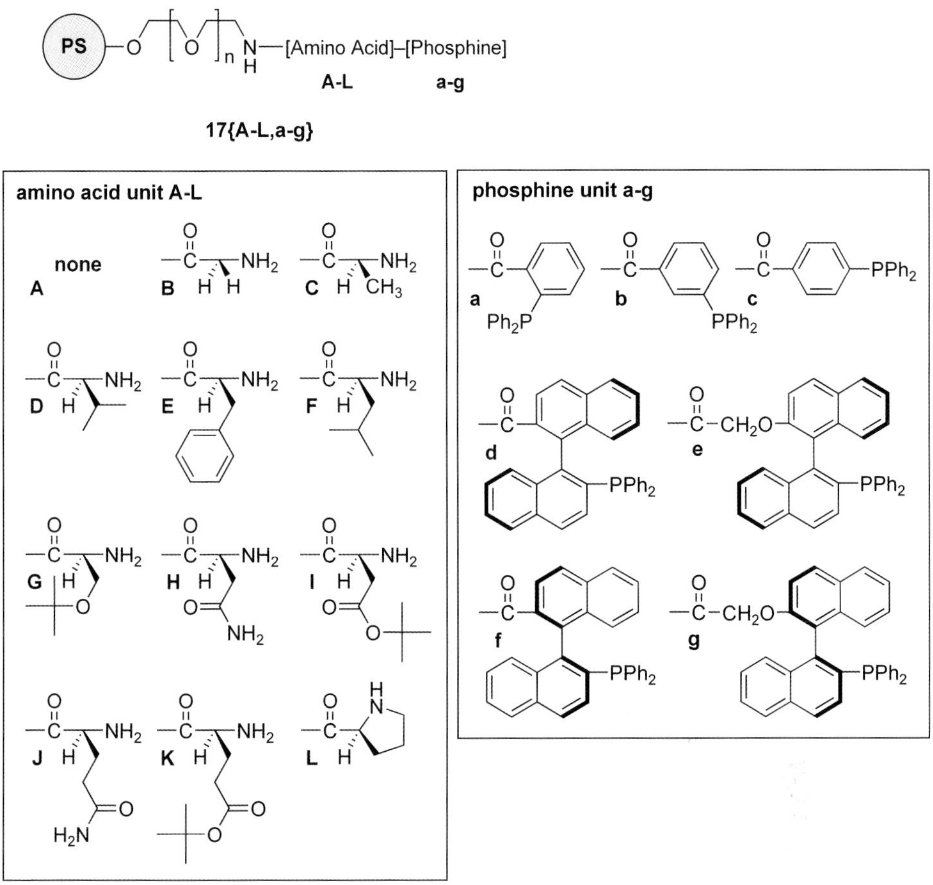

Figure 6.3 A library of PS-PEG resin-supported chiral phosphine ligands.

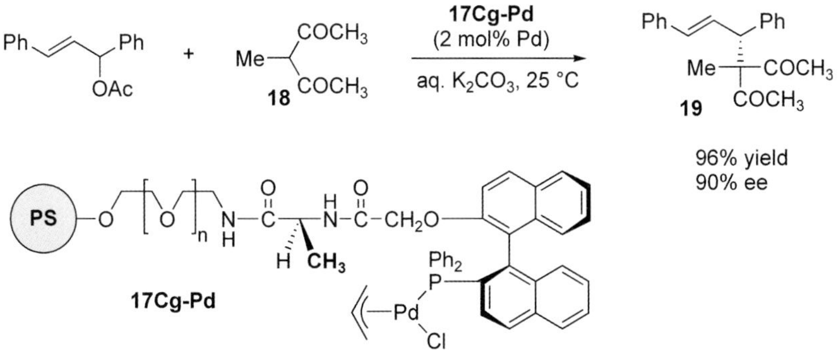

Scheme 6.2 Asymmetric aquacatalytic allylic substitution with a polymeric chiral palladium–phosphine complex.

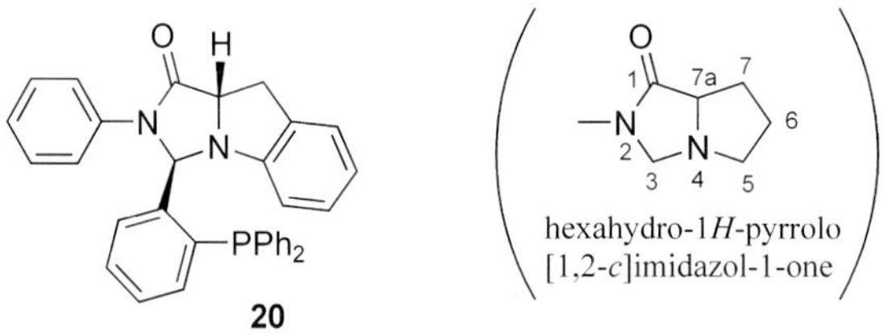

Figure 6.4 Upper: (3R,9aS)-(2-phenyl-3-(2-diphenylphosphino)-phenyl)-tetrahydro-1H-imidazo[1,5-a]indole-1-one (**20**); Lower: ORTEP drawing of the **20**-PdCl$_2$ complex.

6.2.2
Imidazoindole Phosphine

6.2.2.1 Design and Preparation

Highly functionalized optically active bicyclic amines having a pyrrolo[1,2-c]imidazolone framework were identified as effective chiral agents through a diversity-based approach to new chiral amine catalysts [9]. The results indicated that a novel P,N-chelate chiral ligand having the pyrrolo[1,2-c]imidazolone skeleton as a basic chiral unit [10] would be readily immobilized on the PS-PEG resin to achieve highly enantioselective heterogeneous catalysis in water (Figure 6.4). (3R,9aS)-(2-Aryl-3-(2-diphenylphosphino)phenyl)tetrahydro-1H-imidazo[1,5-a]indole-1-one (**23**), which was readily prepared from (S)-indoline-2-carboxylic acid, 4-{3-methoxycarbonyl}propyl}aniline, and 2-(diphenylphosphino)benzaldehyde by a sequence of reactions outlined in Scheme 6.3, was immobilized on PS-PEG-NH$_2$ resin to give the PS-PEG resin-supported chiral P,N-chelate ligand *(R,S)-***23** (Scheme 6.3). The formation of a palladium complex of the P,N-chelate ligand was performed by mixing [PdCl(η^3-C$_3$H$_5$)]$_2$ in toluene at room temperature for 10 min to give the PS-PEG supported P,N-chelate complex **23-Pd** in quantitative yield.

6.2.2.2 Allylic Alkylation

In order to explore the enantiocontrolling potential of the resin-supported complexes in water, Uozumi initially elected to study the palladium-catalyzed asymmetric allylic substitution of cyclic substrates, which is still a major challenge even using homogeneous chiral catalysts [11]. It is interesting to note that high stereoselectivity was achieved in water when the PS-PEG resin-supported catalyst **23-Pd** was used for allylic substitution of the cyclic substrates with dialkyl malonate (Table 6.1) [12]. Thus, the reactions of the methyl cycloalkenyl carbonates **26–30** and dialkyl malonate were carried out in water with 2–10 mol% palladium of the polymeric complex **23-Pd** in the presence of lithium carbonate at 40 °C to give good to excellent yields of the corresponding allylic malonates **33–37** with high stereoselectivities of up to 99% ee. The PS-PEG supported catalyst **23-Pd** was effective for the asymmetric allylic alkylation of both cyclic and acyclic substrates in water. The reactions of 1,3-diphenylpropenyl acetate (**31**) and pivalate (**32**) were catalyzed by **23-Pd** under the same reaction conditions to give **38** with 91% and 94% ee,

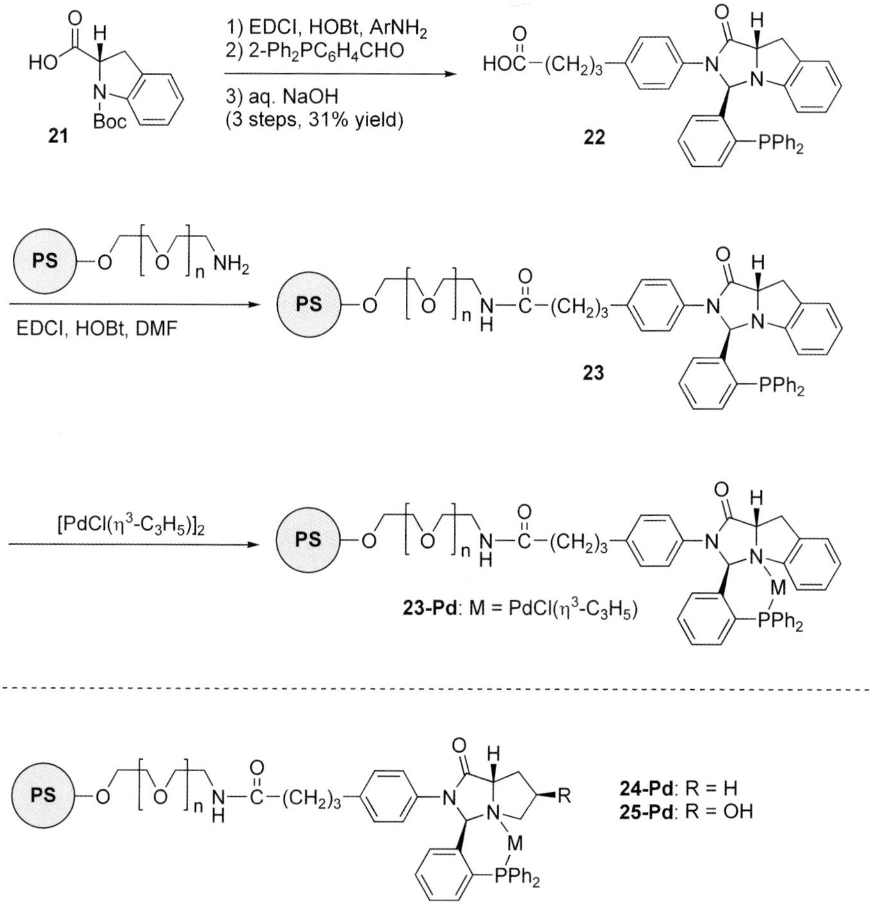

Scheme 6.3 Preparation of PS-PEG resin-supported chiral palladium complexes.

respectively (Table 6.1; entries 14 and 15). Alkylation of the racemic cis-5-carbomethoxy-2-cyclohexenyl methyl carbonate (**28**) gave 92% ee of **35b** in 90% yield as a single diastereoisomer having the cis-configuration (entries 8 and 9), demonstrating that the reaction pathway of allylic substitution in water is essentially the same as that of the homogeneous counterpart in organic solvent. Other PS-PEG supported catalysts, **24-Pd** and **25-Pd**, which lack the fused aromatic moiety on their pyrroloimidazolone ring system, exhibited much lower selectivity. Interestingly, the immobilized complex **23-Pd** is less catalytically active in an organic solvent. Thus, the alkylation of **28** with diethyl malonate in dichloromethane in the presence of N,O-bis(trimethylsilyl)acetamide (BSA) and lithium acetate gave a 27% yield of the adduct **35b** with 87% ee, whereas the reaction proceeded smoothly in aqueous lithium carbonate (Table 6.1; entries 9 and 10). In the aqueous media, the organic substrates (e.g. **28** and malonate) must diffuse into the hydrophobic PS matrix to form a highly concentrated reaction sphere which

Table 6.1 Asymmetric allylic alkylation in water with a chiral polymeric palladium complex.

a

MeOOCO⌒⌒X + ⌒⌒X OCOOMe + NuH → Nu⌒⌒X
racemic
Pd* (2-10 mol% Pd), H_2O, Li_2CO_3, 40 °C, 12 h

Pd* ≡ PS-O-[O]$_n$-NH-C(=O)-(CH$_2$)$_3$-C$_6$H$_4$-N-... (chiral ligand)

23-Pd: M = $PdCl(\eta^3-C_3H_5)$

Entry	Allylic ester	NuH	Product	Yield (%)	ee (%)
1	cyclopentenyl-OCOOMe	$CH_2(COOMe)_2$	cyclopentenyl-CH(COOMe)$_2$	65	92
2		$CH_2(COOEt)_2$	cyclopentenyl-CH(COOEt)$_2$	74	92
3	cyclohexenyl-OCOOMe	$CH_2(COOMe)_2$	cyclohexenyl-CH(COOMe)$_2$	88	90
4	cyclohexenyl-CH(COOEt)$_2$	$CH_2(COOEt)_2$		90	92
5	(cat: 2nd use)	$CH_2(COOEt)_2$	34b	89	90
6	(cat: 3rd use)	$CH_2(COOEt)_2$	34b	90	92
7	(cat: 4th use)	$CH_2(COOEt)_2$	34b	91	91

b

Entry	Allylic ester	NuH	Product	Yield (%)	ee (%)
8	cyclohexenyl-OCOOMe (with COOMe)	$CH_2(COOMe)_2$	cyclohexenyl-CH(COOMe)$_2$ (with COOMe)	71	90
9		$CH_2(COOEt)_2$	cyclohexenyl-CH(COOEt)$_2$ (with COOMe)	90	92
10	[BSA, $Li_2(OAc)_2$ in CH_2Cl_2]	$CH_2(COOEt)_2$	35b	27	87
11	cycloheptenyl-OCOOMe	$CH_2(COOMe)_2$	cycloheptenyl-CH(COOMe)$_2$	85	97

Table 6.1 Continued

Entry	Allylic ester	NuH	Product	Yield (%)	ee (%)
12		$CH_2(COOEt)_2$	(cycloheptenyl)–CH(COOEt)$_2$	91	98
13	(tetrahydropyridyl)–OCOOMe, N-COO-tBu	$CH_2(COOMe)_2$	(tetrahydropyridyl)–CH(COOEt)$_2$, N-COO-tBu	95	99
14	Ph–CH=CH–CH(Ph)–OCOOMe	$CH_2(COOMe)_2$	Ph–CH=CH–CH(Ph)–CH(COOMe)$_2$	88	91
15	Ph–CH=CH–CH(Ph)–OCOO-tBu	$CH_2(COOMe)_2$	38	60	94

should react with the ionic species (e.g. aqueous alkaline) through the interfacial PEG region to afford higher reactivity than that in an organic solvent.

6.2.2.3 Allylic Amination

Amination of the methyl cyclohexenyl carbonate (**27**) with 5 equivalents of dibenzylic amines (**a, b** and **c**) was carried out in water at 25 °C with shaking for 24 h in the presence of 4 or 8 mol% of the palladium complex of **23-Pd** [13]. The reaction mixture was filtered and the catalyst resin rinsed with super critical CO_2 (scCO_2) or tetrahydrofuran (THF) to extract the desired product. The crude mixture obtained from the extract was analyzed chromatographically to give the corresponding (S)-N,N-dialkyl(cyclohexen-2-yl)amine (**39a–c**), the enantiomeric purities of which were determined to be 94%, 92% and 90% ee, respectively (Table 6.2; entries 1–3). The cyclohexenyl ester **28** bearing a methoxycarbonyl group with the *cis* configuration also underwent π-allylic substitution with the dibenzylamines **a–c** to give the corresponding allylamines **40a–c** having the *cis* configuration with high enantioselectivity ranging from 90% to 96% ee (entries 4–5). A higher stereoselectivity was observed when cycloheptenyl carbonate (**29**) was used as the substrate (entries 7–9). Thus, amination of **29** gave N,N-dibenzyl(cycloheptenyl)amine (**41a**), N-benzyl-N-4-methoxy benzyl(cyclohepten-2-yl)amine (**41b**), and N,N-di(4-m ethoxybenzyl)(cyclohepten-2-yl)amine (**41c**), the chemical yields and enantiomeric purities of which were 91% yield, 98% ee (**41a**); 82% yield, 97% ee (**41b**); and 89% yield, 96% ee (**41c**), respectively. The aminopiperidines **42a–c** were also prepared in optically active form from the tetrahydropyridyl carbonate **30** under similar conditions (Table 6.2; entries 10–12) with 93–95% enantiomeric excess.

As observed in the aquacatalytic asymmetric alkylation mentioned above, under these conditions, the π-allylic amination does not take place in organic solvents.

Table 6.2 Asymmetric allylic amination in water with 23-Pd.

Entry	Allylic ester	Amine (HNRR')	Product	Yield (%)	ee (%)
1	(cyclohexenyl OCOOMe)	a	(cyclohexenyl NRR') 39a	90	94
2		b	39b	77	92
3		c	39c	75	90
4	(cyclohexenyl OCOOMe, COOMe)	a	(cyclohexenyl NRR', COOMe) 40a	61	90
5		b	40b	80	96
6		c	40c	85	95
7	(cycloheptenyl OCOOMe)	a	(cycloheptenyl NRR') 41a	91	98
8		b	41b	82	97
9		c	41c	89	96
10	(N-Boc dihydropyridinyl OCOOMe)	a	(N-Boc NRR') 42a	89	95
11		b	42b	99	93
12		c	42c	59	94

HNRR':
a: PhCH$_2$-NH-CH$_2$Ph
b: MeO-C$_6$H$_4$-CH$_2$-NH-CH$_2$Ph
c: MeO-C$_6$H$_4$-CH$_2$-NH-CH$_2$-C$_6$H$_4$-OMe

Thus, the reaction of the cycloheptenyl carbonate **29** with 5 equivalents of dibenzylamine in the presence of 8 mol% palladium of the PS-PEG resin-supported **23-Pd** complex was carried out in THF or dichloromethane and showed no catalytic activity at 25 °C, whereas the same system in water proceeded smoothly to give 91% yield of the cycloheptenylamine **41a** (Scheme 6.4).

The recycling experiments were examined for the amination of the cycloheptenyl ester **29** with 1 equivalent of dibenzylamine. After the first use of the polymeric chiral catalyst to give 98% ee of **41a**, the recovered resin catalyst was taken on to second and third uses, without any additional charge of palladium, and exhibited no loss of its catalytic activity or stereoselectivity.

6.2.2.4 Allylic Etherification

Although a vast amount of research has been devoted to the asymmetric π-allylic substitution of acyclic esters (e.g. 1,3-diphenylpropenyl esters) with carbon and nitrogen nucleophiles, studies on the catalytic asymmetric substitution of cyclic

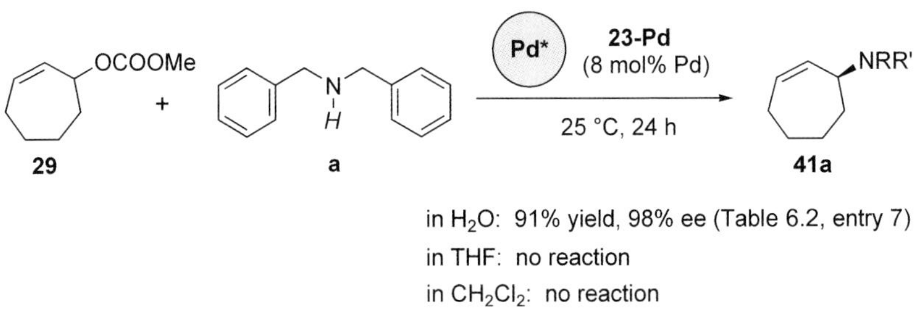

Scheme 6.4

29 (cycloheptenyl OCOOMe) + **a** (dibenzylamine) → **41a** (cycloheptenyl-NRR')

Conditions: Pd* 23-Pd (8 mol% Pd), 25 °C, 24 h

- in H_2O: 91% yield, 98% ee (Table 6.2, entry 7)
- in THF: no reaction
- in CH_2Cl_2: no reaction

substrates with oxygen nucleophiles have been limited to well-developed, albeit isolated, reports. Uozumi and Kimura reported that the heterogeneous aquacatalytic asymmetric etherification of cycloalkenyl esters with phenolic nucleophiles, which is catalyzed by the PS-PEG resin-supported palladium-imidazoindolephosphine complex **23-Pd**, gave optically active aryl(cycloalkenyl) ethers with up to 94% ee (Table 6.3) [14]. Thus, for example, the reaction of methyl cyclohexenylcarbonate (**27**) and 1.0 equivalent of 4-methoxyphenol (**a**) was carried out in the presence of 2 mol% palladium of the polymeric complex **23-Pd** and K_2CO_3 (1 mol. equiv) in water at 25 °C with shaking for 12 h to give 3-(4-methoxyphenoxy)cyclohexene (**43a**) in 89% yield with 86% ee (S-configuration) (entry 1). The results obtained for the asymmetric etherification of various cycloalkenylcarbonates with the phenols **a–c** are summarized in Table 6.3. With the racemic cis-5-carbomethoxy-2-cyclohexenyl methyl carbonate **28**, the 4-methoxyphenyl ether **44a**, the 4-benzyloxyphenyl ether **44b** and the 2-benzyloxyphenyl ether **44c** were obtained in 93% ee (entry 4), 93% ee (entry 5) and 94% ee (entry 6), respectively, while the reaction of **27** with the phenols **2a–c**, which lacks the carbomethoxy substituent at the 5 position, resulted in lower enantioselectivity ranging from 84% to 86% ee (Table 6.3; entries 1–3). The reaction using the cycloheptenyl carbonate **29** gave the cycloheptenyl aryl ethers, **45a, 45b** and **45c**, in 92%, 89% and 93% ee, respectively (entries 7–9). The exclusive formation of the cycloalkenyl ethers **44** having the cis-configuration from the cis-allylic ester **28** revealed that the π-allylic etherification proceeds via a double-inversion pathway (stereoinversive π-allylpalladium formation and stereoinversive nucleophilic attack with the phenol) in water under the present conditions. The catalytic asymmetric introduction of oxygen functionalities to a piperidine framework also took place with high stereoselectivity. The tetrahydropyridyl carbonate **30** reacted with the phenols **a–c** under similar conditions to afford the phenoxypiperidines **46a–c** with 92–94% enantiomeric excesses (entries 10–12).

6.2.2.5 Synthetic Application

A multi-step asymmetric synthesis of a hydrindane framework was achieved in water via asymmetric allylic alkylation, propargylation and aquacatalytic cycloisomerization of a 1,6-enyne, where all three steps were performed in water with the recyclable polymeric catalysts. The racemic cyclohexenyl ester **27** reacted with diethyl malonate under the conditions mentioned in Table 6.1 to give 90–92% ee of **34b**. The polymeric chiral palladium complex **23-Pd** was reused four times

Table 6.3 Asymmetric allylic etherification in water with 23-Pd.

Entry	Allylic ester	Phenol (HOAr)	Product		Yield (%)	ee (%)
1	OCOOMe (cyclohexenyl)	a	OAr (cyclohexenyl)	43a	89	86
2		b		43b	92	84
3		c		43c	80	86
4	OCOOMe (cyclohexenyl, COOMe)	a	OAr (cyclohexenyl, COOMe)	44a	93	93
5		b		44b	93	93
6		c		44c	88	94
7	OCOOMe (cycloheptenyl)	a	OAr (cycloheptenyl)	45a	90	92
8		b		45b	94	89
9		c		45c	90	93
10	OCOOMe (N-Boc piperidine)	a	OAr (N-Boc piperidine)	46a	80	94
11		b		46b	72	94
12		c		46c	62	92

HOAr:
a: 4-methoxyphenol (HO–C₆H₄–OMe)
b: 4-(benzyloxy)phenol (HO–C₆H₄–OCH₂Ph)
c: 2-(benzyloxy)phenol (HO–C₆H₄–OCH₂Ph, ortho)

without any loss of stereoselectivity (Table 6.1; entries 4–7). Propargylation of the cyclohexenylmalonate **34b** with propargyl bromide was performed with PS-PEG ammonium hydroxide **47** [15], which has been developed as an immobilized phase-transfer catalyst (PTC) base with a view towards its use in water to give a quantitative yield of the 1,6-enyne *(S)*-**48**. The polymeric ammonium reagent was recovered as its bromide salt, which was reactivated by washing with aqueous KOH and reused. The enyne *(S)*-**48** underwent cycloisomerization (see Scheme 6.1, Equation 6.4) with the amphiphilic polymeric palladium **3** to afford the hydrindane (3a*R*,7a*S*)-**49** in 94% yield (Scheme 6.5).

Asymmetric aquacatalytic allylic C_1-substitution was also achieved with nitromethane as a C_1 nucleophile by using the PS-PEG resin-supported **23-Pd** catalyst in which, under water-based conditions, nitromethane did not explode even under basic conditions (Scheme 6.6) [16]. To the best of the present author's knowledge,

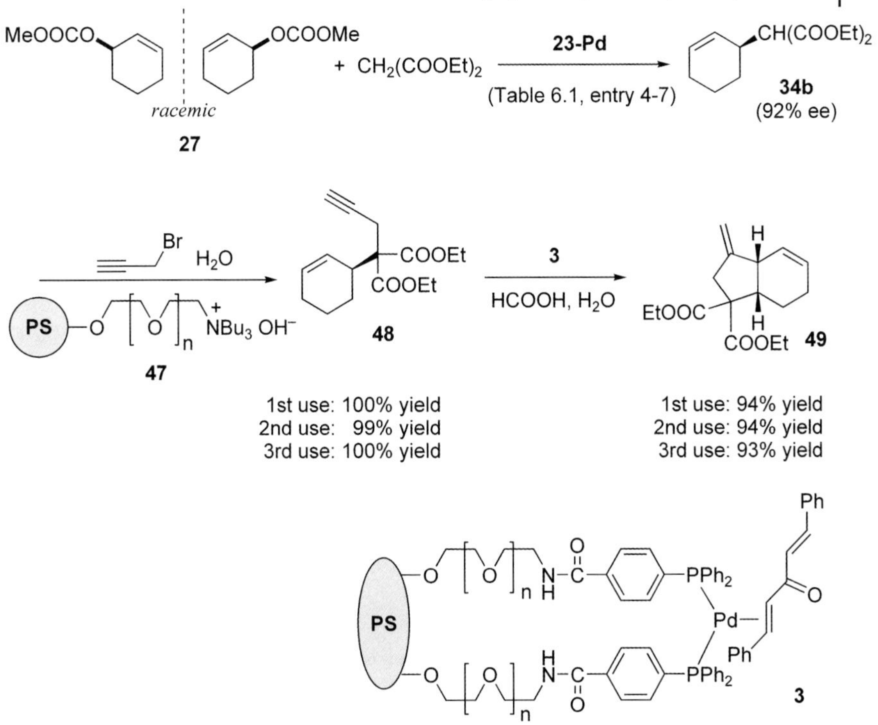

Scheme 6.5 Asymmetric multistep synthesis of a hydrindane framework in water.

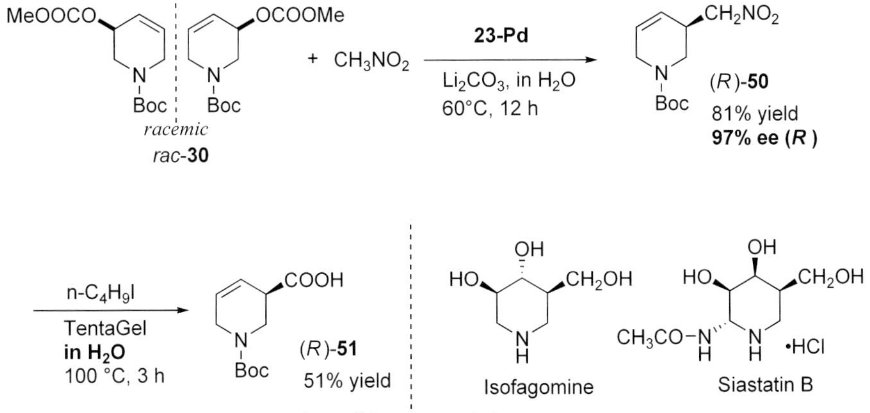

Scheme 6.6 Asymmetric aquacatalytic allylic nitromethylation.

research into the catalytic asymmetric substitution of cyclic substrates with C_1 nucleophiles has been limited to Trost's well-developed report [17] citing the asymmetric π-allylic nitromethylation of cycloalkenyl esters in dichloromethane. Although high enantioselectivities of up to 98% ee were achieved with Trost's chiral bisphosphine ligand, the risk of explosion remains a serious problem.[1] Clearly, although pioneering strides have been made, additional studies on water-based safe protocols are still warranted.

The tetrahydropyridyl ester *rac*-30 underwent nitromethylation in water with the polymeric chiral catalyst 18 under similar conditions to give an 81% isolated yield of *(R)*-nitromethyl(tetrahydro)pyridine 50 with 97% ee. The resultant nitromethyl(tetrahydro)pyridine 50 was readily converted to the tetrahydropyridyl carboxylic acid 51 also in water by the modified Carreira conditions [18], which is a promising synthetic intermediate for Isofagomine and Siastatin B [19].

6.3
Heterogeneous- and Aqueous-Switching of Asymmetric Catalysis

6.3.1
BINAP Catalysts

Heterogeneous-switching and aqueous-switching of the asymmetric catalysis with transition-metal complexes of the axially chiral 2,2'-bis(diphenylphosphino)-1,1'-binaphthyl (BINAP) derivatives has been examined often by introduction of solid supports [20–23] and hydrophilic functional groups [24–26], respectively, onto the BINAP skeleton [27]. Representative examples are shown in Figure 6.5. However, compared to the impressive development of the aqueous- and heterogeneous-switching of BINAP-based asymmetric catalysis, to the best of the present author's knowledge the previously reported successful studies on the development of the BINAP catalyst which realized heterogeneous- and aqueous-switching at the same time in one system have been limited to the pioneering investigations of Hayashi and coworkers [28].

The PS-PEG resin-supported BINAP ligand was prepared and used successfully for the rhodium-catalyzed asymmetric 1,4-addition reaction in water [29]. BINAP bearing a carboxylic group at the 6-position was immobilized by an amide bond on an amphiphilic PS-PEG NH_2 resin (Scheme 6.7). The PS-PEG BINAP 59 was treated with $Rh(acac)(C_2H_4)_2$ to form the PS-PEG–BINAP–Rh complex. The polymeric BINAP–Rh complex showed high catalytic activity and high enantioselectivity in water for the 1,4-addition of phenylboronic acid to α,β-unsaturated ketones (Scheme 6.8).

---

1) Trost's asymmetric nitromethylation was reproduced in the author's laboratory with the cycloheptenyl ester (25 °C, 96 h, 71% yield, 96% ee), whereas the reaction mixture exploded at 40 °C.

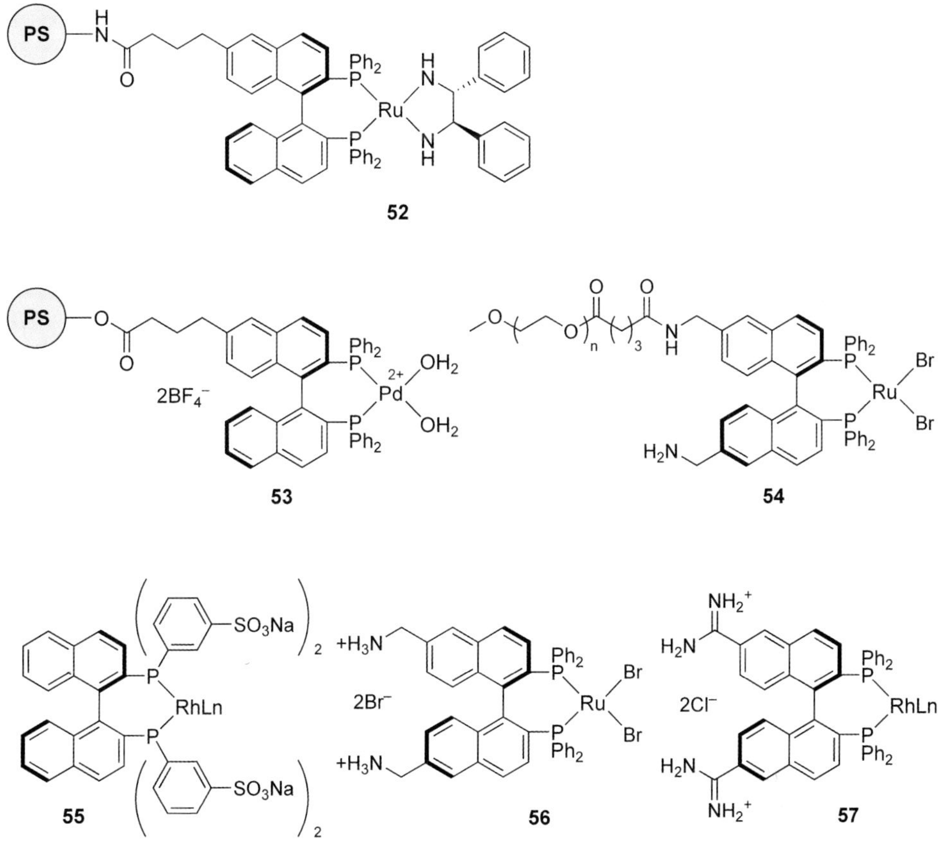

Figure 6.5 Typical examples of heterogeneous BINAP complexes and aquacatalytic BINAP complexes.

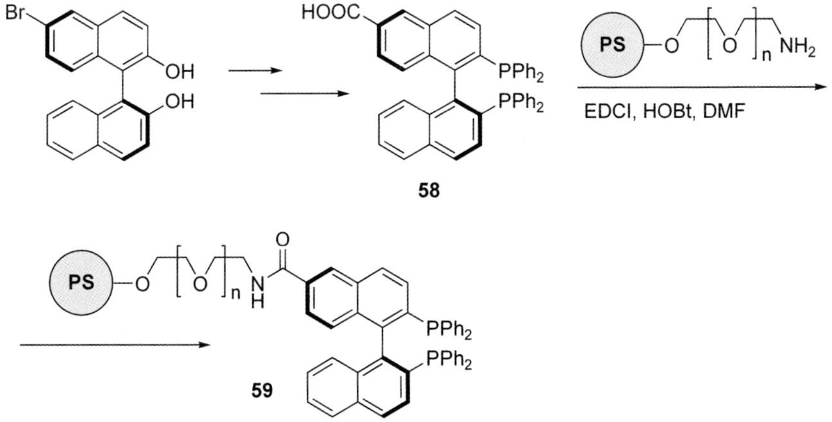

Scheme 6.7 Preparation of PS-PEG-supported BINAP.

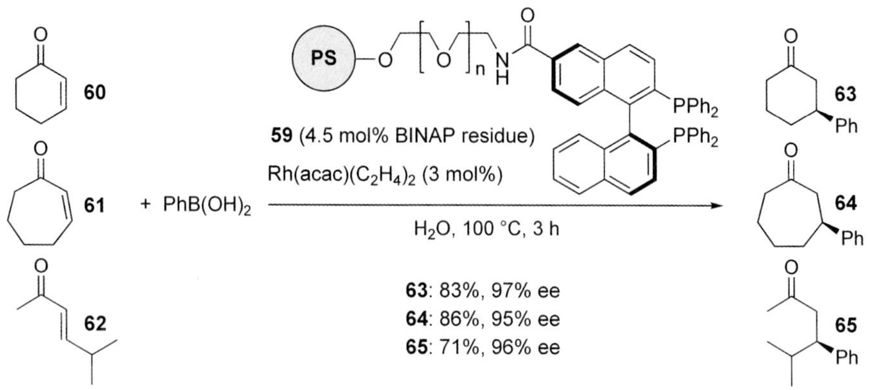

Scheme 6.8 Asymmetric 1,4-addition in water with PS-PEG–BINAP–Rh complex.

Reagents and conditions: 59 (4.5 mol% BINAP residue), Rh(acac)(C$_2$H$_4$)$_2$ (3 mol%), H$_2$O, 100 °C, 3 h.

63: 83%, 97% ee
64: 86%, 95% ee
65: 71%, 96% ee

6.3.2
DPEN Catalysts

(R,R)- or (S,S)-N-(p-Toluenesulfonyl)-1,2-diphenylethylenediamine (TsDPEN, **66**) and its derivatives constitute a most important class of chiral ligands, most notably for ruthenium-catalyzed asymmetric transfer hydrogenations [30]. Thus, as a typical example, the Ru–TsDPEN complex **67**, which was developed by Ikariya, Noyori and coworkers, efficiently promoted the transfer hydrogenation of a wide range of prochiral ketones with HCOOH-Et$_3$N (an azeotropic mixture) to afford the corresponding secondary alcohols with high to excellent enantiomeric selectivity [31]. The structures of the recently developed water-soluble TsDPEN ligands (**68** and **69**) are shown in Figure 6.6; the parent ligand **66** and its ruthenium–cymene complex **67** are also included in this figure [32, 33].

Heterogeneous- and aqueous-switching of asymmetric catalysis has been realized via immobilization of the TsDPEN unit onto silica or polymer supports (Scheme 6.9). The silica-supported TsDPEN ligands **73**, **74** and **75** were developed by Tu and coworkers, where the ligand unit was immobilized via an N-arylsulfonyl group bearing the trimethoxysilylethyl group at the 4-position [34]. N-Arylsulfonylated DPEN **70**, prepared from (R,R)-1,2-diphenylethylenediamine (**68**) and the arylsulfonyl chloride **69**, was treated with amorphous silica gel, mesoporous MCM-4 and SBA-15 in refluxing toluene to give the corresponding silica-supported ArSO$_2$DPEN ligands **73–75** (Scheme 6.9). The Ru-catalyzed asymmetric transfer hydrogenation of acetophenone was examined with the immobilized ArSO$_2$DPEN ligands in water with sodium formate in the presence of various surfactants (Scheme 6.10). Of the immobilized chiral ruthenium complexes, **73-Ru** of the amorphous silica-supported ArSO$_2$DPEN ligands exhibited high catalytic activity and enantioselectivity under water-based conditions, although the reactivity decreased in the recycling experiments. Thus, the reaction of acetophenone (**76**) was carried out with HCOONa, **73-Ru** and PEG-1000 (acetophenone/

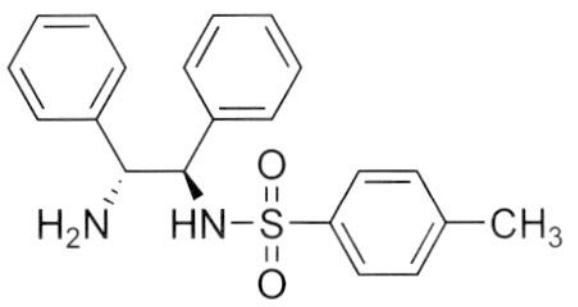

(*R,R*)-TsDPEN (**66**)

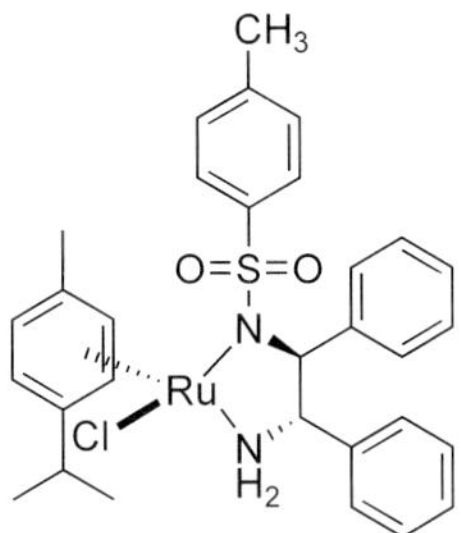

67

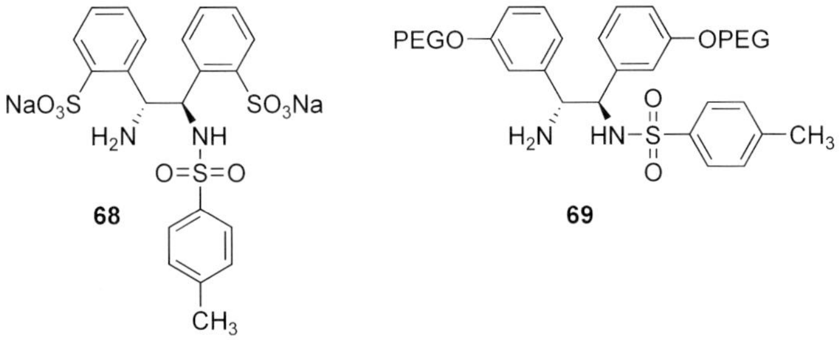

Figure 6.6 Water-soluble TsDPEN ligands.

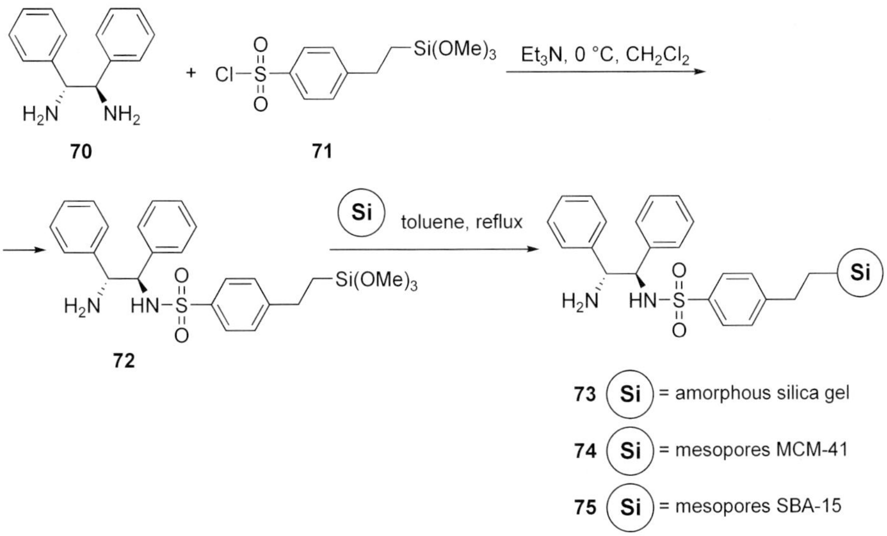

Scheme 6.9 Heterogeneous-switching of the DPEN ligand.

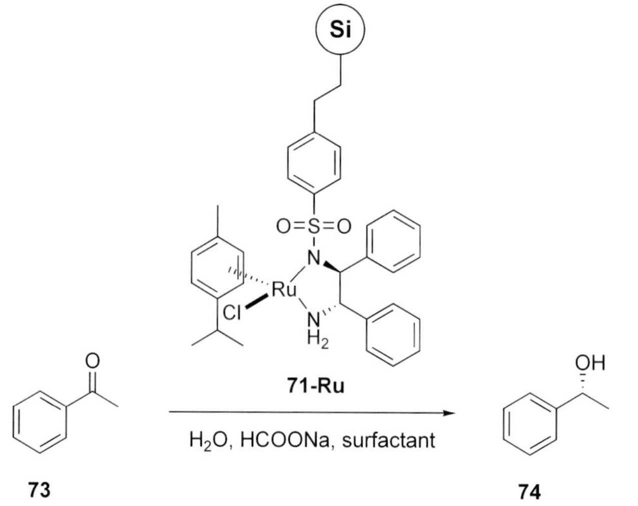

Scheme 6.10 Asymmetric transfer hydrogenation in water with Si-supported DPEN–Ru complexes.

SDS	1st run:	9 h,	>99% conv.; 94% ee
	2nd run:	21 h,	>99% conv.; 95% ee
	3rd run:	72 h,	70% conv.; 94% ee
PEG1000	1st-2nd run:	2-5 h,	>99% conv.; 95% ee
	2rd-6th run:	3.5-35 h,	>99-95% conv.; 95% ee

catalyst/HCOONa/PEG = 1/0.01/500/4) at 80 °C to give 95% ee of phenethyl alcohol **77**.

Itsuno and coworkers have developed a series of PS-supported TsDPEN ligands, as shown in Figure 6.7, for use in water [35]. The enantioselective potential of the polymeric ligands in water was examined for the Ru-catalyzed transfer hydrogenation of acetophenone. Among these ligands, the amphiphilic PS-ArSO$_2$DPEN **82** having an ammonium sulfonate residue was found to bring about high enantioselectivity. The reaction of acetophenone with sodium formate in neat water in the presence of 1 mol% ruthenium of the polymeric DPEN–Ru complex generated *in situ* by mixing **82** and [RuCl$_2$(*p*-cymene)] exhibited a high catalytic performance (40 °C, 3 h, 100% conversion) to afford 98% ee *(R)* of phenethyl alcohol.

6.3.3
Miscellaneous

Several miscellaneous polymer-supported chiral phosphine ligands and oxazoline ligands are depicted in Figure 6.8 and Figure 6.9, respectively. Binaphthyl monophosphine MOP [36] was anchored onto PS-PEG resin at the 2′-, 6- and 6′-positions of its binaphthyl backbone (compounds **84**, **85** and **86**) [37]. The polymeric phos-

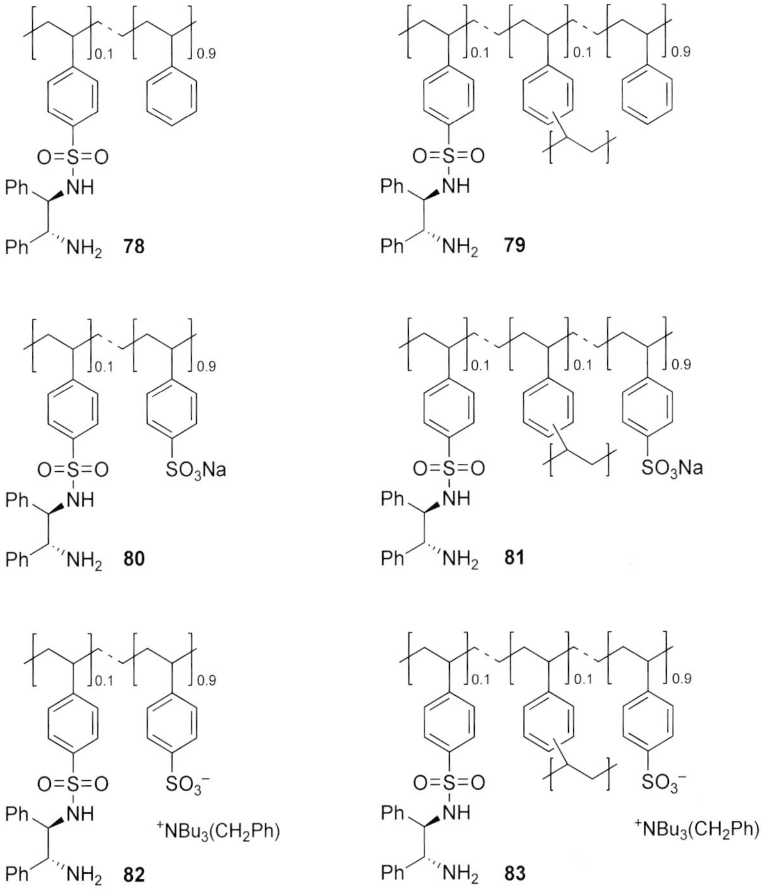

Figure 6.7 Polymer-supported TsDPEN ligands.

phines **87** and **88** were Trost-type bisphosphine ligands immobilized on PS-PEG resin [38]. As discussed in Section 6.2, the amphiphilic PS-PEG supports realize aqueous- and heterogeneous-switching simultaneously. Although these polymeric phosphine ligands have not been investigated for their asymmetric controlling potency in water, they should exhibit aquacatalytic activity with stereoselectivity.

Some typical examples of PS-PEG resin-supported chiral oxazoline ligands [39], which have aquacatalytic potential as well as stereoselective abilities, are illustrated in Figure 6.9.

The rhodium complex **93** coordinated with the chiral phosphoramide ligand **92** was ionically immobilized on mesoporous aminosilicate AITUD (Figure 6.10) [40]. AITUD-**93** was used in water, and showed excellent enantiomeric selectivity and activity in asymmetric hydrogenation.

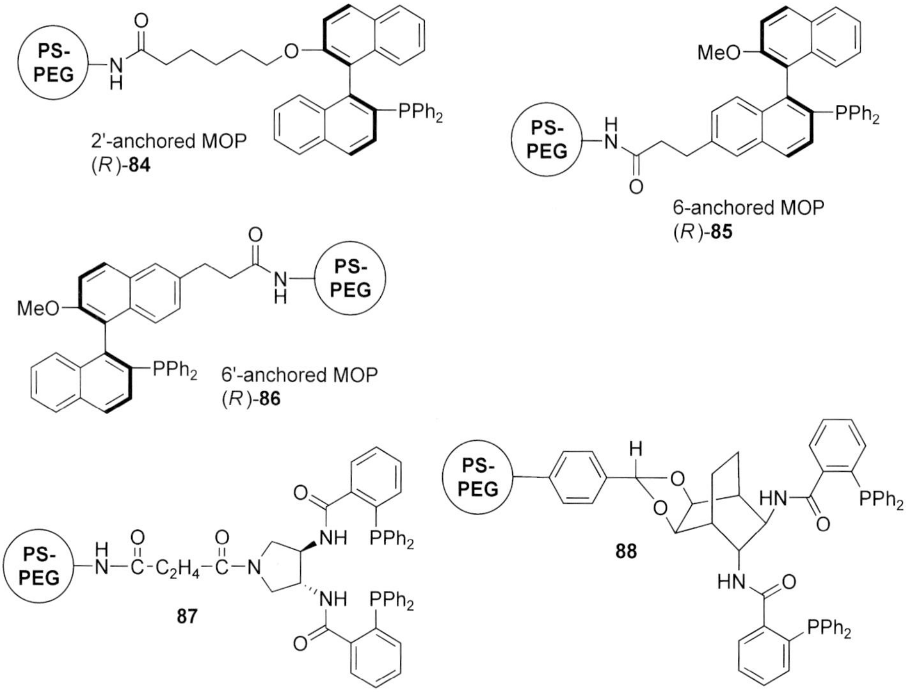

Figure 6.8 PS-PEG-supported chiral phosphine ligands.

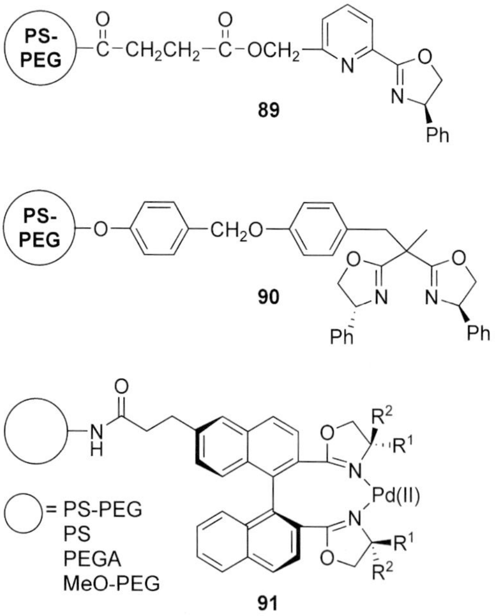

Figure 6.9 PS-PEG-supported chiral oxazoline ligands.

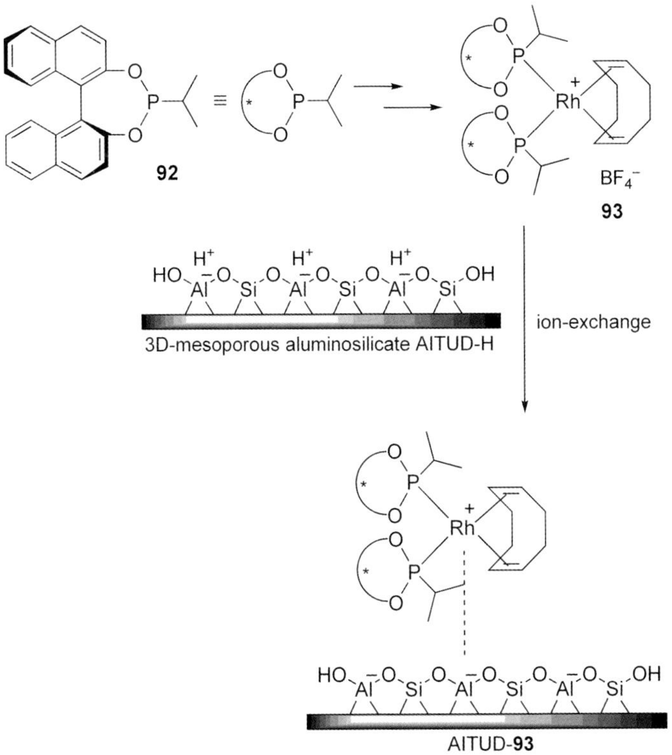

Figure 6.10 AITUD-immobilization of a Rh–phosphoramidite complex.

References

1. (a) Bailey, D.C. and Langer, S.H. (1981) *Chemical Reviews*, **81**, 109.
   (b) Shuttleworth, S.J., Allin, S.M. and Sharma, P.K. (1997) *Synthesis*, 1217.
   (c) Shuttleworth, S.J., Allin, S.M., Wilson, R.D. and Nasturica, D. (2000) *Synthesis*, 1035.
   (d) Dörwald, F.Z. (2000) *Organic Synthesis on Solid Phase*, Wiley-VCH Verlag GmbH, Weinheim.
   (e) Leadbeater, N.E. and Marco, M. (2002) *Chemical Reviews*, **102**, 3217.
   (f) McNamara, C.A., Dixon, M.J. and Bradley, M. (2002) *Chemical Reviews*, **102**, 3275.
   (g) De Vos, D. E., Vankelecom, I.F.J. and Jacobs, P.A. (2000) *Chiral Catalyst Immobilization and Recycling*, Wiley-VCH Verlag GmbH, Weinheim.
   (h) Fan, Q.-H., Li, Y.-M. and Chan, A.S.C. (2002) *Chemical Reviews*, **102**, 3385.

2. For a recent review of solid-phase reactions using palladium catalysts, see: (a) Uozumi, Y. and Hayashi, T. (2002) Solid-phase palladium catalysis for high-throughput organic synthesis, in *Handbook of Combinatorial Chemistry* (eds K.C. Nicolaou, R. Hanko and W. Hartwig), Wiley-VCH Verlag GmbH, Weinheim, Chapter 19, pp. 531–584.
   (b) Uozumi, Y. (2004) *Topics in Current Chemistry*, **242**, 77.

3. For reviews on aqueous-switching, see: (a) Li, C.-J. and Chan, T.-H. (1997) *Organic Reactions in Aqueous Media*, Wiley-VCH Verlag GmbH, New York.
   (b) Grieco, P.A. (1997) *Organic Synthesis in Water*, Kluwer Academic Publishers, Dordrecht.

(c) Herrmann, W.A. and Kohlpaintner, C.W. (1993) *Angewandte Chemie–International Edition in English*, **32**, 1524.
(d) Lindström, U.M. (2002) *Chemical Reviews*, **102**, 2751.
(d) Lindström, U.M. (2007) *Organic Reactions in Water*, Blackwell Publishing, Oxford.

4 For reviews on aqueous-asymmetric catalysis, see: (a) Manabe, K. and Kobayashi, S. (2002) *Chemistry–A European Journal*, **8**, 4095. (b) Sinou, D. (2002) *Advanced Synthesis Catalysis*, **344**, 221.

5 (a) Brogan, A.P., Dickerson, T.J. and Janda, K.D. (2006) *Angewandte Chemie–International Edition*, **45**, 8100.
(b) Hayashi, Y. (2006) *Angewandte Chemie–International Edition*, **45**, 8103.

6 For studies on organic transformations with polymer-supported complex catalysts in water, see:(a) Uozumi, Y., Danjo, H. and Hayashi, T. (1999) *Journal of Organic Chemistry*, **64**, 3384 (cross-coupling).
(b) Uozumi, Y. and Watanabe, T. (1999) *Journal of Organic Chemistry*, **64**, 6921 (carbonylation reaction).
(c) Shibatomi, K., Nakahashi, T. and Uozumi, Y. (2000) *Synlett*, 1643 (Michael addition).
(d) Uozumi, Y. and Nakai, Y. (2002) *Organic Letters*, **4**, 2997 (Suzuki–Miyaura coupling).
(e) Uozumi, Y., Kimura, T. and T. (2002) *Synlett*, 2045 (Heck reaction).
(f) Uozumi, Y. and Nakazono, M. (2002) *Advanced Synthesis Catalysis*, **344**, 274 (rhodium catalysis).
(g) Uozumi, Y. and Kobayashi, Y. (2003) *Heterocycles*, **59**, 71 (Sonogashira reaction).
(h) Uozumi, Y. and Nakao, R. (2003) *Angewandte Chemie–International Edition.*, **42**, 194 (oxidation).
(i) Uozumi, Y. and Nakao, R. (2003) *Angewandte Chemie*, **115**, 204 (oxidation).
(j) Uozumi, Y. and Kikuchi, M. (2005) *Synlett*, 1775 (Suzuki–Miyaura coupling).
(k) Nakao, R., Rhee, H., Uozumi, H. and Y. (2005) *Organic Letters*, **7**, 163 (reduction).
(l) Yamada, Y.M.A. and Uozumi, Y. (2006) *Organic Letters*, 1375 (alkylation).
(m) Uozumi, Y., Danjo, H. and Hayashi, T. (1997) *Tetrahedron Letters*, **38**, 3557.
(n) Uozumi, Y., Danjo, H. and Hayashi, T. (1998) *Tetrahedron Letters*, **39**, 8303.
(o) Danjo, H., Tanaka, D., Hayashi, T. and Uozumi, Y. (1999) *Tetrahedron*, **55**, 14341.
(p) Uozumi, Y. and Shibatomi, K. (2001) *Journal of the American Chemical Society*, **123**, 2919.
(q) Uozumi, Y., Tanaka, H. and Shibatomi, K. (2004) *Organic Letters*, **6**, 281.
(r) Nakai, Y. and Uozumi, Y. (2005) *Organic Letters*, **7**, 291.
(s) Uozumi, Y. and Kimura, M. (2006) *Tetrahedron: Asymmetry*, **17**, 161.

7 (a) Bayer, E. (1991) *Angewandte Chemie–International Edition*, **30**, 113–129.
(b) Rapp, W. (1996) PEG Grafted Polystyrene Tentacle Polymers: Physico-Chemical Properties and Application in Chemical Synthesis, in *Combinatorial Peptide and Nonpeptide Libraries* (ed. G. Jung), Wiley-VCH Verlag GmbH, Weinheim, p. 425.
(c) Du, X. and Armstrong, R.W. (1997) *Journal of Organic Chemistry*, **62**, 5678.
(d) Gooding, O.W., Baudert, D., Deegan, T.L., Heisler, K., Labadie, J.W., Newcomb, W.S., Porco, J.A.Jr and Eikeren, P.J. (1999) *Journal of Combinatorial Chemistry*, **1**, 113.

8 Uozumi, Y., Danjo, H. and Hayashi, T. (1998) *Tetrahedron Letters*, **39**, 8303.

9 (a) Uozumi, Y., Mizutani, K. and Nagai, S.-I. (2001) *Tetrahedron Letters*, **42**, 407.
(b) Uozumi, Y., Yasoshima, K., Miyachi, T. and Nagai, S.-I. (2001) *Tetrahedron Letters*, **42**, 411.

10 Shibatomi, K. and Uozumi, Y. (2002) *Tetrahedron: Asymmetry*, **13**, 1769.

11 For a recent review on asymmetric π-allylic substitution, see: (a) Acemoglu, L. and Williams, J.M.J. (1945) *Handbook of Organopalladium Chemistry* (ed. E. Negishi), John Wiley & Sons, Ltd, New York, 2002.
(b) Trost, B.M. and Crawley, M.L. (2003) *Chemical Reviews*, **103**, 2921.

12 Preliminary communication, see: Uozumi, Y. and Shibatomi, K. (2001) *Journal of the American Chemical Society*, **123**, 2919. Data

shown in Table 6.1 were optimized after the preliminary report.
13. Uozumi, Y., Tanaka, H. and Shibatomi, K. (2004) *Organic Letters*, **6**, 281.
14. Uozumi, Y. and Kimura, M. (2006) *Tetrahedron: Asymmetry*, **17**, 161.
15. Shibatomi, K., Nakahashi, T. and Uozumi, Y. (2000) *Synlett*, 1643.
16. Uozumi, Y. and Suzuka, T. (2006) *Journal of Organic Chemistry*, **71**, 8644.
17. Trost, B.M. and Surivet, J.P. (2000) *Angewandte Chemie – International Edition*, **39**, 3122.
18. Czekelius, C. and Carreira, E.M. (2005) *Angewandte Chemie – International Edition*, **44**, 612.
19. (a) Jakobsen, P., Lundbeck, J.M., Kristiansen, M., Breinholt, J., Demuth, H., Pawlas, J., Candela, M.P.T., Andersen, B., Westergaard, N., Lundgren, K. and Asano, N. (2001) *Bioorganic & Medicinal Chemistry*, **9**, 744.
   (b) Knapp, S. and Zhao, D. (2000) *Organic Letters*, **2**, 2037.
20. Baytoston, D.J., Fraser, J.L., Ashton, M.R., Baxter, A.D., Polywka, M.E.C. and Moses, E. (1998) *Journal of Organic Chemistry*, **63**, 3137.
21. Ohkuma, T., Takeno, H., Honda, Y. and Noyori, R. (2001) *Advanced Synthesis Catalysis*, **343**, 369.
22. (a) Hagiwara, E., Fujii, A. and Sodeoka, M. (1998) *Journal of the American Chemical Society*, **120**, 2474.
   (b) Fujii, A., Hagiwara, E. and Sodeoka, M. (1999) *Journal of the American Chemical Society*, **121**, 5450.
23. Guerreiro, P., Ratovelomanana-Vidal, V., Genet, J.P. and Dellis, P. (2001) *Tetrahedron Letters*, **42**, 3423.
24. (a) Wan, K.T. and Davis, M.E. (1993) *Journal of the Chemical Society D – Chemical Communications*, 1262.
   (b) Wan, K.T. and Davis, M.E. (1993) *Tetrahedron: Asymmetry*, **4**, 2461.
   (c) Wan, K.T. and Davis, M.E. (1994) *Journal of Catalysis*, **148**, 1.
25. (a) Lamouille, T., Saluzzo, C., ter Halle, R., Le Guyader, F. and Lemaire, M. (2001) *Tetrahedron Letters*, **42**, 663.
   (b) Berthod, M., Saluzzo, C., Mignani, G. and Lemaire, M. (2004) *Tetrahedron: Asymmetry*, **15**, 639.
26. Amengual, R., Michelet, V. and Genet, J.-P. (2002) *Synlett*, 1791.
27. (a) Aquacatalytic asymmetric allylic alkylation was also examined with a combination system of a BINAP–palladium complex and an external surfactant, see: Sinou, D., Rabeyrin, C. and Nguefack, C. (2003) *Advanced Synthesis Catalysis*, **345**, 357.
   (b) Rabeyrin, C. and Sinou, D. (2003) *Tetrahedron: Asymmetry*, **14**, 3891.
28. Otomaru, Y., Senda, T. and Hayashi, T. (2004) *Organic Letters*, **6**, 3357.
29. For reviews, see: Hayashi, T. and Yamasaki, K. (2003) *Chemical Reviews*, **103**, 2829.
30. For recent reviews, see: (a) Blaser, H.-U., Malan, C., Pugin, B., Spindler, F., Steiner, H. and Studer, M. (2003) *Advanced Synthesis Catalysis*, **345**, 103.
   (b) Everaere, K., Mortreux, A. and Carpentier, J.-F. (2003) *Advanced Synthesis Catalysis* **345**, 66.
   (c) Saluzzo, C. and Lemaire, M. (2002) *Advanced Synthesis Catalysis*, **344**, 915.
31. (a) Hashiguchi, S., Fujii, A., Takehara, J., Ikariya, T. and Noyori, R. (1995) *Journal of the American Chemical Society*, **117**, 7562.
   (b) Fujii, A., Hashiguchi, S., Uematsu, N., Ikariya, T. and Noyori, R. (1996) *Journal of the American Chemical Society*, **118**, 2521.
   (c) Noyori, R. and Hashiguchi, S. (1997) *Accounts of Chemical Research*, **30**, 97.
32. Sulfonated TsDPEN, see: (a) Ma, Y., Liu, H.L., Chen, L., Cui, X., Zhu, J. and Deng, J. (2003) *Organic Letters*, **5**, 2103.
   (b) Wu, J., Wang, F., Ma, Y., Cui, X., Cun, L., Zhu, J., Deng, J. and Yu, B. (2006) *Chemical Communications*, 1766.
33. PEG-modified TsDPEN, see: (a) Li, X., Chen, W., Hems, W., King, F. and Xiao, J. (2004) *Tetrahedron Letters*, **45**, 951.
   (b) Li, X., Wu, X., Chen, W., Hancock, F.E., King, F. and Xiao, J. (2004) *Organic Letters*, **5**, 3321. The asymmetric transfer hydrogenation of various acetophenones was catalyzed in water by a ruthenium complex of PEG–TsDPEN (0.1–1.0 mol%) with sodium formate, and the polymeric catalyst was readily recovered via precipitation, and reused 14 times without significant loss of either catalytic activity or stereoselectivity.

34 (a) Liu, P.N., Deng, J.G., Tu, Y.Q. and Wang, S.H. (2004) *Chemical Communications*, 2070.
(b) Liu, P.N., Gu, P.-M., Deng, J.G., Tu, Y.Q. and Ma, Y.-P. (2005) *European Journal of Organic Chemistry*, 3221.

35 Arakawa, Y., Haraguchi, N. and Itsuno, S. (2006) *Tetrahedron Letters*, **47**, 3239.

36 (a) Uozumi, Y. and Hayashi, T. (1991) *Journal of the American Chemical Society*, **113**, 9887.
(b) Uozumi, Y., Tanahashi, A., Lee, S.Y. and Hayashi, T. (1993) *Journal of Organic Chemistry*, **58**, 1945.
(c) Uozumi, Y., Suzuki, N., Ogiwara, A. and Hayashi, T. (1994) *Tetrahedron*, **50**, 4293.
(d) Uozumi, Y., Kawatsura, M. and Hayashi, T. (2002) *Organic. Synthesis*, **78**, 1–13.

37 (a) Hocke, H. and Uozumi, Y. (2004) *Tetrahedron*, **60**, 9297.
(b) Hocke, H. and Uozumi, Y. (2003) *Tetrahedron*, **59**, 619.

38 (a) Song, C.E., Yang, J.W., Roh, Y.E., Lee, S.-G., Ahn, J.H. and Han, H. (2002) *Angewandte Chemie–International Edition*, **41**, 3852.
(b) Trost, B.M., Pan, Z., Zambrano, J. and Kujat, C. (2002) *Angewandte Chemie–International Edition*, **41**, 4691.

39 (a) Hallman, K., Macedo, E., Nordström, K. and Moberg, C. (1999) *Tetrahedron: Asymmetry*, **10**, 4037.
(b) Hallman, K. and Moberg, C. (2000) *Tetrahedron: Asymmetry*, **12**, 1475.
(c) Glos, M. and Reiser, O. (2000) *Organic Letters*, **2**, 2045.
(d) Hocke, H. and Uozumi, Y. (2000) *Synlett*, 2049.

40 Simons, C., Hanefeld, U., Arends, I.W.C.E., Minnaard, A.J., Maschmeyer, T. and Sheldon, R. (2004) *Chemical Communications*, 2830.

7
Enantioselective Catalysis in Ionic Liquids and Supercritical CO_2

Sang-gi Lee[1] and Yong Jian Zhang[2]

7.1
Introduction

Although homogeneous asymmetric catalysis very often progresses through the development of highly effective chiral catalysts, the solvent may also serve as a key ingredient in an asymmetric catalytic system. In addition to interactions between the catalysts and substrates, interactions with solvents may also exert a significant impact on the catalyst stability, activity and selectivity. Hence, the choice of solvent may also be critically important in asymmetric catalysis. Moreover, the judicious choice of solvent can often play a crucial role in solving the problems associated with catalyst separation and recycling. One very promising approach is that catalysis may be conducted in either a solvent or a biphasic system, in which the catalyst is immobilized and with the possibility of its being separated from the reaction products and then recycled. However, many organic solvents are both hazardous and volatile, and may also be deleterious to human health as well as causing environmental problems. In contrast, the direct immobilization of a chiral catalyst in common organic solvents is not possible. Consequently, the search for a new 'green' medium either to replace or to minimize the use of organic solvents, and simultaneously to immobilize chiral catalysts, is highly desirable from both economic and environmental points of view.

During the past decade, ionic liquids have attracted increasing attention as a new class of reaction medium [1]. Ionic liquids are organic salts which are composed of an organic cation and an inorganic or organic anion, and have a low melting point (generally below 100 °C). Due to their negligible vapor pressure and nonflammable nature, ionic liquids are currently regarded as being eco-friendly alternatives to volatile organic solvents in chemical processes. Moreover, the properties of ionic liquids such as melting temperature and hydrophilicity/hydrophobicity can be fine-tuned by changing the structure of cation and anion. In this way they can be tailored to be immiscible with water or with certain organic solvents, thus making them more useful for facilitating catalyst recovery from the reaction mixture. In 1995, Chauvin *et al.* reported for the first time on the use of an ionic

Handbook of Asymmetric Heterogeneous Catalysis. Edited by K. Ding and Y. Uozumi
Copyright © 2008 WILEY-VCH Verlag GmbH & Co. KGaA, Weinheim
ISBN: 978-3-527-31913-8

liquid as a liquid support in catalytic asymmetric hydrogenation [2]. Since then, a range of enantioselective catalytic reactions have been transposed into ionic liquids and, in many cases, the use of ionic liquids has offered many advantages over the reactions in organic solvents in terms of activity, enantioselectivity and stability, as well as the reusability of the ionic solvent-catalyst systems [3]. However, the mutual solubility of ionic liquids (including chiral catalysts) in reaction products (or extraction solvents) and the solubility of products (or extraction solvent) in ionic liquids may render the separation process more complex. In order to solve the problems associated with contamination of the extract phase with ionic liquids and catalyst losses, the use of supercritical carbon dioxide (scCO$_2$) to remove the reaction products from the ionic liquid phase has recently been proposed. The feasibility of this concept has been successfully demonstrated in several types of asymmetric catalysis [4]. In contrast to ionic liquids, scCO$_2$ possesses the characteristics of nonpolar solvents and high volatility, and hence can be considered as solvents that are complementary to each other. Although a significant amount of CO$_2$ dissolves in ionic liquids, the ionic liquids do not dissolve in CO$_2$. Moreover, in the presence of scCO$_2$ the viscosity of ionic liquid solutions is reduced, and this facilitates mass-transfer. Apart from these potential benefits for product separation from ionic liquids, scCO$_2$ also has a prodigious potential as an environmentally benign reaction medium for sustainable chemical synthesis. The properties of scCO$_2$ may also be beneficial in asymmetric catalysis; for example, the gas-like mass-transfer properties can facilitate exchange between the catalyst and substrate phases. Due to the high volatility and relatively poor solubility of scCO$_2$ for chiral catalysts, the catalyst can be easily separated with reduced leaching as compared to conventional organic solvents. Since the first report on asymmetric catalysis in scCO$_2$ in 1995 [5] a variety of asymmetric catalyses – notably asymmetric hydrogenation and hydroformylation – have been investigated in scCO$_2$ [6].

7.2
Enantioselective Catalysis in Ionic Liquids

One of the major advantages of using ionic liquids in asymmetric catalysis is the easy immobilization of expensive chiral catalysts. Due to the high affinity of ionic liquids for ionic species, many cationic and anionic chiral catalysts can be immobilized without structural modification. For neutral catalysts, leaching of the catalyst into the organic phase can be reduced by increasing their affinity to ionic liquids through the incorporation of an ionic or a polar tag onto the chiral ligands. The miscibility of ionic liquids with organic solvents (or compounds) may be adjustable through the variation of the structures of cations and/or anions. In general, ionic liquids tend to be immiscible with less-polar solvents (i.e. hydrocarbons, diethyl ether) or alcohols higher than ethanol (particularly isopropanol), and these solvents are often used for biphasic catalytic reaction and/or to extract the products from the ionic liquid containing the chiral catalyst at the end of the reaction. If the reaction product is liquid and is immiscible with ionic liquids, the

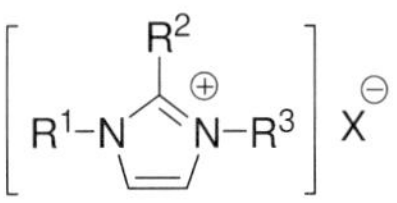

$$\left[R^1\text{-}N \underset{\oplus}{\overset{R^2}{\frown}} N\text{-}R^3 \right] X^{\ominus}$$

X: BF$_4$, PF$_6$, SbF$_6$, NTf$_2$, OTf, OMs
Tf = Trifluoromethanesulfonyl
Ms = Methanesulfonyl

[bmim]: R^1 = Butyl, R^2 = H, R^3 = Me
[emim]: R^1 = Ethyl, R^2 = H, R^3 = Me
[dmpim]: R^1 = Propyl, R^2 = Me, R^3 = Me
[bdmim]: R^1 = Butyl, R^2 = Me, R^3 = Me

[dbim]: R^1 = Butyl, R^2 = H, R^3 = Butyl
[hmim]: R^1 = Hexyl, R^2 = H, R^3 = Me
[moemim]: R^1 = Methoxyethyl, R^2 = H, R^3 = Me
[omim]: R^1 = Octyl, R^2 = H, R^3 = Me

Figure 7.1 Imidazolium-based ionic liquids and their abbreviations.

product can be readily separated by simple decantation. An alternative approach to the product isolation is to distill the volatile liquid products from the nonvolatile ionic liquid phase, but this method may sometimes cause catalyst decomposition. In order to facilitate the separation and reduce the use of ionic liquids, the 'supported aqueous-phase catalysis' (SAPC) concept was transposed to ionic liquids, in which the ionic liquid phase containing catalyst is impregnated onto an inorganic high-surface solid to generate the so-called 'supported ionic liquid phase' (SILP) catalyst. This approach combines the advantages of ionic liquids with those of heterogeneous catalysis. Of course there is no universal immobilization method, and the choice of immobilization method will depend upon the type of reaction and catalyst, as well as on the nature of the ionic liquids employed. To date, the imidazolium cation-based ionic liquids bearing BF$_4$, SbF$_6$, PF$_6$, NTf$_2$ (Tf = trifluoromethanesulfonyl) and OMs anions are the most widely employed in asymmetric catalysis. The most commonly used imidazolium cation-based ionic liquids used in asymmetric catalysis are depicted in Figure 7.1 and discussed in this chapter.

7.2.1
Asymmetric Reductions in Ionic Liquids

7.2.1.1 Asymmetric Hydrogenations of the C=C Bond

Homogeneous catalytic asymmetric hydrogenations have been established as one of the most versatile and powerful methods for the synthesis of optically pure organic compounds. During the twentieth century, a variety of innovative chiral catalysts for asymmetric catalytic hydrogenation have been developed to provide extremely high activity and enantioselectivity that are difficult to realize with heterogeneous catalysts [7]. However, as with the most homogeneous catalysts, product separation and catalyst recovery tend to be less easy than with their heterogeneous counterparts. Some initial studies performed by Chauvin with an ionic rhodium–DIOP complex demonstrated the feasibility of the ionic liquids as liquid supports for ionic chiral catalysts (Scheme 7.1) [2]. Since then, a wide range of ionic chiral complexes composed of transition metal (Rh, Ru, Ir, etc.) and chiral phosphine ligands (Figure 7.2) have been investigated for the hydrogenation of various substrates including α-acetamidocinnamic acid, methyl α-acetamidoacrylate, simple aryl enamide, functionalized olefins and carbonyls in ionic liquids. In most cases, the catalytic activities (reactivity and enantioselectivity) in ionic liquids

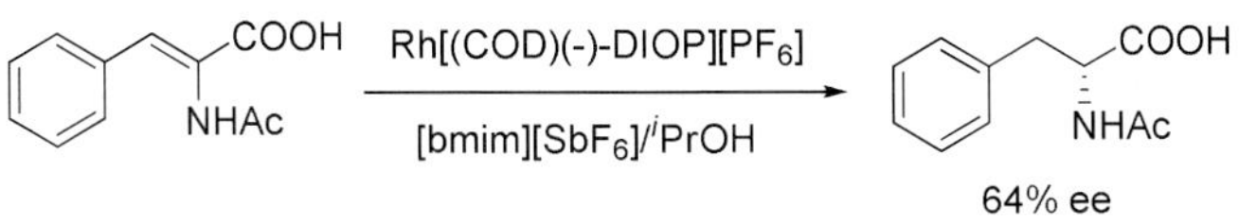

7 Enantioselective Catalysis in Ionic Liquids and Supercritical CO_2

Scheme 7.1

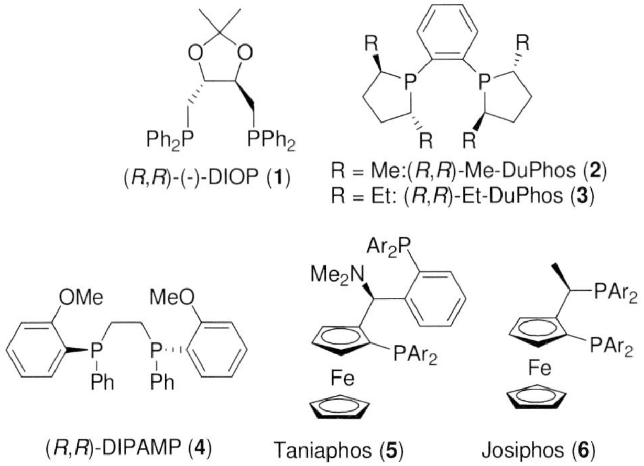

Ph-CH=C(NHAc)-COOH → (with Rh[(COD)(-)-DIOP][PF$_6$], [bmim][SbF$_6$]/iPrOH) → Ph-CH$_2$-CH(NHAc)-COOH, 64% ee

(R,R)-(-)-DIOP (**1**)

R = Me: (R,R)-Me-DuPhos (**2**)
R = Et: (R,R)-Et-DuPhos (**3**)

(R,R)-DIPAMP (**4**)   Taniaphos (**5**)   Josiphos (**6**)

Figure 7.2 Chiral bisphosphine ligands used in asymmetric hydrogenations in ionic liquids without structural modifications.

were comparable to or even better than those obtained in common organic solvents. Many of the reactions were carried out in ionic liquid/organic solvent (mostly *i*PrOH) biphasic solvent systems in which the catalysts were immobilized in ionic liquids. The reaction product, when dissolved in the organic phase, could be separated using simple phase separation or by extraction with less-polar solvents. The recovered ionic liquid containing the active catalyst could be readily reused several times, without any significant loss of catalytic activity.

One interesting finding was that the hydrogen concentration in ionic liquid layer was shown to have a greater effect on both the conversion and the enantioselectivity than the hydrogen pressure in gas phase [8]. As shown in Table 7.1, in the asymmetric hydrogenation of (Z)-α-acetamidocinnamic acid using an ionic catalyst, [Rh(COD)(R,R)–Et–DuPhos][CF$_3$SO$_3$], both the conversion and the enantioselectivity increased with the enhancement of hydrogen solubility in the ionic liquid. At the same hydrogen pressure (50 atm), a relatively lower conversion and enantioselectivity were obtained in ionic liquid solvents in comparison with those obtained in an organic solvent, *i*PrOH. This effect can be ascribed to the higher hydrogen concentration in *i*PrOH than in ionic liquids. The solubility of hydrogen in a hydrophobic ionic liquid, [bmim][PF$_6$], is lower than in hydrophilic [bmim][BF$_4$], in which a higher conversion and enantioselectivity have been obtained. When the asymmetric hydrogenation of enamides, methyl α-acetamidoacrylate or methyl (Z)-α-acetamidocinnamate was performed in a biphasic [bmim][PF$_6$]/*i*PrOH solvent system using [Rh(COD)](R,R)-Me-DuPhos][CF$_3$SO$_3$] as a chiral catalyst, the

Table 7.1 Asymmetric hydrogenation of (Z)-α-acetamidocinnamic acid: the effect of hydrogen concentration in the liquid phase on conversion and enantioselectivity.

$$\text{PhCH=C(NHAc)COOH} \xrightarrow[\text{ionic liquid or }^i\text{PrOH}]{\text{H}_2 \text{ (pressure)}, [\text{Rh(COD)(R,R)-Et-DuPhos}][\text{CF}_3\text{SO}_3]} \text{PhCH}_2\text{CH(NHAc)COOH}$$

Entry	Solvent	P(atm)	H$_2$ solubility/mol l^{-1}	Conv. (%)	ee (%)
1	[bmim][PF$_6$]	5	4.4×10^{-3}	7	66
2	[bmim][PF$_6$]	50	4.4×10^{-2}	26	81
3	[bmim][PF$_6$]	100	8.9×10^{-1}	41	90
4	[bmim][BF$_4$]	50	1.5×10^{-1}	73	93
5	^{i}PrOH	50	129.3	99	94

enantioselectivities (80–96% ee) were similar to those obtained in iPrOH as a single solvent, and the catalyst was recovered and readily recycled. However, the conversion decreased upon the reuse of the recovered ionic liquid/catalyst (first run: 83%, second run: 64%), which suggested that leaching of the catalyst into the iPrOH had occurred during phase separation [9].

The catalytic activity of the Rh–Et–DuPhos complex in the asymmetric hydrogenation was also largely dependent on the miscibility of the ionic liquid with the organic solvent [10]. The turnover frequency (TOF) of the catalyst in a pure ionic liquid, [bmim][PF$_6$], was almost zero, but was dramatically increased by the addition of an organic cosolvent. For example, in the asymmetric hydrogenation of methyl α-acetamidoacrylate, a much higher TOF was observed in a [bmim][PF$_6$]-MeOH solvent system (MeOH was completely miscible with [bmim][PF$_6$]) (entry 5 in Table 7.2) than in [bmim][PF$_6$]-EtOH (solubility of EtOH in [bmim][PF$_6$]: 17 wt% at 20 °C) (entry 6 in Table 7.2) and [bmim][PF$_6$]-iPrOH (iPrOH is immiscible with [bmim][PF$_6$]) (entry 7 in Table 7.2). The increased catalytic efficiency may be attributed to the increased concentrations of the reagents, hydrogen and substrate in the system composed of a miscible organic solvent and an ionic liquid. Other chiral phosphine ligands such as DIPAMP, Taniaphos and Josiphos, Walphos and BPPM were also applied to the Rh-catalyzed asymmetric hydrogenation of methyl α-acetamidoacrylate or methyl α-acetamidocinnamate in ionic liquid-based solvent systems [11, 12]. Asymmetric hydrogenations using chiral Rh-complexes in an ionic liquid/H$_2$O binary solvent system, a so-called 'wet ionic liquid', exhibited an improved catalytic activity. One promising result reported by Feng and coworkers [12] was that the Rh–Taniaphos complex in the [omim][BF$_4$]/H$_2$O system exhibited very high catalytic activity (100% conversion) and enantioselectivities (99% ee), and the catalyst immobilized in the ionic liquid layer could be reused.

In general, the catalytic activities of the Rh catalysts in the ionic liquid-based solvent systems are largely dependent not only on the chiral ligands but also on the solvent system. Mostly ionic liquid/iPrOH biphasic solvent systems are

Table 7.2 Asymmetric hydrogenation of methyl α-acetamidoacrylate with Rh(I)–Et–DuPhos complex: the effect of the miscibility of ionic liquid with an organic solvent on catalytic activity.

$$\underset{NHAc}{\overset{COOMe}{\diagup}}\!\!=\!\! \quad \xrightarrow[\text{ionic liquid/alcohol}]{\text{H}_2 \;\; \text{Rh(I)-(R,R)-Et-DuPhos}} \quad \underset{NHAc}{\overset{COOMe}{\diagup}}\!\!-\!\!$$

Entry	Solvent	TOF (h^{-1})	ee (%)
1	MeOH	3225	97
2	EtOH	3220	98
3	iPrOH	2950	96
4	[bmim][PF$_6$]	0	0
5[a]	[bmim][PF$_6$]/MeOH	3012 (2953)	97 (98)
6[a]	[bmim][PF$_6$]/EtOH	758 (755)	98 (99)
7[a]	[bmim][PF$_6$]/iPrOH	460 (453)	95 (94)

a Data in parentheses are for the second run with the recovered catalyst immobilized in ionic liquid.

employed but, due to the solubility of chiral Rh-complexes and/or ionic liquids in iPrOH, leaching of the catalyst into the iPrOH layer could not be completely avoided, and this caused a decrease in catalytic activity when reusing the ionic liquid layer containing the Rh-catalyst. To solve the problem associated with catalyst leaching, imidazolium ionic tags have been attached to the chiral diphosphine (Figure 7.3). The incorporation of the ionic tags avoided catalyst leaching, and increased the stability of the catalyst in the ionic liquid [13]. The Rh complex of diphosphine Me-BDPMI **7** [14], which showed excellent catalytic activity in asymmetric hydrogenations, was modified task-specifically with the imidazolium salt to give a chiral diphosphine–Rh complex **8** (ILG-BDPMI), and applied as a catalyst for the asymmetric hydrogenation of an enamide in an ionic liquid/iPrOH biphasic system. As shown in Table 7.3, the catalytic efficiency of **7** in an ionic liquid fell significantly after two runs (entries 1–3); hence, the conversion was decreased significantly in the third (78%, entry 3) and fourth (51%, entry 4) runs. The reaction was not completed even after a prolonged reaction time (entry 5). In contrast, the imidazolium ion-tagged Rh-complex **8** was successfully immobilized in the ionic liquid, and could be reused three times without any loss of catalytic activity. Although in the fourth run, the catalytic activity was slightly decreased, the reaction was completed when the reaction time was extended to 8 h. The enantioselectivity remained almost constant. Inductively coupled plasma atomic emission spectroscopy (ICP-AES) analyses of the iPrOH layer separated from the first runs revealed little leaching of Rh and phosphorus within detection limits, which in turn indicated that the appended imidazolium ion tag had dramatically improved the preferential solubility of the Rh-complex in the ionic liquid.

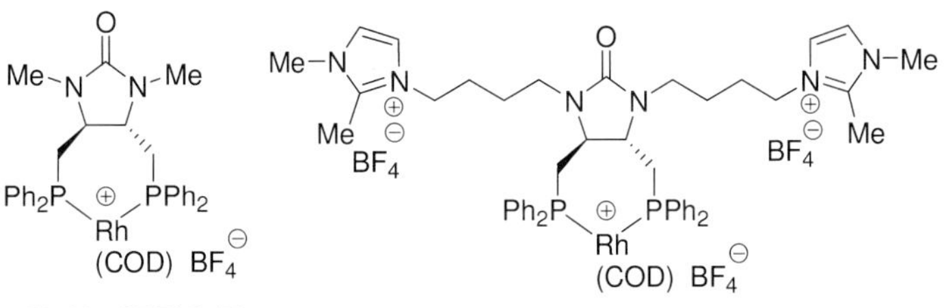

Figure 7.3 Rh-complexes of Me-BDPMI and ILG-BDPMI.

Table 7.3 Asymmetric hydrogenation of N-acetylphenylethenamine using Rh–Me–BDPMI and Rh–ILG–BDPMI in [bmim][SbF$_6$]/iPrOH biphasic system.

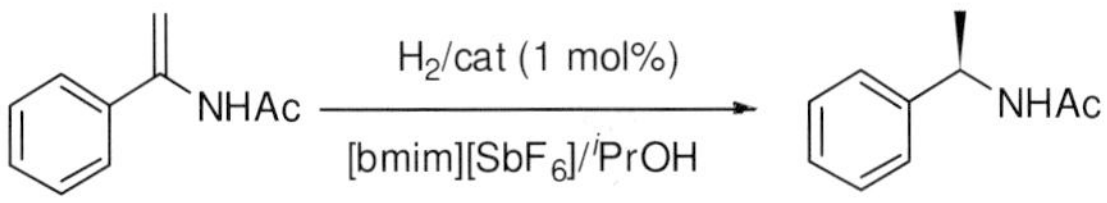

Entry	Catalyst (1 mol%)	Run	Time (h)	Conv. (%)	ee (%)
1	Rh–Me–BDPMI	1	1	100	96
2	Rh–Me–BDPMI	2	1	100	95
3	Rh–Me–BDPMI	3	1	78	94
4	Rh–Me–BDPMI	4	1	51	91
5	Rh–Me–BDPMI	4	12	85	88
6	Rh–ILG–BDPMI	1	1	100	97
7	Rh–ILG–BDPMI	2	1	100	97
8	Rh–ILG–BDPMI	3	1	100	96
9	Rh–ILG–BDPMI	4	1	82	95
10	Rh–ILG–BDPMI	4	8	100	95

One new strategy for overcoming the catalyst/ionic liquid leaching problem is organic solvent nanofiltration (OSN); this has been used recently in the effective separation of reaction products from ionic liquids [15]. In the asymmetric hydrogenation of dimethyl itaconate using the Ru–BINAP complex 9 in a mixed solvent of MeOH and ionic liquids (imidazolium ionic liquids, and phosphonium and ammonium ionic liquids), the catalyst and ionic liquid could be separated by using a solvent-resistant polyimide OSN membrane (Starmem 122, molecular weight cut-off: 220 g mol^{-1}), and recycled several times (Scheme 7.2). Although the OSN-based approach is effective for the separation of catalyst/ionic liquid, the choice of substrate, catalyst, ionic liquids and molecular weight cut-off range of the nanomembrane is crucial.

7.2.1.2 Asymmetric Hydrogenations of the C=O Bond

In contrast to air-sensitive chiral Rh–phosphine complexes, the chiral Ru complexes of BINAP or BINAP analogues were known to be more effectively

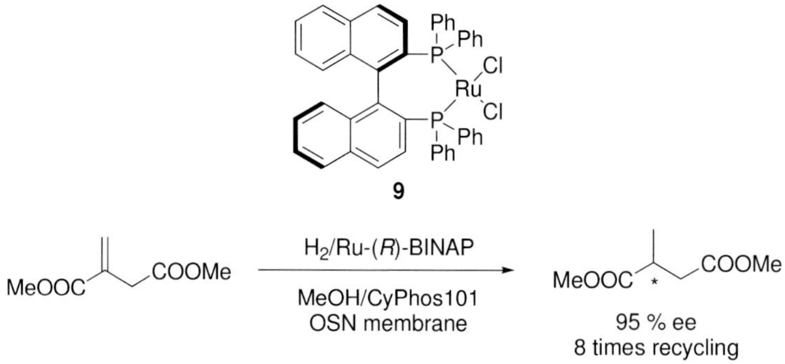

Scheme 7.2

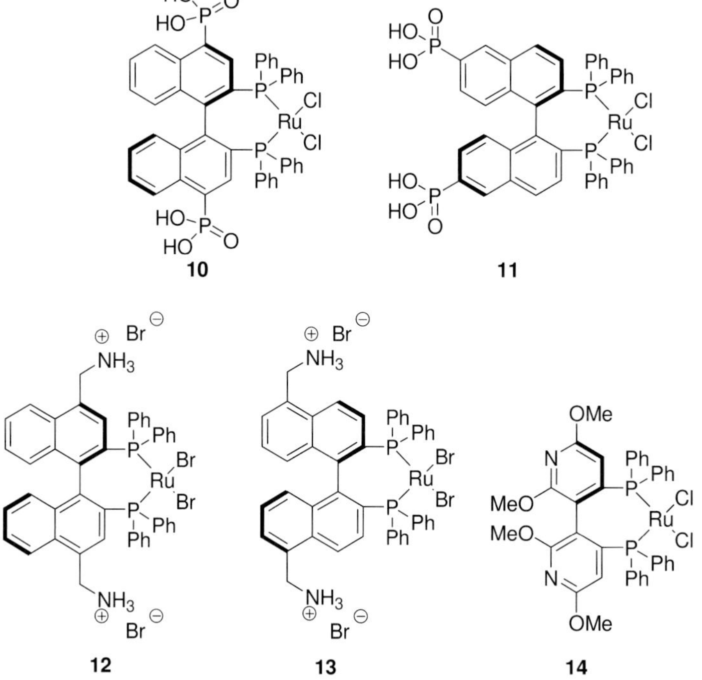

Figure 7.4 Ru complexes of BINAP analogues for asymmetric hydrogenations of the C=O bond.

immobilized in ionic liquids, which have been used in asymmetric hydrogenation of carbonyl group [8, 16]. Nevertheless, the leaching of catalysts from the ionic liquid phase is very often observed. To overcome these obstacles, Lin and coworkers attached a polar phosphonic acid group onto the BINAP ligand, and the polar Ru–BINAP complexes were then used in the catalytic asymmetric hydrogenation of β-keto esters in ionic liquids (Figure 7.4) [17]. In the [dmpim][NTf$_2$]/MeOH solvent system, 4,4′-bisphosphonic acid-derived catalyst **10** was separated by simple extraction, and reused four times without any deterioration in activity and enanti-

7.2 Enantioselective Catalysis in Ionic Liquids

Table 7.4 Ru-catalyzed asymmetric hydrogenation β-keto esters in ionic liquids.

$$R\overset{O}{\underset{}{\text{-COOEt}}} \xrightarrow[\text{ionic liquid}]{H_2/\text{Ru-ligand}} R\overset{OH}{\underset{}{\text{-COOEt}}}$$

Entry	Catalyst	R¹	R²	Solvent	Run	Conv. (%)	ee (%)
1	Ru-BINAP	Me	Et	[dmpim][NTf₂]/MeOH	1	>99	99
2	10	Me	Et	[dmpim][NTf₂]/MeOH	1	>99	99
3					2	>99	99
4					3	>99	98
5					4	>99	98
6	11	Me	Et	[bmim][BF₄]/MeOH	1	>99	97
7					2	97	98
8					3	94	97
9					4	75	95
10	10	Ph	Et	[bmim][BF₄]/MeOH	1	>98	97
11					2	94	94
12					3	81	86
13					4	62	95
14	12	Me	Et	[bmim][BF₄]	1	100	76
15					2	100	90
16	13	Me	Et	[bmim][BF₄]	1	100	80
17	14	Me	Me	[bmim][PF₆]/MeOH	1	>99	98
18					2	99	98
19					3	96	98
20					4	97	96
21					5	96	93
22					6	96	86
23					7	96	78
24					8	94	57
25					9	94	49

oselectivity (Table 7.4; entries 2–5). However, the catalytic efficiency and reusability of 6,6'-bisphosphonic acid-derived catalyst **11** in [bmim][BF₄]/MeOH declined gradually (entries 6–9). For the hydrogenation of ethyl benzoylacetate with catalyst **10** in the [bmim][BF₄]/MeOH system, the conversion and ee-values declined upon recycling of the catalyst (entries 10–13). Similarly, Lemaire and coworkers [18] developed ammonium salt-derived complexes **12** and **13** as polar and ionic liquid-soluble Ru-BINAP analogues for the asymmetric hydrogenation of ethyl acetoacetate in ionic liquids; however, the enantioselectivity was not sufficiently high (entries 14–16). Chan and coworkers [19] also performed catalytic asymmetric hydrogenation of methyl acetoacetate using ruthenium catalyst **14** in ionic liquids. As shown in Table 7.4, in the [bmim][PF₆]/MeOH system, the Ru–P-Phos could be reused and recycled with good activity and excellent enantioselectivity in the first four runs of hydrogenation reactions (entries 17–20). These results clearly demonstrated the effects of the 'ionic tag' on the catalyst recycling in ionic liquids.

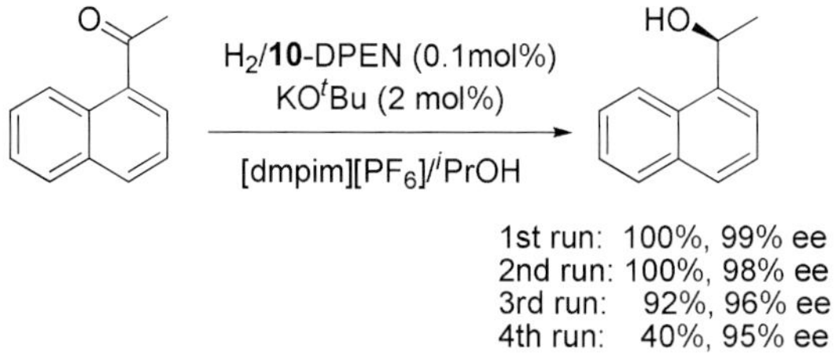

Scheme 7.3

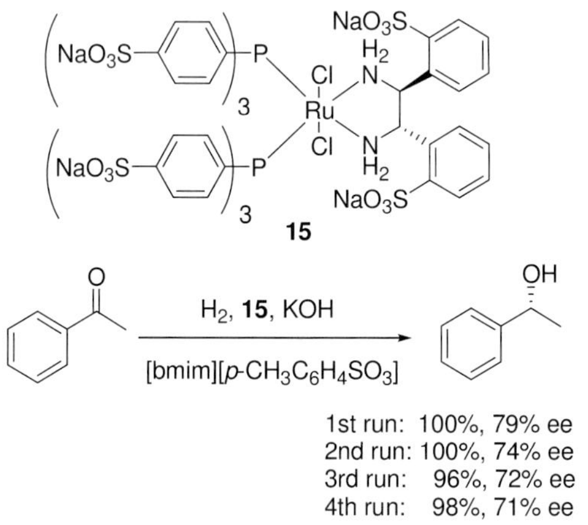

```
O                H₂/10-DPEN (0.1 mol%)        HO
‖                KO^tBu (2 mol%)               |
Naphthyl-C-CH₃  ─────────────────────→    Naphthyl-CH-CH₃
                 [dmpim][PF₆]/iPrOH
```

1st run: 100%, 99% ee
2nd run: 100%, 98% ee
3rd run: 92%, 96% ee
4th run: 40%, 95% ee

15

Ru complex with two (NaO₃S-C₆H₄-)₃P ligands, two Cl, and a DPEN-type diamine bearing NaO₃S groups on both phenyl rings.

Scheme 7.4

```
O                                        OH
‖            H₂, 15, KOH                 |
Ph-C-CH₃  ──────────────────→         Ph-CH-CH₃
          [bmim][p-CH₃C₆H₄SO₃]
```

1st run: 100%, 79% ee
2nd run: 100%, 74% ee
3rd run: 96%, 72% ee
4th run: 98%, 71% ee

However, the conversion and enantioselectivity were slowly decreased from the fifth run onwards (entries 21–25). Nevertheless, these results clearly demonstrated the effects of "ionic tag" on the catalyst recycling in ionic liquids.

Noyori-type Ru-complexes, prepared by the combination of polar phosphonic acid-derived **10** and **11** with (R,R)-DPEN (1,2-diphenylethylenediamine), were also used to catalyze the asymmetric hydrogenation of aromatic ketones in an ionic liquid/iPrOH biphasic system (Scheme 7.3) [20]. The conversions and enantioselectivities were quite comparable to those obtained from homogeneous reaction in iPrOH. The Ru complex **10**-DPEN immobilized in the ionic liquid layer could be recycled six times without any significant loss of enantioselectivity, albeit with a dramatic drop in activity after the third run.

Li and coworkers [21] recently reported a polar Noyori-type catalyst **15** containing tetrasulfonate salts, and used the polar catalyst to asymmetric hydrogenation of aromatic ketones in [bmim][p-CH₃C₆H₄SO₃] (Scheme 7.4). The introduction of sodium sulfonate salts onto the Noyori-type catalyst significantly reduced the leaching to improve the reusability of the catalytic system. A synergistic effect was observed between (S,S)-DPENDS (1,2-diphenyl-1,2-ethylenediamine sulfonate disodium) and KOH in the ionic liquid solution, and hence the reaction rate was significantly accelerated with enhanced enantioselectivity. When the catalytic

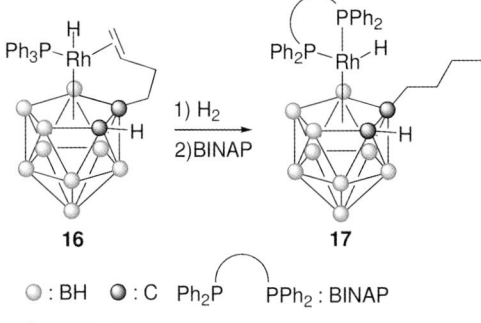

Scheme 7.5

Ph$_3$P–Rh ... (16) → 1) H$_2$ 2) BINAP → Ph$_2$P–Rh–PPh$_2$ ... (17)

○ : BH   ● : C   Ph$_2$P⌒PPh$_2$: BINAP

Table 7.5 Rh-catalyzed asymmetric hydrogenation of acetophenone using rhodacarborane catalyst **17**.

Entry	Catalyst	Solvent	Conv. (%)	ee (%)	TOF (h^{-1})
1	17	[omim][BF$_4$]	100	97	194
2	16	[bmim][PF$_6$]	100	98	207
3	16	[BP][CB$_{10}$H$_{12}$]	100	99	239
4	16	THF	82	91	96
5	[Rh(COD)Cl]$_2$/BINAP	[omim][BF$_4$]	17	10	16
6	[Rh(COD)Cl]$_2$/BINAP	THF	8	6	11

system was prepared with the Ru complex, *(S,S)*-DPENDS and KOH with a ratio of 1:6:36, the hydrogenation in [bmim][*p*-CH$_3$C$_6$H$_4$SO$_3$] provided excellent catalytic efficiency, with 100% conversion and 79% ee. The immobilized catalyst **15** in the ionic liquid was reused four times with a slight decrease in both activity and enantioselectivity (Scheme 7.4).

Zhu and coworkers [22] reported an asymmetric hydrogenation of aromatic ketones in the presence of a rhodacarborane-based chiral catalyst **17**, which was derived from rhodacarborane precursor **16** and *(R)*-BINAP, in ionic liquids (Scheme 7.5). The hydrogenations of acetophenone in the presence of the catalyst precursor **16** and *(R)*-BINAP (0.001 : 0.001 5 for acetophenone) were performed in the ionic liquid medium such as [omim][BF$_4$], [bmim][PF$_6$], or a new liquid salt comprised of 1-carbododecaborate ions and *N-n*-butylpyridinium (BP) ions, [BP][CBIX$_{10}$H$_{12}$], and tetrahydrofuran (THF) at 50 °C under H$_2$ (12 atm) for 12 h. As shown in Table 7.5, the catalytic activities and enantioselectivities in ionic liquids (entries 1–3) were higher in comparison with those obtained in THF (entry 4). Moreover, it was found that the replacement of **16** with [Rh(COD)Cl]$_2$ as the catalyst precursor resulted in a decreased catalytic performance in both [omim][BF$_4$] and THF (entries 5–6). Based on these observations, it has been suggested that the hydrogenation process with rhodacarborane-based **16** may differ from that with [Rh(COD)Cl]$_2$ in terms of its mechanistic pathways. The most efficient ionic

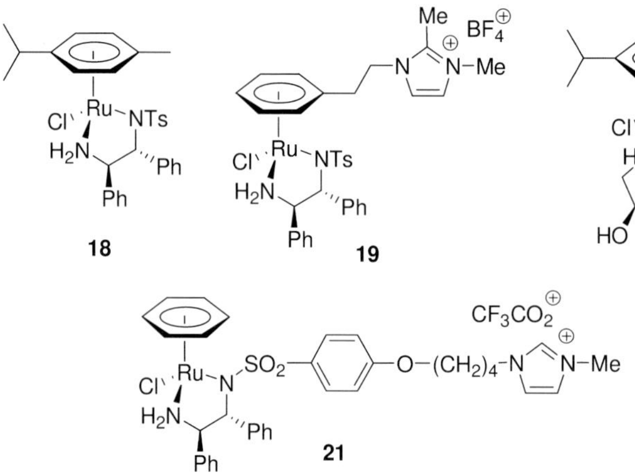

Figure 7.5 Ru complexes for asymmetric transfer hydrogenations.

liquid reaction medium, [BP][CBIX$_{10}$H$_{12}$], has a large number of B–H bonds that may be involved in the reaction, to play an important role under hydrogenation conditions.

7.2.1.3 Asymmetric Transfer Hydrogenations

Dyson and Geldbach [23] utilized Ru complex **19** bearing η^6-arene with an imidazolium salt tag for asymmetric transfer hydrogenations in ionic liquids (Figure 7.5). Asymmetric transfer hydrogenations of acetophenone in an ionic liquid, [bdmim][PF$_6$], using the well-known Ru complex **18** [24] and ionic liquid-grafted complex **19** as catalysts showed that the reusability of the catalyst was largely dependent on the reaction conditions (Table 7.6). When the transfer hydrogenations were conducted in iPrOH/KOH conditions, the neutral complex **18** immediately lost its activity during recycling (compare entry 2 with entry 8), while the cationic analogue **19** was recycled three times with fall in activity (entries 1–2 versus entries 7–10), demonstrating the positive effects of the imidazolium salt tag. Interestingly, when a formic acid/triethylamine azeotrope was employed as the reductant, the reusability of the catalyst system composed of **18**/[bdmim][PF$_6$] was much superior to that of **19**/[bdmim][PF$_6$]. However, as the formic acid/triethylamine azeotrope forms a homogeneous phase together with [bdmim][PF$_6$], the solution containing the catalyst required washing with water after product extraction with hexane or Et$_2$O, and drying in vacuum prior to the next catalytic run. In this manner, an ionic liquid solution containing **18** could be reused five times without any significant decrease in activity (entries 3–6). Moreover, the catalyst solution could be stored for days without any loss of activity or enantioselectivity. However, this procedure was not applicable to complex **19** (entries 11–14). The Ru complex with proline derivative amino amide **20** was also examined in an ionic liquid for the asymmetric transfer hydrogenation of acetophenone using the formic acid/triethylamine azeotrope as a hydrogen source [25]. In terms of results, the catalyst **20** in [bmim][PF$_6$] could be reused five times without any significant decrease in activity; nevertheless, the enantioselectivities were moderate (entries

7.2 Enantioselective Catalysis in Ionic Liquids

Table 7.6 Asymmetric transfer hydrogenation of acetophenone in ionic liquids.

Entry	Catalyst	Ionic liquid	Run	Proton source	Conv. (%)	ee (%)
1	18	[bdmim][PF_6]	1	iPrOH/KOH	95	98
2			2	iPrOH/KOH	5	
3	18	[bdmim][PF_6]	1	HCO_2H/Et_3N	>99	99
4			2	HCO_2H/Et_3N	>99	99
5			3	HCO_2H/Et_3N	>99	99
6			4	HCO_2H/Et_3N	99	99
7	19	[bdmim][PF_6]	1	iPrOH/KOH	80	98
8			2	iPrOH/KOH	66	
9			3	iPrOH/KOH	57	
10			4	iPrOH/KOH	21	
11	19	[bdmim][PF_6]	1	HCO_2H/Et_3N	>99	99
12			2	HCO_2H/Et_3N	68	
13			3	HCO_2H/Et_3N	19	
14			4	HCO_2H/Et_3N	1	
15	20	[bmim][PF_6]	1	HCO_2H/Et_3N	99	72
16			2	HCO_2H/Et_3N	99	71
17			3	HCO_2H/Et_3N	99	71
18			4	HCO_2H/Et_3N	99	70
19	21	[bmim][PF_6]	1	HCO_2H/Et_3N	98	92
20			2	HCO_2H/Et_3N	>99	93
21			3	HCO_2H/Et_3N	99	93
22			4	HCO_2H/Et_3N	92	93
23			5	HCO_2H/Et_3N	75	90

15–18). Quite recently, Ohta and coworkers [26] also developed a task-specifically modified chiral Ru-catalyst **21** by attaching an imidazolium salt unit onto TsDPEN (*N*-(*p*-toluenesulfonyl)-1,2-diphenylethylene diamine), and investigated its reactivity and reusability in the asymmetric transfer hydrogenation of aryl ketones in ionic liquids. As shown in entries 19–23 of Table 7.6, catalyst **21** was fully soluble in [bmim][PF_6] and easily recovered after extraction of the product 1-phenylethanol by the addition of an organic solvent. This catalyst was reused five times without any significant decrease in activity and enantioselectivity.

7.2.2
Asymmetric Oxidations in Ionic Liquids

7.2.2.1 Asymmetric Dihydroxylation
The Sharpless Os-catalyzed asymmetric dihydroxylation (AD) of olefins is a well-established and robust methodology for the synthesis of a wide range of

enantiomerically pure vicinal diols [27]. Although the AD reaction offers a number of processes that could be applied to the synthesis of chiral drugs, natural products and fine chemicals, the high cost and toxicity of osmium, together with the possible contamination of products with the osmium catalyst, have restricted its industrial application. In order to explore the possibility of the recycling and reuse of the osmium/ligand catalyst system, several attempts have been made to immobilize the catalyst [28–31]. Early approaches to immobilizing OsO_4 on solid-supported alkaloid ligands suffered from a number of disadvantages, including the need for a complicated synthesis of the supported ligand system and reduced catalytic efficiency [28]. Recently, alternative methods for immobilizing the osmium catalyst by the microencapsulation of OsO_4 in a polymer matrix [29], by using an ion-exchange technique [30] or by the osmylation of macroporous resins bearing residual vinyl groups such as Amberlite XAD-4 [31], have been reported. Although recycling experiments using this type of immobilized osmium catalyst have been successfully performed in some cases, a higher loading (1–5 mol%) of immobilized osmium catalyst was generally required than with homogeneous AD systems (0.2 mol%).

Recently, a simple and practical approach to a recyclable catalytic system of the AD reaction in ionic liquids has been reported [32, 33]. To investigate the effect of an ionic liquid on the AD reaction and the reusability of the catalytic components, AD reactions were initially carried out with the well-known ligand, 1,4-bis(9-O-dihydroquininyl)phthalazine [(DHQ)$_2$PHAL] (Figure 7.6), using standard Upjohn conditions [34] (using N-methylmorpholine-N-oxide (NMO) as a co-oxidant) in the presence of [bmim][PF$_6$] at 20 °C The results were comparable to those obtained without an ionic liquid. For example, the AD reaction of trans-stilbene afforded the corresponding diol in 94% yield with 97% ee. As the

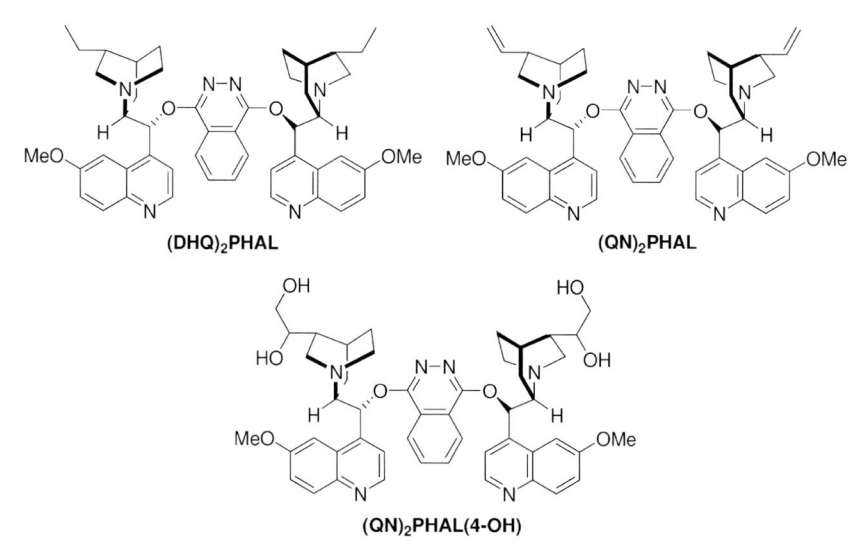

Figure 7.6 (DHQ)PHAL, (QN)2PHAL and (QN)2PHAL(4-OH) ligands for AD reactions.

complex formation between OsO$_4$ and an alkaloid ligand is expected to be reversible, lowering the concentration of the chiral ligand in the ionic liquid phase might result in more OsO$_4$ leaching from the ionic liquid phase. Therefore, it was presumed that the use of an alkaloid ligand that could be strongly immobilized in an ionic liquid might minimize Os leaching during product extraction. To prove this hypothesis, the 1,4-bis(9-O-quininyl)phthalazine [(QN)$_2$PHAL] [35] was converted to the alkaloid [(QN)$_2$PHAL(4-OH)] bearing highly polar residues (four hydroxy groups) under dihydroxylation conditions (Figure 7.6). Such polar residues increase the preferential solubility of chiral ligands in the ionic liquids. The recovered ionic liquid phase containing both osmium and ligand [(QN)$_2$PHAL(4-OH)] was recycled several times, even using 0.1 mol% of OsO$_4$ (Scheme 7.6).

The AD reaction using K$_3$Fe(CN)$_6$ as the co-oxidant in ionic liquids has been screened in a variety of ionic liquids [33, 36]. Among these ionic liquids, [bmim][PF$_6$] and [omim][PF$_6$] proved superior to others, affording comparable or higher yields and enantioselectivities than those in the conventional H$_2$O/tBuOH solvent system for the AD reaction of various olefins, such as styrene, α-methylstyrene and 1-hexene. For the AD reaction of 1-hexene in [bmim][PF$_6$]/H$_2$O or [bmim][PF$_6$]/tBuOH/H$_2$O, it was possible to reuse the catalytic system for nine runs with only a slight decrease in yield (Scheme 7.7). The hydrogen peroxide-based AD reaction of styrene in [bmim][PF$_6$], utilizing the osmium-VO(acac)$_2$ catalytic system, afforded the dihydroxylated product in 80% yield with 75% ee (Scheme 7.8) [37].

Most recently, a new chiral ionic liquid has been prepared by combining the tetra-*n*-hexyl-dimethylguanidinium cation with readily available chiral anions, and

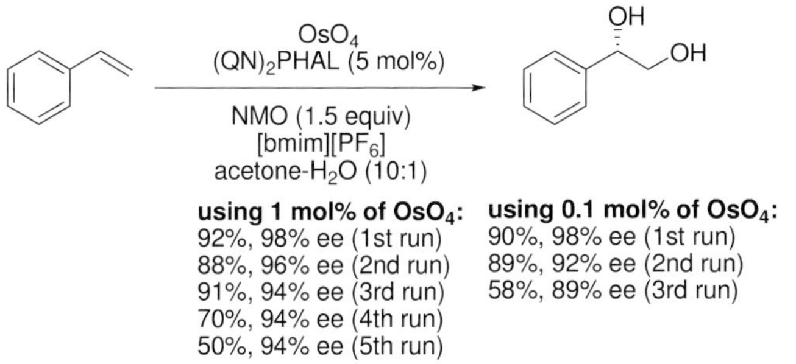

OsO$_4$
(QN)$_2$PHAL (5 mol%)

NMO (1.5 equiv)
[bmim][PF$_6$]
acetone-H$_2$O (10:1)

using 1 mol% of OsO$_4$:
92%, 98% ee (1st run)
88%, 96% ee (2nd run)
91%, 94% ee (3rd run)
70%, 94% ee (4th run)
50%, 94% ee (5th run)

using 0.1 mol% of OsO$_4$:
90%, 98% ee (1st run)
89%, 92% ee (2nd run)
58%, 89% ee (3rd run)

Scheme 7.6

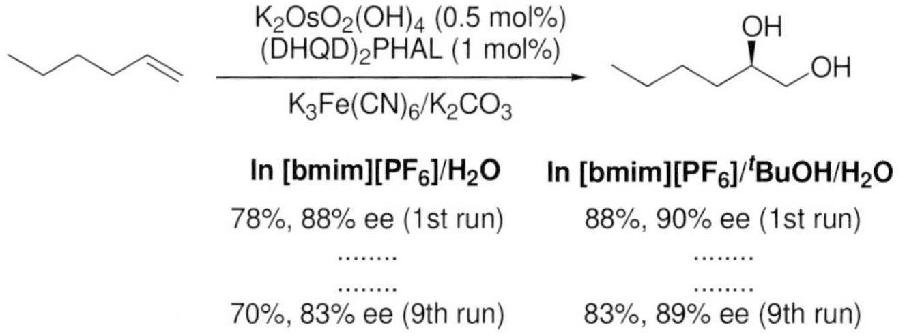

K$_2$OsO$_2$(OH)$_4$ (0.5 mol%)
(DHQD)$_2$PHAL (1 mol%)

K$_3$Fe(CN)$_6$/K$_2$CO$_3$

In [bmim][PF$_6$]/H$_2$O
78%, 88% ee (1st run)
....
....
70%, 83% ee (9th run)

In [bmim][PF$_6$]/^{t}BuOH/H$_2$O
88%, 90% ee (1st run)
....
....
83%, 89% ee (9th run)

Shceme 7.7

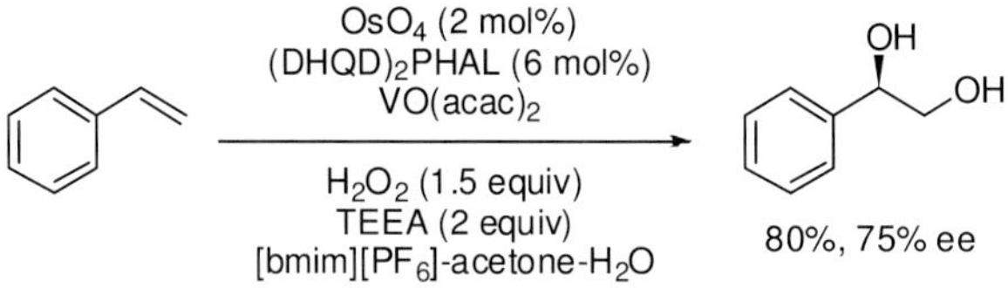

Scheme 7.8

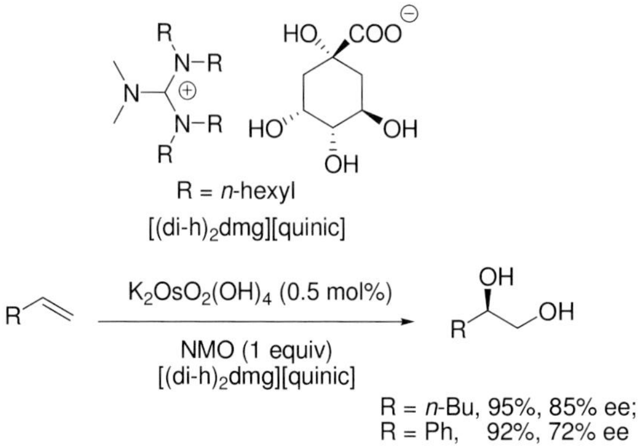

Styrene →(OsO$_4$ (2 mol%), (DHQD)$_2$PHAL (6 mol%), VO(acac)$_2$; H$_2$O$_2$ (1.5 equiv), TEEA (2 equiv), [bmim][PF$_6$]-acetone-H$_2$O)→ 1-phenyl-1,2-ethanediol, 80%, 75% ee

Scheme 7.9

[(di-h)$_2$dmg][quinic], R = n-hexyl

R-CH=CH$_2$ →(K$_2$OsO$_2$(OH)$_4$ (0.5 mol%), NMO (1 equiv), [(di-h)$_2$dmg][quinic])→ R-CH(OH)-CH$_2$OH

R = n-Bu, 95%, 85% ee; R = Ph, 92%, 72% ee

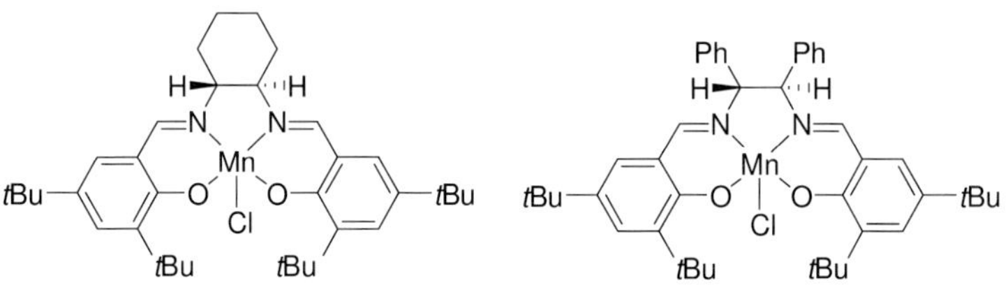

Figure 7.7 Mn(III)(salen) complexes for asymmetric epoxidations.

22, 23

used as an asymmetric inducing agent in the AD reaction. Without using any chiral ligands, the AD reactions of 1-hexene and styrene in the chiral ionic liquid, [(di-h)$_2$dmg][quinic], in the presence of NMO as co-oxidant, provided the corresponding diols in high yields and good enantiomeric excesses (Scheme 7.9) [38].

7.2.2.2 Asymmetric Epoxidations

The catalytic asymmetric epoxidation of alkenes offers a powerful strategy for the synthesis of enantiomerically enriched epoxides. Among the several existing catalytic methods, the asymmetric epoxidation of unfunctionalized alkenes catalyzed by chiral Mn(III)(salen) complexes such as homochiral [(N,N)-bis(3,5-di-*tert*-butylsalicylidene)-1,2-cyclohexanediamine]manganese(III) chloride (**22**) (Figure 7.7), as developed by Jacobsen and coworkers, represents one of the most reliable methods [39].

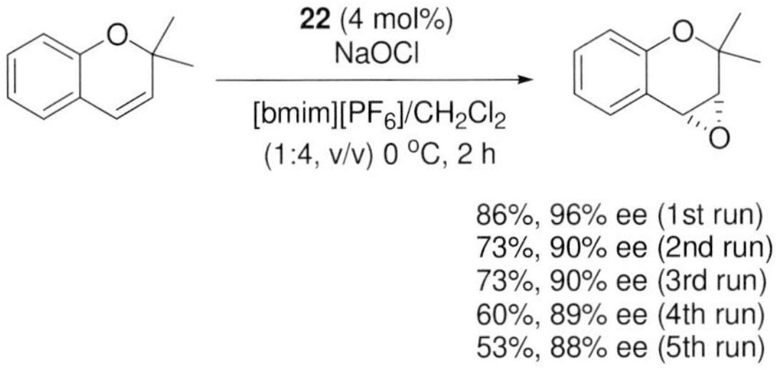

Scheme 7.10

22 (4 mol%)
NaOCl

[bmim][PF$_6$]/CH$_2$Cl$_2$
(1:4, v/v) 0 °C, 2 h

86%, 96% ee (1st run)
73%, 90% ee (2nd run)
73%, 90% ee (3rd run)
60%, 89% ee (4th run)
53%, 88% ee (5th run)

Song and Roh demonstrated for the first time the usefulness of an ionic liquid in this important reaction [40]. The epoxidations of various olefins were carried out using 22 as the catalyst and NaOCl as the co-oxidant in a mixture of [bmim][PF$_6$] and CH$_2$Cl$_2$ (1:4, v/v). Good conversion and enantioselectivity were observed for all tested substrates. Interestingly, an enhancement in catalytic activity was obtained by adding the ionic liquid to the organic solvent. The epoxidation of 2,2-dimethylchromene using 4 mol% of Mn(III)(salen) catalyst 22 in the presence of [bmim][PF$_6$] was completed in 2 h, whereas the same reaction without the ionic liquid required 6 h to achieve complete conversion. This rate acceleration effect induced by the ionic liquid was shown even more dramatically when the amount of the catalyst was reduced to 0.5 mol%. Moreover, the use of an ionic liquid solvent allows for easier catalyst recycling, without the need for catalyst modification. However, the enantioselectivity and activity for the reaction using the recovered catalyst decreased upon reuse. After five cycles, the yield and enantioselectivity fell from 83% to 53% and from 96% to 88% ee, respectively (Scheme 7.10), although this deterioration may have been due to degradation of the salen catalyst under oxidation conditions. Gaillon and Bedioui reported the electroassisted biomimetic activation of molecular oxygen using a chiral Mn(salen) complex in [bmim][PF$_6$], where the highly reactive manganese-oxo intermediate [(salen)Mn(V)=O]$^+$ is capable of transferring its oxygen to an olefin [41]. This method has provided a potential in electrocatalytic asymmetric epoxidations using molecular oxygen in ionic liquid media.

Liu and coworkers [42] reported the Mn(III)(salen)-catalyzed asymmetric epoxidation of unfuctionalized olefins in an MCM-48-supported ionic liquid phase (SILP) (Figure 7.8). The Mn(III)(salen) catalysts *(S,S)*-22 or *(S,S)*-23 were immobilized by mixing of their acetone solutions with ionic liquid-anchored MCM-48 and ionic liquid [bmim][PF$_6$]. The catalytic activity of the SILP catalyst was evaluated for the epoxidation of unfunctionalized olefins with m-CPBA/NMO as oxidant in CH$_2$Cl$_2$ at 273 K for 2 h. High yields and enantioselectivities were obtained for epoxidation of α-methylstyrene (99%, 99% ee) and 1-phenylcyclohexene (95%, 92% ee). The immobilized catalysts were recycled at least three times, without any significant decrease in either activity or enantioselectivity (Figure 7.8).

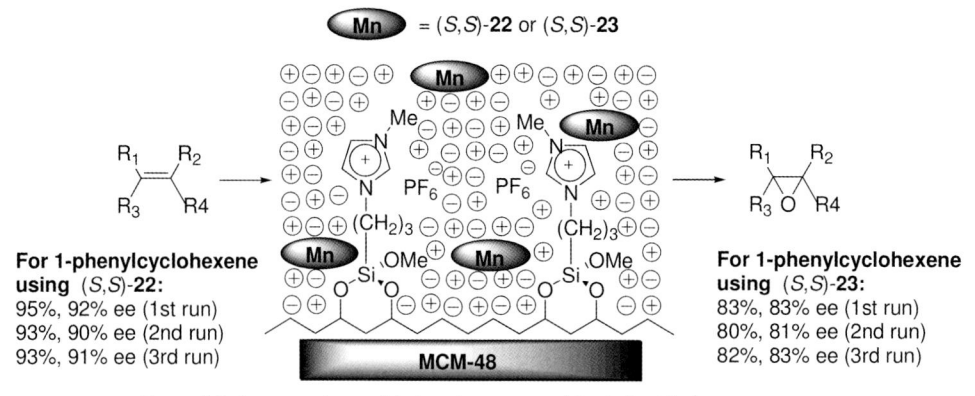

Figure 7.8 Asymmetric epoxidations in supported ionic liquid phase.

For 1-phenylcyclohexene using (S,S)-**22**:
95%, 92% ee (1st run)
93%, 90% ee (2nd run)
93%, 91% ee (3rd run)

For 1-phenylcyclohexene using (S,S)-**23**:
83%, 83% ee (1st run)
80%, 81% ee (2nd run)
82%, 83% ee (3rd run)

in CH_2CH_2, endo:exo = 79:21, **26:27** = 76:24, yield = 4 %
in [dbim][BF_4], endo:exo = 93:7, **26:27** = 96:4, yield = 65 %

Scheme 7.11

7.2.3
Asymmetric Carbon–Carbon and Carbon–Heteroatom Bond Formation in Ionic Liquids

7.2.3.1 Asymmetric Diels–Alder Reactions

Oh and Meracz were the first to report asymmetric Diels–Alder reaction in ionic liquids [43]. It has also been shown that Diels–Alder reactions in ionic liquids give unusually high stereoselectivities at room temperature as compared to those in conventional organic solvents, where a low temperature was required to achieve good stereoselectivities. For example, the Diels–Alder reaction of cyclopentadiene and dienophile **25** with Cu(II)-bisoxazoline complex **24** as catalyst in [dbim][BF_4] showed a higher endoselectivity (*endo/exo* = 93/7) and regioselectivity (**26/27** = 96/4) than in CH_2Cl_2 (*endo/exo* = 79/21, **26/27** = 76/24) (Scheme 7.11).

Doherty and coworkers [44] reported asymmetric Diels–Alder reactions of **25** with cyclopentadiene (shown in Scheme 7.11) using platinum complexes of BINAP

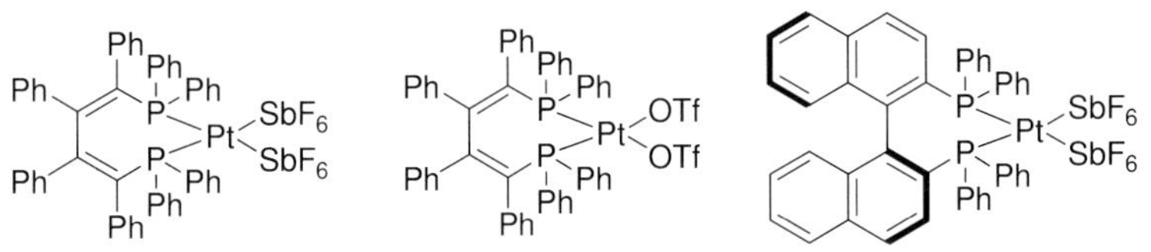

Figure 7.9 Pt-complexes for asymmetric Diels–Alder reactions.

Table 7.7 Asymmetric Diels–Alder reaction between cyclopentadiene and dienophile **25** in ionic liquids using δ-**28**, δ-**29** and (S)-**30** as catalysts.

Entry	Solvent	Catalyst (mol%)	Run	Conv. (%) (time) (h)	% endo	endo ee (%)
1	CH$_2$Cl$_2$	δ-**28** (20)		45 (20)	80	67
2	[bmim][PF$_6$]	δ-**28** (20)	1	100 (1)	89	93
3			2	96 (1)	87	90
4			3	90 (1)	84	90
5	[emim][NTf$_2$]	δ-**28** (20)	1	100 (1)	87	90
6			2	100 (1)	85	91
7			3	96 (1)	82	91
8	[emim][NTf$_2$]	δ-**29** (10)	1	71 (1)	82	90
9			2	64 (1)	79	88
10			3	65 (1)	79	89
11	[emim][NTf$_2$]	(S)-**30** (20)	1	100 (1)	80	92
12			2	100 (1)	69	94
13			3	96 (1)	82	91

(**30**) and conformationally flexible NUPHOS-type diphosphines δ-**28** and δ-**29** in ionic liquids (Figure 7.9). All reactions conducted in ionic liquids were found to be significantly faster than those performed in dichloromethane (DCM) (Table 7.7; entries 1, 2, 5 and 8). Moreover, significantly improved enantioselectivities were also obtained in ionic liquids compared with those in DCM; for example, the reaction using δ-**28** in [bmim][PF$_6$] achieved 93% ee (Table 7.7, entry 2), whereas the corresponding reaction in DCM gave only 67% ee (Table 7.7, entry 1). For catalyst recycling, the extractions were performed in air and no significant change in terms of ee-value or conversion were found in the successive reactions (Table 7.7).

Quite recently, palladium-phosphinooxazolidine catalysts have been used for asymmetric Diels–Alder reactions in ionic liquids [45]. By screening various counterions of the Pd catalysts and ionic liquids, a combination of cationic Pd-POZ catalyst **31c** with SbF$_6$ counterion and [bmim][BF$_4$] as the ionic liquid was found to be the most effective. The Diels–Alder reaction of cyclopentadiene with dienophile **32** using **31c** as a catalyst provided DA adduct **33** with high chemical and optical yields. The catalyst **31c** immobilized in [bmim][BF$_4$] was successfully recycled eight times to give **33** in 89–99% yields and 88–99% ee-values (Scheme 7.12).

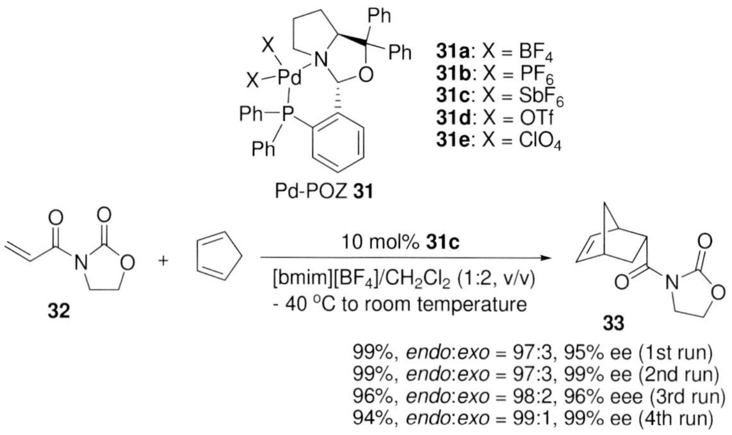

Scheme 7.12

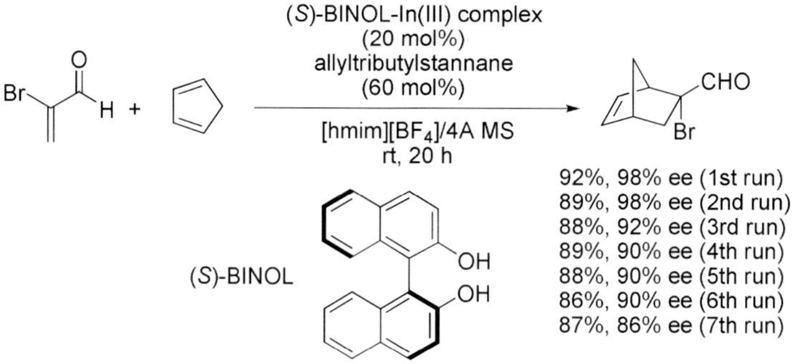

Pd-POZ 31
- 31a: X = BF₄
- 31b: X = PF₆
- 31c: X = SbF₆
- 31d: X = OTf
- 31e: X = ClO₄

32 + cyclopentadiene → 33

Conditions: 10 mol% **31c**, [bmim][BF$_4$]/CH$_2$Cl$_2$ (1:2, v/v), −40 °C to room temperature

- 99%, endo:exo = 97:3, 95% ee (1st run)
- 99%, endo:exo = 97:3, 99% ee (2nd run)
- 96%, endo:exo = 98:2, 96% eee (3rd run)
- 94%, endo:exo = 99:1, 99% ee (4th run)

Scheme 7.13

(S)-BINOL-In(III) complex (20 mol%)
allyltributylstannane (60 mol%)

[hmim][BF$_4$]/4A MS, rt, 20 h

(S)-BINOL

- 92%, 98% ee (1st run)
- 89%, 98% ee (2nd run)
- 88%, 92% ee (3rd run)
- 89%, 90% ee (4th run)
- 88%, 90% ee (5th run)
- 86%, 90% ee (6th run)
- 87%, 86% ee (7th run)

Catalytic asymmetric Diels–Alder reactions in ionic liquids, using an air- and moisture-stable chiral BINOL–indium complex prepared from (S)-BINOL and InCl$_3$, have also reported [46]. The reactions of various dienes with 2-methacrolein or 2-bromoacrolein proceeded in an ionic liquid with good yields and excellent enantioselectivities (up to 98% ee). The chiral In(III) catalyst immobilized in ionic liquid [hmim][PF$_6$] could be reused seven times without any significant loss in catalytic activity (Scheme 7.13).

7.2.3.2 Asymmetric Ring Opening of Epoxides

The asymmetric ring-opening (ARO) reaction of epoxides by trimethylsilyl azide (TMSN$_3$) catalyzed by the chiral Cr(salen) complex has been recognized as an attractive approach to the synthesis of optically enriched β-amino alcohols [47]. In particular, the chiral Cr(salen) catalyst **34** exhibits remarkable stability under catalytic conditions, which allows its repeated recycling. Jacobsen and coworkers reported that this reaction could be run without solvents, and that the catalyst could be recycled several times without any loss of activity or enantioselectivity

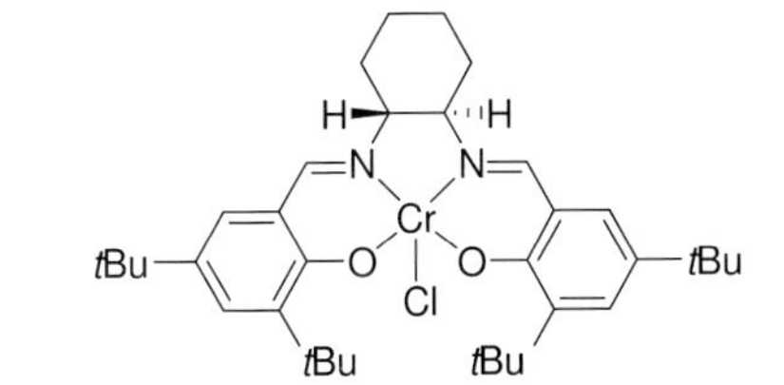

7.2 Enantioselective Catalysis in Ionic Liquids

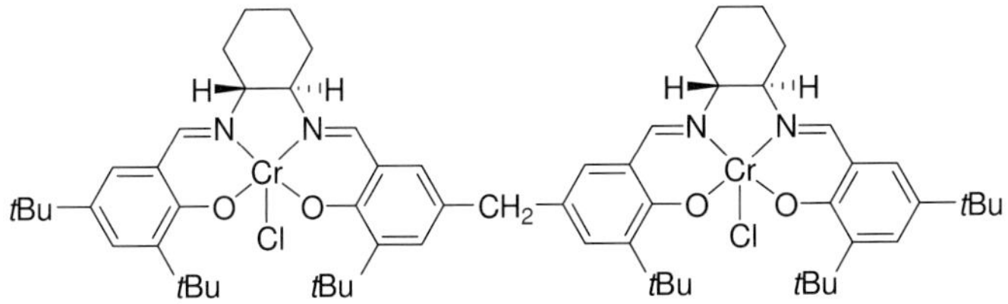

Scheme 7.14

Reaction conditions: (R,R)-Cr(salen) (3 mol%), TMSN$_3$, [bmim][PF$_6$]/[bmim][OTf] (5:1, v/v), 20 °C, 28 h

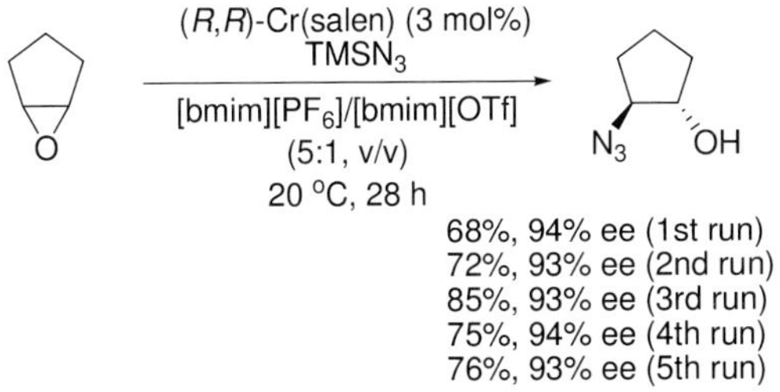

- 68%, 94% ee (1st run)
- 72%, 93% ee (2nd run)
- 85%, 93% ee (3rd run)
- 75%, 94% ee (4th run)
- 76%, 93% ee (5th run)

Figure 7.10 Dimeric Cr(salen) complex for asymmetric ring opening reaction of epoxides.

[48]. However, this catalyst recycling procedure involves the potentially hazardous distillation of neat liquid azides, which cannot be applied to large-scale applications. Song and coworkers were the first to develop a highly practical recycling procedure for the chiral Cr(salen) complex using ionic liquids [49]. This consisted of running a reaction of *meso*-epoxides with TMSN$_3$ in the presence of catalytic amounts of (R,R)-Cr(salen) complex **34** dissolved in the [bmim] salts. The yield and enantioselectivity of this reaction were found to depend heavily on the nature of the counteranion of the ionic liquids. While the reactions performed in hydrophobic [PF$_6$] and [SbF$_6$] salts afforded the product in high yields and enantioselectivities (similar to those obtained in organic solvents), the reactions were not successful in hydrophilic [BF$_4$] and [OTf] salts. Significantly, the best recyclable catalytic system was obtained by immobilizing the catalyst in a 5/1 (v/v) mixture of hydrophobic [bmim][PF$_6$] and hydrophilic [bmim][OTf]. The recovered ionic liquid phase containing the catalyst was reused several times without any loss of activity and enantioselectivity (Scheme 7.14).

The ARO reaction of epoxides with TMSN$_3$ in the SILP phase using dimeric Cr(salen) complex **35** (Figure 7.10) has been developed as one of the robust

Table 7.8 Asymmetric ring opening of epoxides catalyzed by **35** immobilized in a supported ionic liquid phase, the dimeric complex impregnated on silica and homogeneous reactions with the monomeric Cr(salen) complex **34**.

Entry	Catalyst (run)	Reaction (cat/mol%)	Conv. (%) (time/h)	% ee of epoxide	% ee of product	Leaching (%)
1	SILP-35 (1st)	1 (1.5)	52 (3)	95	84	0.4
2	SILP-35 (2nd)	1 (1.5)	48 (3)	96	87	0.5
3	SILP-35 (1st)	2 (1.5)	93 (20)	—	75	1.1
4	SILP-35 (2nd)	2 (1.5)	92 (20)	—	73	1.1
5	Impregnation-35	1 (3)	59 (10)	96	66	1.0
6	Impregnation-35	2 (3)	98 (70)	—	65	1.1
7	Homogeneous-34	1 (2)	45 (27)	—	97	1.0
8	Homogeneous-34	2 (2)	83 (18)	—	84	—

recyclable catalytic systems [50]. The SILP catalyst was prepared by dissolving the dimeric Cr(salen) catalyst **35** together with the [bmim][PF$_6$] in acetone. Following addition of the silica support material, the resultant suspension was opened to the air and stirred until all the solvent had evaporated. As shown in Table 7.8, compared with the impregnated dimeric complex [51] (entries 5 and 6) and homogeneous reactions [48, 52] (entries 7 and 8), the use of a SILP to immobilize the dimeric Cr(salen) catalyst **35** offered several benefits. An increased reactivity and selectivity were observed (entries 1–4 in Table 7.8), which could be attributed to the methods of immobilization. The impregnated catalyst was adsorbed onto the silica surface which limited its activity, whereas the catalyst in the SILP was more accessible for the reaction. The leaching rate for the silica-supported ionic liquid phase was comparable to that observed for the impregnated catalyst (3.1% over four runs).

7.2.3.3 Hydrolytic Kinetic Resolution of Epoxides

The hydrolytic kinetic resolution (HKR) of racemic epoxides using Jacobsen's chiral Co(III)(salen)–OAc complex **36a** as a catalyst is one of the most practical approaches to the preparation of enantiopure terminal epoxides (Scheme 7.15) [47, 53]. Although the chiral catalyst is readily accessible and displays high enantioselectivity, it provides only relatively low turnover numbers. Thus, in order to facilitate catalyst separation and reuse, several attempts were made to anchor Jacobsen's catalyst onto insoluble supports [54]. Although these heterogeneous

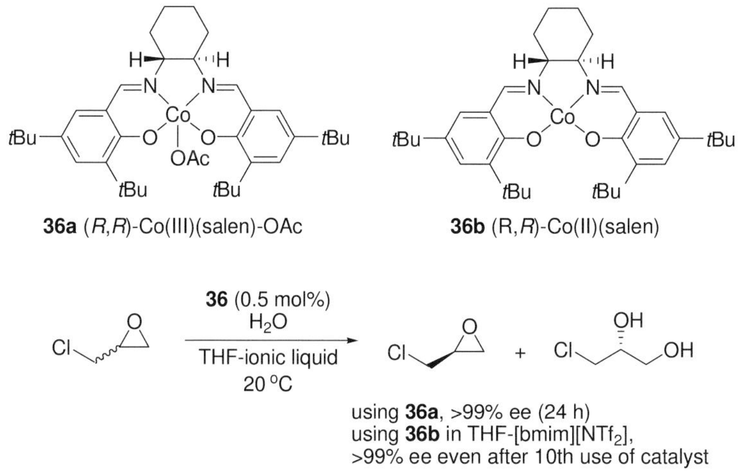

7.2 Enantioselective Catalysis in Ionic Liquids

Scheme 7.15

36a (*R,R*)-Co(III)(salen)-OAc

36b (R,*R*)-Co(II)(salen)

using **36a**, >99% ee (24 h)
using **36b** in THF-[bmim][NTf$_2$],
>99% ee even after 10th use of catalyst

analogues of Co(III)(salen)-OAc produced almost the same enantioselectivities as their homogeneous counterparts, complicated synthetic manipulations were required for their preparation. Moreover, during the reaction the Co(III)-complexes in either solid-bound [54] or homogeneous systems [53a] were reduced to the Co(II)(salen) complex **36b**, which is known to be inactive for HKR. Thus, it was necessary to re-oxidize the recovered catalyst to the Co(III) complex with acetic acid under air before it could be used in the next run. Recently, it was found that ionic liquids caused an interesting effect in this catalytic reaction [55]. For example, in the HKR of racemic epoxides using catalytic amounts of *(R,R)*-Co(III)(salen)-OAc **36a** in a mixture of THF and an ionic liquid, [bmim][X] (X = PF$_6$, NTf$_2$) (4/1, v/v) at 20 °C, the yields and ee-values were comparable to those obtained without ionic liquids [53a]. Interestingly, it was demonstrated by both ultraviolet and X-ray photoelectron spectrometry (XPS) analyses of the recovered ionic liquid phase that the oxidation state of the Co(salen) complex dissolved in the recovered ionic liquid phase remained as Co(III). More interestingly, it was also found that catalytically inactive Co(II)(salen) complex **36b** could be used directly as a catalyst precursor instead of Co(III)(salen)-OAc (**36a**) in the presence of the ionic liquid. This implies that the Co(II)(salen) complex **36b** was oxidized, in the absence of acetic acid, to the catalytically active Co(III) complex during the HKR reactions, which was not possible in conventional organic solvents. Thus, all HKRs of racemic epichlorohydrin using catalytic amounts of *(R,R)*-Co(II)(salen) complex **36b** in [bmim][PF$_6$] or [bmim][NTf$_2$] proceeded smoothly, even when only 0.025 mol% of *(R,R)*-Co(II)(salen) **36b** was used. This catalytic system involving the ionic liquid [bmim][NTf$_2$] was reusable up to ten times without any loss of activity and enantioselectivity (>99% ee) (Scheme 7.15).

7.2.3.4 Asymmetric Cyanosilylation of Aldehydes

Optically pure cyanohydrins are versatile synthetic intermediates in the synthesis of a wide range of homochiral products such as α-hydroxy acids and β-hydroxy

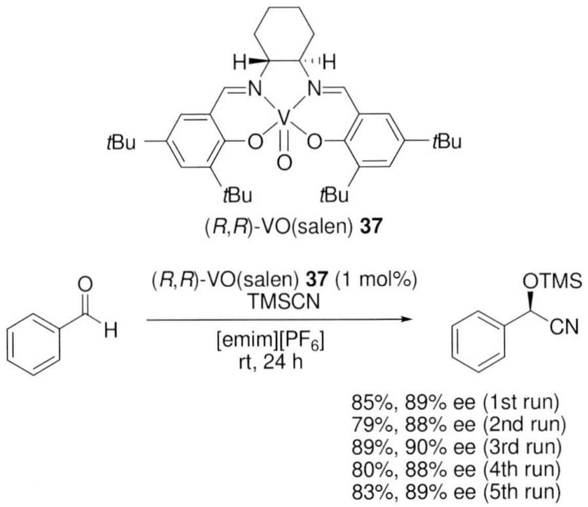

Scheme 7.16

(R,R)-VO(salen) **37**

(R,R)-VO(salen) **37** (1 mol%)
TMSCN
[emim][PF$_6$]
rt, 24 h

85%, 89% ee (1st run)
79%, 88% ee (2nd run)
89%, 90% ee (3rd run)
80%, 88% ee (4th run)
83%, 89% ee (5th run)

amines. Many enantioselective catalytic hydrocyanations and cyanosilylations of aldehydes and ketones have been reported [56]. Recently, Corma and coworkers [57] undertook the enantioselective cyanosilylation of aldehydes using chiral vanadium(salen) complex **37** as a catalyst in various ionic liquids, and found that both yield and enantioselectivity were heavily dependent upon the nature of the counteranion. While the reaction performed in hydrophobic [PF$_6$] salts gave yields and degrees of enantioselectivity similar to those obtained in CH$_2$Cl$_2$, much lower yields and ee-values were obtained in hydrophilic ionic liquids bearing [BF$_4$] or [Cl] salts. Moreover, the products were easily separated from the reaction mixture by hexane extraction. The remaining [bmim][PF$_6$] ionic liquid phase containing the catalyst was found to be reusable for at least four further cycles, without any loss of activity or selectivity (Scheme 7.16).

A series of supported chiral VO(salen) complexes anchored on silica, single-wall carbon nanotube, activated carbon or ionic liquids have been prepared through the simple methods based on the addition of mercapto groups to terminal C=C double bonds (Scheme 7.17) [58]. The four recoverable catalysts and the standard VO(salen) complex **37** were tested for the enantioselective cyanosilylation of benzaldehyde using trimethylsilyl cyanide (Table 7.9). It should be noted that the ionic liquid-supported IL-VO(salen) showed the highest catalytic activity, though the ee-value was considerably reduced compared to the soluble **37** in [bmim][PF$_6$] (entries 4 and 5).

7.2.3.5 Asymmetric Allylic Substitution

A number of homogeneous chiral ligands have been developed for the Pd-catalyzed asymmetric allylic substitution reaction, which is one of the useful methods for asymmetric C–C and C–X bond formation [59]. Although several of Pd-catalyzed asymmetric allylic substitutions have been investigated in ionic

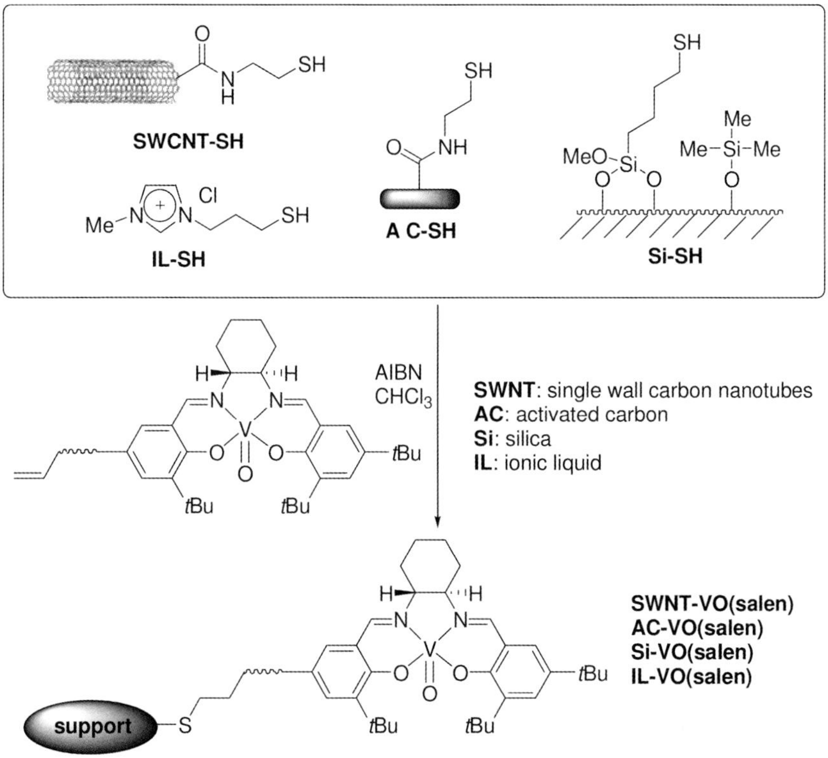

7.2 Enantioselective Catalysis in Ionic Liquids

Scheme 7.17

SWNT: single wall carbon nanotubes
AC: activated carbon
Si: silica
IL: ionic liquid

SWNT-VO(salen)
AC-VO(salen)
Si-VO(salen)
IL-VO(salen)

Table 7.9 Asymmetric cyanosilylation of benzaldehyde in ionic liquids using supported VO(salen) complexes.

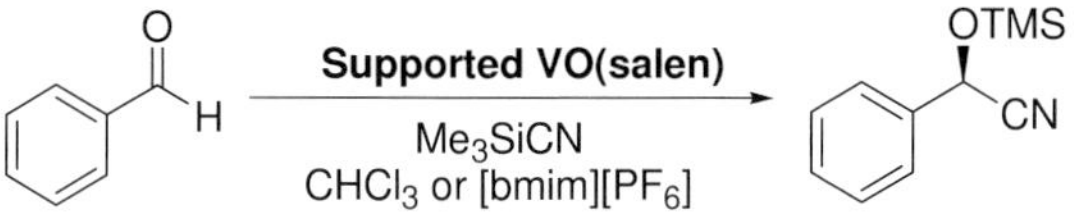

Entry	Catalyst	Conditions	Conv. (%)	TOF (h^{-1})	ee (%)
1	Si-VO(salen)	CHCl$_3$, 0.24 mol%, 0 °C	78	2.7	85
2	SWCNT-VO(salen)	CHCl$_3$, 0.3 mol%, 0 °C	67	3.1	66
3	AC-VO(salen)	CHCl$_3$, 0.3 mol%, 0 °C	81	3.75	48
4	IL-VO(salen)	[bmim][PF$_6$], 0.2 mol%, rt	88	18.3	57
5	37	[bmim][PF$_6$], 1 mol%, rt	85	3.5	89

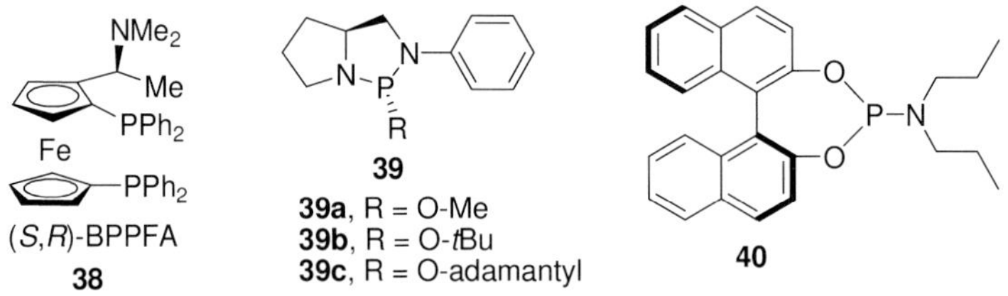

7 Enantioselective Catalysis in Ionic Liquids and Supercritical CO_2

Figure 7.11 Chiral phosphorous ligands for asymmetric allylic substitution reactions.

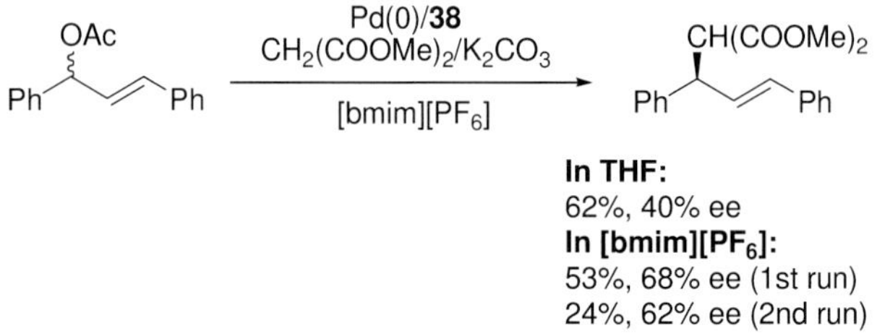

Compounds shown: (S,R)-BPPFA **38**; **39** (39a, R = O-Me; 39b, R = O-tBu; 39c, R = O-adamantyl); **40**.

Scheme 7.18

Ph-CH(OAc)-CH=CH-Ph + $CH_2(COOMe)_2$/K_2CO_3 → Ph-CH(CH(COOMe)$_2$)-CH=CH-Ph

Pd(0)/**38**, [bmim][PF$_6$]

In THF:
62%, 40% ee
In [bmim][PF$_6$]:
53%, 68% ee (1st run)
24%, 62% ee (2nd run)

liquids, the immobilization of chiral catalysts has been less successful [60, 61]. The Pd-complex of ferrocenylphosphine BPPFA **38** (Figure 7.11) catalyzed the allylic substitution of (rac)-1,3-diphenyl-2-propenyl acetate in [bmim][PF$_6$] giving the product with 68% ee, which was higher than that (40% ee) observed in THF. The catalytic activity of the recovered catalyst immobilized in ionic liquid decreased significantly upon reuse, as shown in Scheme 7.18; this finding was ascribed to the leaching of catalyst during product extraction with toluene from the ionic liquid (Scheme 7.18) [60a]. Similarly, the palladium-catalyzed asymmetric allylic amination of 1,3-diphenyl-2-propenyl acetate with di-n-propylamine has also been carried out in an ionic liquid. As shown in Table 7.10, the allylic amination in [bdmim][BF$_4$] using **39** and **40** as chiral ligands gave the aminated product in high yields, but with moderate enantioselectivities (entries 2, 5 and 7). Unfortunately, the catalytic activities fell significantly in the recycle (entries 3, 4, 6, 8 and 9), most likely due to partial leaching of the catalyst [61].

7.2.3.6 Asymmetric Allylic Addition

The In(III)-catalyzed asymmetric allylation of aldehydes and ketones using PYBOX-type ligands such as **41** in an ionic liquid has recently been reported [62]. The allylation of benzaldehyde with allyltributylstannane was performed using 20 mol% In(III)-**41** complex in the presence of 4 Å molecular sieve in a [hmim][PF$_6$]/CH_2Cl_2 system at −60 °C for 30 h, affording the corresponding homoallylic alcohol in 74% yield with 89% ee. The immobilized catalyst in the ionic liquid layer could be recycled four times with comparable enantioselectivities and yields (Scheme 7.19) [62a]. The asymmetric allylation of ketones was also performed using In(III)-**41** complex as a catalyst in a [hmim][PF$_6$]/CH_2Cl_2 system with good catalytic activities

Table 7.10 Palladium catalyzed asymmetric allylic amination in [bdmim][BF$_4$].

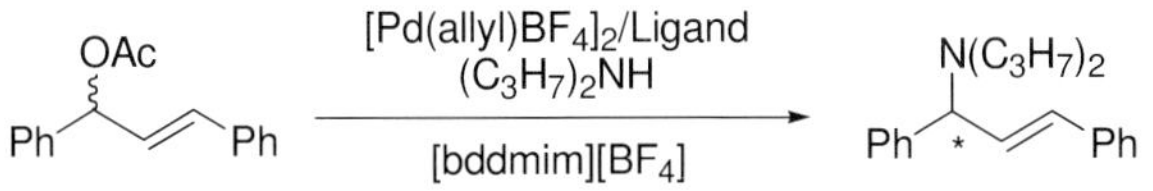

Ph~~~OAc~~~Ph → [Pd(allyl)BF$_4$]$_2$/Ligand, (C$_3$H$_7$)$_2$NH, [bddmim][BF$_4$] → Ph~*~N(C$_3$H$_7$)$_2$~~Ph

Entry	Ligand	Cycle	Conv. (%)	ee (%)
1	39a	1	100	3
2	39b	1	100	77
3	39b	2	71	75
4	39b	3	45	76
5	39c	1	100	84
6	39c	2	10	68
7	40	1	97	50
8	40	2	40	51
9	40	3	10	51

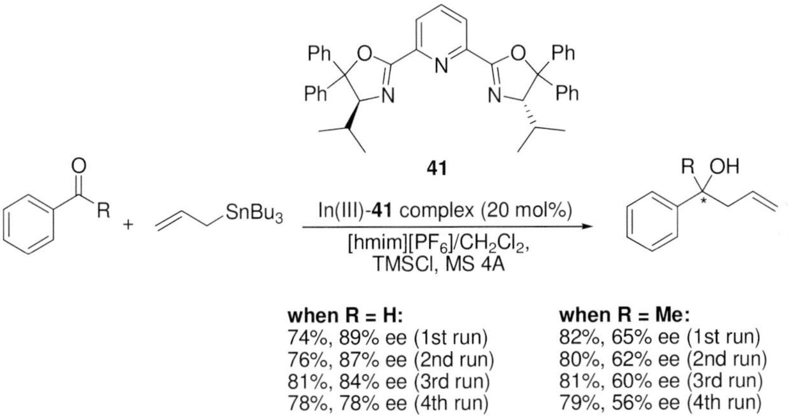

41

PhCOR + ~~~SnBu$_3$ → In(III)-**41** complex (20 mol%), [hmim][PF$_6$]/CH$_2$Cl$_2$, TMSCl, MS 4Å → Ph-C*(R)(OH)-CH$_2$-CH=CH$_2$

when R = H:
74%, 89% ee (1st run)
76%, 87% ee (2nd run)
81%, 84% ee (3rd run)
78%, 78% ee (4th run)

when R = Me:
82%, 65% ee (1st run)
80%, 62% ee (2nd run)
81%, 60% ee (3rd run)
79%, 56% ee (4th run)

Scheme 7.19

and enantioselectivities; here, the catalyst was reused four times with moderate yields and ee-values (Scheme 7.19) [62b].

7.2.3.7 Asymmetric Cyclopropanation

Cu(II) complexes of bisoxazolines **42a**, **42b** and **43** (Figure 7.12) have been reported to catalyze the asymmetric cyclopropanation of olefins with diazo-compounds in ionic liquids (Table 7.11). The catalytic activities increased in ionic liquids compared with that in organic solvent (compare entry 1 with 9; entry 2 with entries 3 and 6) [63]. One important finding here was that catalytically less-active but cheaper and moisture-stable CuCl$_2$ could be activated in ionic liquids, in which the anion of the ionic liquid may exchange with chloride to generate the more reactive

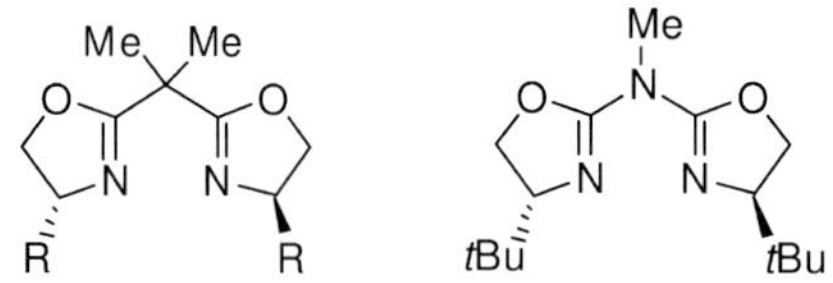

7 Enantioselective Catalysis in Ionic Liquids and Supercritical CO$_2$

Figure 7.12 Chiral Bis(oxazoline) ligands for Cu-catalyzed asymmetric cyclopropanations.

42a: R = Ph
42b: R = tBu

Table 7.11 Asymmetric cyclopropanation of styrene with ethyl diazoacetate catalyzed by bisoxazoline–copper complexes in an ionic liquid.

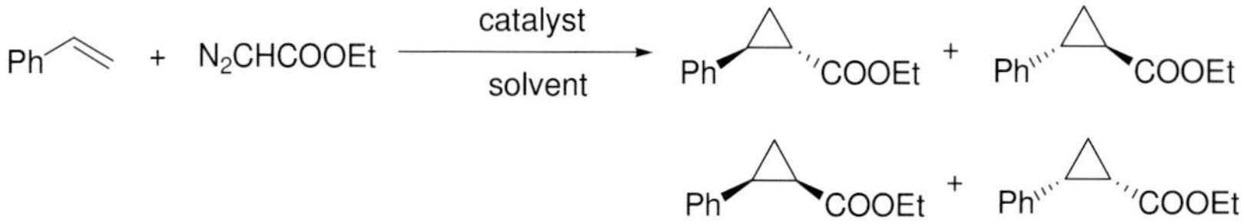

Entry	Ligand	CuX$_2$	Solvent	Run	Yield (%)	ee trans (%)	ee cis (%)
1	42a	Cu(OTf)$_2$	CH$_2$Cl$_2$	1	33	60	51
2	42a	CuCl$_2$	CH$_2$Cl$_2$	1	19	17	13
3	42a	CuCl$_2$	[emim][NTf$_2$]	1	34	55	47
4	42a	CuCl$_2$	[emim][NTf$_2$]	2	32	53	45
5	42a	CuCl$_2$	[emim][NTf$_2$]	3	33	53	44
6	42b	CuCl	[emim][OTf]	1	51	85	78
7	42b	CuCl	[emim][OTf]	2	56	85	78
8	42b	CuCl	[emim][OTf]	3	45	77	71
9	42b	Cu(OTf)	[bmim][BF$_4$]	1	88	97	94
10	42b	Cu(Otf)	[bmim][BF$_4$]	2	88	95	93
11	42b	Cu(Otf)	[bmim][BF$_4$]	3	90	93	91
12	42b	Cu(Otf)	[bmim][BF$_4$]	5	69	76	77
13	42b	Cu(Otf)	[bmim][BF$_4$]	8	51	77	78
14	43	CuCl	[emim][OTf]	1	62	91	82
15	43	CuCl	[emim][OTf]	2	40	91	82

Cu(NTf$_2$)$_2$ or Cu(OTf)$_2$. Moreover the complex, when dissolved in [emim][NTf$_2$], was successfully recycled twice without any loss of activity and enantioselectivity (entries 3–5) [63a]. A similar catalyst activation in ionic liquids was also observed in Cu(I)-catalyzed cyclopropanation. The catalytic activity and enantioselectivity of the cyclopropanation using CuCl-**42b** as a catalyst in dry [emim][OTf] was much more efficient than that using CuCl$_2$ as a copper source in [emim][NTf$_2$] (Table 7.11; entries 6–9) [63b]. The azabisoxazoline ligand **43**, when complexed with CuCl, also catalyzed cyclopropanation effectively in [emim][OTf] (entries 14 and 15) [63c]. These advantages were most likely due to the electron-donating property of the aza bridge, which increases the stability of the copper complex and improves

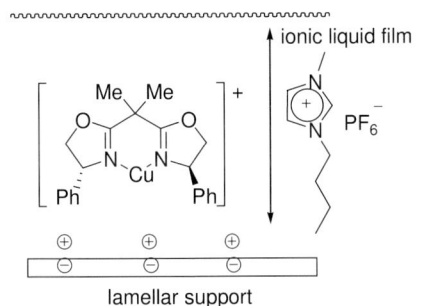

Figure 7.13 Asymmetric cyclopropanation using Cu-bis(oxazoline) complex **42a** in supported ionic liquid film (SILF).

Table 7.12 Effect of the nature of the support on cyclopropanation reaction between styrene and ethyl diazoacetate catalyzed by SILF.

Entry	Support	IL/support (ml g^{-1})	Yield (%)	ee trans (%)	ee cis (%)
1	Homogeneous	—	42	54	45
2	Laponite	0.134	12	32	−36
3	Laponite	0.067	3	27	−53
4	Bentonite	0.05	76	47	4
5	Bentonite	0.03	36	6	−62
6	Graphite	0.05	26	54	45
7	Graphite	0.01	19	56	48
8	Zeolite Y	0.05	7	38	20
9	Silica	0.05	26	54	38
10	K10	0.03	24	27	−42
11	K10	0.01	26	5	−61

the reusability of the immobilized catalyst. Cyclopropanation of styrene with diazoacetate catalyzed by Cu(OTf) complex of bisoxazoline **42b** in various ionic liquids gave the cyclopropanated products with high yields and enantioselectivities [64]. The catalyst could be recycled at least three times before a significant drop in yield and ee- values (entries 9–13).

In 2007, it was found that the supported ionic liquid films (SILF) of nanometer-thickness containing bisoxazoline-copper complexes could be used as recoverable catalysts for enantioselective cyclopropanation reactions [65]. The SILF was prepared by mixing CuCl, bisoxazoline ligand **42a** and the ionic liquid, [bmim][PF$_6$], and a clay (e.g. laponite) in CH$_2$Cl$_2$ (Figure 7.13). The resultant free-flowing powders were used as heterogeneous catalysts in solvent-free cyclopropanation between alkenes and ethyldiazoacetate. The heterogeneous catalytic system was seen to behave as a near-two-dimensional nanoreactor in which the restrictions in rotational mobility and close proximity to the surface support produced variations in stereoselectivity and enantioselectivity, leading to a complete reversal of the overall selectivity of the reaction. As shown in Table 7.12, the reaction efficiency

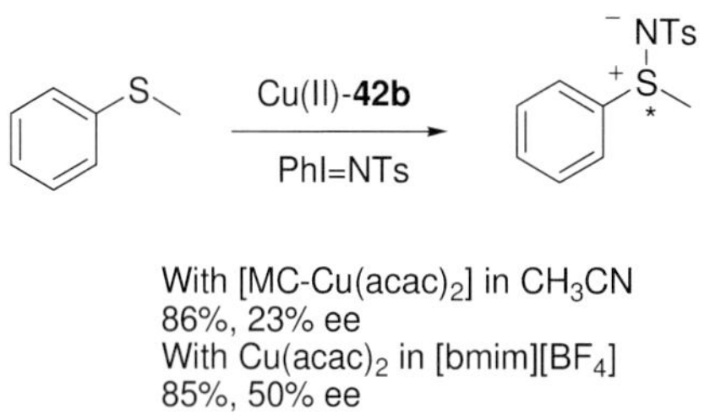

Scheme 7.20

PhS-C₆H₅ → (with Cu(II)-**42b**, PhI=NTs) → Ph-S(+)(NTs)-C₆H₅ (chiral)

With [MC-Cu(acac)$_2$] in CH$_3$CN
86%, 23% ee
With Cu(acac)$_2$ in [bmim][BF$_4$]
85%, 50% ee

Scheme 7.20

was largely dependent on the ionic nature of the support surface and the thickness of the ionic liquid film. The change – and even reversal – in both *trans/cis* ratio and enantioselectivity were observed with different supports and the thickness of the ionic liquid film (Table 7.12; entries 2–11).

7.2.3.8 Asymmetric Sulfimidation

Sulfimides are the nitrogen equivalents of sulfoxides, and their use in organic synthesis is rather limited due to a lack of convenient synthetic methods for their optically active forms. Evans and coworkers reported that PhI=Ts was an effective asymmetric nitrene transfer reagent to alkenes in the presence of a catalytic amount of Cu(I) salt together with a chiral bisoxazoline ligand [66]. Mn(salen) [67] and Ru(salen) [68] complexes have also been used as efficient catalysts for the same reaction. However, the scope of the reaction is rather limited when using these catalysts as a longer reaction time is required and it is difficult to recycle the catalyst. Kantam and coworkers [69] reported the asymmetric heterogeneous sulfimidation of various sulfides using microencapsulated copper(II) acetylacetonate, [MC-Cu(acac)$_2$] [70] or Cu(acac)$_2$ immobilized in ionic liquids with PhI=NTs as the nitrene donor. It was found that, in an ionic liquid [bmim][BF$_4$], both reaction rate and enantioselectivity were enhanced compared to that in acetonitrile (Scheme 7.20). The immobilized Cu(II)-**42b** complex in an ionic liquid could be reused for several cycles, with constant activity and enantioselectivity.

7.2.3.9 Asymmetric Diethylzinc Addition

Chiral ionic liquids **44** derived from α-pinene have been used as either additive or cosolvent in the copper-catalyzed enantioselective addition of diethylzinc to various enones (Scheme 7.21) [71]. The diethylzinc addition of cyclohex-2-enone using 3 mol% Cu(OTf)$_2$ in the presence of a chiral ionic liquid achieved up to 76% ee, thus providing a potential application of chiral ionic liquids as asymmetric inducing agents in asymmetric reactions.

7.2.3.10 Asymmetric Fluorination

Chiral organofluorine compounds containing a fluorine atom bonded directly to a stereogenic center have been used in a variety of research investigations, including mechanistic studies of enzymes and intermediates in the asymmetric synthesis of bioactive compounds. The development of effective methodologies for the

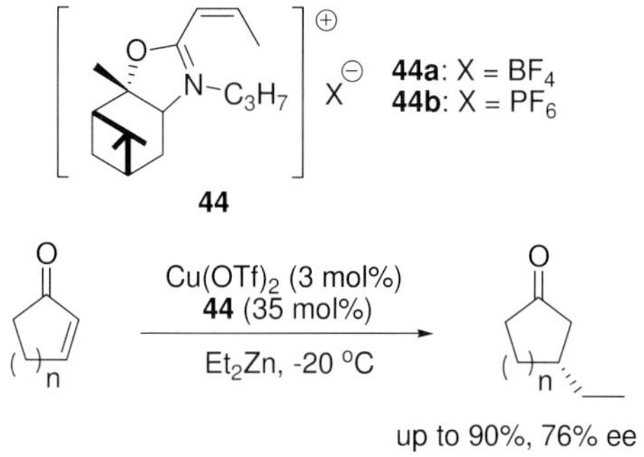
44a: X = BF4
44b: X = PF6
44
Cu(OTf)2 (3 mol%)
44 (35 mol%)
Et2Zn, -20 °C
up to 90%, 76% ee

Scheme 7.21

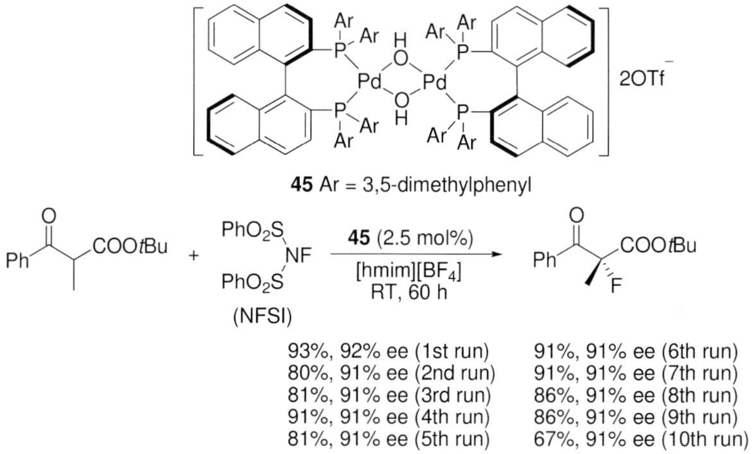
2OTf
45 Ar = 3,5-dimethylphenyl
COOtBu
(NFSI)
45 (2.5 mol%)
[hmim][BF4]
RT, 60 h
93%, 92% ee (1st run)
80%, 91% ee (2nd run)
81%, 91% ee (3rd run)
91%, 91% ee (4th run)
81%, 91% ee (5th run)
91%, 91% ee (6th run)
91%, 91% ee (7th run)
86%, 91% ee (8th run)
86%, 91% ee (9th run)
67%, 91% ee (10th run)

Scheme 7.22

preparation of chiral fluorine compounds is a challenging issue in fluorine chemistry [72]. The first asymmetric fluorination of β-ketoesters in ionic liquids were successfully conducted using palladium complex **45** as a catalyst [73]. The fluorination of β-ketoester with *N*-fluorobenzenesulfonimide (NFSI) (1.5 equiv.) was performed in the presence of 2.5 mol% of palladium complex **45** in [hmim][BF_4] (Scheme 7.22). Although the chemical yield and enantioselectivity were almost the same as those obtained in EtOH (40 h, 92%, 91% ee), a longer reaction time (60 h) was necessary for completion of the reaction. The catalyst **45** could be recycled more than 10 times, with excellent enantioselectivity (91–92% ee).

Kim and coworkers reported the asymmetric fluorination of β-keto phosphonates in ionic liquids using palladium complex **46** as a catalyst [74]. The fluorination of β-keto phosphonate with NFSI in the presence of 5 mol% of **46** in [bmim][BF_4] at room temperature produced a fluorinated product in 95% yield with 93% ee. The chemical yield and enantioselectivity were quite similar to those obtained in MeOH (8 h, 93%, 97% ee). Moreover, the catalyst immobilized in the ionic liquid layer was recycled seven times without any obvious loss in yield or

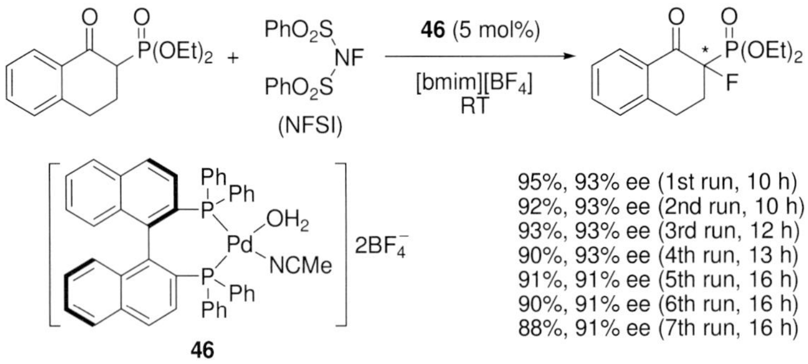
P(OEt)2
PhO2S
NF
PhO2S
(NFSI)
46 (5 mol%)
[bmim][BF4]
RT
P(OEt)2
F
Ph
Ph
P
OH2
Pd
NCMe
P
Ph
Ph
2BF4−
46
95%, 93% ee (1st run, 10 h)
92%, 93% ee (2nd run, 10 h)
93%, 93% ee (3rd run, 12 h)
90%, 93% ee (4th run, 13 h)
91%, 91% ee (5th run, 16 h)
90%, 91% ee (6th run, 16 h)
88%, 91% ee (7th run, 16 h)

Scheme 7.23

enantioselectivity, although a rather longer reaction time was required to complete the reactions in recycling runs. Here again, leaching of the catalyst/ionic liquid during extraction may be the major reason for the decline in activity (Scheme 7.23).

7.2.4
Enantioselective Organocatalysis

For a long time, metal- and enzyme-based asymmetric catalysis dominated the field of asymmetric catalysis. However, quite recently a new-type of asymmetric catalysis – asymmetric organocatalysis – in which an enantioselective transformation is promoted by a catalytic amount of a chiral organic molecule, has emerged as the third discipline. The research area of asymmetric organocatalysis has grown rapidly to become one of the most exciting research fields in organic chemistry [75]. The use of organocatalysts has provided several clear advantages over metal-based chiral catalysts, most notably the safety and availability of the organic catalysts. During recent years the scope of asymmetric organocatalytic reactions has rapidly expanded, and many typical transition-metal-catalyzed asymmetric reactions have now been transposed into the metal-free organocatalytic reactions. However, the turnover numbers of organocatalysts are generally not very high, and hence recycling of the catalysts might become an important issue for the practical application of asymmetric organocatalysis, particularly in the case that the catalyst has been synthesized by multistep transformations using expensive starting materials. Moreover, immobilization would simplify the separation of the organic catalyst from the organic products. Several organocatalysts have been immobilized by covalent bonding to the solid supports, and recycled efficiently in some cases [76]. However, in many cases the activity and/or enantioselectivity of the immobilized organocatalysts were reduced largely due to the unsuitable structural modification of the parent organic catalyst or steric hindrance by support materials or mass transfer limitations, and so on. In order to circumvent these problems, room-temperature ionic liquids – particularly 1,3-dialkylimidazolium-based ionic liquids – have recently been utilized for the immobilization of organocatalysts. Although studies on the immobilization of organocatalysts in ionic liquids are still in their infancy, in many cases the organocatalysts immobilized in ionic liquids may be readily recovered and reused several times without any significant loss in catalytic activities. In some cases, increased reaction rates and enantioselectivities have also been observed when the organocatalysis were conducted in ionic liquids. To date, most of the asymmetric organocatalysis studied have focused on chiral secondary amine-catalyzed reactions such as aldol, Mannich, Michael addition or Diels–Alder reactions. Thus, the improved catalytic performance of organocatalysts might be due to the stabilization of the iminium intermediate formed from the carbonyl and the secondary amine of the catalyst.

7.2.4.1 Asymmetric Aldol Reactions

The asymmetric aldol reaction is one of the most efficient synthetic methods for obtaining optically active β-hydroxy carbonyl compounds, which are important building blocks in organic synthesis [77]. Independent pioneering studies by List and Barbas have each shown that L-proline is able to catalyze the direct asymmetric intra- and intermolecular aldol reactions with high enantioselectivity and yield [78]. Usually, polar solvents such as dimethyl sulfoxide (DMSO), N,N-dimethylformamide (DMF), or water are used for this reaction. Early attempts involved either the use of chloroform as a solvent, in which proline is insoluble, while the catalyst could be recovered quantitatively by filtration; an alternative approach involved the covalent immobilization of L-proline on silica gel. Unfortunately, both approaches resulted in a significant reduction in enantioselectivity as compared to the reaction performed homogeneously in DMSO. The first asymmetric organocatalysis in ionic liquids was reported by Loh in 2002 [79a], since when several research groups have investigated L-proline-catalyzed direct aldol reactions in ionic liquids, and shown the yields and enantioselectivities to be comparable with those obtained in organic solvents [79]. For example, the aldol reaction of benzaldehyde with acetone in [bmim][PF$_6$] afforded the corresponding chiral hydroxy ketone with 71% ee [79a] (76% ee in Ref. [75b]), which is somewhat higher than that obtained in DMSO (60% ee). Moreover, in many of the cases studied either less or no dehydration product was detected. On completion of the reaction the products could be extracted with diethyl ether, and the immobilized proline in the ionic liquid phase simply recovered and reused in subsequent reactions without any significant loss of activity or enantioselectivity (Scheme 7.24).

The catalytic efficiency of L-proline in ionic liquid was enhanced by the addition of DMF as cosolvent, which may be largely due to the increased mass transfer in the presence of DMF [79c]. Thus, the use of only 5 mol% L-proline was sufficient to accomplish the cross-aldol reactions of aliphatic aldehydes, affording α-alkyl-β-hydroxyaldehydes with extremely high enantioselectivities (>99% ee) in moderate to high diastereoselectivities (diastereomeric ratio 3:1 ~ >19:1). However, under the same reaction conditions, much lower ee-values and yields were observed in a one-pot synthesis of pyranose derivatives by sequential cross-aldol reactions. The L-proline immobilized in the ionic liquid layer could be recovered and reused without any deterioration in catalytic efficiency, with the diastereoselectivity,

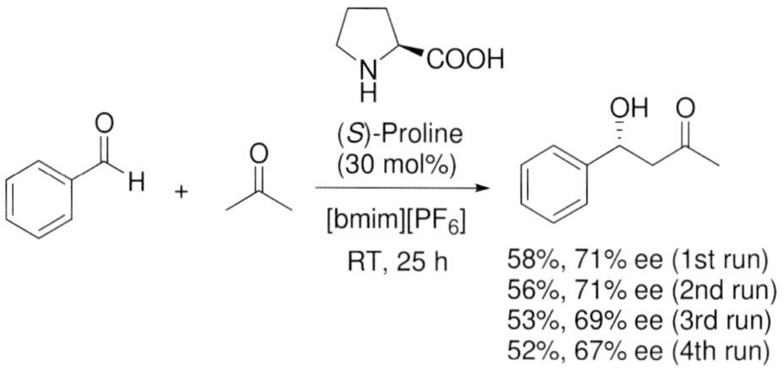

Scheme 7.24

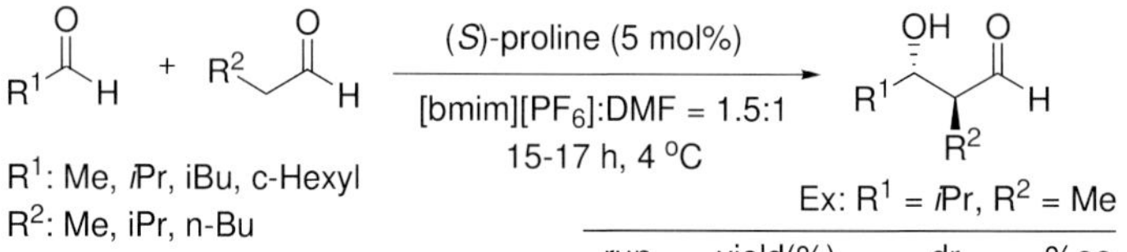

benzaldehyde + acetone → (S)-Proline (30 mol%), [bmim][PF$_6$], RT, 25 h → β-hydroxy ketone product

58%, 71% ee (1st run)
56%, 71% ee (2nd run)
53%, 69% ee (3rd run)
52%, 67% ee (4th run)

Scheme 7.25

R^1CHO + R^2CH$_2$CHO → (S)-proline (5 mol%), [bmim][PF$_6$]:DMF = 1.5:1, 15–17 h, 4 °C → aldol product

R^1: Me, iPr, iBu, c-Hexyl
R^2: Me, iPr, n-Bu

Ex: R^1 = iPr, R^2 = Me

run	yield(%)	dr	%ee
1st	76	>19:1	>99
2nd	74	>19:1	>99
3rd	78	>19:1	>99
4th	76	>19:1	>99
5th	75	>19:1	>99

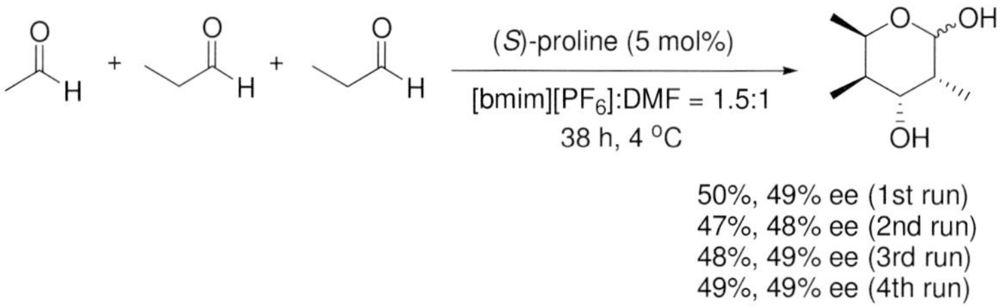

acetaldehyde + propanal + propanal → (S)-proline (5 mol%), [bmim][PF$_6$]:DMF = 1.5:1, 38 h, 4 °C → pyranose product

50%, 49% ee (1st run)
47%, 48% ee (2nd run)
48%, 49% ee (3rd run)
49%, 49% ee (4th run)

enantioselectivity and yields being retained upon reuse of the L-proline/ionic liquid system (Scheme 7.25).

By using of a modified proline, L-prolinamide **47** (which is known to be a more reactive catalyst than L-proline in cross-aldol reactions [80]), the enantioselectivity of the direct aldol reactions in ionic liquid [bmim][BF$_4$] was remarkably increased as compared with the reaction carried out in acetone (69% ee) (Scheme 7.26) [81]. However, the reusability of the recovered **47** when immobilized in the ionic liquid layer was somewhat inferior to that of the L-proline catalyst; this effect could be ascribed to the increased solubility of the organocatalyst **47** in the extracting organic solvents (not provided in the literature), leading to an increased leaching of the catalyst.

The catalyst leaching problem could be solved by incorporating an 'ionic tag' onto the 4-hydroxyproline [82]. Interestingly, the imidazolium-tagged organo-

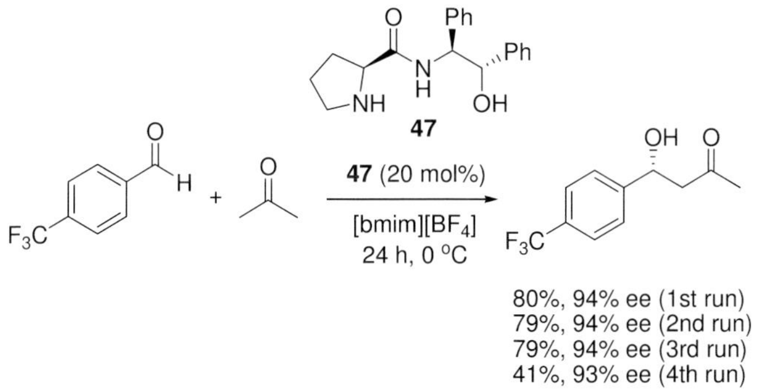

7.2 Enantioselective Catalysis in Ionic Liquids

Scheme 7.26

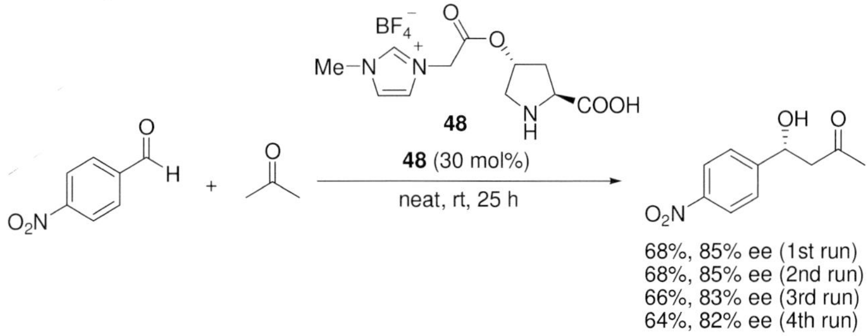

Reaction: F₃C-C₆H₄-CHO + acetone, catalyst **47** (20 mol%), [bmim][BF₄], 24 h, 0 °C → aldol product (F₃C-aryl, OH).

Catalyst **47**: proline-derived amide with NH-CH(Ph)-CH(Ph)-OH.

- 80%, 94% ee (1st run)
- 79%, 94% ee (2nd run)
- 79%, 94% ee (3rd run)
- 41%, 93% ee (4th run)

Scheme 7.27

Reaction: O_2N-C₆H₄-CHO + acetone, catalyst **48** (30 mol%), neat, rt, 25 h → aldol product.

Catalyst **48**: Me–N(imidazolium)–N–CH₂–C(O)O–pyrrolidine–COOH, BF₄⁻ counterion.

- 68%, 85% ee (1st run)
- 68%, 85% ee (2nd run)
- 66%, 83% ee (3rd run)
- 64%, 82% ee (4th run)

catalyst **48** was not soluble in methylene chloride, and hence the catalyst **48** could be separated simply by washing the ionic liquid-tagged catalysts with methylene chloride. Importantly, the reaction rate of the aldol reactions between *p*-nitrobenzaldehyde and acetone using catalyst **48** was dramatically increased. A nuclear magnetic resonance (NMR) study of the aldol reaction in d_6-DMSO revealed an extremely high initial rate and, as a result, about 87% of the conversion was achieved within 10 min. The increased reaction rate with ionic liquid-tagged catalyst **48** was most likely due to the Lewis acidity of the imidazolium moiety, which may facilitate reaction of the enamine intermediate with aldehyde in the proximity of the ionic moiety. The recovered catalyst **48** could be reused four times without any loss of catalytic activity (Scheme 7.27) [82a].

Other ionic liquid-tagged organocatalysts such as **49a** also showed high catalytic efficiency in terms of yield and reusability in aldol reactions [82b]. However, dramatic decreases in both diastereoselectivity and enantioselectivity were observed compared with the unmodified, proline-catalyzed aldol reactions. Moreover, a significant amount of dehydrated product was also observed. In contrast, the addition of H_2O and acetic acid as additives could significantly inhibit the formation of the dehydrated product. For example, an aldol reaction of *p*-nitrobenzaldehyde with an excess of acetone using 20 mol% **49a** as a catalyst afforded the aldol adduct in

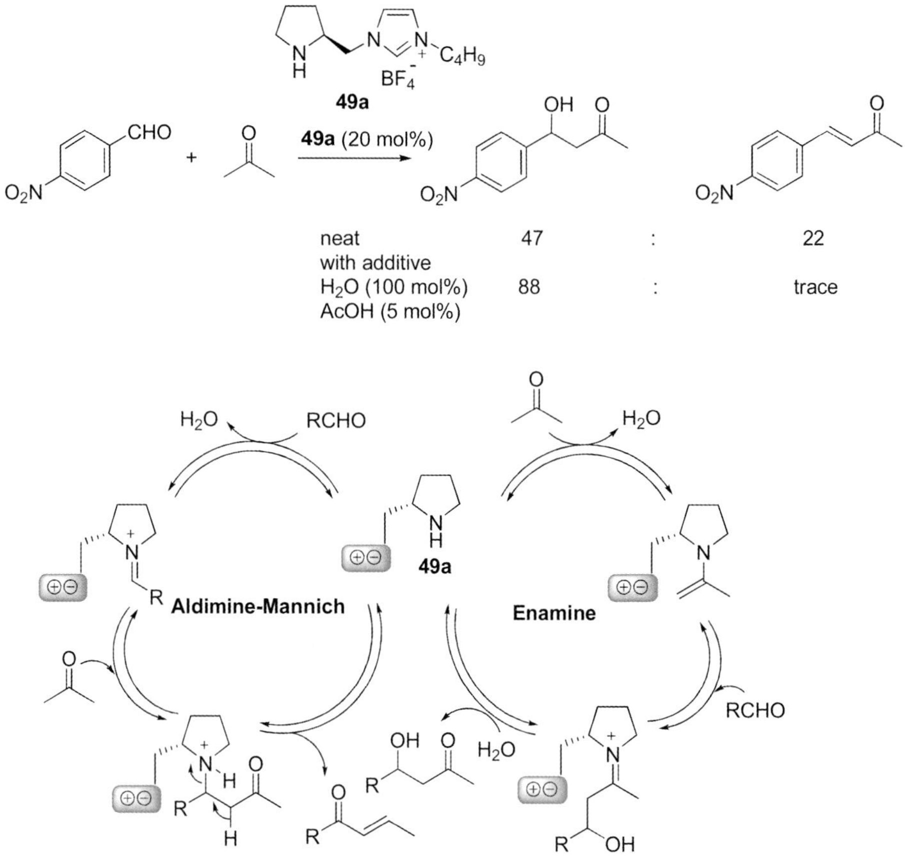

Scheme 7.28

47% yield, along with 22% of the dehydrated product. The addition of H_2O (100 mol%) and acetic acid (5 mol%) completely inhibited formation of the dehydrated product, affording the aldol adduct in 88% yield. Although the role of additive is not yet clear, it has been suggested that the acidic additives might promote the enamine catalytic cycle, thereby accelerating the reaction rate and suppressing the general base-mediated aldimine–Mannich condensation pathway (Scheme 7.28). Although the diastereoselectivity (syn/anti = 2.1/1–4.8/1) and enantioselectivity (<5–11% ee for syn, 26–37% ee for anti) of the aldol reaction with the ionic liquid-supported catalyst **49a** were not sufficiently high, the catalyst could be recovered and reused several times without any significant loss of activity (89–96% yields).

The L-proline-catalyzed direct asymmetric aldol reactions between acetone and aldehydes in a SILP have also been investigated (Scheme 7.29) [83]. The major advantages of SILP catalyst systems are that the amount of expensive ionic liquid used as the reaction medium can be reduced, the catalyst can be recovered by simple filtration, and it can be used in homogeneous form. The results obtained

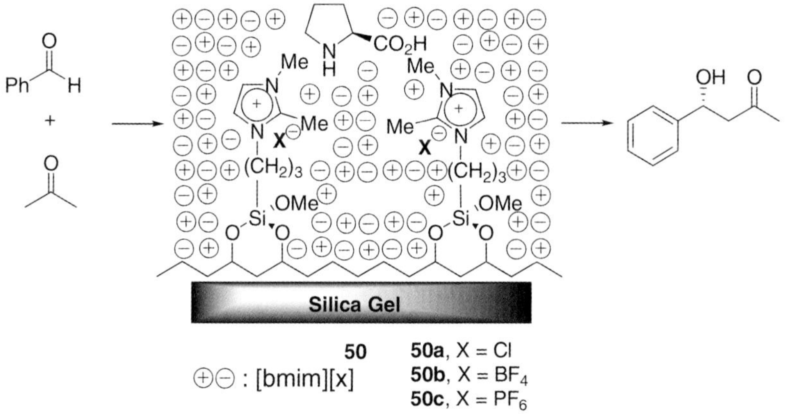

Scheme 7.29

50a, X = Cl
50b, X = BF$_4$
50c, X = PF$_6$

⊕⊖ : [bmim][x]

Table 7.13 Direct aldol reaction between benzaldehyde and acetone with different L-proline catalyst forms.

Entry	Catalyst	Yield (%)	ee (%)	Reference
1	50b/[bmim][BF$_4$]	51	64	[77]
2	SiO$_2$/[bmim][BF$_4$]	38	12	[77]
3	50c/[bmim][PF$_6$]	15	52	[77]
4	proline/DMSO	62	60	[73]
5	proline/[bmim][PF$_6$]	58	71	[75a]
6	proline/[bmim][PF$_6$]	55	76	[75b]
7	PEG-proline/DMF	45	59	[74]

from the L-proline-catalyzed direct aldol reaction between benzaldehyde and acetone with various catalytic systems are summarized in Table 7.13. The catalytic efficiency of the supported ionic liquid-phase system was found to be comparable with that of other catalytic systems, and was largely dependent on the anion of the supported ionic liquid phase. For example, the aldol reaction in a catalytic system composed of **50b**/[bmim][BF$_4$] afforded the aldol product in 51% yield with 64% ee (entry 1), whereas the **50c**/[bmim][PF$_6$] system afforded the product in only 15% yield with 52% ee (entry 3). An important role of the imidazolium moiety impregnated onto silica gel was also observed by a comparison of the result with that obtained using the unmodified silica gel/[bmim][BF$_4$] catalytic system, where the catalytic efficiency was dramatically decreased (entry 2). The **50b**/[bmim][BF$_4$] catalytic system could be reused three times without any loss of catalytic activity.

7.2.4.2 Asymmetric Michael Addition

The asymmetric Michael addition is one of the most frequently studied reactions in asymmetric organocatalysis, and impressive levels of conversion and enantiose-

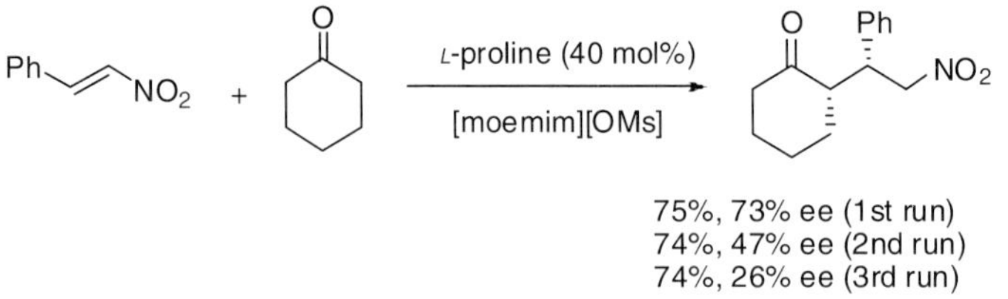

Scheme 7.30

Ph–CH=CH–NO$_2$ + cyclohexanone →[L-proline (40 mol%)][[moemim][OMs]] cyclohexanone-CH(Ph)–CH$_2$NO$_2$

75%, 73% ee (1st run)
74%, 47% ee (2nd run)
74%, 26% ee (3rd run)

lectivities have been achieved using a variety of organocatalysts. In contrast to aldol reactions, only a few Michael additions have been conducted in ionic liquids [84–87]. The initial studies on the asymmetric organocatalysis using L-proline or other amino acids and amines as catalysts in ionic liquids showed little promise [84–86], and in most cases decreased enantioselectivities were observed in comparison with the reaction conducted in organic solvents. Moreover, the enantioselectivity of the recovered catalyst immobilized in an ionic liquid layer was also dramatically decreased upon reuse. For example, the Michael addition of cyclohexanone to β-nitrostyrene catalyzed by L-proline in a hydrophilic ionic liquid, [moemi][OMs] (1-methoxyethyl-3-methylimidazolium methanesulfonate), afforded the Michael adduct in 75% yield with 73% ee. However, upon reusing the recovered catalyst/ionic liquid the enantioselectivity decreased to 47% ee, even though the yield was retained (Scheme 7.30) [87].

Very successful asymmetric Michael additions have been achieved by using imidazolium salt-tagged pyrrolidines **49** as organocatalysts [88, 89]. In contrast to the aldol reaction shown in Scheme 7.28, imidazolium-tagged pyrrolidines **49** showed extremely high catalytic efficiency in Michael addition (Table 7.14). For example, the Michael addition of cyclohexanone to β-nitrostyrene using 15 mol% of chiral ionic liquids **49a** in the present of 5 mol% of trifluoroacetic acid (TFA) proceeded quantitatively with excellent diastereoselectivity (*syn/anti* = 99/1) and enantioselectivity (99% ee) within 8 h (entry 1) [88]. Moreover, the imidazolium-tagged catalyst **49a** could be recovered by precipitation using ethyl ether, and reused four times with slightly decreased catalytic efficiency (entries 1–4). Similar catalytic efficiency and reusability have been observed when the reaction was conducted in an ionic liquid (entries 5–14). However, the catalytic efficiencies were dramatically decreased in organic solvents (entries 15–16).

7.2.4.3 Asymmetric Mannich, α-Aminoxylation, and Diels–Alder Reaction

In contrast to the aldol reaction, L-proline immobilized in ionic liquids showed an extremely high catalytic efficiency as well as a good reusability in Mannich reactions [90]. The reaction rate of the proline-catalyzed direct asymmetric Mannich reaction of N-*p*-methoxyphenyl (PMP)-protected α-imino ethyl glyoxylate with cyclohexanone in [bmim][BF$_4$] was enhanced between four- and 50-fold faster than the reaction in standard organic solvents. Accordingly, the reaction was completed within 30 min, providing the Mannich adduct in quantitative yield with excellent enantioselectivity and diastereoselectivity (99% ee, dr = 19:1). For recycling of the ionic liquid containing L-proline, there was no diminution in ee-value and a slight

Table 7.14 Asymmetric Michael addition of cyclohexanone to β-nitrostyrene using ionic liquid-supported chiral pyrrolidines as a catalyst.

49a: R = Bu, X = BF$_4$
49b: R = Bu, X = PF$_6$
49c: R = Bu, X = Br
49d: R = (CH$_2$)$_2$OH, X = Br
49e: R = Me, X = Br
49f: R = Et, X = Br
49g: R = Et, X = BF$_4$
49h: R = Et, X = PF$_6$

Entry	Catalyst (mol%)	Additive or solvent	T (h)	Yield (%)	syn/anti	ee (%)
1	49a (1st run) (15)	TFA (5 mol%)	8	100	99:1	99
2	49a (2nd run) (15)	TFA (5 mol%)	8	97	97:3	94
3	49a (3rd run) (15)	TFA (5 mol%)	24	99	96:4	91
4	49a (4th run) (15)	TFA (5 mol%)	48	96	97:3	93
5	49b (15)	TFA (5 mol%)	12	86	98:2	87
6	49c (15)	TFA (5 mol%)	10	99	99:1	98
7	49d (15)	TFA (5 mol%)	18	86	97:3	89
8	49e (20)	[bmim][PF$_6$]	24	97	93:7	95
9	49f (1st run) (20)	[bmim][PF$_6$]	24	98	92:8	97
10	49f (2nd run) (20)	[bmim][PF$_6$]	32	97	90:10	97
11	49f (3rd run) (20)	[bmim][PF$_6$]	40	97	91:9	97
12	49f (4th run) (20)	[bmim][PF$_6$]	100	90	90:10	96
13	49g (20)	[bmim][PF$_6$]	96	55	92:8	80
14	49h (20)	[bmim][PF$_6$]	96	78	90:10	88
15	49f (20)	DMSO	96	53	91:9	79
16	49f (20)	iPrOH	96	45	93:7	98

decrease in yield for the four consecutive reaction cycles when the reaction time was strictly controlled within 30 min (Scheme 7.31). The reaction rates of the three-component Mannich reactions of aldehyde, ketone and p-methoxyaniline in ionic liquids were also significantly faster than those of the reactions conducted in organic solvents.

Recently, some extensive research has been devoted to exploring a diastereoselective and enantioselective route for the synthesis of α-hydroxyaldehydes or α-hydroxyketones because they are important building blocks for the construction of complex natural products and biologically active molecules [91]. In parallel with the transition-metal-catalyzed asymmetric nitroso-aldol reaction [92], much interest has also been expressed towards the proline-catalyzed direct asymmetric α-aminoxylation of aldehydes or ketones for the synthesis of optically active α-hydroxyladehydes and α-hydroxyketones [93]. Wang [94] and Huang [95] independently reported an L-proline-catalyzed asymmetric α-aminoxylation reaction in ionic liquids, whereby it was found that aldehydes and ketones could undergo

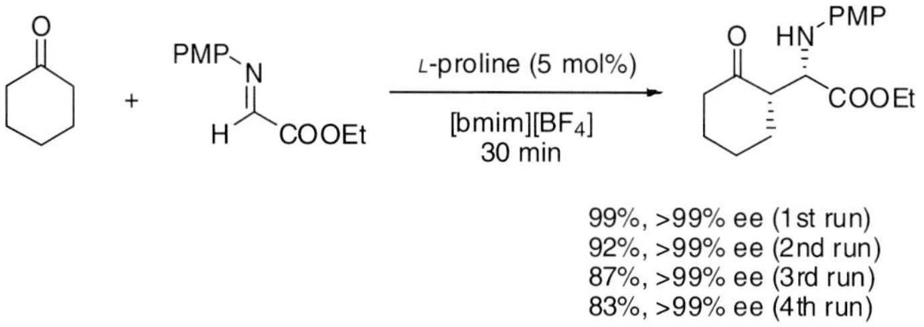

Scheme 7.31

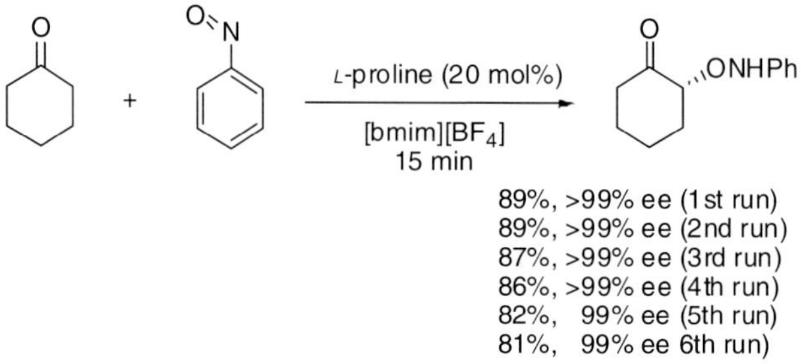

Cyclohexanone + PMP-N=CH-COOEt → (with L-proline (5 mol%), [bmim][BF$_4$], 30 min) → α-substituted cyclohexanone with HN-PMP and COOEt group

99%, >99% ee (1st run)
92%, >99% ee (2nd run)
87%, >99% ee (3rd run)
83%, >99% ee (4th run)

Scheme 7.32

Cyclohexanone + nitrosobenzene (O=N-Ph) → (with L-proline (20 mol%), [bmim][BF$_4$], 15 min) → 2-(ONHPh)-cyclohexanone

89%, >99% ee (1st run)
89%, >99% ee (2nd run)
87%, >99% ee (3rd run)
86%, >99% ee (4th run)
82%, 99% ee (5th run)
81%, 99% ee 6th run)

asymmetric aminoxylation smoothly in the presence of 20 mol% L-proline with nitrosobenzene in various ionic liquids at room temperature. This provided the corresponding aminoxy aldehydes and ketones in good yields (67–98%), with excellent enantioselectivities (97 to >99% ee). The reusability of the ionic liquid containing L-proline was examined for the direct α-aminoxylation reaction of cyclohexanone with nitrosobenzene. The catalyst could be recycled six times without any significant loss of catalytic efficiency (Scheme 7.32) [95] (77–50% yield in Ref. [90]).

The addition reaction of aldehydes and ketones to diethyl azodicarboxylate (DEAD) using organocatalysts in ionic liquids was also investigated [96]. In order to determine the enantioselectivity of the adduct, the unstable primary reaction products were converted to stable N-(ethoxycarbonylamino)oxazolidinone derivatives by reduction with NaBH$_4$ and subsequent treatment with aqueous NaOH (Scheme 7.33). Remarkably, very high yields (up to 92%) were obtained in [bmim][PF$_6$] using only a slight excess (1.1 equiv.) of aldehydes and 5 mol% of L-proline. However, the reactions involving ketones as substrates gave less satisfactory results. Among the tested ionic liquids for the addition of 3-methylbutanal to DEAD, [bmim][BF$_4$] proved to be the best with respect to both yield (85%) and enantioselectivity (84% ee). Recycling of the catalytic system was possible, although the catalytic activity fell considerably, most likely due to the partial loss of the organocatalyst during the extraction.

Asymmetric Diels–Alder reactions in ionic liquids have been conducted using the privileged MacMillian's catalyst imidazolidin-4-one 51 [97]. The Diels–Alder reactions between cyclohexadiene and acrolein were carried out using 5 mol% of

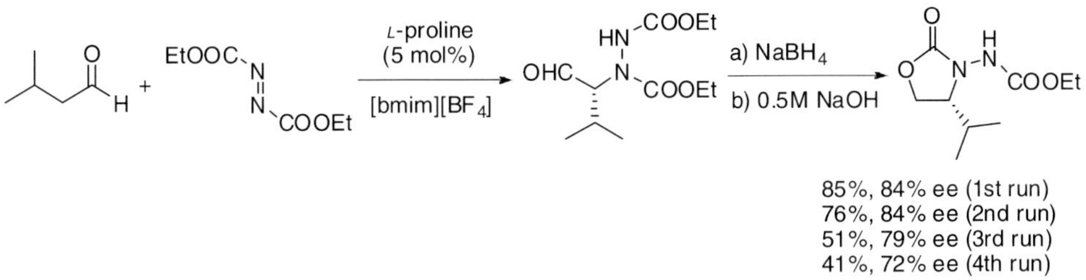

7.2 Enantioselective Catalysis in Ionic Liquids

Scheme 7.33

85%, 84% ee (1st run)
76%, 84% ee (2nd run)
51%, 79% ee (3rd run)
41%, 72% ee (4th run)

Table 7.15 Asymmetric Diels–Alder reaction between cyclohexadiene and acrolein using organocatalyst **51** in ionic liquids.

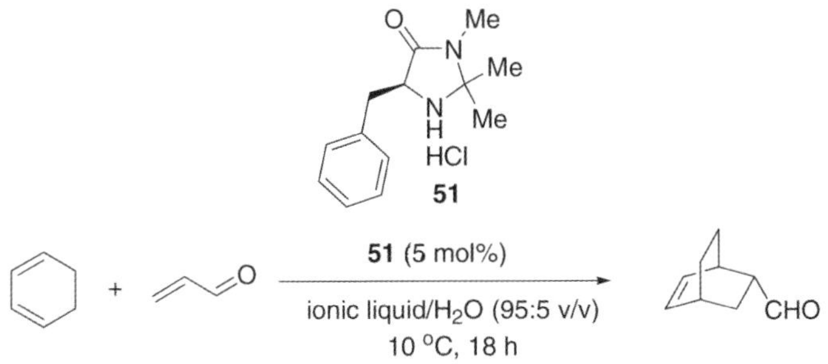

Entry	Solvent	Run	Yield (%)	endo:exo	ee endo (%)
1	[bmim][PF$_6$]	1	76	17:1	93
2	[bmim][PF$_6$]	2	72	17:1	91
3	[bmim][PF$_6$]	3	70	17:1	87
4	[bmim][SbF$_6$]	1	74	17:1	92
5	[bmim][OTf]	1	7	17:1	0
6	[bmim][BF$_4$]	1	5	17:1	0
7	CH$_3$CN	1	82	14:1	94

51 in a mixture of ionic liquid and water (95:5, v/v) (Table 7.15). The counteranion of an ionic liquid was found to play an important role in catalytic efficiency; for example, when [bmim][PF$_6$] was used a 17:1 endo/exo mixture of the Diels–Alder adduct was obtained in 76% yield with a 93% ee-value for the major *endo*-diastereomer (entry 1), which is comparable to that obtained in acetonitrile (entry 7). The ionic liquid containing catalyst **51** could be recycled three times without any significant loss of the catalytic efficiency (entries 1–3). Reactions employing [bmim][SbF$_6$] gave similar results to that obtained in [bmim][PF$_6$] (entry 4). However, when the counteranion was OTf$^-$ or BF$_4^-$, the yields decreased dramatically and no enantiomeric enrichment for the *endo*-isomer was observed (entries 5 and 6).

Although asymmetric organocatalysis in ionic liquids are still in many ways in their infancy, the results of the studies described here have shown that ionic liquids can be used successfully as alternative solvents to immobilize organocatalysts. The application of ionic liquids often allows facile recycling of the organic catalysts, and sometimes also provides an environment which is helpful for the stabilization of ionic intermediates. Consequently, higher turnover frequency and/or turnover numbers of the reaction can be achieved under favorable circumstances. In particular, the incorporation of ionic moieties into the organocatalysts has provided more opportunities not only for the development of new chiral organocatalysts but also for catalyst immobilization in ionic liquids. The combination of asymmetric organocatalysis with ionic liquids has extended the scope of investigations in a very interesting manner. Clearly, the constant accumulation of results of asymmetric organocatalysis in ionic liquids will increase the prospects of their application in practical processes.

7.3
Enantioselective Catalysis in Supercritical Carbon Dioxide (scCO$_2$)

Above its critical point (T_c = 304 K, P_c = 7.4 MPa), both gas and liquid carbon dioxide coexist, and this is referred to as supercritical carbon dioxide (scCO$_2$). scCO$_2$ has high miscibility with other gases such as hydrogen and carbon monoxide, and high solvation properties for many organic molecules, and this may be advantageous in many catalytic asymmetric reactions such as hydrogenation and hydroformylation. For example, the gas-like mass-transfer properties of scCO$_2$ can facilitate exchange between the catalyst and substrate phases. Moreover, the relatively poor solubility of most organometallic compounds in scCO$_2$ generally leads to a reduced leaching of organometallic catalysts compared with common organic solvents. Hence, the combination of a high solubility of organic substrates and products and a low solubility of organometallic catalysts in scCO$_2$ can allow the easy separation of catalysts from the reaction mixture. During recent years, the application of scCO$_2$ as a reaction medium in asymmetric catalysis has been investigated and, in some cases, increased enantioselectivities have been observed in scCO$_2$ compared to those in organic solvents.

7.3.1
Asymmetric Hydrogenation

In 1995, Burk and Tumas reported for the first time the asymmetric hydrogenation of several α-enamides in scCO$_2$ using a cationic Rh complex with Et-DuPHOS [5]. The reaction proceeded homogeneously under 5000 psi of the supercritical phase (H$_2$ partial pressure 200 psi) at 40 °C. The enantioselectivities obtained in scCO$_2$ were comparable to or higher than those obtained in either methanol or hexane (Scheme 7.34).

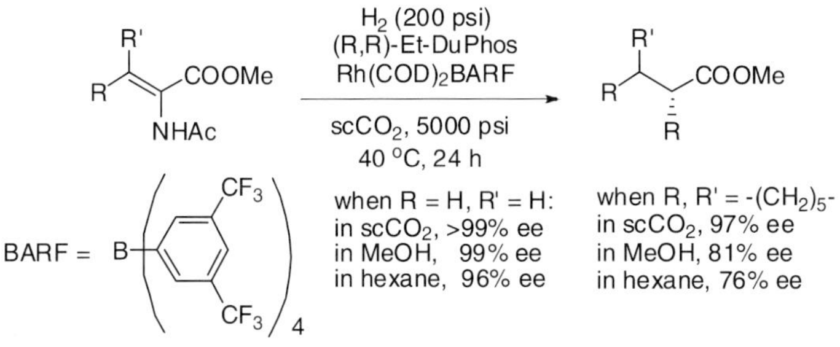

Scheme 7.34

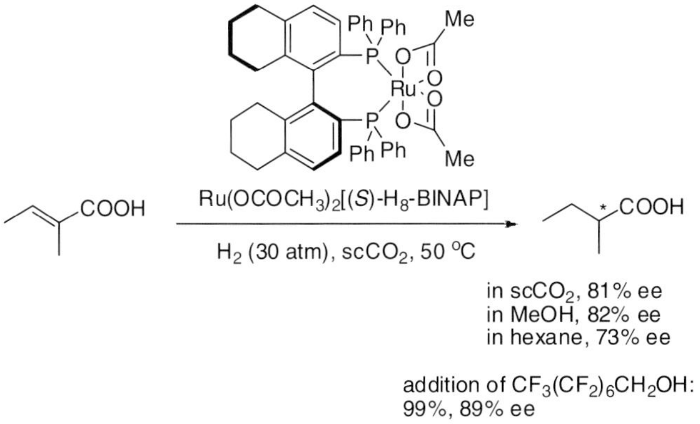

R'R C(=)COOMe with NHAc substituent

H₂ (200 psi)
(R,R)-Et-DuPhos
Rh(COD)₂BARF
───────────────→
scCO₂, 5000 psi
40 °C, 24 h

R'RCH–CH(COOMe)(R̄)

BARF = B[3,5-(CF₃)₂C₆H₃]₄⁻

when R = H, R' = H:
in scCO₂, >99% ee
in MeOH, 99% ee
in hexane, 96% ee

when R, R' = -(CH₂)₅-
in scCO₂, 97% ee
in MeOH, 81% ee
in hexane, 76% ee

Scheme 7.35

CH₃–C(CH₃)=CH–COOH

Ru(OCOCH₃)₂[(S)-H₈-BINAP]
───────────────→
H₂ (30 atm), scCO₂, 50 °C

CH₃–CH(CH₃)–*CH(COOH)

in scCO₂, 81% ee
in MeOH, 82% ee
in hexane, 73% ee

addition of CF₃(CF₂)₆CH₂OH:
99%, 89% ee

Subsequently, Noyori and coworkers used Ru(OCOCH₃)₂(H₈-BINAP) as a catalyst for the hydrogenation of tiglic acid in scCO₂ (Scheme 7.35) [98]. The enantioselectivity in scCO₂ (81% ee) was comparable to that in methanol (82% ee) and greater than that in hexane (73% ee). It has been also found that the addition of a fluorinated alcohol CH₃(CF₂)₆CH₂OH to the reaction in scCO₂ increased both the conversion (99%) and enantioselectivity (89% ee). Although the role of the alcohol is not yet clear, it has been proposed that solubility of the catalyst may be increased by the added fluorinated alcohol.

Wang and Kienzle [99] reported that the asymmetric hydrogenation of 2-(4-fluorophenyl)-3-methylbutenoic acid in scCO₂ using MeOH as a cosolvent gave the optically active carboxylic acid with 63–84% ee; this compound is an important intermediate in the synthesis of a new type of calcium antagonist (Scheme 7.36). In comparison with hydrogenation in MeOH (99%, 93% ee), however, the results were disappointing. The decreased catalytic efficiency might be largely due to the relatively poor solubility of the catalyst and the reactant in scCO₂. Similarly, the asymmetric hydrogenation of the naproxen precursor, 2-(6′-methoxy-2′-naphthyl)propenoic acid, using Ru–BINAP in scCO₂ medium also proved to be less effective, and a decreased product enantioselectivity was obtained compared to that achieved in MeOH [100].

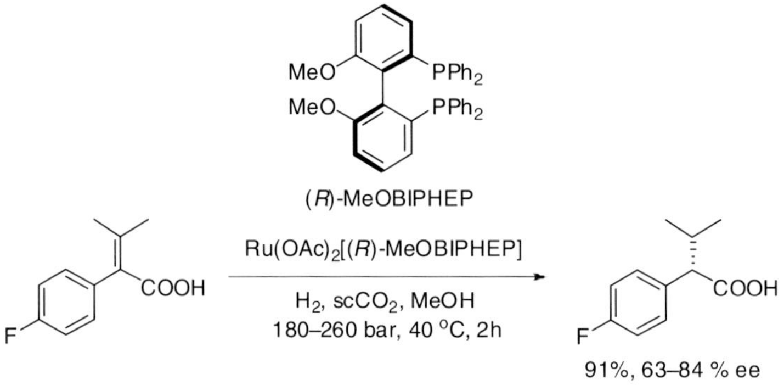

Scheme 7.36

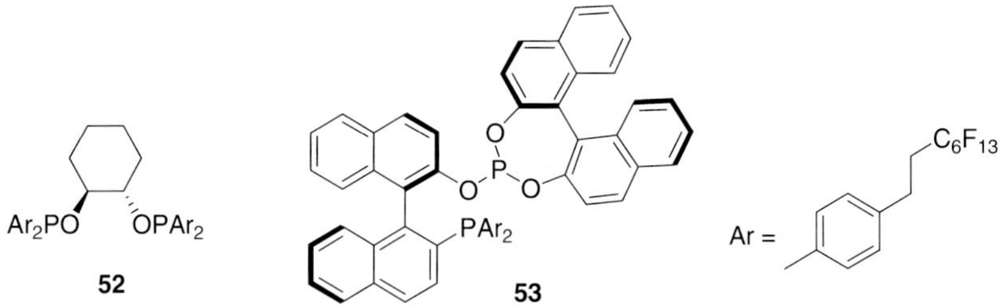

(R)-MeOBIPHEP

Ru(OAc)$_2$[(R)-MeOBIPHEP]
H$_2$, scCO$_2$, MeOH
180–260 bar, 40 °C, 2h

91%, 63–84 % ee

Figure 7.14 'CO$_2$-philic' phosphorus ligands bearing perfluoroalkyl chains for asymmetric hydrogenations in scCO$_2$.

In order to increase the solubility of chiral catalysts in scCO$_2$, several research groups have independently developed chiral phosphorus ligands bearing 'CO$_2$-philic' groups, such as fluorinated substituents. For example, Leitner and coworkers developed two CO$_2$-philic phosphorus ligands containing perfluorinated chains, and investigated their catalytic activities for asymmetric hydrogenation in scCO$_2$ [101]. The asymmetric hydrogenation of dimethyl itaconate in scCO$_2$ using the **52**-Rh(COD)$_2$BF$_4$ complex (Figure 7.14) under 200–240 bar total pressure (H$_2$ partial pressure 30–45 bar) at 40–45 °C for 20 h exhibited only 13% ee [101a]. However, addition of the perfluorinated alcohol C$_6$F$_{13}$CH$_2$CH$_2$OH improved the enantioselectivity to 73% ee. By changing the BF$_4$ (the counterion of the catalyst) to the BARF anion, enantioselectivity was increased to 72% ee without any addition of perfluorinated alcohol (Scheme 7.37). More recently, asymmetric hydrogenation reactions in scCO$_2$/water biphasic system using the Rh complex with perfluorinated chiral phosphine/phosphite ligand *(R,S)*-BINAPHOS **53** as a catalyst have also been investigated [101b]. Full conversion and an ee-value of 97% were achieved for the hydrogenation of methyl 2-acetamidoacrylate using 0.5 mol% catalyst in a scCO$_2$/H$_2$O biphasic system under an H$_2$ partial pressure of 30 bar at 56 °C. The protected amino acid was found to be partitioned almost exclusively into the aqueous phase, and the product was isolated in quantitative yield directly from the water layer. The catalyst recovered from scCO$_2$ could be recycled five times

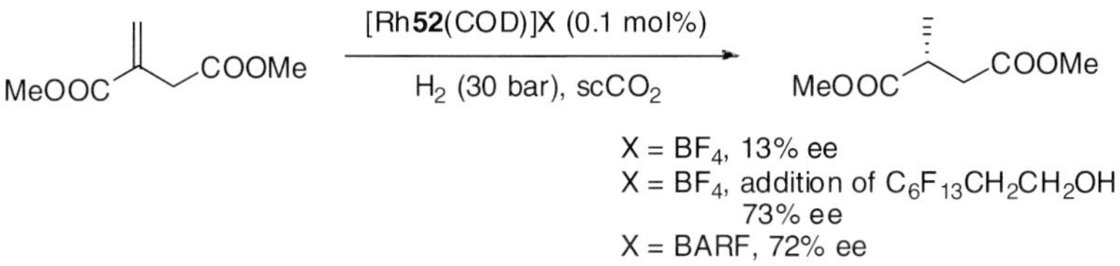

7.3 Enantioselective Catalysis in Supercritical Carbon Dioxide (scCO2) | 277

Scheme 7.37

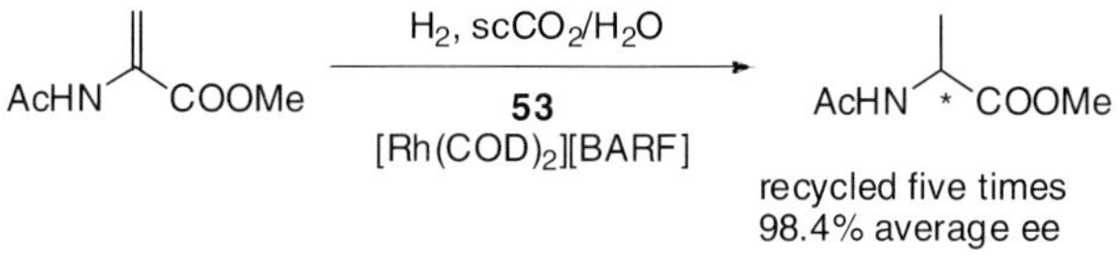

MeOOC-CH=CH-COOMe →[Rh**52**(COD)]X (0.1 mol%), H$_2$ (30 bar), scCO$_2$]→ MeOOC-CH(*)-CH$_2$-COOMe

X = BF$_4$, 13% ee
X = BF$_4$, addition of C$_6$F$_{13}$CH$_2$CH$_2$OH 73% ee
X = BARF, 72% ee

Scheme 7.38

AcHN-C(=CH$_2$)-COOMe →[H$_2$, scCO$_2$/H$_2$O, **53** [Rh(COD)$_2$][BARF]]→ AcHN-CH(*)-COOMe

recycled five times
98.4% average ee

54: Ar = p-CF$_3$OC$_6$H$_4$, BINAP with PAr$_2$/PAr$_2$

55: BINAP with C$_6$F$_{13}$ substituents on 6,6'-positions, PPh$_2$/PPh$_2$

56: BINAP with C$_6$F$_{13}$ substituents on 5,5'-positions, PPh$_2$/PPh$_2$

57: BINAP with C$_6$F$_{13}$H$_2$CH$_2$C- substituents, PPh$_2$/PPh$_2$

58: BINAP with C$_6$F$_{13}$ substituents, phosphite O-P-R
- 58a, R = Ph
- 58b, R = NMe$_2$
- 58c, R = OPh
- 58d, R = OC$_6$H$_4$-p-C$_6$F$_{13}$

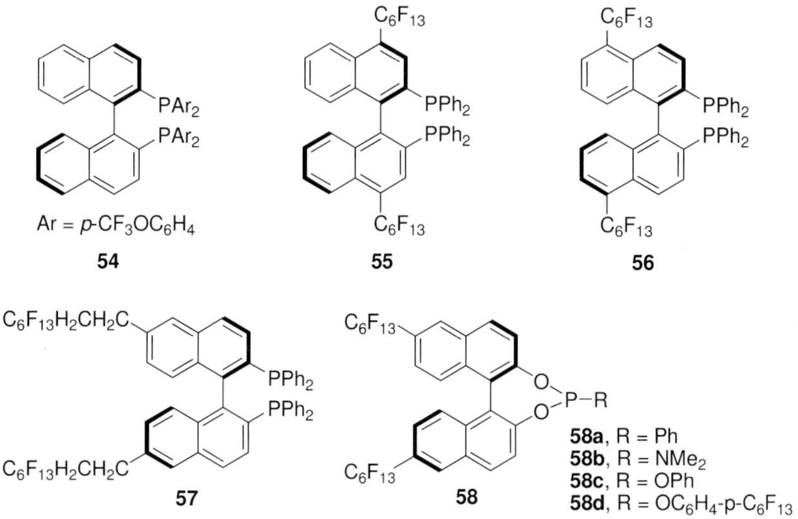

Figure 7.15 Fluoroalkylated BINAP analogs for Ru-catalyzed asymmetric hydrogenations in scCO$_2$.

with an average ee-value of 98.4% and without any significant loss of catalytic efficiency (Scheme 7.38). The leaching of rhodium metal and the chiral ligand into the aqueous layer was 1.4 and 5.2 ppm, respectively, during the first cycle, but for all of the following cycles the contamination of the aqueous phase with the metal and the phosphorus species was below the limits of detection.

Erkey and Dong developed the fluorinated BINAP analogue **54** (Figure 7.15) to confer solubility to their Ru-complexes in dense scCO$_2$ [102]. However, the enantioselectivity of the asymmetric hydrogenation of tiglic acid in scCO$_2$ using the **54**-Ru-complex as a catalyst was only modest (25% ee), and much lower than that obtained in neat MeOH (82% ee). In scCO$_2$ containing MeOH, the

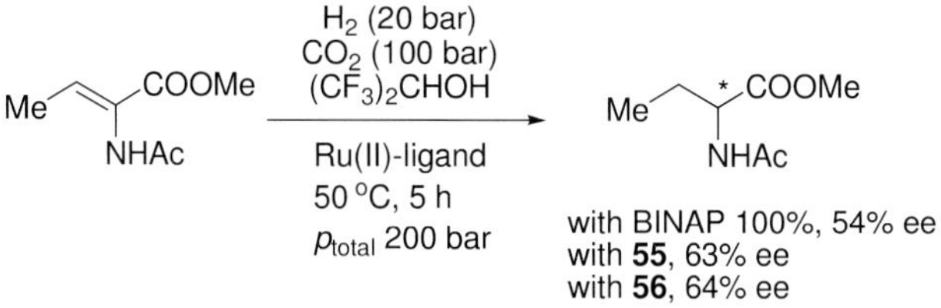

Scheme 7.39

Me–CH=C(COOMe)(NHAc)
$\xrightarrow[\text{Ru(II)-ligand, 50 °C, 5 h, } P_{total}\ 200\ \text{bar}]{\text{H}_2\ (20\ \text{bar}),\ \text{CO}_2\ (100\ \text{bar}),\ (\text{CF}_3)_2\text{CHOH}}$
Me–CH_2–*CH(COOMe)(NHAc)

with BINAP 100%, 54% ee
with **55**, 63% ee
with **56**, 64% ee

Table 7.16 Catalytic asymmetric hydrogenation of dimethyl itaconate using chiral perfluoronated phosphorus ligands in $scCO_2$.

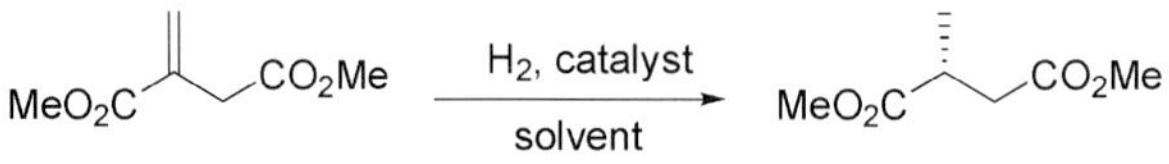

$MeO_2C-C(=CH_2)-CH_2-CO_2Me \xrightarrow[\text{solvent}]{\text{H}_2,\ \text{catalyst}} MeO_2C-\overset{*}{C}H(CH_3)-CH_2-CO_2Me$

Entry	Catalyst	Solvent	Additive	Conv. (%)	ee (%)
1	Ru-57	MeOH	—	100	96
2	Ru-57	$scCO_2$	—	100	74
3	Ru-57	$scCO_2$/MeOH	—	100	96
4	Rh-(58a)$_2$	$scCO_2$	—	9	8
5	Rh-(58b)$_2$	$scCO_2$	—	11	10
6	Rh-(58b)$_2$	$scCO_2$	NaBARF	84	31
7	Rh-(58b)$_2$	CH_2Cl_2	—	100	>99
8	Rh-(58c)$_2$	$scCO_2$	—	8	7
9	Rh-(58d)$_2$	$scCO_2$	—	21	34
10	Rh-(58d)$_2$	$scCO_2$	NaBARF	28	65
11	Rh-(58d)$_2$	CH_2Cl_2	—	14	Racemic

enantioselectivity was increased to 54% ee. Lemaire and coworkers reported BINAP analogues **55** and **56** containing perfluoroalkyl groups in the backbone, and their applications in the Ru-catalyzed asymmetric hydrogenation of methyl-2-acetamidoacrylate in $scCO_2$ [103]. However, when the reaction was conducted in $scCO_2$ without any cosolvents it did not proceed at all. By incorporating 1,1,1,3,3,3-hexafluoro-2-propanol as the cosolvent, complete conversions were obtained with good enantioselectivities in all cases (Scheme 7.39). The perfluoro BINAP **55** and **56** gave an ee-value for the product which was slightly higher than that of BINAP, and this was most likely due to the presence of perfluoroalkyl groups increasing the solubility of the catalyst and thus making it more effective. Various chiral fluoroalkylated BINAP ligands **57** and **58** have also been applied for the catalytic asymmetric hydrogenation of dimethyl itaconate in $scCO_2$ [104]. As shown in Table 7.16, with Ru-**57** complex as a catalyst, the enantioselectivity in $scCO_2$ was lower than that obtained in MeOH (entries 1 and 2), but with the addition of MeOH to the $scCO_2$ medium the ee-value was increased to 96% (entry 3) [104a]. Similar results were observed in the Rh-catalyzed asymmetric hydrogenation of dimethyl itaconate using monophosphorus ligand **58** [104b]. The catalytic activities and enantioselectivities in scCO$_2$ were remarkably lower than those obtained in CH$_2$Cl$_2$ (entries 4–8). When using ligand **58d** in the presence of NaBARF as an additive, a higher catalytic efficiency was observed (entries 8–11).

Most recently, Poliakoff and coworkers reported a continuous catalytic asymmetric hydrogenation reaction in scCO$_2$ using chiral catalysts supported on the alumina [105]. The Rh-skewphos [skewphos = (S,S)-2,4-bis(diphenylphosphino)pentane] complex was immobilized on γ-alumina using a phosphotungstic acid linker, H$_3$O$_{40}$PW$_{12}$. In the hydrogenation of dimethyl itaconate, the transport of both the substrate and product were shown to be affected by flowing scCO$_2$ (60–120 bar). Neither rhodium nor tungsten (<1 ppm) was detected in the product when separated by decompression of the scCO$_2$, although catalyst leaching (Rh 7 ppm, W 1 ppm) did occur at a higher temperature (100 °C) [105a]. Various immobilized Rh-chiral phosphine complexes have also been examined in the continuous asymmetric hydrogenation of dimethyl itaconate in scCO$_2$ medium, but only moderate enantioselectivities were achieved [105b].

7.3.2
Asymmetric Hydroformylation

Asymmetric hydroformylation of alkenes is one of the most important catalytic processes in the production of chiral aldehydes. Currently, there is a growing market for special chemicals produced via the asymmetric hydroformylation of alkenes, with applications of the resultant aldehydes in the production of pharmaceuticals, fragrances and pesticides [106]. In particular, the branched aldehydes obtained via the hydroformylation of styrenes can be converted to a variety of nonsteroidal anti-inflammatory agents. The chiral phosphine/phosphate ligand (R,S)-BINAPHOS allows rhodium-catalyzed asymmetric hydroformylation of vinyl arenes with outstanding levels of enantiocontrol [107]. The first study of asymmetric hydroformylation in an scCO$_2$ medium was reported by Leitner and Kainz in 1998 [108]. The asymmetric hydroformylation of styrene using an Rh–BINAPHOS complex as catalyst proceeded smoothly in compressed CO$_2$ to show an appreciable asymmetric induction (66% ee). A breakthrough in the asymmetric hydroformylation of vinylarene substrates in scCO$_2$ has been achieved by using CO$_2$-philic perfluoroalkylated BINAPHOS **53** as a ligand, with high conversions (>90%) and excellent enantioselectivities (generally >90% ee) [109]. In some cases, higher enantioselectivity has been achieved in scCO$_2$ than in organic solvents; for example, the hydroformylation of vinyl acetate in scCO$_2$ showed 95% ee, but 90% ee in benzene. The catalyst was readily separated from the reactants and products by scCO$_2$ extraction, and the recovered catalysts were reused eight times without any significant loss of catalytic activity (Table 7.17). These reduced ee-values may not be due to the decreased catalytic activity, but rather to the partial racemization of the product over the longer reaction time. The 70% ee of run 7 (see Table 7.17) could be restored to 88.2% (run 8) by the addition of one more equivalent to rhodium of the ligand. Based on this observation, it could be inferred that although

Table 7.17 Rhodium-catalyzed asymmetric hydroformylation of styrene in $scCO_2$.

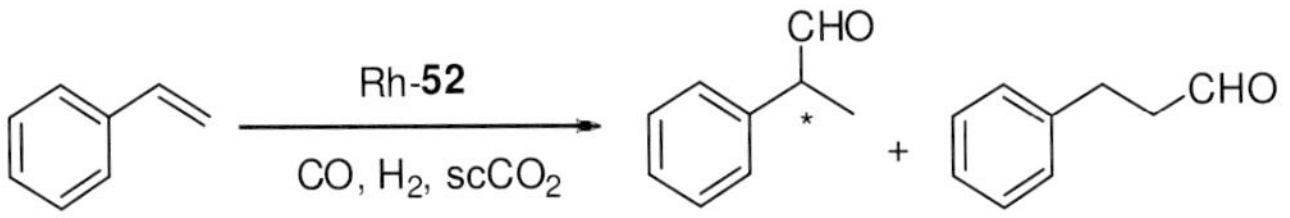

Run	S/Rh3	T (°C)	PH2/CO (bar)	t (h)	CO2 for extraction			Conv. (%)	ee (%)	Rh leaching (ppm)
					T (°C)	P	V (l)			
1	1000	55	19	17	63	94	200	83	87	0.97
2	1000	55	34	15	60	100	170	97	90	0.78
3	2000	50	40	65	61	105	270	>99	90	0.45
4	1500	54	40	15	61	103	200	92	82	0.36
5	1400	55	38	15	59	107	128	96	78	1.94
6	3000	40	61	114	60	115	213	99	66	0.95
7	1500	50	50	16	60	104	173	95	70	0.41
8	1000	—	—	41	—	—	—	89	88	—

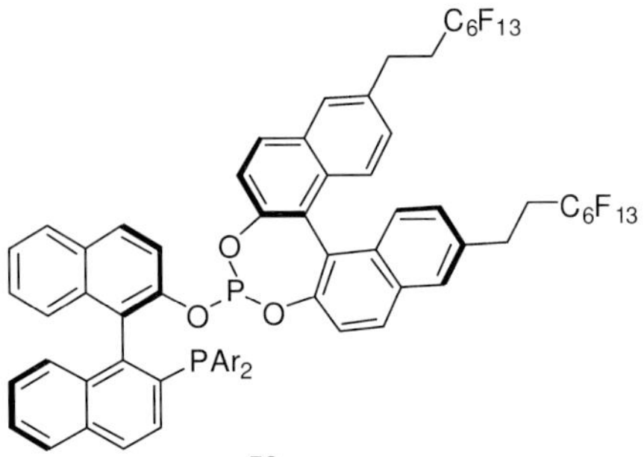

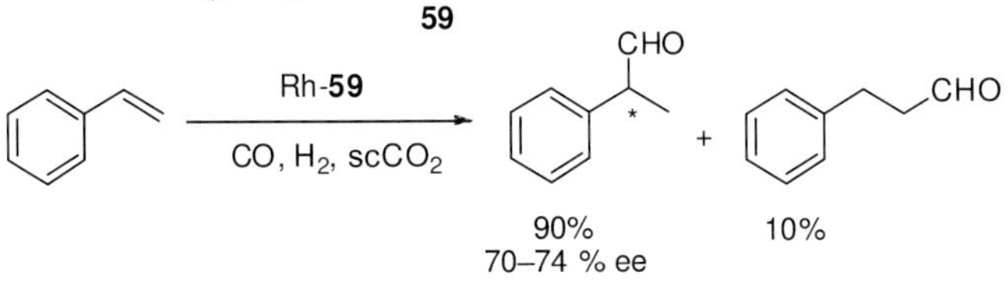

Scheme 7.40

the leaching of Rh metal is not significant (<2 ppm in each run), an amount of ligand may have been leached into the $scCO_2$ phase during the extraction [109].

Ojima and coworkers developed $scCO_2$-soluble perfluoronated BINAPHOS **59** (Scheme 7.40) [110], but their Rh-complexes gave lower ee-values for the asymmetric hydroformylation of styrene than did the Rh-**53** complex described above (Figure 7.14). The catalytic performance of the Rh-**59** complex in $scCO_2$ was inferior to that in benzene; this poorer performance may be attributed to the rapid racemization of the product aldehyde under the reaction conditions.

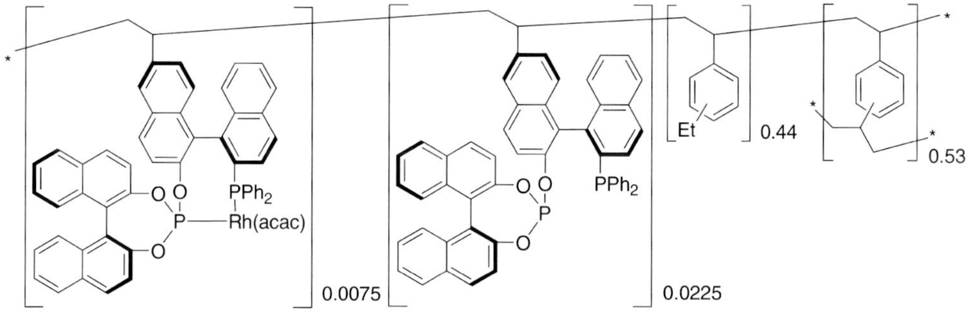

Figure 7.16 Cross-linked polystyrene-supported (R,S)-BINAPHOS-Rh(I) catalyst.

Table 7.18 Sequential asymmetric hydroformylation of alkenes catalyzed by crosslinked polystyrene-supported Rh–BINAPHOS **60** in pulsed flowing scCO$_2$.

Cycle	Alkene	Conv. (%)	Ratio b/l	ee (%)
1	Styrene	49	82:18	77
2	Vinyl acetate	5	70:30	74
3	1-Octene	47	21:79	73
4	1-Hexene	40	21:79	60
5	Styrene	36	81:19	82
6	2,3,4,5,6-Pentafluorostyrene	27	89:11	88
7	CF$_3$(CF$_2$)$_5$CH=CH$_2$	21	91:9	78
8	Styrene	54	80:20	80

b/l = branched/linear.

Nozaki and coworkers reported a highly crosslinked polystyrene-supported (R,S)-BINAPHOS-Rh(I) catalyst **60** for the asymmetric hydroformylation of olefins under solvent-free conditions (Figure 7.16) [111]. Gaseous alkenes could be hydroformylated in a simple flow system, but less-volatile substrates such as vinyl acetate and styrene were hydroformylated in flowing scCO$_2$. Following incubation of the catalyst under CO/H$_2$ in scCO$_2$ for 15 min, styrene was injected and allowed to react for 30 min. The products were then swept from the reactor using a higher pressure of CO$_2$ (120 bar) for 15 min. This process was repeated up to seven times, whereupon conversion, selectivities to the branched isomer (80%) and ee-values (80–85%) were all reasonably constant, with the conversion being increased (90%) at 120 bar total pressure compared to 80 bar (85%). The scCO$_2$ flow system can be used sequentially for the asymmetric hydroformylation of a variety of substrates, with good results in most cases (Table 7.18). The injection of styrene in runs 1, 5 and 8 clearly showed that the catalytic activity had not deteriorated with continuous use.

7.3.3
Asymmetric Carbon–Carbon Bond Formation

Fukuzawa and coworkers reported the rare earth (III) salt-catalyzed asymmetric Diels–Alder reaction of cyclopentadiene with a chiral dienophile in scCO$_2$ [112]. The Diels–Alder reaction in scCO$_2$ proceeded rapidly to give the corresponding adducts with higher diastereoselectivities than those in CH$_2$Cl$_2$ (Scheme 7.41).

The reaction of cyclopentadiene with 3-crotonoyl-2-oxazolidinone in scCO$_2$ (40 °C, 10 MPa) catalyzed by the Sc(III) complex of chiral pybox gave the Diels–Alder adduct in 71% yield with 83% ee (Scheme 7.42), which was comparable to the result obtained in CH$_2$Cl$_2$ [112c]. The reaction proceeded more rapidly in scCO$_2$, being completed in a shorter time period (0.5 h) than in CH$_2$Cl$_2$ (18 h).

An asymmetric Mukaiyama aldol reaction in supercritical fluids using a binaphthol-based titanium complex has also been reported (Scheme 7.43) [113].

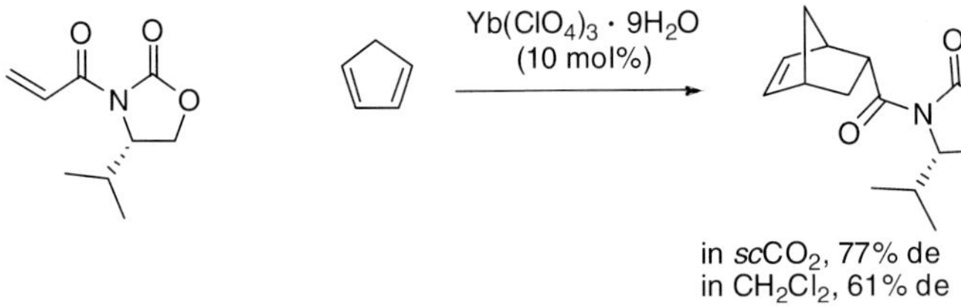

Scheme 7.41

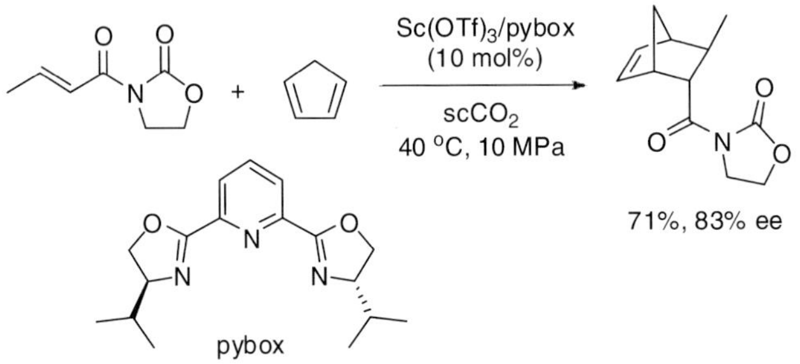

Scheme 7.42

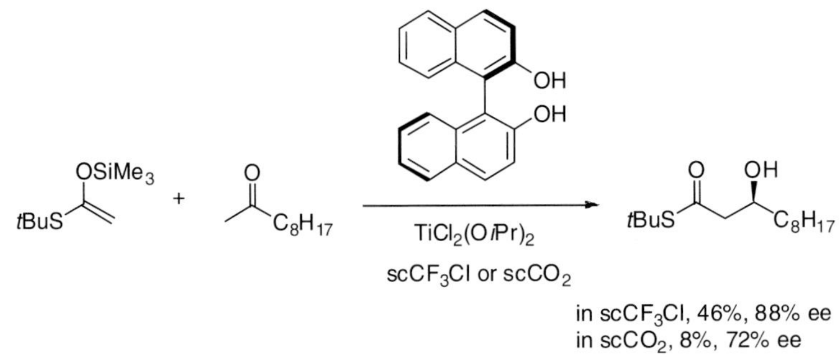

Scheme 7.43

Under the high dilution conditions (0.2 mM catalyst, 4.0 mM aldehyde), the reaction did not occur in conventional solvents (toluene or CH_2Cl_2), but proceeded to give the product in the supercritical solvents. Although the conversions were modest (20–46% in $scCF_3Cl$, 8% in $scCO_2$), the enantioselectivities were as high as 88% ee in $scCF_3Cl$ (72% ee in $scCO_2$).

7.4
Enantioselective Catalysis in the Combined Use of Ionic Liquids and Supercritical CO_2

As described in Section 7.3, the problems associated with the mutual solubility of ionic liquids (including chiral catalysts) in reaction products (or extraction solvents) and the solubility of products (or extraction solvent) in ionic liquids, and contamination of the extraction phase with ionic liquids to cause catalyst losses, may be overcome by the use of $scCO_2$. Brennecke and coworkers [114] showed that it was possible to extract a solute from an ionic liquid using $scCO_2$, without any contamination of the ionic liquid, and therefore the extraction of solutes from ionic liquids using $scCO_2$ could be combined in reaction operations. In contrast, as mentioned above, ionic liquids and $scCO_2$ can each dramatically affect reaction rate and selectivity, and so a different reaction performance could be expected in the combined reaction medium of ionic liquid and $scCO_2$.

Jessop and coworkers investigated the asymmetric hydrogenation of tiglic acid using Ru–tolBINAP as a catalyst in wet [bmim][PF$_6$] [115, 116]. Extraction of the product with $scCO_2$ from the ionic liquid containing the catalysts provided the extremely pure product from the CO_2 effluent, in which neither the ionic liquid nor catalyst was contaminated at all. In this way a conversion of up to 99% and an ee-value of 90% were obtained. The recovered ionic liquid catalytic solution was reused up to four times without any reduction of the conversion and enantioselectivity (Scheme 7.44).

Most recently, Leitner and coworkers studied the asymmetric hydrogenation of imines using cationic iridium complexes of chiral phosphinooxazoline ligands, for example, complex **61**, in an ionic liquid and $scCO_2$ medium [117]. As shown in Table 7.19, in the absence of CO_2, the conversion was only marginal under standard conditions (P_{H2} = 30 bar, T = 40 °C), and an acceptable conversion required 100 bar of hydrogen pressure (entries 1 and 2). Fortunately, the addition of CO_2 improved the conversion at lower hydrogen pressure, and thus quantitative formation of the hydrogenation product was observed at a hydrogen partial pressure of 30 bar (entries 3). A similar beneficial effect of added CO_2, albeit less dramatic, was also observed with other ionic liquids (entries 4–6). These remarkable effects were observed on the enantioselectivity of the reaction upon variation of the anion in the ionic liquid, with the ee-values varying from 30% with BF_4^- to 78% with $BARF^-$. The catalyst recycling in the ionic liquid/CO_2 system was also examined in a solution of [bmim][PF$_6$] containing the catalyst **61** in the presence of $scCO_2$ (d = 0.68 g ml^{-1}) under 40 bar of hydrogen partial pressure at 40 °C. $scCO_2$ was shown to be effective for the quantitative extraction of hydrogenated product from the

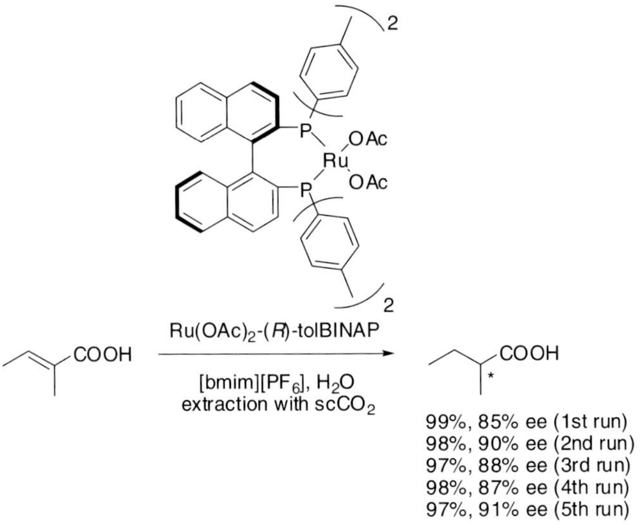

Scheme 7.44

Reaction: CH₂=C(Me)COOH → CH₃-CH(Me)-COOH (with *) via Ru(OAc)₂-(R)-tolBINAP, [bmim][PF₆], H₂O, extraction with scCO₂

99%, 85% ee (1st run)
98%, 90% ee (2nd run)
97%, 88% ee (3rd run)
98%, 87% ee (4th run)
97%, 91% ee (5th run)

Table 7.19 Asymmetric hydrogenation of N-(1-phenylethylidene)aniline in ionic liquid/CO_2 systems.

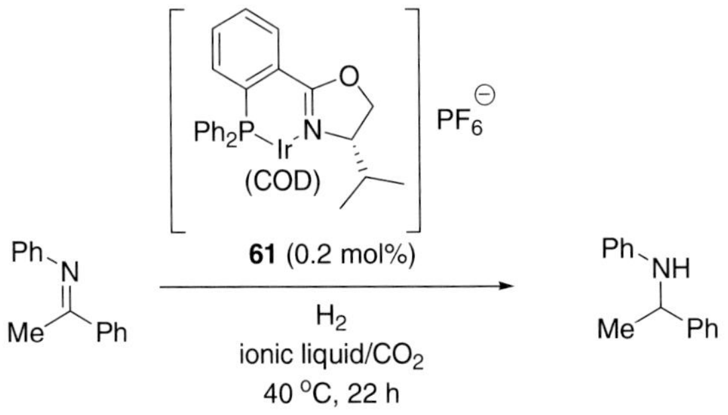

Ph-N=C(Me)Ph → Ph-NH-CH(Me)Ph via **61** (0.2 mol%), H₂, ionic liquid/CO_2, 40 °C, 22 h

Entry	Ionic liquid	P_{H2} (bar)	CO_2 (g)	Conv. (%)	ee (%)
1	[emim][NTf₂]	30	—	3	—
2	[emim][NTf₂]	100	—	97	58
3	[emim][NTf₂]	30	8.9	>99	56
4	[pmim][PF₆]	30	8.0	>99	65
5	[bmim][BF₄]	30	7.6	92	30
6	[emim][BARF]	30	8.9	>99	78

Table 7.20 Mass of extracted product, yield, ee-values and osmium content for recycling experiments of osmium catalyst for asymmetric dihydroxylation.

$$\text{Ph}\diagup\!\!\!\diagdown\text{COOMe} \xrightarrow[\text{extraction with scCO}_2]{\substack{\text{K}_2\text{OsO}_2(\text{OH})_4 \text{ (0.5 mol\%)} \\ (\text{DHQD})_2\text{PYR (1 mol\%)} \\ \text{NMO, [omim][PF}_6]}} \text{Ph}\diagup\!\!\!\diagdown\text{COOMe}\ (\text{OH, OH})$$

Run	Extract (mg)	Yield (%)	ee (%)	Os content[a]
1	198	91	77	<0.05
2	188	87	83	<0.05
3	193	89	85	<0.05
4	191	88	84	<0.03
5	196	90	80	<0.03
6	185	85	84	<0.03

a Percentage of osmium relative to initial amount detected by ICP in the combined organic phase (detection limit 0.05%).

catalyst solution. Moreover, the catalyst/ionic liquid solution showed constant activity and enantioselectivity under these conditions for at least seven recycling runs.

A combination of an ionic liquid with scCO$_2$ was applied in osmium-catalyzed asymmetric dihydroxylation [118]. The asymmetric dihydroxylation of methyl *trans*-cinnamate in [omim][PF$_6$] was carried out using K$_2$OsO$_2$(OH)$_4$ (0.5 mol%) and (DHQD)$_2$PYR (1 mol%) in the presence of 1.3 equiv of NMO, followed by scCO$_2$ extraction under 125 bar at 40 °C. As shown in Table 7.20, most of the product could be extracted when 60 ml of CO$_2$ at 125 bar was passed through the reaction mixture at 40 °C, without any significant contamination with osmium.

7.5
Summary and Outlook

Clearly, catalyst and solvent recycling are highly desirable from both economic and environmental points of view, and the results of the studies described in this chapter have shown that ionic liquids can be used successfully as alternative solvents to immobilize not only metal-based chiral catalysts but also chiral organocatalysts. The application of ionic liquids in a variety of asymmetric catalyses allows the facile recycling of chiral catalysts, whilst it would also appear that in many cases ionic liquids can cause a stabilization of the catalysts and prevent their decomposition/deactivation. As a result, increased turnover numbers can be achieved. Although, in some cases much higher enantioselectivities are obtained when the asymmetric catalyses are conducted in ionic liquids, the latter approach does not represent a 'magic solution' to the problems associated with catalyst leaching. Many chiral catalysts may leach into the organic solvent, leading to decreased catalytic activity when recycling the recovered catalyst/ionic liquids. This effect may be due largely due to a mutual solubility between the ionic liquid and organic solvent used for product extraction. Although ionic liquid-tagging strategies have provided new ways to solve this problem, multi-step synthetic efforts are often required. Many organic substrates and products are significantly soluble in $scCO_2$, whereas the majority of metal-based chiral catalysts are very poorly soluble. However, whilst the insolubility of metallic chiral catalysts in $scCO_2$ has led to a decrease in catalytic efficiency, this problem may be overcome by the introduction of perfluoroalkyl groups into the chiral ligand, thus providing $scCO_2$-soluble chiral catalysts. Although the combination of ionic liquids with $scCO_2$ extractions or SILPs shows great promise, further investigations along these lines are imperative. The constant accumulation of information relating to factors which affect not only catalytic efficiency but also catalyst recycling will raise the prospects of their industrial development.

References

1 (a) Dupont, J., de Souza, R.F. and Suarez, P.A.Z. (2002) *Chemical Reviews*, **102**, 3667.
(b) Wasserscheid, P. and Welton, T. (2003) *Ionic Liquids in Synthesis*, Wiley-VCH Verlag GmbH, Weinheim, Germany.
(c) Dyson, P.J. and Geldbach, T.J. (2005) *Metal-Catalysed Reactions in Ionic Liquids*, Springer, The Netherlands.

2 Chauvin, Y., Mussmann, L. and Oliver, H. (1995) *Angewandte Chemie – International Edition*, **34**, 2698.

3 (a) Lee, S.-g. (2006) *Chemical Communications*, 1049.
(b) Song, C.E. (2004) *Chemical Communications*, 1033.
(c) Baudequin, C., Brégeon, D., Levillain, J., Guillen, F., Plaquevent, J.-C. and Gaumont, A.-C. (2005) *Tetrahedron: Asymmetry*, **16**, 3921.
(d) Baudequin, C., Baudoux, J., Levillain, J., Cahard, D., Gaumont, A.-C. and Plaquevent, J.-C. (2003) *Tetrahedron: Asymmetry*, **14**, 3081.
(e) Fan, Q.-H., Li, Y.-M. and Chan, A.S.C. (2002) *Chemical Reviews*, **102**, 3385.

4 Dzyuba, S.V. and Bartsch, R.A. (2003) *Angewandte Chemie – International Edition*, **42**, 148.

5 Burk, M.J., Feng, S., Gross, M.F. and Tumas, W. (1995) *Journal of the American Chemical Society*, **117**, 8277.

6 (a) Jessop, P.G., Ikariya, T. and Noyori, R. (1999) *Chemical Reviews*, **99**, 475.
(b) Leitner, W. (2002) *Accounts of Chemical Research*, **35**, 746.
(c) Jessop, P.G. (2006) *The Journal of Supercritical Fluids*, **38**, 211.
(d) Cole-Hamilton, D.J. (2006) *Advanced Synthesis Catalysis*, **348**, 1341.
(e) Rayner, C.M. (2007) *Organic Process Research & Development*, **11**, 121.

7 (a) Jacobsen, E.N., Pfaltz, A. and Yamamoto, H. (1999) *Comprehensive Asymmetric Catalysis*, Vol. 1, Springer, Berlin, pp. 121–318.
(b) Ohkuma, T., Kitamura, M. and Noyori, R. (2000) Asymmetric hydrogenation, in *Catalytic Asymmetric Synthesis* (ed. I. Ojima), Wiley-VCH Verlag GmbH, New York, p. 1.
(c) Lin, G., Li, Y. and Chan, A.S.C. (2001) *Principles and Applications of Asymmetric Synthesis*, John Wiley & Sons, Ltd, New York.
(d) Blaser, H.-U., Malan, C., Pugin, B., Spindler, F., Steiner, H. and Studer, M. (2003) *Advanced Synthesis Catalysis*, **345**, 103.

(e) Tang, W. and Zhang, X. (2003) *Chemical Reviews*, **103**, 3029.

8 Berger, A., de Souza, R.F., Delgado, M.R. and Dupont, J. (2001) *Tetrahedron: Asymmetry*, **12**, 1825.

9 Guernik, S., Wolfson, A., Herskowitz, M., Greenspoon, N. and Geresh, S. (2001) *Chemical Communications*, 2314.

10 Wolfson, A., Vankelecom, I.F.J. and Jacobs, P.A. (2005) *Journal of Organometallic Chemistry*, **690**, 3558.

11 Fráter, T., Gubicza, L., Szöllosy, Á. and Bakos, J. (2006) *Inorganica Chimica Acta*, **359**, 2756.

12 Pugin, B., Studer, M., Kuesters, E., Sedelmeier, G. and Feng, X. (2004) *Advanced Synthesis Catalysis*, **346**, 1481.

13 Lee, S.-g., Zhang, Y.J., Piao, J.Y., Yoon, H., Song, C.E., Choi, J.H. and Hong, J. (2003) *Chemical Communications*, 2624.

14 (a) Lee, S.-g., Zhang, Y.J., Song, C.E., Lee, J.K. and Choi, J.H. (2002) *Angewandte Chemie – International Edition*, **41**, 847.
(b) Lee, S.-g. and Zhang, Y.J. (2002) *Organic Letters*, **4**, 2429.
(c) Zhang, Y.J., Kim, K.Y., Park, J.H., Song, C.E., Lee, K., Lah, M.S. and Lee, S.-g. (2005) *Advanced Synthesis Catalysis*, **347**, 563.

15 Wong, H.-T., See-Toh, Y.H., Ferreira, F.C., Crook, R. and Livingston, A.G. (2006) *Chemical Communications*, 2063.

16 Monterio, A.L., Zinn, F.K., de Souza, R.F. and Dupont, J. (1997) *Tetrahedron: Asymmetry*, **8**, 177.

17 (a) Hu, A., Ngo, H.L. and Lin, W. (2004) *Angewandte Chemie – International Edition*, **43**, 2501.
(b) Ngo, H.L. Hu, A. and Lin, W. (2003) *Chemical Communications*, 1912.

18 Berthod, M., Joerger, J.-M., Mignani, G., Vaultier, M. and Lemaire, M. (2004) *Tetrahedron: Asymmetry*, **15**, 2219.

19 Lam, K.H., Xu, L., Feng, L., Ruan, J., Fan, Q. and Chan, A.S.C. (2005) *Canadian Journal of Chemistry – Revue Canadienne de Chimie*, **83**, 903.

20 Ngo, H.L. Hu, A. and Lin, W. (2005) *Tetrahedron Letters*, **46**, 595.

21 Xiong, W., Lin, Q., Ma, H., Zheng, H., Chen, H. and Li, X. (2005) *Tetrahedron: Asymmetry*, **16**, 1959.

22 Zhu, Y., Carpenter, K., Bun, C.C., Bahnmueller, S., Ke, C.P., Srid, V.S., Kee, L.W. and Hawthorne, M.F. (2003) *Angewandte Chemie – International Edition*, **42**, 3792.

23 (a) Geldbach, T.J. and Dyson, P.J. (2004) *Journal of the American Chemical Society*, **126**, 8114.
(b) Geldbach, T.J., Brown, M.R.H., Scopelliti, R. and Dyson, P.J. (2005) *Journal of Organometallic Chemistry*, **690**, 5055.

24 (a) Noyori, R. and Hashiguchi, S. (1997) *Accounts of Chemical Research*, **30**, 97.
(b) Palmer, M.J. and Wills, M. (1999) *Tetrahedron: Asymmetry*, **10**, 2045.
(c) Everaere, K., Mortreux, A. and Carpentier, J.-F. (2003) *Advanced Synthesis Catalysis*, **345**, 67.

25 Joerger, J.-M., Paris, J.-M. and Vaultier, M. (2006) *ARKIVOC*, **iv**, 152.

26 Kawasaki, I., Tsunoda, K., Tsuji, T., Yamaguchi, T., Shibuta, H., Uchida, N., Yamashita, M. and Ohta, S. (2005) *Chemical Communications*, 2134.

27 (a) Kolb, H.C., VanNieuwenhze, M.S. and Sharpless, K.B. (1994) *Chemical Reviews*, **94**, 2483.
(b) Markó, I.E. and Svendsen, J. (1999) Dihydroxylation of carbon-carbon double bond, in *Comprehensive Asymmetric Catalysis*, vol. **2** (eds E.N. Jacobsen, A. Pfaltz and H. Yamamoto), Springer Verlag, Berlin, p. 713.
(c) Johnson, R.A., Sharpless, K.B. (2000) Catalytic asymmetric dihydroxylation-discovery and development, in *Catalytic Asymmetric Synthesis* (ed. I. Ojima), Wiley-VCH Verlag GmbH, New York, p. 357.

28 (a) Salvadori, P., Pini, D. and Petri, A. (1999) *Synlett*, 1181.
(b) Bolm, C. and Gerlach, A. (1998) *European Journal of Organic Chemistry*, **21**, 21.

29 (a) Nagayama, S., Endo, M. and Kobayashi, S. (1998) *Journal of Organic Chemistry*, **63**, 6094.
(b) Kobayashi, S., Endo, M. and Nagayama, S. (1999) *Journal of the American Chemical Society*, **121**, 11229.
(c) Kobayashi, S., Ishida, T. and Akiyama, R. (2001) *Organic Letters*, **3**, 2649.

30 (a) Choudary, B.M., Chowdari, N.S., Kantam, M.L. and Raghavan, K.V. (2001) *Journal of the American Chemical Society*, **123**, 9220.
(b) Choudary, B.M., Chowdari, N.S., Jyothi, K. and Kantam, M.L. (2002) *Journal of the American Chemical Society*, **124**, 5341.

31 Yang, J.W., Han, H., Roh, E.J., Lee, S.-g. and Song, C.E. (2002) *Organic Letters*, **4**, 4685.

32 Song, C.E., Jung, D.-u., Roh, E.J., Lee, S.-g. and Chi, D.Y. (2002) *Chemical Communications*, 3038.

33 Branco, L.C. and Afonso, C.A.M. (2002) *Chemical Communications*, 3036.

34 Van Rheenen, V., Kelly, R.C. and Cha, P.Y. (1976) *Tetrahedron Letters*, **23**, 1973.

35 Song, C.E., Yang, J.W., Ha, H.J. and Lee, S.-g. (1996) *Tetrahedron: Asymmetry*, **7**, 645.

36 Branco, L.C. and Afonso, C.A.M. (2004) *Journal of Organic Chemistry*, **69**, 4381.

37 Johansson, M., Lindén, A.A. and Bäckvall, J.-E. (2005) *Journal of Organometallic Chemistry*, **690**, 3614.

38 Branco, L.C., Gois, P.M.P., Lourenco, N.M.T., Kurteva, V.B. and Afonso, C.A.M. (2006) *Chemical Communications*, 2371.

39 Jacobsen, E.N., Wu, M.H. (1999) Epoxidation of alkenes other than allylic alcohols, in *Comprehensive Asymmetric Catalysis*, Vol. 2 (eds E. N. Jacobsen, A. Pfaltz and H. Yamamoto), Springer Verlag, Berlin, p. 649.

40 Song, C.E. and Roh, E.J. (2000) *Chemical Communications*, 837.

41 Gaillon, L. and Bedioui, F. (2001) *Chemical Communications*, 1458.

42 Lou, L.-L., Yu, K., Ding, F., Zhou, W., Peng, X. and Liu, S. (2006) *Tetrahedron Letters*, **47**, 6513.

43 Meracz, I. and Oh, T. (2003) *Tetrahedron Letters*, **44**, 6465.

44 Doherty, S., Goodrich, P., Hardacre, C., Luo, H.-K., Rooney, D.W., Seddon, K.R. and Styring, P. (2004) *Green Chemistry*, **6**, 63.

45 Takahashi, K., Nakano, H. and Fujita, R. (2007) *Chemical Communications*, 263.

46 Fu, F., Teo, Y.-C. and Loh, T.-P. (2006) *Organic Letters*, **8**, 5999.

47 (a) Jacobsen, E.N. (2000) *Accounts of Chemical Research*, **33**, 421.
(b) Katsuki, T. (2000) Asymmetric epoxidation of unfunctionalized olefins and related reactions, in *Catalytic Asymmetric Synthesis* (ed. I. Ojima), Wiley-VCH Verlag GmbH, p. 287.
(c) Jacobsen, E.N. and Wu, M.H. (1999) Ring opening of epoxides and related reactions, in *Comprehensive Asymmetric Catalysis*, vol. 3 (eds E.N. Jacobsen, A. Pfaltz and H. Yamamoto), Springer Verlag, Berlin, p. 1309.

48 Martínez, L.E., Leighton, J.L., Carsten, D.H. and Jacobsen, E.N. (1995) *Journal of the American Chemical Society*, **117**, 5897.

49 Song, C.E., Oh, C.R., Roh, E.J. and Choo, D.J. (2000) *Chemical Communications*, 1743.

50 Dioos, B.M.L. and Jacobs, P.A. (2006) *Journal of Catalysis*, **243**, 217.

51 Dioos, B.M.L. and Jacobs, P.A. (2005) *Applied Catalysis A: General*, **282**, 181.

52 Hansen, K.B., Leighton, J.L. and Jacobsen, E.N. (1996) *Journal of the American Chemical Society*, **118**, 10924.

53 (a) Tokunaga, M., Larrow, J.F., Kakiuchi, F. and Jacobsen, E.N. (1997) *Science*, **277**, 936. (b) Schaus, S.E., Brandes, B.D., Larrow, J.F., Togunaga, M., Hansen, K.B., Gould, A.E., Furrow, M.E. and Jacobsen, E.N. (2002) *Journal of the American Chemical Society*, **124**, 1307.

54 (a) Annis, D.A. and Jacobsen, E.N. (1999) *Journal of the American Chemical Society*, **121**, 4147.
(b) Kim, G.-J. and Park, D.-W. (2000) *Catalysis Today*, **63**, 537.

55 Oh, C.R., Choo, D.J., Shim, W.H., Lee, D.H., Roh, E.J., Lee, S.-g. and Song, C.E. (2003) *Chemical Communications*, 1100.

56 North, M. (1993) *Synlett*, 807.

57 Baleizão, C., Gigante, B., Garcia, H. and Corma, A. (2002) *Green Chemistry*, **4**, 272.

58 (a) Baleizão, C., Gigante, B., García, H. and Corma, A. (2004) *Tetrahedron*, **60**, 10461.
(b) Baleizão, C., Gigante, B., Garcia, H. and Corma, A. (2003) *Tetrahedron Letters*, **44**, 6813.

59 Trost, B.M., Lee, C.B. (2000) Asymmetric allylic alkylation reactions, in *Catalytic Asymmetric Synthesis*, 2nd edn (ed. I.

Ojima), Wiley-VCH Verlag GmbH, p. 593.
60 (a) Toma, Š., Gotov, B., Kmentová, I. and Solčániová, E. (2000) *Green Chemistry*, **2**, 149.
(b) Kmentová, I., Gotov, B., Solčániová, E. and Toma, Š. (2002) *Green Chemistry*, **4**, 103.
61 Lyubimov, S.E., Davankov, V.A. and Gavrilov, K.N. (2006) *Tetrahedron Letters*, **47**, 2721.
62 (a) Lu, J., Ji, S.-J. and Loh, T.-P. (2005) *Chemical Communications*, 2345.
(b) Lu, J., Ji, S.-J., Teo, Y.-C. and Loh, T.-P. (2005) *Tetrahedron Letters*, **46**, 7435.
63 (a) Fraile, J.M., García, J.I., Herrerías, C.I., Mayoral, J.A., Carrié, D. and Vaultier, M. (2001) *Tetrahedron: Asymmetry*, **12**, 1891.
(b) Fraile, J.M., García, J.I., Herrerías, C.I., Mayoral, J.A., Gmough, S. and Vaultier, M. (2004) *Green Chemistry*, **6**, 93.
(c) Fraile, J.M., García, J.I., Herrerías, C.I., Mayoral, J.A., Reiser, O. and Vaultier, M. (2004) *Tetrahedron Letters*, **45**, 6765.
64 Davies, D.L., Kandola, S.K. and Patel, R.K. (2004) *Tetrahedron: Asymmetry*, **15**, 77.
65 Castillo, M.R., Fousse, L., Fraile, J.M., García, J.I. and Mayoral, J.A. (2007) *Chemistry—A European Journal*, **13**, 287.
66 Evans, D.A., Faul, M.M., Bilodeau, M.T., anderson, B.A. and Barnes, D.M. (1993) *Journal of the American Chemical Society*, **115**, 5328.
67 (a) Nishikori, H., Ohta, C., Oberlin, E., Irie, R. and Katsuki, T. (1999) *Tetrahedron*, **55**, 13937.
(b) Ohta, C. and Katsuki, T. (2001) *Tetrahedron Letters*, **42**, 3885.
68 (a) Marakami, M., Uchida, T. and Katsuki, T. (2001) *Tetrahedron Letters*, **42**, 7071.
(b) Marakami, M., Uchida, T., Saito, B. and Katsuki, T. (2003) *Chirality*, **15**, 116.
69 Kantam, M.L., Kavita, B., Neeraja, V., Haritha, Y., Chaudhuri, M.K. and Dehury, S.K. (2005) *Advanced Synthesis Catalysis*, **347**, 641.
70 Kantam, M.L., Kavita, B., Neeraja, V., Haritha, Y., Chaudhuri, M.K. and Dehury, S.K. (2003) *Tetrahedron Letters*, **44**, 9029.
71 Malhotra, S.V. and Wang, Y. (2006) *Tetrahedron: Asymmetry*, **17**, 1032.
72 (a) Ramachandran, P.V. (2000) *Asymmetric Fluoroorganic Chemistry: Synthesis, Application and Future Directions*, ACS Symposium Series 746, American Chemical Society, Washington, DC.
(b) Soloshonok, V.A. *et al.* (1999) *Enantiocontrolled Synthesis of Fluoro-Organic Compounds*, John Wiley & Sons, Ltd, Chichester.
73 Hamashima, Y., Takano, H., Hotta, D. and Sodeoka, M. (2003) *Organic Letters*, **5**, 3225.
74 Kim, S.M., Kang, Y.K., Lee, K.S., Mang, J.Y. and Kim, D.Y. (2006) *Bulletin of the Korean Chemical Society*, **27**, 423.
75 (a) List, B. (2002) *Tetrahedron*, **58**, 5573.
(b) List, B. (2004) *Accounts of Chemical Research*, **37**, 548.
(c) Notz, W., Tanaka, F. and Barbas, C.F. III (2004) *Accounts of Chemical Research*, **37**, 580.
(d) Dalko, P.I. and Moisan, L. (2004) *Angewandte Chemie—International Edition*, **43**, 5138.
(e) Berkessel, A. and Groger, H. (2005) *Asymmetric Organocatalysis*, Wiley-VCH Verlag GmbH, Weinheim.
76 (a) Sakthivel, K., Notz, W., Bui, T. and Barbas, C.F. III (2001) *Journal of the American Chemical Society*, **123**, 5260.
(b) Benaglia, M., Cinquini, M., Cozzi, F., Puglisi, A. and Celentano, G. (2002) *Advanced Synthesis Catalysis*, **344**, 533.
77 (a) Trost, B.M. (1991) *Science*, **254**, 1471.
(b) Trost, B.M. (1995) *Angewandte Chemie—International Edition*, **34**, 259.
78 List, B., Lerner, R.A. and Barbas, C.F. III (2000) *Journal of the American Chemical Society*, **122**, 2395.
79 (a) Loh, T.-P., Feng, L.-C., Yang, H.-Y. and Yang, J.-Y. (2002) *Tetrahedron Letters*, **43**, 8741.
(b) Kotrusz, P., Kmentová, I., Gotov, B., Toma, Š. and Solčániová, E. (2002) *Chemical Communications*, 2510.
(c) Córdova, A. (2004) *Tetrahedron Letters*, **45**, 3949.
80 (a) Tang, Z., Jiang, F., Yu, L.-T., Cui, X., Gong, L.-Z., Mi, A.-Q., Jiang, Y.-Z. and

Wu, Y.-D. (2003) *Journal of the American Chemical Society*, **125**, 5263.
(b) Tang, Z., Jiang, F., Cui, X., Gong, L.-Z., Mi, A.-Q., Jiang, Y.-Z. and Wu, Y.-D. (2004) *Proceedings of the National Academy of Sciences of the United States of America*, **101**, 5755.

81 Guo, H.-M., Cun, L.-F., Gong, L.-Z., Mi, A.-Q. and Jiang, Y.-Z. (2005) *Chemical Communications*, 1450.

82 (a) Miao, W. and Chan, T.H. (2006) *Advanced Synthesis Catalysis*, **348**, 1711.
(b) Luo, S., Mi, X., Zhang, L., Liu, S., Xu, H. and Cheng, J.-P. (2007) *Tetrahedron*, **63**, 1923.

83 Gruttadauria, M., Riela, S., Meo, P.L., D'Anna, F. and Noto, R. (2004) *Tetrahedron Letters*, **45**, 6113.

84 Kotrusz, P., Toma, Š., Schmalz, H.-G. and Adler, A. (2004) *European Journal of Organic Chemistry*, 1577.

85 Mečiarová, M., Toma, Š. and Kotrusz, P. (2006) *Organic and Biomolecular Chemistry*, **4**, 1420.

86 Hagiwara, H., Okabe, T., Hoshi, T. and Suzuki, T. (2004) *Journal of Molecular Catalysis A–Chemical*, **214**, 167.

87 Rasalkar, M.S., Potdar, M.K., Mohile, S.S. and Salunkhe, M.M. (2005) *Journal of Molecular Catalysis A–Chemical*, **235**, 267.

88 Luo, S., Mi, X., Zhang, L., Liu, S., Xu, H. and Cheng, J.-P. (2006) *Angewandte Chemie–International Edition*, **45**, 3093.

89 Xu, D., Luo, S., Yue, H., Wang, L., Liu, Y. and Xu, Z. (2006) *Synlett*, 2569.

90 (a) Chowdari, N.S., Ramachary, D.B. and Barbas, C.F. III (2003) *Synlett*, 1906.
(b) Notz, W., Chowdari, S.-i., Watanabe, N.S., Zhong, G., Betancort, J.M., Tanaka, F. and Barbas, C.F. III (2004) *Advanced Synthesis Catalysis*, **346**, 1131.

91 Davis, F.A. and Chen, B.C. (1995) Oxygenation of enolates, in *Houben-Weyl: Methods of Organic Chemistry*, Vol. **E21** (eds G. Helmchen, R.W. Hoffmann, J. Mulzer and E. Schaumann), Georg Thieme: Stuttgart, Germany, p. 4497.

92 Momiyama, N. and Yamamoto, H. (2003) *Journal of the American Chemical Society*, **125**, 6038.

93 (a) Zhong, G. (2003) *Angewandte Chemie–International Edition*, **42**, 4247.
(b) Brown, S.P., Brochu, M.P., Sinz, C.J. and MacMillan, D.W.C. (2003) *Journal of the American Chemical Society*, **125**, 10808.
(c) Hayashi, Y., Yamaguchi, J., Hibino, K. and Shoji, M. (2003) *Tetrahedron Letters*, **44**, 8293.
(d) Hayashi, Y., Yamaguchi, J., Sumiya, T. and Shoji, M. (2004) *Angewandte Chemie–International Edition*, **43**, 1112.
(e) Bøgevig, A., Sundén, H. and Córdova, A. (2004) *Angewandte Chemie–International Edition*, **43**, 1109.

94 Guo, H.-M., Niu, H.-Y., Xue, M.-X., Guo, Q.-X., Cun, L.-F., Mi, A.-Q., Jiang, Y.-Z. and Wang, J.-J. (2006) *Green Chemistry*, **8**, 682.

95 Huang, K., Huang, Z.-Z. and Li, X.-L. (2006) *Journal of Organic Chemistry*, **71**, 8320.

96 Kotrusz, P., Alemayehu, S., Toma, Š., Schmalz, H.-G. and Adler, A. (2005) *European Journal of Organic Chemistry*, 4904.

97 Park, J.K., Sreekanth, P. and Kim, B.M. (2004) *Advanced Synthesis Catalysis*, **346**, 49.

98 Xiao, J., Nefkens, S.C.A., Jessop, P.G., Ikariya, T. and Noyori, R. (1996) *Tetrahedron Letters*, **37**, 2813.

99 Wang, S. and Kienzle, F. (2000) *Industrial and Engineering Chemistry Research*, **39**, 4487.

100 Combes, G., Coen, E., Dehghani, F. and Foster, N. (2005) *The Journal of Supercritical Fluids*, **36**, 127.

101 (a) Lange, S., Brinkmann, A., Trautner, P., Woelk, K., Bargon, J. and Leitner, W. (2000) *Chirality*, **12**, 450.
(b) Burgemeister, K., Francio, G., Hugl, H. and Leitner, W. (2005) *Chemical Communications*, 6026.

102 Dong, X. and Erkey, C. (2004) *Journal of Molecular Catalysis A–Chemical*, **211**, 73.

103 Berthod, M., Mignani, G. and Lemaire, M. (2004) *Tetrahedron: Asymmetry*, **15**, 1121.

104 (a) Hu, Y., Birdsall, D.J., Stuart, A.M., Hope, E.G. and Xiao, J. (2004) *Journal of Molecular Catalysis A–Chemical*, **219**, 57.
(b) Adams, D.J., Chen, W., Hope, E.G., Lange, S., Stuart, A.M., West, A. and Xiao, J. (2003) *Green Chemistry*, **5**, 118.

105 (a) Stephenson, P., Licence, P., Ross, S.K. and Poliakoff, M. (2004) *Green Chemistry*, **6**, 521.
(b) Stephenson, P., Kondor, B., Licence, P., Scovell, K., Ross, S.K. and Poliakoff, M. (2006) *Advanced Synthesis Catalysis*, **348**, 1605.

106 (a) Nozaki, K. and Ojima, I. (2000) Asymmetric carbonylations, in *Catalytic Asymmetric Synthesis* (ed. I. Ojima), Wiley-VCH Verlag GmbH, New York, p. 429.
(b) Nozaki, K. (1999) Hydrocarbonylation of carbon-carbon double bonds, in *Comprehensive Asymmetric Catalysis*, Vol. **1** (eds E.N. Jacobsen, A. Pfaltz and H. Yamamoto), Springer-Verlag, Berlin, p. 381.

107 (a) Nozaki, K., Takaya, H. and Hiayama, T. (1997) *Topics in Catalysis*, **4**, 175.
(b) Nozaki, K., Sakai, N., Mano, S., Higashijima, T., Horiuchi, T. and Takaya, H. (1997) *Journal of the American Chemical Society*, **119**, 4413.

108 Kainz, S. and Leitner, W. (1998) *Catalysis Letters*, **55**, 223.

109 Franciò, G., Wittmann, K. and Leitner, W. (2001) *Journal of Organometallic Chemistry*, **621**, 130.

110 Bonafoux, D., Hua, Z., Wang, B. and Ojima, I. (2001) *Journal of Fluorine Chemistry*, **112**, 101.

111 Shibahara, F., Nozaki, K. and Hiyama, T. (2003) *Journal of the American Chemical Society*, **125**, 8555.

112 (a) Matsuzawa, S.-i., Fukuzawa, H. and Metoki, K. (2001) *Synlett*, 709.
(b) Fukuzawa, S.-I., Metoki, K., Komuro, Y. and Funazukuri, T. (2002) *Synlett*, 134.
(c) Metoki, S.-i., Fukuzawa, K. and Esumi, S.-i. (2003) *Tetrahedron*, **59**, 10445.

113 Mikami, K., Matsukawa, S., Kayaki, Y. and Ikariya, T. (2000) *Tetrahedron Letters*, **41**, 1931.

114 (a) Blanchard, L.A., Hancu, D., Beckman, E.J. and Brennecke, J.F. (1999) *Nature*, **399**, 28.
(b) Blanchard, L.A. and Brennecke, J.F. (2001) *Industrial and Engineering Chemistry Research*, **40**, 287.

115 Brown, R.A., Pollet, P., McKoon, E., Eckert, C.A., Liotta, C.L. and Jessop, P.G. (2001) *Journal of the American Chemical Society*, **123**, 1254.

116 Jessop, P.G., Stanley, R.R., Brown, R.A., Eckert, C.A., Liotta, C.L., Ngo, T.T. and Pollet, P. (2003) *Green Chemistry*, **5**, 123.

117 Solinas, M., Pfaltz, A., Cozzi, P.G. and Leitner, W. (2004) *Journal of the American Chemical Society*, **126**, 16142.

118 (a) Branco, L.C., Serbanovic, A., da Ponte, M.N. and Afonso, C.A.M. (2005) *Chemical Communications*, 107.
(b) Serbanovic, A., Branco, L.C., da Ponte, M.N. and Afonso, C.A.M. (2005) *Journal of Organometallic Chemistry*, **690**, 3600.

8
Heterogenized Organocatalysts for Asymmetric Transformations

Maurizio Benaglia

8.1
Introduction

Among the different possible methodologies available, the use of a chiral catalyst represents, in principle, the most attractive procedure to synthesize enantiomerically enriched compounds, since in a catalytic process a small amount of a 'smart' molecule produces a large quantity of the desired chiral compound [1].

Today, the demand for enantiomerically pure compounds is continuously increasing, not only for use in pharmaceuticals but also in other areas such as agrochemicals, flavor and aroma chemicals, and specialty materials. Recently, strict government regulations that require the individual evaluation of all possible stereoisomers of a compound and the commercialization of a chiral product only as single enantiomer have called for further improvements in the stereoselective synthesis of chiral compounds. While the catalyst usually allows the reaction to be operated under mild conditions, the economic benefits of an efficient catalytic process are also enormous as it is less capital intensive, has lower operating costs, produces higher-purity products and fewer byproducts. In addition, a substoichiometric process may provide important environmental benefits.

In this context it is surprising how relatively few enantioselective catalytic reactions are used on an industrial scale today [2]. This is even more difficult to understand when considering the impressive progress which has been made during the past few years in the field of enantioselective catalysis, where hundreds of catalytic transformations with high chemical and stereochemical efficiencies have been developed. So, the obvious question is: Why has the application of enantioselective catalysis to the fine-chemicals industry, which is of potentially great economic and environmental interest, not been widely pursued on a large scale? It is true that a variety of issues must be addressed: first, not only the cost of the chiral catalyst but also other problems must be considered, such as the general applicability. Many highly selective catalysts have been developed for reactions with selected model substrates, but not tested on differently functionalized molecules. In addition, for many catalysts very little information is available on catalyst selectivity, activity and productivity. The stability of the catalyst and the possibility of its easy separation and recycling are also important aspects that must be considered for an industrial asymmetric catalytic process.

The immobilization of a catalytic species on a solid support may represent a solution to some of these problems [3]. In fact, it is not only the recovery and possible recycling of a catalyst that may be investigated and successfully realized through its immobilization; other issues such as the stability, structural characterization and catalytic behavior of an enantioselective catalyst may also be better conducted on a supported version.

These general considerations are also true for organic catalysts. Whilst the transformation of a stoichiometric process into its catalytic counterpart can be regarded as a significant step towards the development of a truly 'green chemistry' [4], catalytic reactions are amenable to a variety of improvements that can make them increasingly greener. Among these reactions, the replacement of metal-based catalysts with equally efficient metal-free counterparts – the so-called 'organic catalysts' – can be extremely important [5]. Indeed, the possibility of using catalytic amounts of an organic compound of relatively low molecular weight and simple structure to promote reactions that previously required a costly (and possibly toxic) transition metal-based catalyst, has spurred an incredibly intense and productive line of research over the past few years, and this has led to the establishment of 'organocatalysis' as a viable alternative to organometallic catalysis. A number of chemically robust organocatalysts, each capable of displaying enzyme-like activity, has thus become available, and their application which encompasses many fundamental organic reactions that can now be run in the presence of nontoxic, cheap and more environmentally friendly promoters. In this context, the term 'organic' is synonymous with 'metal-free', and includes all the advantages of performing a reaction under metal-free conditions. These advantages might also include, *inter alia*, the possibility of working in wet solvents and under an aerobic atmosphere, dealing with a stable and robust catalyst and avoiding the problem of a (possibly) expensive and toxic metal leaching into the organic product [6].

The immobilization of a catalyst on a support – with the aim of facilitating separation of the product from the catalyst, and thus the recovery and recycling of the latter – can also be regarded as an important improvement for a catalytic process [7]. In this regard, the immobilization of organic catalysts seems particularly attractive, because the metal-free nature of these compounds avoids from the outset the problem of metal leaching that often negatively affects and practically prevents the efficient recycling of a supported organometallic catalyst [8]. Furthermore, a simple organic compound will be less affected by the connection to a support than more structurally complex (and somehow more 'delicate') enzymes, from which they are conceptually derived and to which they are often compared [9]. In this very hectic area of organocatalysis the immobilization of chiral catalytic species may play (and in part it has already played) a decisive role in further developments of the field. In addition, it is important to note how immobilization on a support can endow the catalyst with special properties (e.g. a different solubility profile or an

8.2
General Considerations on the Immobilization Process

enhanced catalytic activity) that can be fine-tuned by careful selection of the support such that the range of application of the catalyst is expanded.

8.2
General Considerations on the Immobilization Process

The main goals when immobilizing a catalyst on a support are to simplify the reaction work-up, the recovery and hopefully the recycling of the precious chiral catalytic species. However, besides the recovery and reuse of the catalyst, other reasons may lead to the development of a supported version of a catalytic species.

Catalyst instability can be a problem that may be tackled by developing an immobilized catalyst. Organic catalysts do exist that slowly decompose under the conditions necessary for their reaction and release trace amounts of byproducts that must be separated from the products. For example, in photooxygenation reactions catalyzed by porphyrin the release of highly colored materials derived from the photosensitizer may cause major problems with the product purification. By immobilizing the catalyst this problem can be solved because the decomposed materials are also supported and can be removed from the reaction medium during the work-up process.

In this context, soluble polymers have recently become a subject of an intense research activity; by allowing the reaction to be carried out in a homogeneous solution they would secure higher chemical and stereochemical efficiencies than would insoluble polymers. Among the soluble polymeric matrices employed, poly(ethylene glycol) (PEG)s are the most successful [10]. These polymers, with a molecular weight (M_w) greater than 2000 Da, are readily functionalized and commercially available inexpensive supports that feature convenient solubility properties. Indeed, while they are generally soluble in many common organic solvents they are insoluble in a few other solvents, such as diethyl ether, hexanes and *t*-butyl-methyl ether. Therefore, the correct choice of solvent system will allow a reaction to be run under homogeneous catalysis conditions (where the PEG-supported catalyst is expected to perform at its best), after which the catalyst may be recovered under heterogeneous conditions, as if it were bound to an insoluble matrix.

Starting from the commercially available monomethylether of PEG (MeOPEG), 5,10,15,20-tetrakis-(4-hydroxyphenyl)-porphyrin (also available commercially) was immobilized to give the PEG-supported porphyrin **1** (M_W = 2000; Scheme 8.1).

The irradiation of a 0.01 *M* methylene chloride solution of bisdialine **2** with a 100 W halogen lamp, in the presence of 3 mol% of PEG-supported tetrahydroxyphenyl-porphyrin (PEG-TPP, **1**) as a sensitizer gave a 82/18 mixture of *supra* and *antara* diastereoisomeric endoperoxides **3** in quantitative yield after 1 h (see Equation a in Scheme 8.1). The polymer-bound catalyst not only showed a similar activity to that of the nonsupported species, but also greatly simplified the product isolation [11]. On completion of the reaction the reaction mixture was concentrated

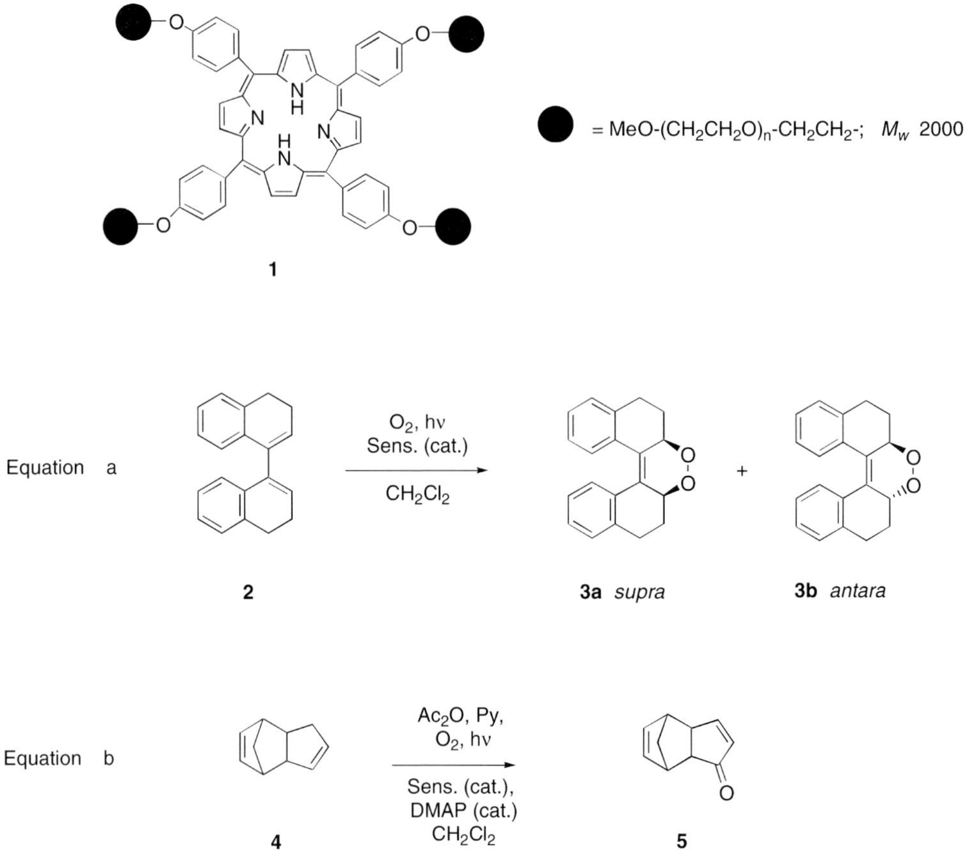

Scheme 8.1 PEG-supported porphyrin as a sensitizer for photooxidation reactions.

in vacuo, diethyl ether was added to the residue, and the precipitated PEG-supported porphyrin was quantitatively recovered by filtration. From the concentrated filtrate solution the endoperoxides were easily isolated by crystallization from ethanol. The one-pot oxidation of olefin to α,β-unsaturated ketones was applied on a gram-scale to convert dicyclopentadiene **4** into the corresponding dicyclopentadienone **5**, a useful starting material for the preparation of enantio-enriched diols; following filtration of the supported catalyst, the product was isolated by simple evaporation of the organic solvent as an analytically pure compound, that required no further purification. (Equation b in Scheme 8.1). It should be noted that the PEG-supported sensitizer was recycled six times, with no appreciable loss of either chemical or stereochemical efficiency.

The immobilization of an organic catalyst can also be used to facilitate the process of the catalyst's optimization. Surprisingly, only a single example of the application of this methodology has been reported to date; this deserves special mention as it represents a remarkable exception in which the development of the immobilized catalyst preceded and was crucial to that of the nonsupported

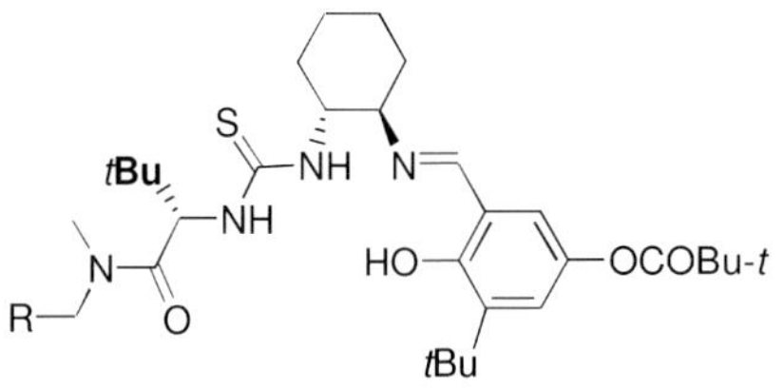

8.2 General Considerations on the Immobilization Process | 297

6a R = 1% crosslinked polystyrene
6b R = H

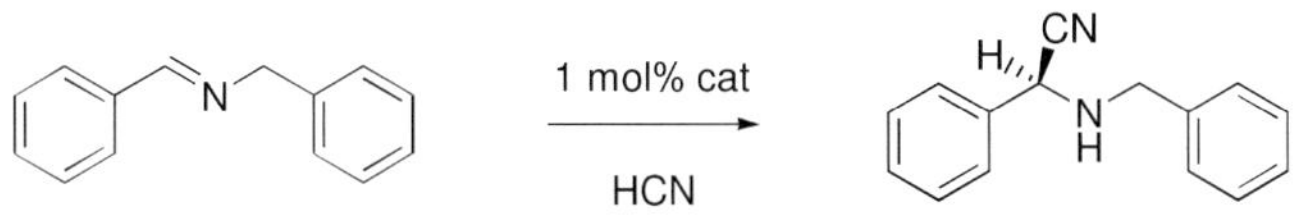

Scheme 8.2 Supported thiourea Jacobsen's catalysts.

counterpart [12]. Jacobsen developed a fully organic catalyst for the Strecker reaction using a thiourea-based chiral Brønsted acid that was found eventually to be extremely chemically active, stereoselective, and broad in terms of its application. Optimization of the catalyst structure was realized through a series of modifications of the salen-based structure carried out on an insoluble polystyrene support and using the principles of combinatorial chemistry to identify the best amino acid, diamine and diamine–amino acid linker combination. The screening of three successive libraries led to the identification of the supported thiourea catalyst **6a** as the best one from which the nonsupported counterpart **6b** was derived (Scheme 8.2).

At a loading as low as 1 mol%, **6b** promoted the hydrocyanation of N-allyl or -benzyl imines derived from aromatic and aliphatic aldehydes and of some ketones with very high yield and almost complete stereoselectivity. It is of interest to note that the soluble and resin-bound catalysts performed equally well. Moreover, recovery and recycling of the supported catalyst was shown to occur without any erosion of chemical and stereochemical efficiency over 10 reaction cycles.

Given these excellent results, it is surprising that this approach has not been used more extensively for chiral organic catalyst discovery. The success of this methodology is even more significant if it is considered that catalysts **6** are among the few chiral organocatalysts to be currently employed at the industrial level [12].

The preparation of supported catalysts for use in environmentally friendly or green solvents, as a part of the drive towards developing a more 'green chemistry', is also becoming increasingly widespread.

PEG-supported metal-free catalysts have shown to perform well in water (Scheme 8.3). For example, the synthesis of a PEG-supported TEMPO (2,2,6,6-tetramethyl-piperidine-1-oxyl), and its use as a highly efficient, recoverable and recyclable catalyst in oxidation reactions were described [13].

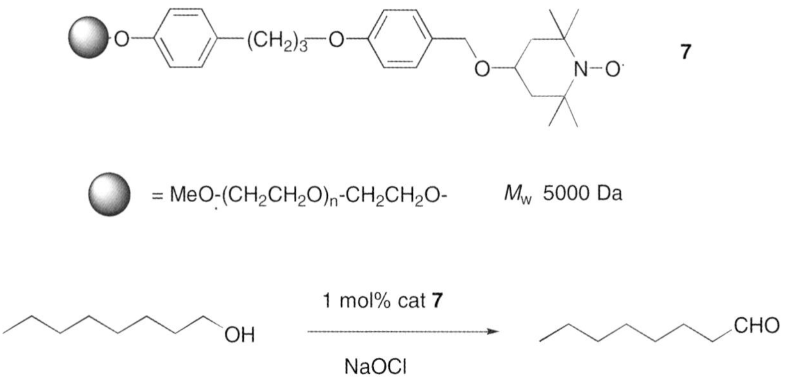

Scheme 8.3 PEG-supported TEMPO as a catalyst for oxidation reactions.

The use of oxoammonium ions such as those derived from TEMPO, in combination with cheap and easy-to-handle terminal oxidants in the conversion of alcohols into aldehydes, ketones and carboxylic acids, represents a significant example of how it is possible to develop a safer and greener chemistry, by avoiding the use of environmentally unfriendly toxic metals. Unfortunately, the separation of products from TEMPO can be problematic, especially when the reactions are run on a large scale. However, immobilization on a solid support may offer a solution to this problem [14].

The TEMPO-catalyzed oxidation of alcohols to carbonyl compounds with buffered aqueous NaOCl has found broad application even in large-scale operations. Indeed, this selective methodology involves the use of safe and inexpensive inorganic reagents under mild reaction conditions. A supported TEMPO **7**, which is soluble in CH_2Cl_2 and acetic acid but insoluble in ethers and hexane, was prepared and proved to be an effective catalyst for the selective oxidation of 1-octanol with various stoichiometric oxidants. When **7** was employed at 1 mol% as a catalyst with a stoichiometric amount of NaOCl, the aldehyde was obtained in 95% yield after only 30 min of reaction. The recycling of catalyst **7** was shown to be possible for seven reaction cycles in the oxidation of 1-octanol, that occurred in undiminished conversion and selectivity under similar reaction conditions.

Besides the simplification of the reaction work-up, it is clear that the recovery and recycling of the precious chiral catalyst should represent an even more important issue for the immobilization of an enantiomerically pure organocatalyst. Until now, within the field of organocatalysis research efforts have mainly been focused on catalyst discovery, while the development of immobilized chiral catalysts has been less intensively pursued. Based on recently published reviews describing the immobilization of *achiral* and *chiral* organic catalysts [8], this chapter will outline the more relevant achievements reported in the field of immobilized chiral organic catalysts, with special attention being paid to those investigations that have been conducted during the past few years. Whenever possible, comparisons between the behavior of supported versus nonsupported catalytic species will be discussed. Neither catalysis by molecularly imprinted polymers or dendrimeric supports will be considered here, nor ion liquid-based technologies and recoverable fluorous systems, all of which are described in the other chapters of this Handbook. Instead, particular attention will be devoted to catalyst recovery and recycling, with discussions on the structure of the polymeric support being limited to those examples for which an influence on catalyst performance has clearly been demonstrated. Some other considerations on the methodologies, the future and problems related to chiral organic catalyst immobilization will also be briefly presented.

8.3
Phase-Transfer Catalysts

Chiral phase-transfer catalysis (PTC) is a very interesting methodology that typically requires simple experimental operations, a mild reaction conditions and inexpensive and/or environmentally benign reagents, and which is amenable to large-scale preparations [15]. The possibility of developing recoverable and recyclable chiral catalysts has attracted the interest of many groups. Indeed, the immobilization of chiral phase-transfer catalysts has provided the first demonstrations of the feasibility of this approach.

The advent during the early 1990s of the O'Donnell–Corey–Lygo protocol for the highly enantioselective alkylation of amino acids imines under PTC conditions, catalyzed by quaternized cinchona alkaloids, led to a series of investigations on the use of supported catalysts in these reactions [15].

Insoluble polymer-anchored ammonium salts prepared from cinchona derivatives, when tested in the standard alkylation of *tert*-butyl glycinate benzophenone imine, often afforded low enantioselectivities and somehow contradictory results. For example, Najera's group has reported that polystyrene (PS)-supported catalyst **8** promoted the benzylation of the benzophenone imine of glycine *i*-propylester (Scheme 8.4, Equation a) carried out in 25% aqueous NaOH and toluene at 0 °C to afford the *(S)*-product in 90% yield and 90% enantiomeric excess (ee) [16]. The use of different ester alkyl residues (ethyl, *tert*-butyl), a higher reaction temperature (25 °C) and, much more surprisingly, very similar alkylating agents (4-bromo-, 4-nitro-, 4-methoxybenzylbromide, and 2-bromomethyl-naphthalene) depressed the ee-value to 40–60%. Even more surprising was the observation that, under the best conditions, catalyst **9** afforded the *(R)*-alkylated product in only 40% ee.

A few years ago Cahard reported a series of studies on the use of immobilized cinchona alkaloid derivatives in asymmetric reactions with phase-transfer catalysts [17]. Two types of polymer-supported ammonium salts of cinchona alkaloids (types A and B in Scheme 8.4) were prepared from PS, and their activity was evaluated. The enantioselectivity was found to depend heavily on the alkaloid immobilized, with the type B catalysts usually giving better results than the type A catalysts. By performing the reaction in toluene at −50 °C in the presence of an excess of solid cesium hydroxide and 0.1 mol equiv of catalyst **10**, benzylation of the *tert*-butyl glycinate-derived benzophenone imine afforded the expected *(S)*-product in 67% yield with 94% ee, a value very close to that observed with the nonsupported catalyst. (Scheme 8.4, Equation b) Unfortunately – and again, inexplicably – the pseudoenantiomer of **10** proved to be much less stereoselective, affording the *(R)*-product in only 23% ee. No mention of catalyst recycling was reported [18].

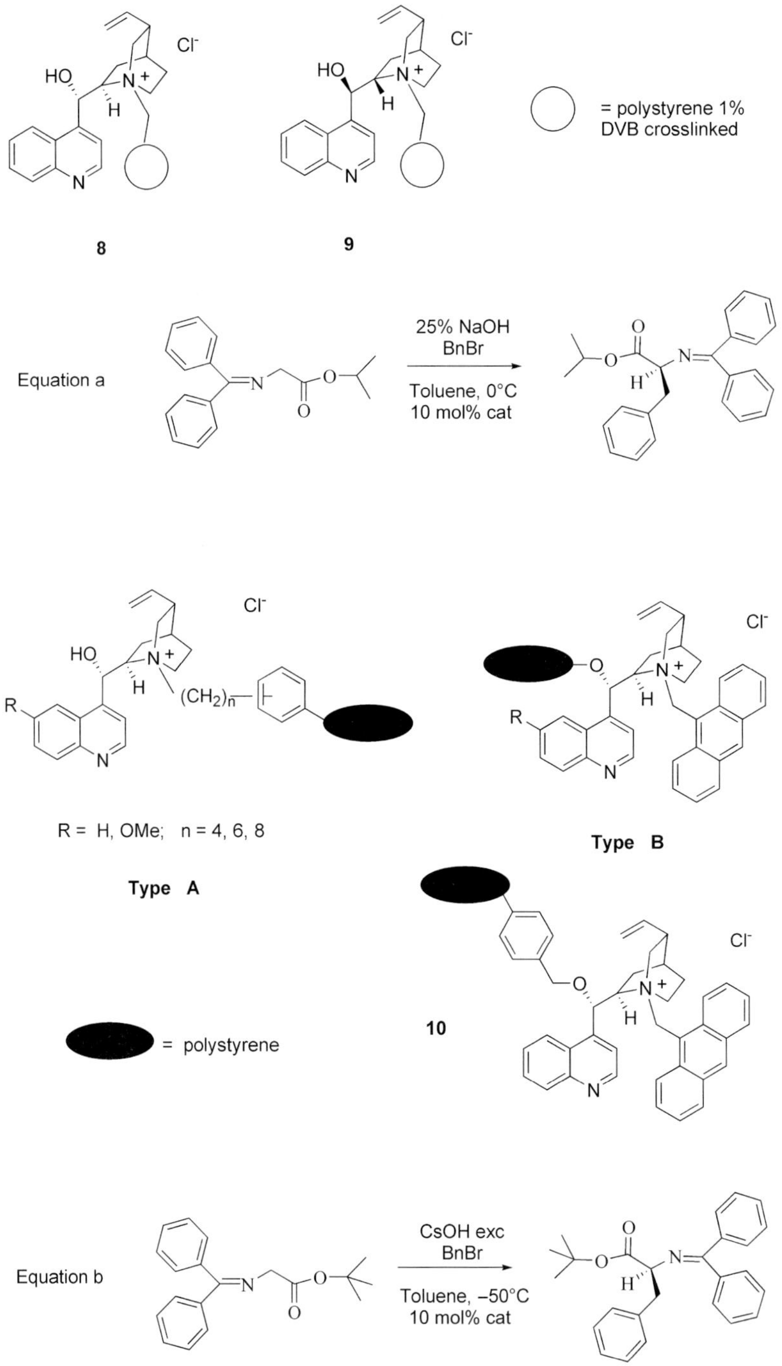

300 | *8 Heterogenized Organocatalysts for Asymmetric Transformations*

8

9

○ = polystyrene 1% DVB crosslinked

Equation a

25% NaOH
BnBr
─────────→
Toluene, 0°C
10 mol% cat

Type A

R = H, OMe; n = 4, 6, 8

Type B

● = polystyrene

10

Equation b

CsOH exc
BnBr
─────────→
Toluene, −50°C
10 mol% cat

Scheme 8.4 Insoluble polymer-supported phase-transfer catalysts.

The *soluble polymer*-supported catalysts **11** and **12** (Scheme 8.5) were prepared by attaching two different MeO-PEG$_{5000}$/spacer fragments to the *N*-anthracenylmethyl salts of *nor*-quinine and cinchonidine, respectively [19]. The behavior of the obtained catalysts, however, fell short of expectations. Whilst with **11** enantioselectivities lower than 12% ee were always obtained, **12** showed good catalytic activity in promoting the benzylation reaction (solid CsOH, DCM, −78 to 23 °C, 22 h, 92% yield) but with only 30% ee. Although this value was increased to 64% by performing the reaction at −78 °C for 60 h, the stereoselectivity remained inferior to that obtained with the nonsupported catalyst.

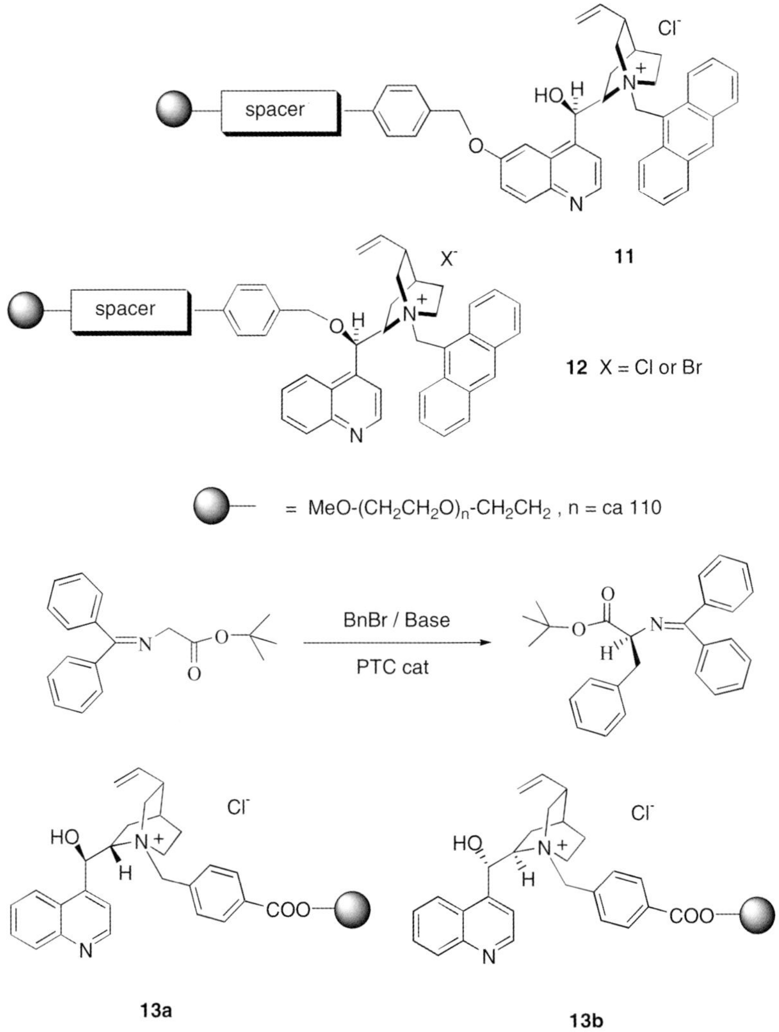

Scheme 8.5 Soluble polymer-supported phase-transfer catalysts.

PEG was considered to be responsible (at least in part) for these results because of the following effects. By increasing the polarity around the catalyst, PEG prevents the formation of a tigh t ion pair between the enolate and the chiral ammonium salt, the formation of which is regarded as crucial for high stereocontrol. Moreover, PEG enhances the solubility of the inorganic cation in the organic phase, leading to a competing nonstereoselective alkylation occurring on the achiral cesium enolate. In order to check the validity of these hypotheses, control experiments were carried out by performing the reaction with the nonsupported catalyst in the presence of the bis-methylether of PEG_{2000}. The observed ee-value of 65% was comparable to that observed with catalyst **12**, but largely inferior to the >90% ee easily achieved with the nonsupported catalyst. During the course of this study it was also found that both the supported and nonsupported catalysts were quite unstable, thus preventing any possibility of recovery and recycling.

In a more recent study, Cahard described the synthesis of two catalysts where cinchonidine (**13a**) and cinchonine (**13b**) were connected through their bridgehead nitrogen atom to $MeOPEG_{5000}$ by an ester linker [20]. These catalysts (10 mol%) afforded the *(S)*-benzylation product in 81% ee and the *(R)*-product in 53% ee, respectively (Scheme 8.5). The relatively large difference in the observed ee-value was quite surprising, as the *quasi*-enantiomeric structure of the alkaloid catalysts should secure virtually identical ee-values for both enantiomers. The chemical yield was about 80% in both cases. The use of other alkylating agents resulted in modest ee-values (20% with *n*-hexyliodide; 34% with benzhydryl bromide). Equally surprising – and without reasonable explanation – was the strong variation in the stereoselectivity of the alkylation observed on changing the reaction solvent: toluene (81% ee), benzene (64% ee), xylene (58% ee), carbon tetrachloride (65% ee), 9 (DCM) (3% ee). Attempts at recycling catalyst **13a** led to a dramatic fall in enantioselectivity, which was ascribed to the instability of the catalyst's ester linkage under the reaction conditions.

More recently, Wang reported a dimeric PEG-supported cinchona ammonium salt for the alkylation of Schiff bases in water as solvent [21]. Soluble polymer-immobilized catalysts **14a** and **14b** were prepared by the reaction of diacetoamido-PEG_{2000} chloride with an excess of cinchonidine and quinine, respectively (Scheme 8.6). The benzylation of the benzophenone imine of *tert*-butyl glycinate in a 1 M aqueous solution of NaOH at 25 °C afforded the *(S)*-product in 75% ee and the *(R)*-enantiomer in 80% ee, respectively. Of note here was the satisfactory enantioselectivity and yields obtained by employing the recovered catalysts three times, thus demonstrating a higher stability of these catalysts than for previously reported systems, most likely due to the acetamido-group connecting the alkaloid moiety to the polymer.

8.4
Nonionic Cinchona-Derived Catalysts

The potential for the use of cinchona alkaloids as organocatalysts has long been recognized [22]. As the alkaloid skeleton offers different sites for polymer attach-

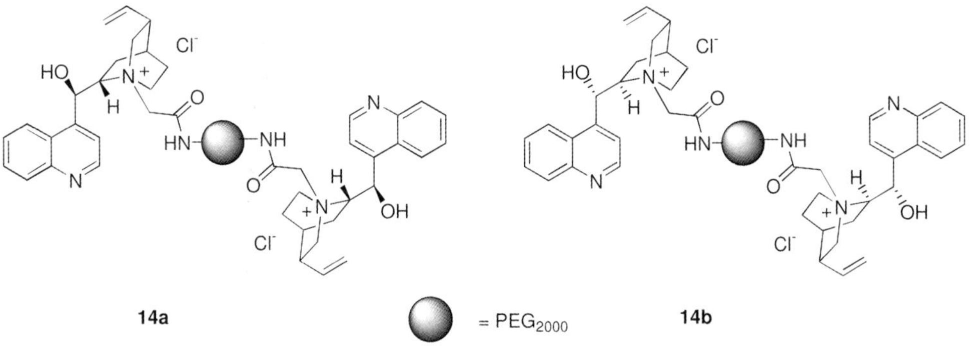

Scheme 8.6 Dimeric PEG-supported phase-transfer catalysts.

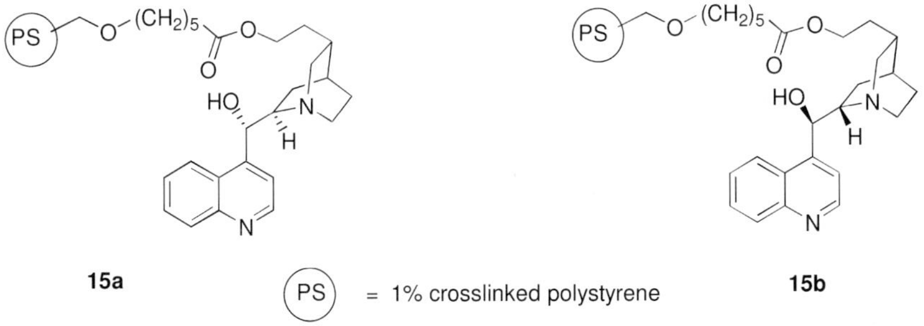

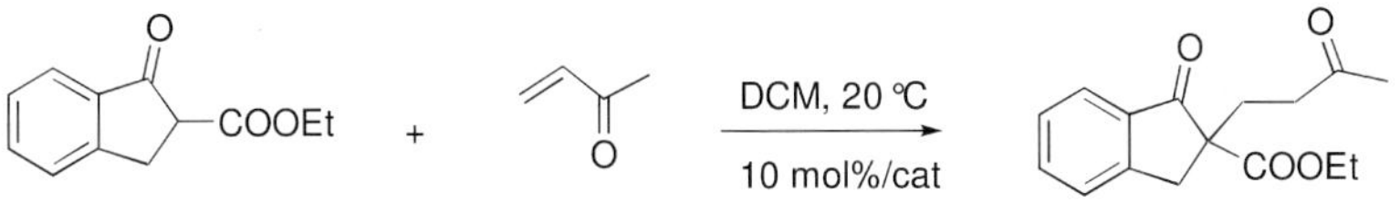

Scheme 8.7 Insoluble polymer-supported cinchona alkaloids.

ment, a variety of immobilized versions of these catalysts has been reported [8a]. In the case of the nonionic catalyst, the two most commonly used connection sites were the double bond of the quinuclidine residue and the oxygen atom at C9, with the former being largely preferred. Following the seminal studies of Kobayashi and Hodge during the 1980s [8a], more recently a hydroxy group was introduced at the terminal atom of the double bond of quinine and quinidine by d'Angelo and coworkers to provide a new mode of attachment of these alkaloids to 1% divinylbenzene (DVB) crosslinked chloromethylated PS [23]. A small collection of ten supported catalysts was thus obtained, featuring different spacers between the polymer and the pseudoenantiomeric alkaloids. These catalysts were tested in the paradigmatic Michael addition described in Scheme 8.7 (0.1 mol equiv of catalyst, DCM, 20 °C). The quinine-derived catalyst **15a** was found to be the more stereoselective, affording the *(R)*-configured product in 85% yield and 87% ee. Surprisingly, the quinidine-based catalyst **15b** was much less selective (39% ee) in the production of the same enantiomer. Even if this behavior has some precedent in the reactions promoted by other polymer-supported cinchona alkaloid catalysts [20], these results demonstrated how relatively small changes in the polymer/catalyst ensemble may produce dramatic and inexplicable effects in the stereochemical outcome of the reaction.

In this field Lectka and coworkers obtained spectacular results in a highly stereoselective synthesis of β-lactam [24]. This outcome was realized in a process which involved the use of solid-phase reagents and catalysts that constituted the packing of a 'series of reaction columns'. The chemical steps in the catalytic asymmetric synthesis of β-lactams are shown in Scheme 8.8, and included a ketene generation step (Supp. Base), the β-lactam formation (Supp. Cat) and a purification step (Supp. Scavenger). The catalyst involved was a quinine derivative, **16**, anchored to Wang resin through an appropriate spacer, and afforded the products with very high stereoselectivity (>90% ee). Of note, the conduction of chemical reactions on sequential columns led to an easy recovery of both catalyst and reagents, and also simplified the purification steps thereby avoiding the need for chromatography.

It is important to note that, under continuous-flow conditions, product isolation, catalyst recovery and recycling are realized in a single operation. The convenience of this approach is demonstrated by the fact that one of the highest recycling numbers ever observed for a chiral organic catalyst has been accomplished. Quite interestingly, several runs (between five and ten) were necessary to obtain a catalyst which was sufficiently aged to afford consistent results, as quinine 'bleeding' from the freshly prepared catalyst was found to occur. A properly aged resin was shown

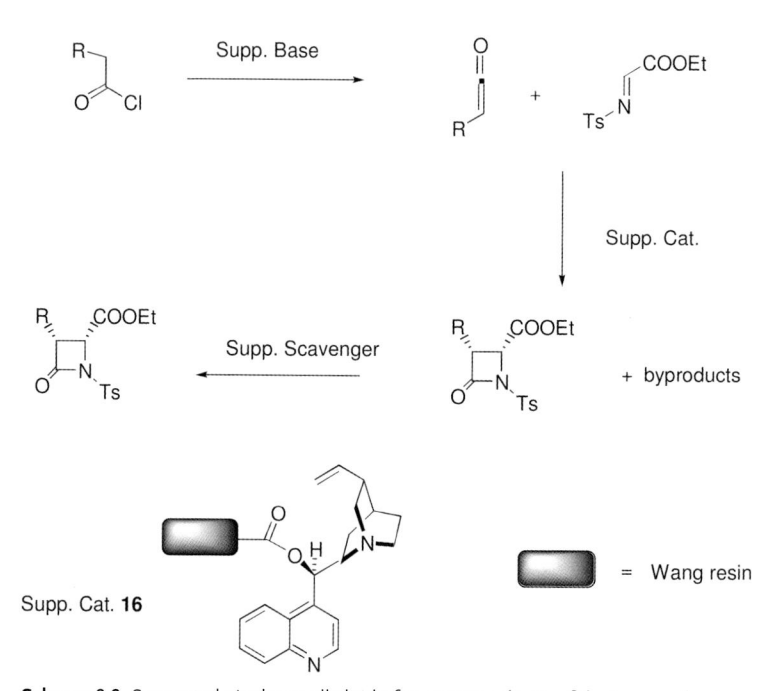

Scheme 8.8 Supported cinchona alkaloids for enantioselective β-lactam synthesis.

to perform as many as 60 reaction cycles, with no erosion in either yield or selectivity.

8.5 Lewis Base Catalysts

Chiral phosphoramides, as developed by Denmark during the late 1990s [25], are efficient catalysts for the allylation of aldehydes with allyl trichlorosilane [26a] or the aldol condensation of trichlorosilyl-enol ethers with aldehydes [26b]. However, the first example of supported chiral phosphoramides on a polymeric matrix was reported only in 2005 [27].

PS-anchored catalysts **17a–c** of different active site content were used as catalysts (10 mol%) to promote the allylation of benzaldehyde with allyl trichlorosilane in the presence of excess diisopropylethylamine (DCM, −78 °C, 6 h), affording the product in 82–84% yield and 62–63% ee (Scheme 8.9). Remarkably, the supported catalysts proved to be more efficient than the corresponding nonsupported derivatives featuring a benzyl group instead of the polymer residue, both in terms of yield and of stereoselectivity. Since bis-phosphoramides have been shown to be more efficient than mono-phosphoramides in promoting the allylation reaction [26a], the better results obtained with **17a–c** were regarded as suggestive that two phosphoramide groups of the supported catalysts could bind the hypervalent octahedrally coordinated silicon atom believed to be involved in the transition structure of the reaction. In other words, the polymer backbone apparently forces the sites of two catalysts into such close proximity that they can behave as bis-phosphoramides [28]. Neither the recycling of **17a–c** nor the extension of their use to the allylation of aldehydes different from benzaldehyde has been described.

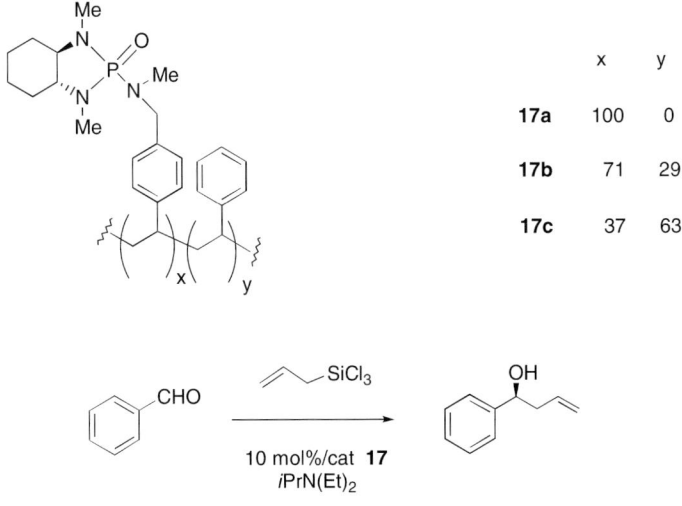

	x	y
17a	100	0
17b	71	29
17c	37	63

Scheme 8.9 Supported Lewis bases.

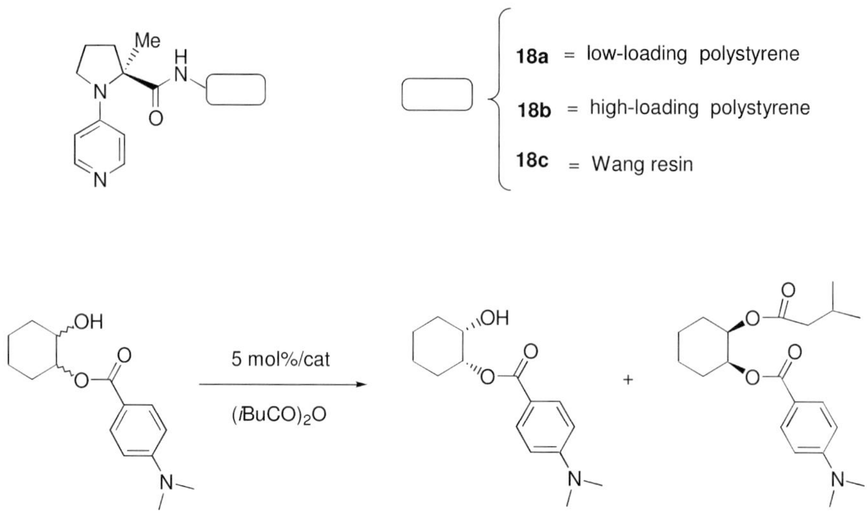

Scheme 8.10 Supported chiral 4-N,N-dialkylaminopyridines.

Although the use of a variety of chiral 4-N,N-dimethylaminopyridine (DMAP) analogues as organic catalysts is very well known [29], only a few examples of supported versions of these compounds have been described to date. A family of chiral acylating agents bearing the N-4'-pyridinyl-α-methylproline unit was recently reported to promote the kinetic resolution of alcohols with a high level of enantioselectivity [30]. The ready availability of these compounds suggested that the easily functionalizable carboxy moiety would be employed to immobilize these catalysts on different polymer supports. Among others, derivatives 18a–c (depicted in Scheme 8.10) were prepared by connecting N-4'-pyridinyl-α-methylproline to low-loading polystyrene (LLPS) or high-loading polystyrene (HLPS), or to Wang resin using standard condensation methods. These compounds were tested as insoluble catalysts (5 mol%) in the kinetic resolution of cis-1,2-cyclohexanediol mono-4-dimethylaminobenzoate carried out with a deficiency of iso-butyric anhydride in DCM (room temperature, 16 h; Scheme 8.10). By stopping the reaction at about 50% conversion, it was possible to recover the unreacted (−)-alcohol at about 75% ee, but this was increased to 93% by allowing the reaction to proceed to 67% conversion. No appreciable difference in either chemical or stereochemical efficiency was observed with the variation of the polymeric support.

Catalyst **18b**, recovered by filtration and thoroughly washed with DCM, was employed in three additional runs to afford the resolved product in slightly higher ee-value at identical conversions. However, the activity of the recycled catalyst was somewhat lower than that of the fresh version, as longer reaction times were necessary to obtain the same conversions. An extension of the use of this catalyst to the kinetic resolution of other secondary alcohols was possible, although the immobilized catalyst performed constantly less efficiently than its best nonsupported analogue [31].

8.6
Catalysts Derived from Amino Acids

Amino acids and their derivatives represent an obvious source of chiral organic catalysts, which has been fully exploited by synthetic organic chemists [32], especially in the field of aminocatalysis, such as enantioselective catalytic processes promoted by enantiomerically pure amines [33].

The ready availability of amino acids and their different functionalizations in the side chains allowed for a number of applications in the field of supported catalysis. While the relatively low cost of many amino acids apparently does not seem to justify the preparation of supported catalysts derived from amino acids, other reasons (as mentioned above) may drive towards the immobilization of chiral catalysts, for example to experiment with different solubilities, the easy separation of the product from the catalyst, and the catalyst's recyclability. The immobilization of these compounds on a support can also be seen as an attempt to develop a minimalist version of an enzyme, with the amino acid playing the role of the enzyme's active site and the polymer that of an oversimplified peptide backbone not directly involved in the catalytic activity [34]. It should be mentioned at this point that, in principle, amine-based catalysts offer also the possibility to be recovered by exploiting their solubility profiles in acids.

8.6.1
Proline Derivatives

One of the most successful and versatile chiral organic catalysts, proline, has been employed in a number of enantioselective syntheses, including aldol and iminoaldol condensations and Michael addition. Proline was immobilized very soon after the initial, seminal studies of List and Barbas [35]. A *soluble polymer*-supported version of this versatile catalyst, **19**, has been prepared by anchoring (2S,4R)-4-hydroxyproline to the monomethyl ether of PEG_{5000} by means of a succinate spacer [34]. In the presence of 0.25–0.35 mol equiv of this catalyst, acetone reacted with enolizable and nonenolizable aldehydes (Scheme 8.11; R≠H, Equation a) in dimethylformamide (DMF) at room temperature (40–60h) to afford the corresponding aldol products in yield (up to 80%) and ee-value (up to >98%) comparable to those obtained using nonsupported proline derivatives as the catalysts (that however gave faster reactions). The condensation of hydroxyacetone with cyclohexanecarboxaldehyde catalyzed by **19** afforded the corresponding *anti*-α,β-dihydroxyketone in 96% ee (*anti/syn* ratio >20:1) (Scheme 8.11; R≠OH, Equation a). The double-loaded catalyst **20** behaved similarly to **19** while allowing the use of only half the weight of catalyst [34, 36].

Replacement of the aldehyde component of the aldol reactions with imines (either preformed or generated *in situ*) opened access to synthetically relevant β-amino- and syn-β-amino-α-hydroxyketones, which were obtained in moderate to good yields (up to 80%) and good to high diastereoselectivity and enantioselectivity (Scheme 8.11; Equation b; ee-value up to 97%) [36]. As far as catalyst recycling is

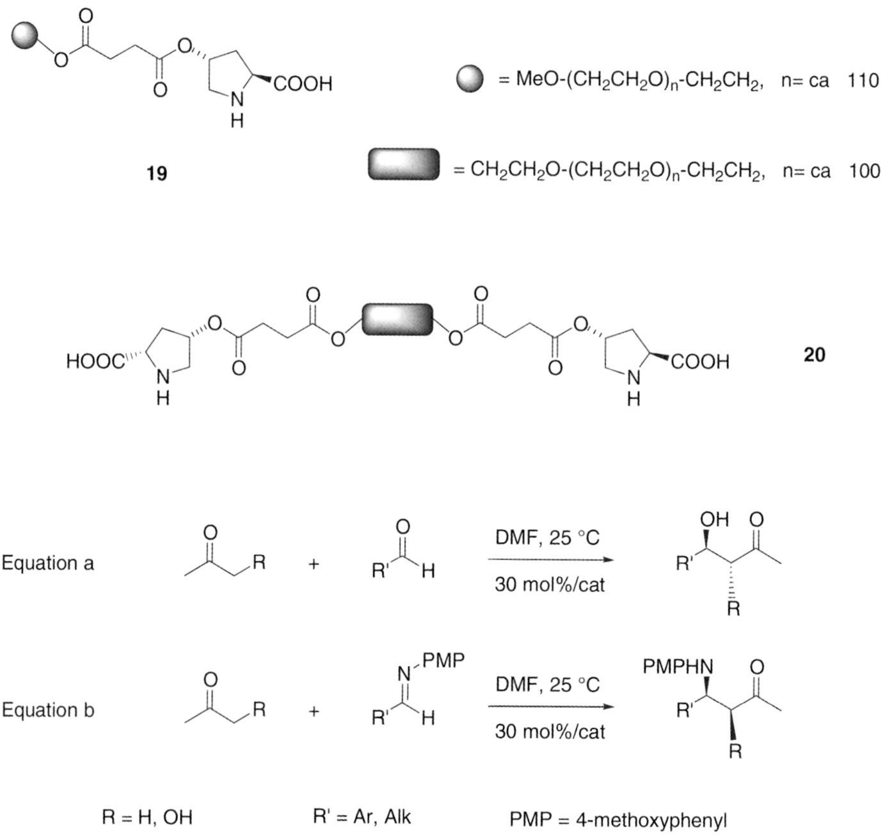

Scheme 8.11 PEG-supported (S)-proline derivatives.

R = H, OH   R' = Ar, Alk   PMP = 4-methoxyphenyl

concerned, it was shown that the supported catalyst **19** could easily be recovered by exploiting its solubility properties and recycled three or four times in all of the above-mentioned reactions. These reactions, however, showed slowly diminishing yields and virtually unchanged ee-values. It should be mentioned at this point that the use of hydroxyproline anchored via a spacer to 1% DVB crosslinked chloromethylated PS as the insoluble catalyst in the aldol step of a Robinson annulation led to the product in only 29% yield and 39% ee [37].

More recently, the modified proline **21** immobilized on the mesoporous siliceous material MCM-41 was prepared, with an active site loading of 0.52 mmol g^{-1} [38]. Large amounts of this catalyst (47–52 mol%) were then employed to promote the condensation between hydroxyacetone and *iso*-butyraldehyde (room temperature, 24 h) or benzaldehyde (90 °C, 24 h) in DMSO or toluene, to afford the products in yields that were only marginally higher than the amount of catalyst used (55–60%) (Scheme 8.12). The reaction times were significantly shortened to 10–30 min by the use of microwave irradiation. Catalyst **21** promoted the exclusive formation of *anti* diols with ee-values >99% in the condensation involving *iso*-butyraldehyde,

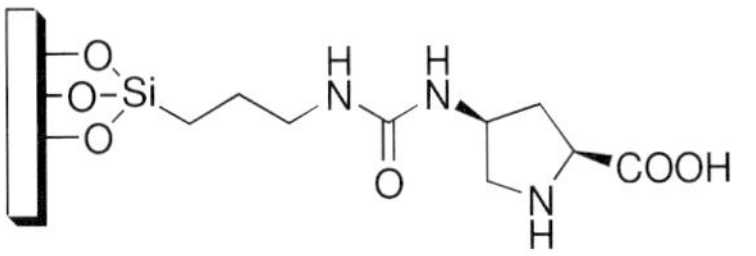

Scheme 8.12 Immobilized (S)-proline derivatives.

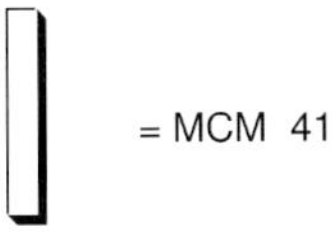

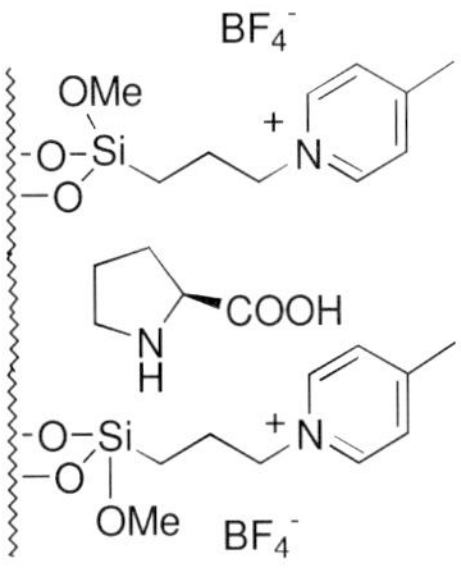

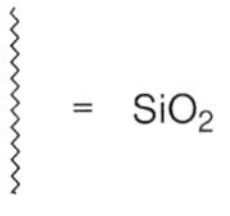

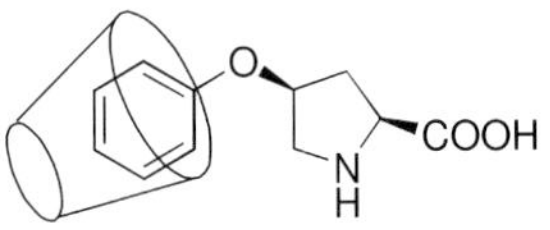

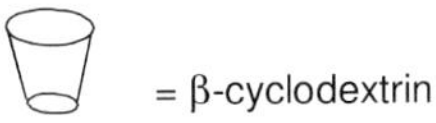

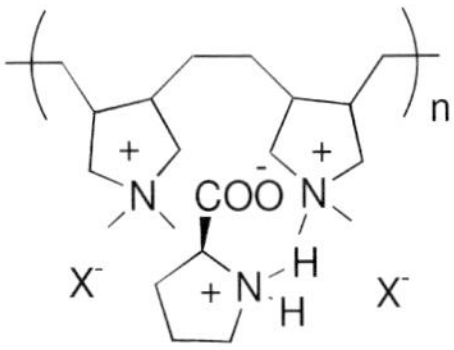

while the reaction with benzaldehyde led to *syn* diols in lower stereocontrol (*syn/anti* ratio 1.4 : 1; ee-value of *syn* 80%). Two recyclings of the catalyst recovered by filtration were shown to occur with slowly decreasing yield and unchanged stereoselectivity.

(S)-Proline has recently been supported on the surface of modified silica gels with a monolayer of covalently attached ionic liquid, with or without additional adsorbed ionic liquid [39]. These materials (e.g. 4-methylpyridinium-modified silica gel, **22**) were able to catalyze the aldol reaction between acetone and several aldehydes, affording the corresponding products in yields and ee-values comparable to those obtained under homogeneous conditions. Moreover, these supported catalysts were easily recovered by simple filtration and reused at least up to seven times.

An interesting noncovalent immobilization technique was exploited by Zhang and coworkers in the synthesis of catalyst **23** (Scheme 8.12). In this case, the nonpolar phenyl ring of 4-phenoxyproline served as the handle for including the amino acid into the β-cyclodextrin cavity. Interestingly, the amount of catalyst actually included was found to depend directly on the temperature of the inclusion reaction, with higher temperatures leading to a higher extent of inclusion. Thus, catalyst samples with different loadings could be obtained [40]. In the aldol addition of acetone to 2-nitrobenzaldehyde, it was shown that in the presence of the highest loaded catalyst (10 mol%/cat) the product was obtained in 90% yield and 83% ee after 16 h at room temperature. The catalyst was recovered by filtration and employed for three subsequent runs with slowly decreasing yields (79%, fourth run) and unchanged enantioselectivity, a behavior that almost perfectly paralleled that observed for PEG-supported proline over the same number of recycling experiments [36]. In this context a *(S)*-proline/polyelectrolyte system has also been recently reported [41], where the amino acid was adsorbed onto the solid support by simply mixing a poly-(diallyldimethylammonium) salt suspension in methanol with the organocatalyst solution in the same solvent (Scheme 8.12). The heterogeneous supported catalytic system **24** was shown to be able to promote the aldol reaction between acetone and aromatic aldehydes in yields and enantioselectivities comparable to those obtained in the proline-catalyzed reactions. The immobilized catalyst was recovered by filtration and reused six times, without any appreciable loss of stereoselectivity.

Very recently, Pericas reported a new strategy to immobilize *trans*-4-hydroxyproline onto an insoluble Merrifield-type polymer by exploiting Cu(I)-catalyzed 1,3-dipolar cycloaddition ('click chemistry') [42]. The supported catalyst **25** was successfully employed in the α-aminoxylation of ketones and aldehydes (Scheme 8.13). Under the optimized reaction conditions (20 mol%/cat, 2 equiv. ketone, DMF, 23 °C, 3 h), the reaction of cyclohexanone with nitrosobenzene catalyzed by **25** gave the product in 60% yield and 98% ee (Scheme 8.13; Equation a). It should be noted that the reaction rates of cyclic ketones with supported catalyst are faster than those reported with *(S)*-proline. The use of a supported catalyst allowed for a simplification of the work-up procedure, as the product could often be obtained after simple filtration of the catalyst and evaporation of the solvents. Furthermore, **25** was recycled up to three times without any decrease in either the chemical and/or stereochemical efficiency.

The same insoluble and recoverable catalyst was shown capable of efficiently promoting the stereoselective aldol reaction in water [43]. The condensation of cyclohexanone with different aromatic aldehydes (Scheme 8.13; Equation b) was efficiently catalyzed in water by **25** in the presence of a substoichiometric amount of water-soluble DiMePEG (M_W 2000) that increased the yields without lowering the enantioselectivity of the process (ee >93% for aromatic and heteroaromatic aldehydes). The high hydrophobicity of the resin and the aqueous environment are believed to be decisive factors for ensuring high stereoselectivity. After filtration, washing the resin with acetic acid and drying, the polymer-supported hydroxyproline was recycled three times without any appreciable loss of performance.

It should be mentioned at this point that, in principle, amine-based catalysts also offer the possibility to be recovered by exploiting their solubility profiles in acids, without the need for functionalization in order to heterogenize the catalytic

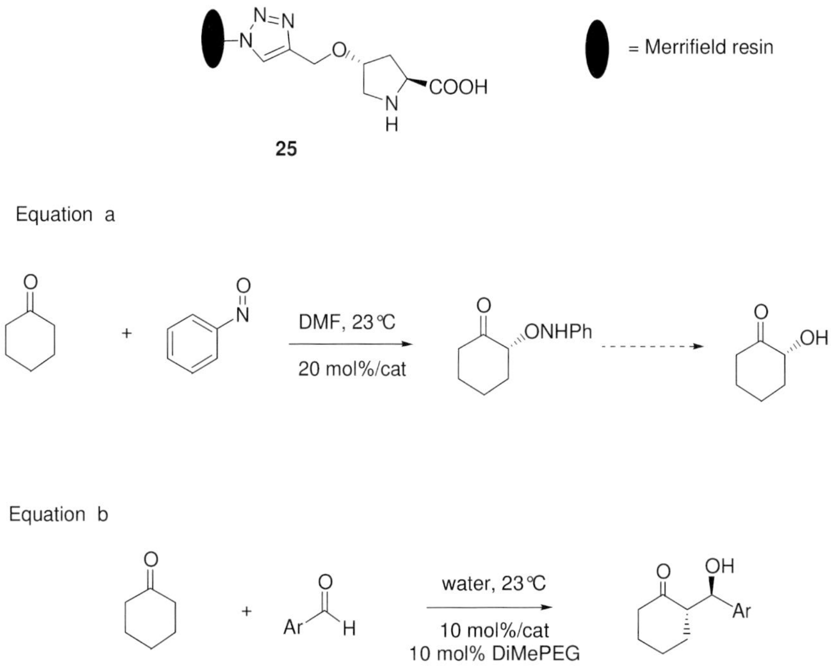

8.6 Catalysts Derived from Amino Acids | 311

25

Equation a

cyclohexanone + PhN=O → (DMF, 23 °C, 20 mol%/cat) → 2-(PhNH-O)-cyclohexanone → 2-hydroxycyclohexanone

Equation b

cyclohexanone + ArCHO → (water, 23 °C, 10 mol%/cat, 10 mol% DiMePEG) → 2-(Ar)(OH)-cyclohexanone

Scheme 8.13 Polystyrene-supported (S)-proline derivatives.

system. Recently, several proline derivatives were prepared and tested in aldol reactions [44]. In this context, the synthesis of new organocatalysts has been realized by connecting the proline moiety to the 1,1′-binaphthyl-2,2′-diamine scaffold, which is easily prepared in a few steps from inexpensive, enantiopure starting materials [45]. In the presence of stearic acid, (S)-prolinamide **26** was able to promote the direct aldol condensation of cyclohexanone and other ketones with different aldehydes in the presence of a massive amount of water in very good yields, high diastereoselectivity, and up to 95% ee (Scheme 8.14) [46].

A preliminary experiment to recover and recycle the catalytic system was attempted for the condensation between cyclohexanone and benzaldehyde catalyzed by **26** and stearic acid. On completion of the reaction pentane was added and the resultant biphasic mixture separated; new reagents were then added to the recovered 'suspension' in aqueous phase, and the reaction was run again. The product was obtained by simple evaporation of the pentane phase. Following this procedure, recycling of the stearic acid/**26** catalytic system provided the product in a 98/2 *anti/syn* ratio and 93% ee (compared to 99/1 *anti/syn* ratio and 93% ee for the first cycle), showing that catalyst reuse is indeed feasible.

Within the field of proline-derivative it should be mentioned that the (2S,4R)-4-hydroxy-N-methylproline derivative **27** supported on JandaJel (a more swellable version of crosslinked PS) was employed by Janda and coworkers to catalyze the kinetic resolution of some cyclic secondary alcohols (Scheme 8.15) [47]. In the presence of 0.15 mol equiv of the catalyst, the benzoylation of racemic *trans*-2-

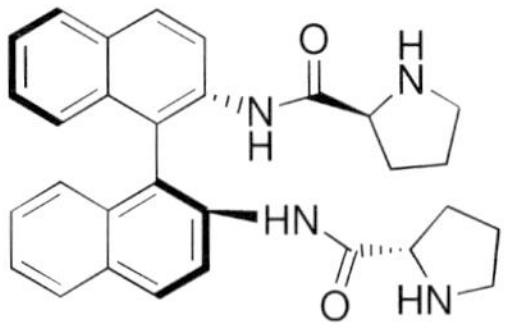

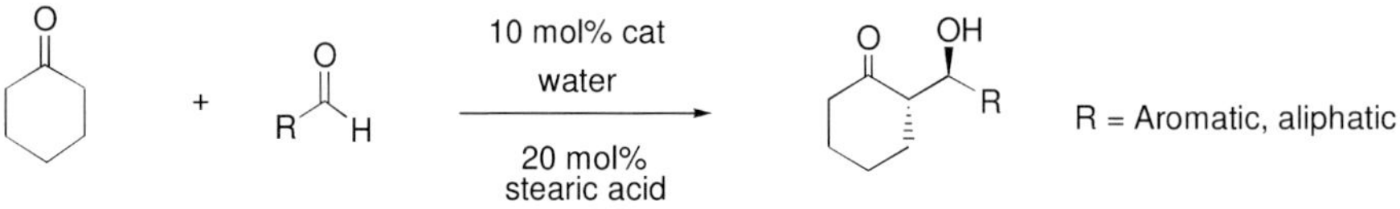

Scheme 8.14 Recoverable prolinamides for aldol reactions in water.

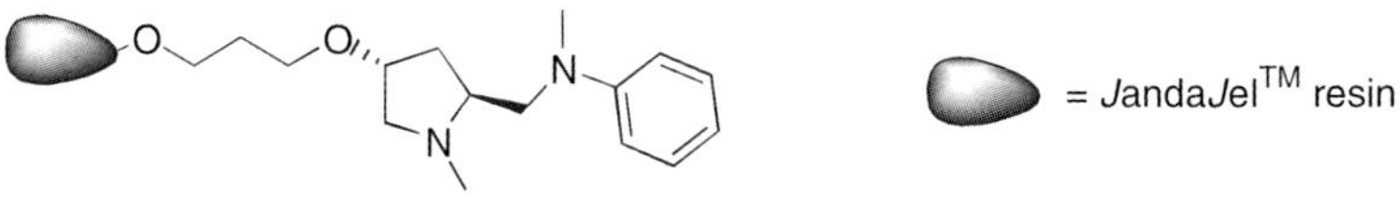

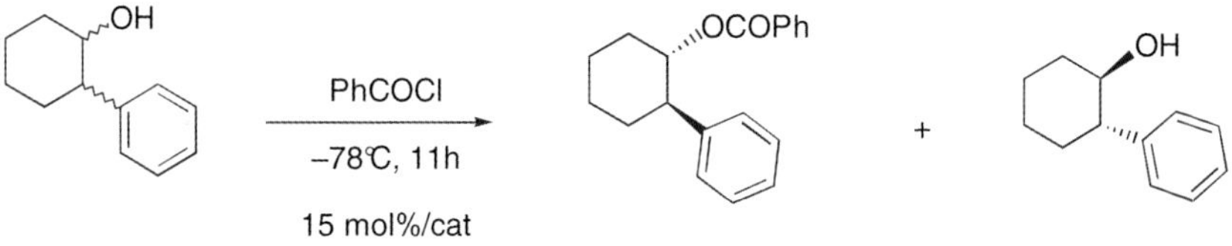

Scheme 8.15 Supported N-methyl proline derivative.

phenylcyclohexanol (benzoyl chloride, triethylamine, dichloromethane, −78 °C, 11 h) afforded a 44% yield of (1S,2R)-(2-phenyl)cyclohexyl benzoate in 96% ee, and a 45% yield of the unreacted (1R,2S)-2-phenylcyclohexanol in 85% ee. An extension of the process to acyclic alcohols proved to be less successful. Recovery and recycling of the insoluble catalyst was demonstrated for up to five reaction cycles, with unchanged yield and enantioselectivity.

8.6.2
Amino Acid-Derived Imidazolinones

Other chiral organic catalysts of major success are the protonated phenylalanine-derived imidazolinones **28** developed by MacMillan [48], that have found widespread use in a number of relevant processes [5, 6]. Immobilized versions of these catalysts have been developed both on soluble (PEG-supported catalyst **29** [49]), and insoluble supports (catalysts **30** and **31** [50]), and employed in enantioselective Diels–Alder cycloadditions of dienes with unsaturated aldehydes (Scheme 8.16).

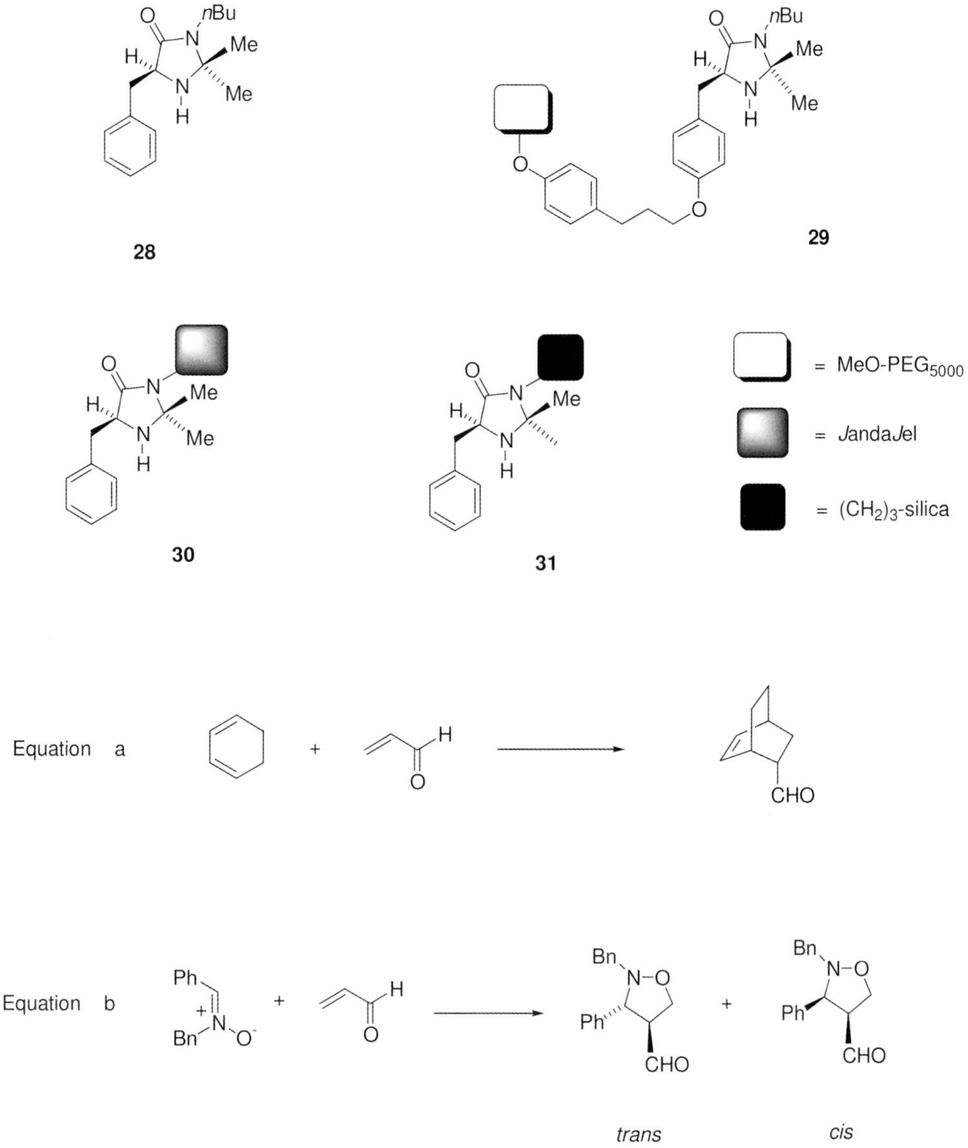

8.6 Catalysts Derived from Amino Acids | 313

Scheme 8.16 Supported chiral imidazolinones.

In the reaction between acrolein and 1,3-cyclohexadiene (Scheme 8.16; Equation a) under the optimized conditions, which involved use of the trifluoroacetate salt of **29** (0.1 mol equiv) in a 95:5 acetonitrile:water mixture at room temperature for 40 h, the product was obtained in 67% yield as a 94:6 mixture of *endo/exo* isomers having 92% and 86% ee, respectively. This result did not differ much from that obtained with 0.05 mol equiv of the nonsupported catalyst (hydrochloride form), that afforded the product in 82% yield and 94% ee under otherwise identical conditions. Unfortunately, catalyst **29** (as well as its nonsupported, commercially available counterpart) proved to be rather unstable under the reaction conditions. Accordingly, catalyst recycling over four cycles showed a marked decrease in the chemical yield of the reaction (from 67% to 38%), while the ee-value was eroded more slowly (from 92% to 85%).

A comparison with insoluble supported systems showed that the JandaJel-supported catalyst **30** performed better not only than its PEG and silica-supported analogues (which behaved almost identically to each other), but also than the nonsupported compound (quite surprisingly).

The PEG-supported imidazolidinone **29** was also employed in 1,3-dipolar cycloadditions [51]. For the reaction of N-benzyl-C-phenylnitrone with acrolein, the reaction outcome was shown to be heavily dependent on the nature of the acid employed to generate the catalyst, and that only the use of HBF_4 allowed reproducible results to be obtained (Scheme 8.16; Equation b). Under the best reaction conditions (20 mol% catalyst, DCM, −20°C, 120 h), the product was obtained in 71% yield as a 85:15 *trans/cis* mixture of isomers, with 87% ee for the *trans* isomer. The major difference between the PEG-supported and nonsupported catalyst resides in the chemical rather than the stereochemical efficiency. Indeed, while the supported catalyst gave *trans/cis* ratios almost identical to and an ee-value only 3–6% lower than those obtained with the nonsupported catalyst, the difference in chemical yields was larger, ranging from 9% to 27%.

The PEG-supported catalyst was recycled twice to afford the product with constant levels of diastereoselectivity and enantioselectivity, but with chemical yields diminishing from 71% to 38%. Several experiments were carried out in order to explain this behavior. For example, after each recovery the supported catalyst was examined using ^1H NMR; this showed degradation which increased after each cycle and was due most likely to an imidazolidinone ring-opening process. Indeed, extensive catalyst degradation in the presence of acrolein, less degradation with crotonaldehyde, and essentially no degradation with cinnamaldehyde, was demonstrated with NMR analysis, while nitrone did not exert any effect on catalyst stability. In agreement with the results of these experiments, the nonsupported catalyst also showed a marked instability and decrease in chemical efficiency when recycled [51].

8.6.3
Other Amino Acids

The use of supported catalysts in the process of chiral catalyst discovery and optimization has already been described for the highly successful Jacobsen's amino acid-derived thioureas employed to promote the hydrocyanation of imines [12].

Another reaction where amino acids play a key role is the Julià–Colonna epoxidation of α,β-unsaturated ketones [52], which involves the use of a catalytic amount of polymeric amino acids, able to catalyze the Weitz–Scheffer epoxidation of chalcone using basic hydrogen peroxide, with high enantioselectivity (Scheme 8.17; Equation a).

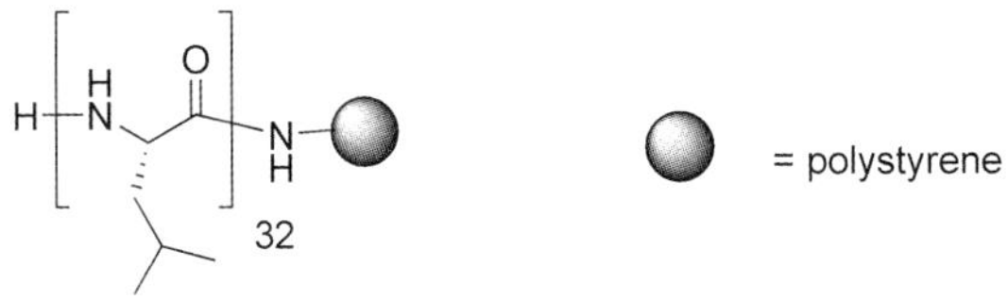

8.6 Catalysts Derived from Amino Acids | 315

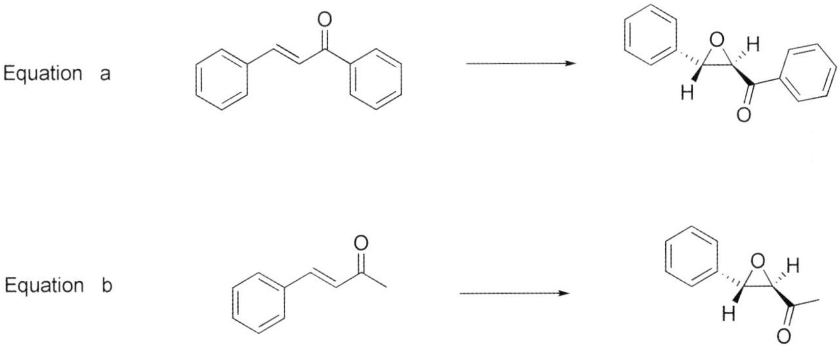

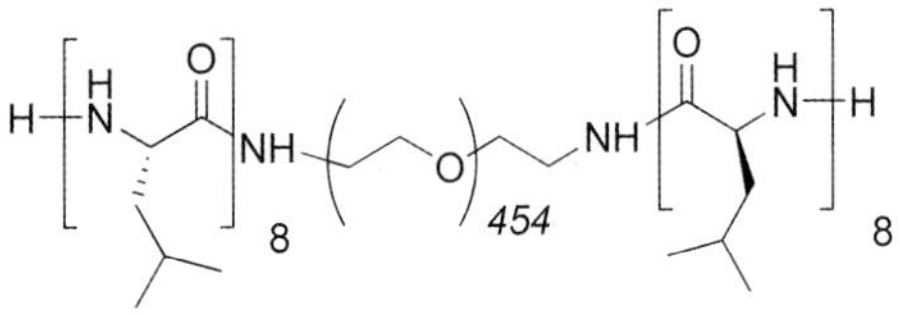

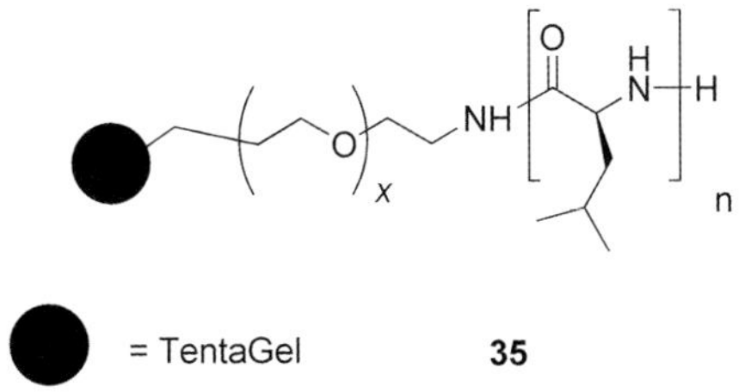

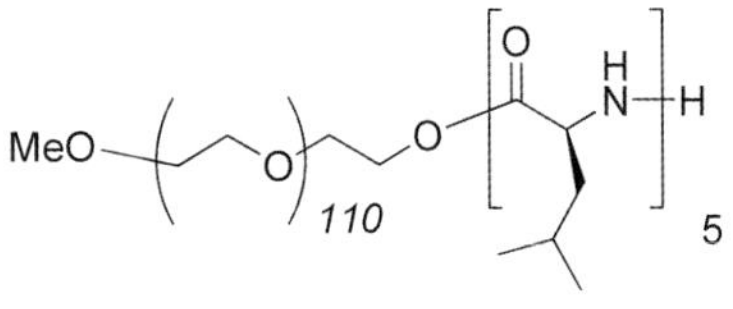

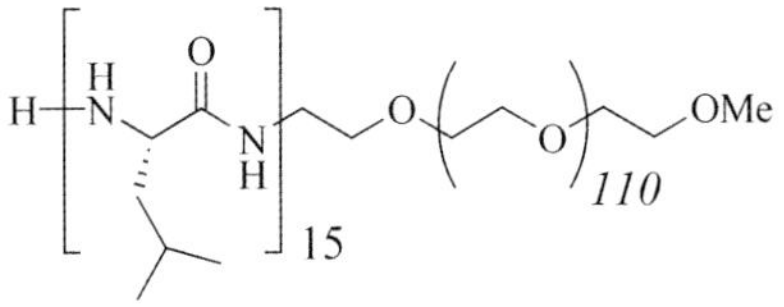

Equation a

Equation b

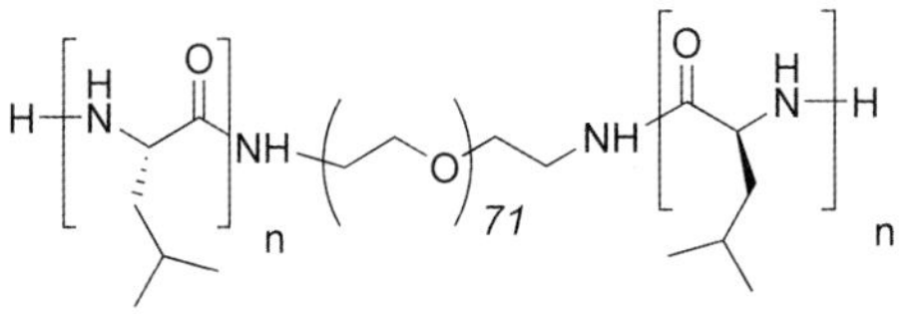

Scheme 8.17 Supported poly(amino acids).

The immobilization of poly(amino acids) on polymeric supports continues to attract much interest. The PS-based supported catalyst **32** developed by Itsuno [53] was used successively by Roberts in an improved procedure for chalcone epoxidation, which involved the use of a urea/hydrogen peroxide complex as the oxidant and 1,8-diazabicyclo[5.4.0]undec-7-ene (DBU) as the base in anhydrous tetrahydrofuran (THF) at room temperature [54]. Under these conditions, fast (30 min), high-yielding (85–100%) and highly stereoselective (>95% ee) reactions were observed. 4-Phenyl-3-buten-2-one was also oxidized in 70% yield and 83% ee under the anhydrous conditions, which suggested that the new protocol could broaden the scope of the Julià–Colonna method by allowing its application to unsaturated ketones other than chalcone (Scheme 8.17; Equation b). However, the recovery and recycling of these catalysts remained a problem. The best results were obtained by Roberts by immobilization of the poly(amino acid) on silica gel; this produced a very active catalyst which was recoverable by filtration and recyclable at least five times without any appreciable loss in activity and stereoselectivity [55].

Following extended studies on insoluble PS-supported polyamino acids, later efforts focused on the immobilization on soluble PEGs to prove the superiority of the latter support. The first soluble version of the Julià–Colonna catalyst was reported in 2001, when Roberts and coworkers described the synthesis of the triblock PEG-polyleucine adducts **33**, which were able to promote the conversion of chalcone to epoxichalcone in up to 97% ee (Scheme 8.17) [56]. The catalysts were prepared by using the commercially available O,O-bis(2-aminoethyl)-poly(ethylene glycol) of average M_W 3350 Da as the initiator of the polymerization of (L)-leucine N-carboxyanhydride. On the basis of their performances, the high-molecular-weight soluble catalyst **34** (Scheme 8.17; average M_W of the PEG fragment 20 000 Da), featuring two terminal leucine octamers, was synthesized [57]. This catalyst was employed to promote the epoxidation of chalcone (99% yield, 94% ee) in a continuously operated membrane reactor, where catalyst retention was achieved by means of a nanofiltration membrane. This equipment allowed 25 reaction cycles to occur with almost unchanged yield and selectivity, after which some decreases in both values were observed.

Berkessel and coworkers prepared a series of catalysts having a general structure **35** (Scheme 8.17) by attaching the C terminus of leucine oligomers of different length to the PEG-modified DVB crosslinked polystyrene TentaGel S NH_2 [58]. These compounds were used to establish that a polymer-supported catalyst containing as few as five amino acid residues was able to catalyze the Julià–Colonna epoxidation of chalcone with up to 98% ee. By replacing the insoluble support of **35** with soluble MeOPEG, adduct **36** was obtained (Scheme 8.17). In this case, a leucine pentamer was also found to be the minimum structural requirement for achieving stereocontrol (>50% ee). Since at least four amino acids residues are required to form one turn of the α-helical structure, it was concluded that one complete turn was required for efficient stereoselectivity. Since in the case of nonsupported polyleucine catalyst a good level of enantioselectivity is observed only at the decamer level, the PEG portions of **35** and **36** could act as a helix surrogate, forcing the oligopeptide to adopt the helical arrangement at shorter chain lengths. More recently, by attaching a polyleucine chain to MeOPEG$_{5000}$NH$_2$, Kelly and Roberts prepared a new soluble catalyst **37** (Scheme 8.17) [59] that promoted the epoxidation of chalcone with a urea/hydrogen peroxide complex in 95% conversion and 97% ee (DBU, THF, room temperature, 3 h). Circular dichroism measurements have shown that the peptide fragments in this catalyst exist mostly in the α-helical structure (86%), once again suggesting the existence of a direct relationship between the content of α-helical structure and catalytic activity and, to a lesser extent, enantioselectivity. Further experiments showed that the minimum number of Leu residues necessary to achieve high stereoselectivity with these PEG-supported catalysts was six; this was in good agreement with the observation that at least four amino acids are required to form a whole turn of the α-helical structure.

8.7
Miscellaneous Catalysts

Dioxiranes derived from chiral ketones have been used extensively to promote the enantioselective epoxidation of alkenes [60]. The first examples of immobilized chiral ketones for this reaction were reported by Sartori and Armstrong [61]. A modified racemic tropinone was supported onto amorphous silica KG-60, mesoporous silica MCM-41, and 2% crosslinked PS, respectively, to obtain the insoluble precatalysts **38a–c** (Scheme 8.18).

In the epoxidation of 1-phenylcyclohexene (40 mol% catalyst, oxone as the terminal oxidant, acetonitrile, aqueous NaHCO$_3$, room temperature, 1 h), it was discovered that the siliceous materials, having different porosity and loading (200 and 670 m^2 g^{-1}, and 0.58 and 0.78 mmol g^{-1}, for **38a** and **38b**, respectively), performed much better than the PS-supported ketone in terms of chemical activity. Modified tropinone samples having 78% ee were then supported on KG-60 and MCM-41

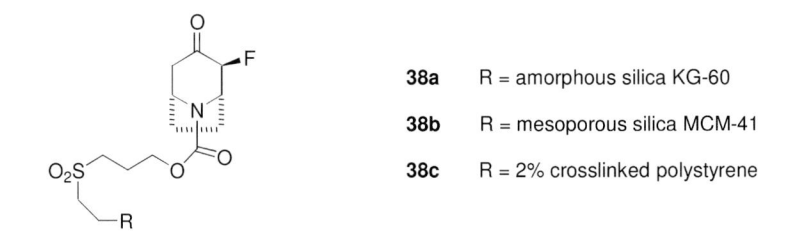

38a R = amorphous silica KG-60
38b R = mesoporous silica MCM-41
38c R = 2% crosslinked polystyrene

40 mol%/cat, oxone, acetonitrile,
aq. NaHCO$_3$, 25 °C, 1 h

Scheme 8.18 Supported catalysts for alkene oxidation.

and employed to promote the epoxidation of a series of trisubstituted alkenes in >93% conversions, >91% yields and 58–80% ee (corrected for the ee-value of tropinone). It is important to note that the observed ee-values were only slightly lower than those obtained with the nonsupported catalyst, although the latter was employed at a lower loading (10 mol%). Finally, the catalyst recovered by filtration was reused three times with unchanged activity and stereoselectivity.

Following the pioneering studies of Jacobsen, several groups have recently investigated the use of ureas and thioureas as chiral Brønsted acid catalysts [62]. Chiral bifunctional catalysts incorporating the thiourea motif have been shown to efficiently promote the aza-Henry and Michael reactions [63]; however, these reactions suffered from the difficulty of recovering the enantiomerically pure organocatalyst. Very recently, Takemoto studied the immobilization of a structurally modified version of the successful bifunctional catalyst **39** [64]. In order to attach the metal-free catalyst to different polymer supports, the ester derivative **40** was prepared. In the diethyl malonate addition to trans-β-nitrostyrene promoted by **40**, the product was obtained after two days in 88% yield and 91% ee (versus 86% yield and 93% ee with **39**; Scheme 8.19). On the basis of these results, a PEG-bound thiourea **41** was prepared and tested in the same reaction to afford the product after 6 days in 71% yield and 86% ee. The recovery and recycle of the supported catalyst was also described.

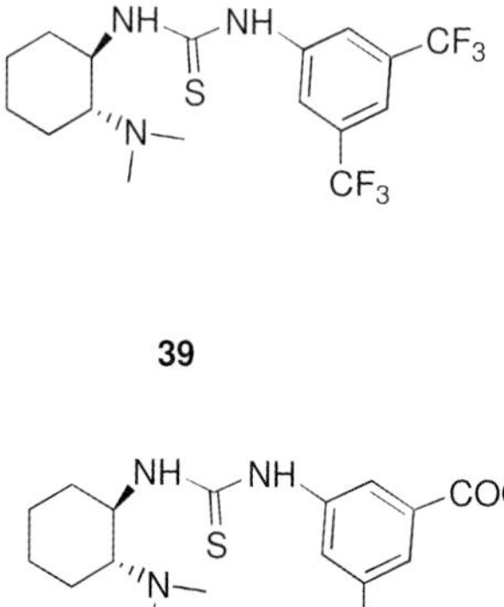

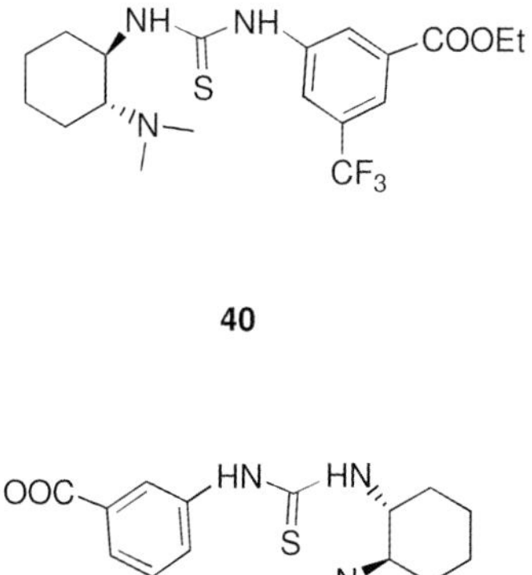

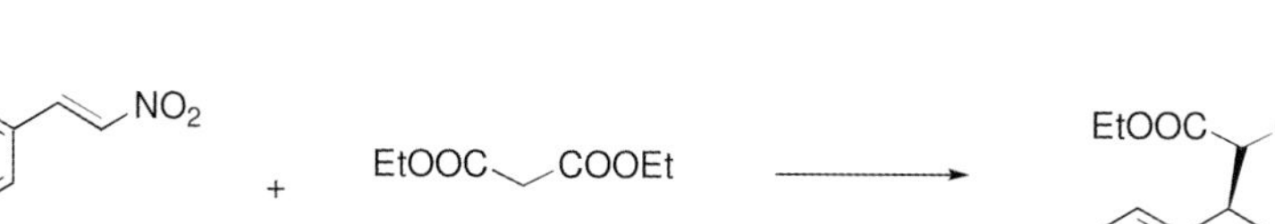

Scheme 8.19 Supported thiourea derivative as chiral bifunctional catalyst.

8.8
Outlook and Perspectives

During the past 20 years, solid-supported organic catalysts have become powerful synthetic tools readily available to the chemical community. The reasons for developing an immobilized version of a chiral catalyst go well beyond the simple – yet still fundamental – aspect of the recovery and recycling of the precious catalytic species. Catalyst stability, structural characterization, catalytic behavior, new or different solubility properties, simplification of the reaction work-up, catalyst discovery and optimization, use in environmentally friendly or 'green' solvents are all issues that may be conveniently addressed working with supported systems.

From a practical perspective, the success or failure of a supported catalyst seems to be independent of the soluble/insoluble nature of the support, or of the class of compounds to which the catalyst belongs. Rather, the support behavior seems fundamental, as the solid support should exert a positive influence or at least not interfere with the reaction course. Therefore, the choice of the support is also crucial, as several of its features may influence the catalyst behavior at every level. The decision to develop an homogeneous or a heterogeneous catalytic system is the first to be made when designing an immobilization. It is impossible to say whether it is better to work with a soluble or an insoluble catalyst, as each has its positive and negative characteristics. While the homogeneous catalytic system is expected to be more reactive, stereochemically more efficient (because it operates in a solution exactly like the nonsupported system) and more reliable in reproducibility than a heterogeneously supported catalyst, the latter is normally believed to be more stable, easier to handle, and simpler to recover and recycle. Although the ideal support does not exist, the best immobilization technique should be selected for each specific catalytic system to be supported. In this sense, significant progress could derive from the development of an interdisciplinary expertise with the contributions of organic, polymer and materials chemists. This may also allow certain problems to be solved which relate to the cost and commercial availability of the support, which is yet another issue of major importance in determining (and somehow limiting) the choice of the support.

Within the context of catalyst separation and recycling, it must be noted that a system where a catalyst needs not to be removed from the reaction vessel is very attractive. One such example occurs in continuous-flow methods [65], by which the immobilized catalyst resides permanently in the reactor, where it transforms the entering starting materials into the exiting products. The retention of a catalyst inside the reaction vessel can be achieved by different techniques, ranging from ultrafiltration through a M_W-selective membrane to immobilization on a silica gel column [66].

The stability of the nonsupported catalyst under the reaction conditions is a prerequisite that must be firmly established both in the presence and in the absence of the intended support. Possible candidates to be supported should be catalysts of great versatility and of wide tolerance of structurally different substrates, possibly with a high catalytic efficiency and with a well-tested stability. The steric and electronic effects exerted by the structural modification of the catalyst required for immobilization should be minimized.

In conclusion, it is clear that stereoselective organocatalysis has today achieved the same relevance to asymmetric synthesis as organometallic catalysis, and each day novel *fully organic* methods are discovered and developed to perform an even wider variety of reactions in the future. A few chiral organocatalysts are already employed in large-scale preparations at the industrial level [12, 62, 67]. In this context, the development of immobilized chiral organic catalysts will play an important role, to further expand the applicability of organic catalysts and even to facilitate the discovery of new chiral organic catalytic species. The immobilization of chiral organic catalysts represents a relatively new field of research which is undergoing great expansion and is open to the interdisciplinary contributions of both organic and materials chemists.

References

1. (a) Jacobsen, E.N., Pfaltz, A. and Yamamoto, H. (1999) *Comprehensive Asymmetric Catalysis*, Springer, Berlin.
   (b) Noyori, R. (1994) *Asymmetric Catalysis in Organic Synthesis*, John Wiley & Sons, Ltd, New York.
2. Blaser, H.U. (2003) *Chemical Communications*, 293–7.
3. Sheldon, R.A. and van Bekkum, H. (2001) *Fine Chemicals through Heterogeneous Catalysis*, Wiley-VCH Verlag GmbH, Weinheim.
4. Anastas, P.T. and Warner, J.C. (1998) *Green Chemistry: Theory and Practice*, Oxford University Press, New York.
5. Berkessel, A. and Groger, H. (2005) *Asymmetric Organic Catalysis*, Wiley-VCH Verlag GmbH, Weinheim.
6. (a) Review: Dalko, I. and Moisan, L. (2004) *Angewandte Chemie – International Edition*, **35**, 5138–75.
   (b) List, B. and Houk, K.N. (eds) (2004) *Accounts of Chemical Research*, **37** (thematic issue 8), 487–631.
   (c) List, B. Bolm, C. (eds) (2004) *Advanced Synthesis Catalysis*, **346** (thematic issue 9-10), 1007–1249.
7. De Vos, D.E., Vankelecom, I.F.J. and Jacobs, P.A. (2000) *Chiral Catalysts Immobilization and Recycling*, Wiley-VCH Verlag GmbH, Weinheim.
8. Reviews on supported organic catalysts: (a) Benaglia, M. Puglisi, A. and Cozzi, F. (2003) *Chemical Reviews*, **103**, 3401–29.
   (b) Cozzi, F. (2006) *Advanced Synthesis Catalysis*, **348**, 1367–90.
   For a perspective on chiral supported metal-free catalysts see: (c) Benaglia, M. (2006) *New Journal of Chemistry*, **30**, 1525–33.
9. Drauz, K. and Waldmann, H. (2002) *Enzyme Catalysis in Organic Synthesis: A Comprehensive Handbook*, 2nd edn, Vol. 3, Wiley-VCH Verlag GmbH, Weinheim.
10. Janda, K.D., Dickerson, T.J. and Reed, N.N. (2002) *Chemical Reviews*, **102**, 3325–244.
11. Benaglia, M., Danelli, T., Fabris, F., Sperandio, D. and Pozzi, G. (2002) *Organic Letters*, **4**, 4229–32.
12. Sigman, M.S. and Jacobsen, E.N. (1998) *Journal of the American Chemical Society*, **120**, 4901–2.
13. Pozzi, G., Cavazzini, M., Quici, S., Benaglia, M. and Dell'Anna, G. (2004) *Organic Letters*, **6**, 441–3.
14. For other supported TEMPO catalysts see: (a) Dijksman, A., Arends, I.W.C.E. and Sheldon, R.A. (2000) *Chemical Communications*, 271–2.
    (b) Benaglia, M., Puglisi, A., Holczknecht, O., Quici, S. and Pozzi, G. (2005) *Tetrahedron*, **61**, 12058–64 and references therein.
15. Review on asymmetric phase-transfer catalysis: Nelson, A. (1999) *Angewandte Chemie – International Edition in English*, **38**, 1583.

16 Chinchilla, R., Mazòn, P. and Najera, C. (2000) *Tetrahedron: Asymmetry*, **11**, 3277.
17 Thierry, B., Plaquevent, J. and Cahard, D. (2001) *Tetrahedron: Asymmetry*, **12**, 983.
18 Thierry, B., Perrard, T., Audouard, C., Plaquevent, J. and Cahard, D. (2001) *Synthesis*, 1742.
19 Danelli, T., Annunziata, R., Benaglia, M., Cinquini, M., Cozzi, F. and Tocco, G. (2003) *Tetrahedron: Asymmetry*, **14**, 461.
20 Thierry, B., Plaquevent, J.-C. and Cahard, D. (2003) *Tetrahedron: Asymmetry*, **14**, 1671–7.
21 Wang, X., Yin, L., Yang, T. and Wang, Y. (2007) *Tetrahedron: Asymmetry*, **18**, 108.
22 For a review on the use of cinchona alkaloids as organic catalysts see: Kacprzak, K. and Gawronski, J. (2001) *Synthesis*, 961–98.
23 Alvarez, R., Hourdin, M.-A., Cavè, C., d'Angelo J. and Chaminade, P. (1999) *Tetrahedron Letters*, **40**, 7091.
24 Hafez, A.M., Taggi, A.E., Dudding, T. and Lectka, T. (2001) *Journal of the American Chemical Society*, **123**, 10853–9 and references cited therein.
25 Denmark, S.E., Nishiogaichi, X., Su, Y., Coe, D.M., Wong, K.-T., Winter, S.B.D. and Choi, J.Y. (1999) *Journal of Organic Chemistry*, **64**, 1958–67 and references cited therein.
26 Allylation: (a) Denmark, S.E., Fu, J. and Lawler, M.J. (2006) *Journal of Organic Chemistry*, **71**, 1523–36.
(b) Denmark, S.E., Fu, J., Coe, D.M., Su, X., Pratt, N.E. and Griedel, B.D. (2006) *Journal of Organic Chemistry*, **71**, 1513–22 and references cited therein.
Aldol condensation: (c) Denmark, S.E. and Bui, T. (2005) *Journal of Organic Chemistry*, **70**, 10393–9.
27 Oyama, T., Yoshioka, H. and Tomoi, M. (2005) *Chemical Communications*, 1857–9.
28 For a recent study describing the determination of site-site distance and site concentration within polymer beads of different polystyrene-type polymers see: Marchetto, R., Cilli, E.M., Jubilut, G.N., Schreier, S. and Nakaie, C.R. (2005) *Journal of Organic Chemistry*, **71**, 4561–8.
29 Murugan, R. and Scriven, E.F.V. (2003) *Aldrichimica Acta*, **36**, 21.
30 Priem, G., Pelotier, B., Macdonald, S.J.F., Anson, M.S. and Campbell, I.B. (2003) *Journal of Organic Chemistry*, **68**, 3844–8.
31 Priem, G., Pelotier, B., Macdonald, S.J.F., Anson, M.S. and Campbell, I.B. (2003) *Synlett*, 679–83.
32 Jarvo, E.R. and Miller, S.J. (2002) *Tetrahedron*, **58**, 2481–95.
33 List, B. (2001) *Synlett*, 1675.
34 Benaglia, M., Celentano, G., Cinquini, M., Puglisi, A. and Cozzi, F. (2002) *Advanced Synthesis Catalysis*, **344**, 149–52.
35 List, B., Lerner, R.A. and Barbas, C.F. III, (2000) *Journal of the American Chemical Society*, **122**, 2395.
36 Benaglia, M., Cinquini, M., Cozzi, F., Puglisi, A. and Celentano, G. (2002) *Advanced Synthesis Catalysis*, **344**, 533–42.
37 Proline supported on 1% crosslinked polystyrene: (a) Kondo, K., Yamano, T. and Takemoto, K. (1985) *Makromolekulare Chemie – Macromolecular Chemistry and Physics*, **186**, 1781. Proline supported on silica gel column: (b) Sakthivel, K., Notz, W., Bui, T. and Barbas, C.F. (2001) *Journal of the American Chemical Society*, **123**, 5260–7.
38 Calderòn, F., Fernandez, R., Sanchez, F. and Fernandez-Mayoralas, A. (2005) *Advanced Synthesis Catalysis*, **347**, 1395–403.
39 (a) Gruttadauria, M., Riela, S., Aprile, C., Lo Meo, P., D'Anna, F. and Noto, R. (2006) *Advanced Synthesis Catalysis*, **348**, 82–92.
(b) See also Giacalone, F., Gruttadauria, M., Marculescu, A.M. and Noto, R. (2007) *Tetrahedron Letters*, **48**, 255–9.
40 Shen, Z., Liu, Y., Jiao, C., Ma, J., Li, M. and Zhang, Y. (2005) *Chirality*, **17**, 556–8.
41 Kucherenko, A.S., Struchkova, M.I. and Zlotin, S.G. (2007) *European Journal of Organic Chemistry*, 2000–4.
42 Font, D., Bastero, A., Sayalero, S., Jimeno, C. and Pericas, M.A. (2007) *Organic Letters*, **9**, 1943–7.
43 Font, D., Jimeno, C. and Pericass, M.A. (2006) *Organic Letters*, **8**, 4653–6.
44 (a) Tang, Z., Yang, Z., Chen, X.-H., Cun, L., Mi, A., Jiang, Y. and Gong, L.-Z. (2005) *Journal of the American Chemical Society*, **127**, 9285.
(b) Samanta, S., Liu, J., Dodda, R. and Zhao, C.-G. (2005) *Organic Letters*, **7**, 5321.

(c) Chen, J.-R., Lu, H., Li, X.-Y., Cheng, L., Wan, J. and Xiao, W.-J. (2005) *Organic Letters*, **7**, 4543.

45 (a) Guillena, G., Hita, M. and Najera, C. (2006) *Tetrahedron: Asymmetry*, **17**, 1027.
(b) Gryko, D., Kowalczyk, B. and Zawadzki, L. (2006) *Synlett*, 1059.
(c) Guizzetti, S., Benaglia, M., Pignataro, L. and Puglisi, A. (2006) *Tetrahedron: Asymmetry*, **17**, 2754.
(d) Guillena, G., Hita, M. and Najera, C. (2006) *Tetrahedron: Asymmetry*, **17**, 1493.

46 Guizzetti, S., Benaglia, M., Raimondi, L. and Celentano, G. (2007) *Organic Letters*, **9**, 1247–50. See also ref. 45d.

47 Clapham, B., Cho, C.-W. and Janda, K.D. (2001) *Journal of Organic Chemistry*, **66**, 868.

48 Ahrendt, K.A., Borths, C.J. and MacMillan, D.W.C. (2000) *Journal of the American Chemical Society*, **122**, 4243–4.

49 Benaglia, M., Celentano, G., Cinquini, M., Puglisi, A. and Cozzi, F. (2002) *Advanced Synthesis Catalysis*, **344**, 149–52.

50 Selkälä, S.A., Tois, J., Pihko, P.M. and Koskinen, A.M.P. (2002) *Advanced Synthesis Catalysis*, **344**, 941–5.

51 Benaglia, M., Celentano, G., Cinquini, M., Puglisi, A. and Cozzi, F. (2004) *European Journal of Organic Chemistry*, 567–73.

52 Colonna, S., Molinari, H., Banfi, S., Julià, S. and Guixer, J. (1984) *Tetrahedron*, **40**, 5207.

53 Itsuno, S., Sakakura, M. and Ito, K. (1990) *Journal of Organic Chemistry*, **55**, 6047.

54 Bentley, P.A., Bergeron, S., Cappi, M.W., Hibbs, D.E., Hursthouse, M.B., Nugent, T.C., Pulido, R., Roberts, S.M. and Wu, L. (1997) *Chemical Communications*, 739.

55 Bentley, P.A., Flood, R.W., Roberts, S.M., Skidmore, J., Smith, C.B. and Smith, J.A. (2001) *Chemical Communications*, 1616–17.

56 Flood, R.W., Geller, T.P., Petty, S.A., Roberts, S.M., Skidmore, J. and Volk, M. (2001) *Organic Letters*, **3**, 683–6.

57 Tsogoeva, S.B., Wöltinger, J., Jost, C., Reichert, D., Künle, A., Krimmer, H.P. and Drauz, K. (2002) *Synlett*, 707–10.

58 Berkessel, A., Gasch, N., Glaubitz, K. and Koch, C. (2001) *Organic Letters*, **3**, 3839–41.

59 Kelly, D.R., Bui, T.T.T., Caroff, E., Drake, A.F. and Roberts, S.M. (2004) *Tetrahedron Letters*, **45**, 3885–8.

60 Denmark, S.E. and Wu, Z. (1999) *Synlett*, 847–59.

61 Sartori, G., Armstrong, A., Maggi, R., Mazzacani, A., Sartorio, R., Bigi, F. and Dominguez-Fernandez, B. (2003) *Journal of Organic Chemistry*, **68**, 3232–7.

62 Connon, S.J. (2006) *Chemistry – A European Journal*, **12**, 5418–27.

63 Takemoto, Y. (2005) *Organic and Biomolecular Chemistry*, **3**, 4299–306.

64 Miyabe, H., Tuchida, S., Yamauchi, M. and Takemoto, Y. (2006) *Synthesis*, 3295–300.

65 Review: Jas, G. and Kirschning, A. (2003) *Chemistry – A European Journal*, **9**, 5708–23.

66 For examples of organocatalyzed reactions carried out under the continuous-flow mode see: (a) Ishiara, K., Hasegawa, A. and Yamamoto, H. (2002) *Synlett*, 1296–8.
(b) Brünjes, M., Sourkouni-Argirusi, G. and Kirschning, A. (2003) *Advanced Synthesis Catalysis*, **345**, 635–42.

67 Ikunaka, M. (2007) *Organic Process Research and Development*, **11**, 495–502.

9
Homochiral Metal–Organic Coordination Polymers for Heterogeneous Enantioselective Catalysis: Self-Supporting Strategy

Kuiling Ding and Zheng Wang

9.1
Introduction

The asymmetric catalysis of organic reactions to provide enantiomerically enriched products is of intense current interest to synthetic chemists in general, and to the pharmaceutical industry in particular [1]. Although significant achievements have been made in the design and application of homogeneous chiral catalysts for asymmetric transformations, most of these have not yet had any major impact on the practical synthesis of fine chemicals [2]. One main reason for this is that the expensive ligands and/or metals used to produce the chiral catalysts (with loadings usually in the range of 1–10 mol%) are often difficult to recover and reuse for homogeneous catalytic processes. Furthermore, the contamination of such products with metal traces that is often associated with a homogeneous process is especially unacceptable in the production of pharmaceuticals. Hence, the immobilization of homogeneous catalysts for heterogeneous asymmetric catalysis, as one of the most favorable methods for resolving these problems, has recently attracted a great deal of attention [3].

Over the past few decades, several approaches have been taken for the immobilization of chiral catalysts, including the use of inorganic materials, organic polymers, dendrimers and membranes as supports. Alternatively, the reactions have been conducted in ionic liquid, fluorous phase and biphasic systems [3, 4]. Both, covalent and noncovalent interactions have been employed to attach the chiral ligand or complex to the support, and many immobilized chiral catalysts have been developed for a wide variety of catalytic asymmetric reactions. Despite the significant achievements made in this field, however, many problems persist during the immobilization of homogeneous catalysts, causing flaws in their performance. In most cases the immobilized catalysts display a lower activity and/or a reduced enantioselectivity compared to their homogeneous counterparts. As a result of the unstable linkage between the catalyst and support, catalyst leaching may occasionally cause problems for an immobilized catalyst. In addition, the synthesis of polymer-anchored chiral ligands/catalysts can be arduous as an additional

Handbook of Asymmetric Heterogeneous Catalysis. Edited by K. Ding and Y. Uozumi
Copyright © 2008 WILEY-VCH Verlag GmbH & Co. KGaA, Weinheim
ISBN: 978-3-527-31913-8

functionalization of ligands is often necessary. When examined from these perspectives, alternative methods for the preparation of recyclable and reusable heterogeneous asymmetric catalysts are clearly highly desirable for practical applications of asymmetric catalysis in fine-chemical synthesis.

Recently, a novel class of immobilized catalysts based on homochiral metal–organic hybrid materials (MOHMs) has been developed, which demonstrated high enantioselectivity and efficiency in a variety of heterogeneous asymmetric catalytic reactions [3b,c, 5–8]. Here, MOHMs are designated as compounds containing both inorganic species and organic molecules as part of an infinite framework, which may be subdivided into metal–organic coordination polymers and organic–inorganic hybrid polymers, depending on their framework compositions (Scheme 9.1) [9, 10]. Metal–organic coordination polymers, alternatively known as metal–organic coordination networks (MOCNs) or metal-organic frameworks (MOFs), can be defined as extended arrays of metal–ligand compounds composed of metal atoms or clusters bridged by polyfunctional (polytopic) organic molecules (Scheme 9.1a) [10, 11]. In the simplest case, the coordination polymer consists of one metal ion and one bridging ligand, both possessing at least two coordination sites. Polymeric chains, layers or networks can be formed depending on the number and orientation of these coordination sites. On the other hand, organic–inorganic hybrid polymers contain infinite metal–X–metal (M—X—M, X = donor atoms O, N, S, Se, etc.) arrays (inorganic units) as a part of their structures, which are modified by organic ligands/functionalities (Scheme 9.1b) [9, 12]. Versatile synthetic approaches have been developed for the assembly of these MOHM structures from molecular building blocks, which can maintain their structural integrity throughout the reactions under mild conditions [7, 9–12]. In principle, such a modular approach in using molecular building units for the assembly of the MOHMs allows systematic engineering of the desired chemical and physical properties, including chirality and catalytic activity, into the resulting extended structures. For example, using a polyfunctional chiral ligand with two (or more) ligating sites and catalytically active metals for the formation of a coordination polymer or modification of an organic–inorganic hybrid polymer (usually a hybrid metal oxide), a MOHM assembly with chiral topology and catalytic activity can be obtained. These homochiral MOHMs usually show low solubility in common organic solvents, and thus may be exploited for applications in heterogeneous asymmetric catalysis [5–8].

Although having a high potential as heterogeneous catalysts in enantioselective synthesis, only a handful of examples using homochiral MOHMs have been reported to date (*vide infra*). The aim of this chapter is to introduce fundamental structural and synthetic aspects in designing homochiral metal–organic coordination polymer (MOCP) catalysts, as well as their applications in heterogeneous asymmetric catalysis. The chapter begins with a brief historical account of the evolution of the concept, listing several selected examples using MOCPs as heterogeneous achiral catalysts. This is followed by discussions on the general design principles of MOCP catalysts, with several issues key to MOCP catalysis being

a) Metal-organic coordination polymers

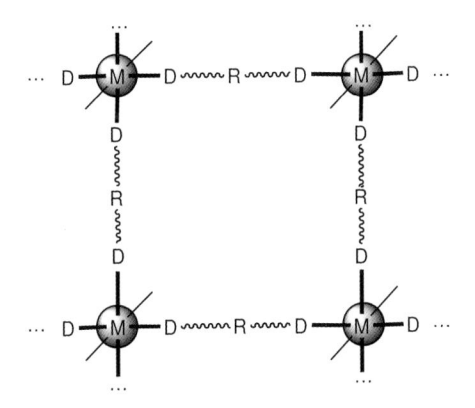

b) Organic-inorganic hybrid Polymers

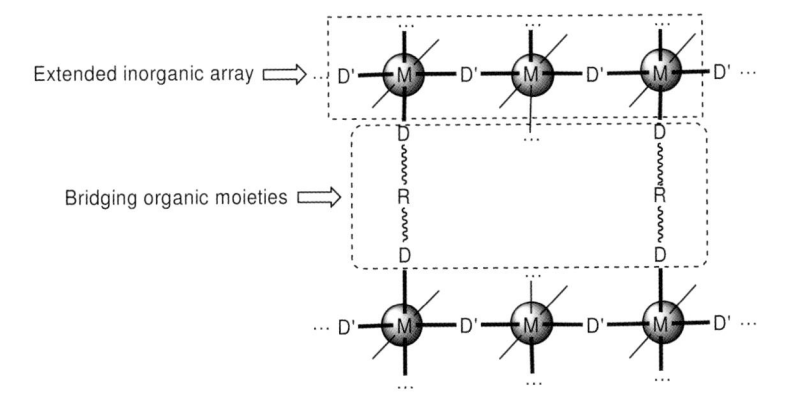

M: Metal atoms
D, D': Donor atoms such as O, N, P, S, Se, etc.
R: Organic spacers, linkers, or functional groups

Scheme 9.1 Schematic representations of metal–organic coordination polymers and organic–inorganic hybrid polymers.

briefly addressed. The following sections provide up-to-date application examples of homochiral MOCP catalysts, with an assessment of both catalytic performance and reusability of MOCP in heterogeneous asymmetric reactions. The chapter concludes with a summary and outlook of the field. As many of the MOCP structures are difficult to depict and require a considerable space to visualize, some will be represented with molecular formulae, except in cases where detailed structures are necessary and appropriate for discussions.

9.2
A Historical Account of Catalytic Applications of MOCPs

Since the 1990s, there has been a surge of research interest in MOCPs owing to their potential applications in catalysis and other areas [7, 9–12]. Some of these coordination polymers not only possess unsaturated metal sites as potentially catalytic metal centers, but also are highly porous and insoluble in organic solvents, thus showing promise for applications in heterogeneous catalysis. Consequently, several excellent reviews on the application of MOCPs as achiral heterogeneous catalysts for organic transformations are available [7, 10, 11a]. In 1994, Fujita and coworkers first reported the heterogeneous catalysis by a porous coordination polymer [13]. A lamellar coordination polymer $[Cd(NO_3)_2(4,4'\text{-bpy})_2]_n$ with cavities surrounded by 4,4'-bpy units, was tested as a Lewis acidic catalyst for the cyanosilation of several aldehydes with cyanotrimethylsilane. Control experiments using powdered $Cd(NO_3)_2$ or 4,4'-bpy alone, or the supernatant from a CH_2Cl_2 suspension of the coordination polymer led to no reaction, thus demonstrating the heterogeneous nature of the catalysis. An interesting shape selectivity of the substrates was observed, which was ascribed to the cavity size of the network. The results of these studies showed that the incorporation of catalytic active metal centers into coordination polymers could be employed as a strategy for designing heterogeneous catalysts. Later, Ohmori and Fujita showed that the same coordination polymer could be used to catalyze the cyanosilylation of imines under heterogeneous conditions [14]. In most cases, the products were isolated quantitatively by simple filtration of the heterogeneous catalyst.

Since the completion of these seminal investigations, explorations on coordination polymers as heterogeneous catalysts have begun to attract increasing attention, and several types of metal poly-phenoxide coordination polymers have now been used as heterogeneous Lewis-acidic catalysts for organic transformations. Aoyama and coworkers demonstrated that Lewis-acidic Zr, Ti and Al coordination polymers with the formulae $[Zr_2(OtBu)_4(L)]_n$, $[Ti_2(OiPr)_2Cl_2(L)]_n$, $[Al_2Cl_2(L)]$, or $[Al_2(OiPr)_2(L)]$ (L = anthracene bisresorcinol derivative), but of unknown structures, could be used as heterogeneous catalysts for the Diels–Alder reaction of acrolein with 1,3-cyclohexadiene [15]. All of the polymers exhibited catalytic activities much higher than their corresponding components, and the Diels–Alder reaction occurs in high stereoselectivities with endo/exo ratios typically >99/1 (Ti, Al) or >95/5 (Zr). Remarkably, the Ti and Zr polymers as solid catalysts could be readily recovered by filtration or decantation and used repeatedly, without any significant deactivation [15a,b]. For the insoluble Zr polymeric catalyst, the reaction could even be conducted in a flow system with a reactant mixture as a mobile phase [15b]. Tanski and Wolczanski tested several Ti aryldioxide coordination polymers $[Ti(OArO)_2(py)_2]$ (Ar = p-C_6H_4-, 2,7-naphthyl, 4,4 -biphenyl; py = pyridine and 4-Me- or 4-Ph-pyridine) for their catalytic activity in Ziegler–Natta polymerization of ethene or propene with methylalumoxane (MAO) as cocatalyst. Poor to mediocre polymerization activities were observed for both olefins, and fragmentation of the coordination polymer occurred during the process [16]. Yamamoto and coworkers introduced an aluminum trisphenoxide coordination polymer as an efficient Lewis acid catalyst for promoting the Diels–Alder reaction of α,β-enals [17a]. In the reaction of acrolein with cyclopentadiene, the polymer catalyst could be recovered quantitatively by simple filtration and reused for seven cycles, without any decrease in activity. Inanaga reported a recyclable coordination polymer, constructed from lanthanide metal ions and biphenyl-4,4'-disulfonic acid, for the heterogeneous catalysis of hetero-Diels–Alder reaction of aldehydes with Danishefsky's diene under solvent-free conditions [17b]. Under these optimized conditions, the corresponding dihydropyranone derivatives were obtained in good yields, while the scandium catalyst could be reused three times without deactivation of the catalyst or leaching of the metal ion.

Several studies have shown that some metal coordination polymers constructed from poly-carboxylate ligands may be used as heterogeneous achiral catalysts. Monge and coworkers found that $In_2(OH)_3(BDC)_{1.5}$ (BDC = 1,4-denzendicarboxylate) was active for the hydrogenation of nitroaromatics and the oxidation of alkylphenylsulfides [18]. The pores in this hybrid material were too small to accommodate a reactant molecule; hence, catalysis was seen to take place on the surface and did not show any rate-dependence on the size of the substrates. A low metal/substrate ratio of 0.1% proved to be effective in the reduction of nitroaromatics, and the In polymer catalyst could be recycled in successive runs, using simple filtration, and reused without any significant loss of activity and selectivity in the oxidation of sulfides. The supernatant liquid phase was demonstrated to be inactive and showed neither the presence of In metal nor the BDC anion, thus confirming the heterogeneous character of the catalysis. Ruiz-Valero et al. showed that 3-D coordination polymers of rare-earth metal succinates, $[Sc_2(OOCC_2H_4COO)_{2.5}(OH)]$, $[Y_2(OOCC_2H_4COO)_3(H_2O)_2]\cdot H_2O$, and $[La_2(OOCC_2H_4COO)_3(H_2O)_2]\cdot H_2O$ [where $(OOCC_2H_4COO)^{2-}$ = succinate], could be used as effective and easily recyclable heterogeneous catalysts in the acetalization of aldehydes and the oxidation of sulfides [19]. Kaskel and coworkers explored the catalytic properties of the desolvated porous metal–organic framework compound $Cu_3(BTC)_2(H_2O)_3\cdot xH_2O$ (BTC = benzene 1,3,5-tricarboxylate) [20]. Although, in this polymer the Lewis acid copper coordination sites are located on the interior of the pore wall and thus can be accessible for catalytic conversions, they are coordinatively saturated by the three weakly bound water molecules and the pores are occupied by the solvent molecules. Removal of the bound water by evacuation in vacuo under elevated temperature prior to the catalysis is necessary to create space vacancy and coordinative unsaturation, leading to $Cu_3(BTC)_2$ which allows easy access of substrate to the copper sites. The chemisorption of benzaldehyde results in its activation, while further cyanosilylation under the heterogeneous catalysis gives a good yield and a high selectivity of product. However, a limited compatibility of the polymer catalyst with organic solvents was noted. Coordinating solvents such as tetrahydrofuran (THF) completely block the Lewis acid sites of the catalyst, and solvents such as CH_2Cl_2 or higher reaction temperatures (353 K) can cause decomposition of the catalyst. The same desolvated coordination polymer $Cu_3(BTC)_2$ was also shown by De Vos et al. to be a highly selective heterogeneous Lewis acid catalyst for the isomerization of terpene derivatives, such as the rearrangement of a-pinene oxide to campholenic aldehyde and the cyclization of citronellal to isopulegol [21]. Catalyst performance, reusability and heterogeneity were critically assessed. Recycling experiments showed that, for each consecutive run, the selectivity remained constant whereas the reaction rate decreased owing to the formation of deposits in the pores. By applying suitable regeneration procedures, however, the catalytic activity of $Cu_3(BTC)_2$ could be brought back close to its original values. Mori et al. reported a 2-D microporous copper carboxylate coordination polymer, $[Cu_2(OOCC_6H_{10}COO)_2]\cdot H_2O$ $[(OOCC_6H_{10}COO)^{2-}$ = trans-1,4-cyclohexanedicarboxylate], which could be used as a highly selective and reusable heterogeneous catalyst for oxidations of various alcohols with hydrogen peroxide [22]. Remarkably, a green-colored active peroxo copper(II) polymeric intermediate, $H_2[Cu_2(OOCC_6H_{10}COO)_2(O_2)]\cdot H_2O$, was isolated and amply characterized by a variety of physical techniques, with its molecular structure established using powder X-ray diffraction (PXRD) analysis. The polymeric peroxo intermediate was stable under 132.8 °C in a solid state, and determined to be a real active species for the heterogeneous oxidation catalysis by the NMR tube reaction of 2-propanol to form acetone in the absence of H_2O_2 in combination with other evidences. It is noteworthy that, as the polymeric peroxo complex is the first example of an active copper peroxo intermediate, the corresponding structure and reactivity study may provide new suggestions that would be useful for the development of heterogeneous oxidation catalysts. Recently, the potential application of metal–organic frameworks (MOF) in heterogeneous catalysis has begun to attract research interest from industry. One R&D group from BASF reported test results using zinc-based coordination polymers with terephthalic acid (BDC) for the catalytic activation of alkynes to prepare methoxypropene from propyne and vinylester from acetylene, respectively [23]. The study results showed clearly that the MOFs may act as heterogeneous catalysts, and that there is wide scope for future research to assess whether the many factors involved in MOF preparation may lead to unusual catalytic properties. It should also become clear whether MOFs can compete with the presently available, well-known heterogeneous industrial catalysts.

Although further examples of achiral coordination polymers used for heterogeneous catalysis are described in some excellent recent reviews, they will not be discussed here for the sake of brevity [10, 11a].

In parallel to the investigations on metal–organic coordination polymer catalysis, another strategy using organic–inorganic hybrid polymers (mainly hybrid metal oxides) as heterogeneous catalysts was also developed [7]. Compared to the coordination polymers, hybrid metal oxides usually possess a higher thermal stability, owing to an increased role of relatively strong metal–oxygen–metal bonds in supporting the frameworks. Due to their tunable pore sizes and easy to functionalize nature, the organically pillared zirconium phosphate/phosphonates have developed into a highly promising class of heterogeneous catalysts [24, 25]. Details of many of these investigations have been included in recent reviews by Clearfield and Wang [24] and Curini [25], and will not be further discussed at this point.

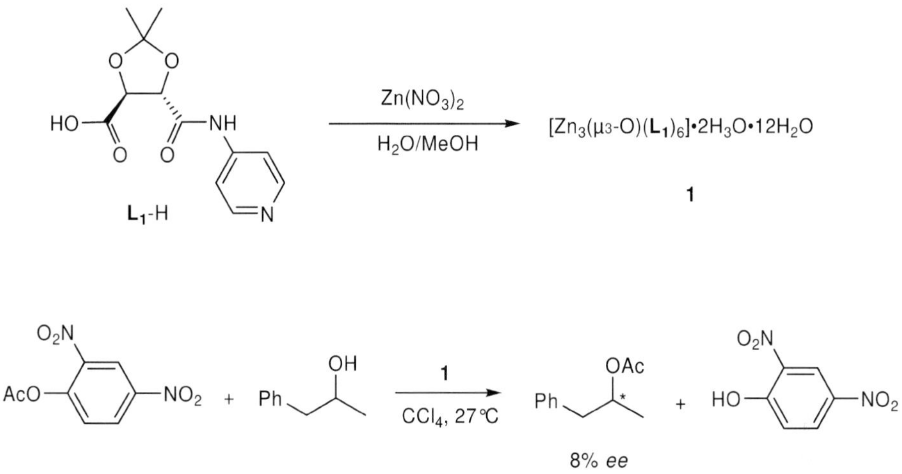

Scheme 9.2 Synthesis of homochiral MOCP **1** and its catalytic application in the enantioselective transesterification of racemic 1-phenyl-2-propanol.

In principle, by the judicious choice of a chiral polytopic ligand in combination with a proper catalytic metal, the mild synthetic conditions should allow for the straightforward construction of homochiral MOCPs, with a chiral environment inside the cavities or on the surface of the solids, with the metal ions acting as the catalytically active centers. Given their usually low solubility, the resulting MOCP may find application as a heterogeneous chiral catalyst for asymmetric transformations. However, despite such clear feasibility this fascinating idea was realized only very recently. In 2000, Kim and coworkers reported the first example of asymmetric catalysis using a crystalline homochiral coordination polymer **1** characterized by trimeric subunits, $\{[Zn_3O(L_1)_6] \cdot 2H_2O \cdot 12H_2O\}_n$ (L_1 = chiral bridging ligand) (Scheme 9.2) [26]. The enantiopure bridging ligand L_1-H with both carboxylic and pyridyl functional groups was easily derived from D-tartaric acid which, upon treatment with $Zn(NO_3)_2$ in a H_2O/MeOH solvent mixture, gave the porous chiral metal–organic polymer with part of the pyridyl groups pointing towards the interiors of chiral 1-D channels. These 'dangling' pyridyl groups in the homochiral polymer were used to catalyze the kinetic resolution of racemic 1-phenyl-2-propanol via transesterification (Scheme 9.2). Although the resulting ee-value was only modest (~8%), this seminal study proved to be the first example using a well-defined homochiral MOCP for enantioselective heterogeneous catalysis.

Considering the numerous coordination polymers developed to date, and the major interest in their potential application as heterogeneous catalytic systems, it seems surprising that only a few studies have demonstrated such catalytic activities. One inherent reason for this phenomenon might be attributed to the fact that most coordination polymers prepared to date have the coordinatively saturated metal centers, and this precludes their involvement in catalytic transformation [20, 21, 27]. Moreover, for some polymers the metal ions are embedded in the polymer matrix and are not freely accessible to reagent molecules. Finally, the compatibility of a coordination polymer with a reaction system may also be problematic, as the leaching of metal ions or organic ligands or even decomposition of the catalyst may all hamper its recovery and reuse [20, 21]. Some general considerations and challenges in the design and synthesis of MOCP for heterogeneous catalysis will be presented in the following section.

9.3
General Considerations on the Design and Construction of Homochiral MOCP Catalysts

Kim's proof-of-concept study triggered further interest in the use of metal–organic hybrid polymers for the immobilization of homogeneous chiral catalysts for asymmetric synthesis which, until then, had been a largely unexplored field of research (*vide infra*). MOCPs are often – but not necessarily – insoluble in common organic solvents, and thus may potentially be used as insoluble supports for the immobilization of homogeneous catalysts, or directly as heterogeneous catalysts themselves. Over the past few years, two general strategies have been developed for incorporating catalytically active moieties into polymeric scaffolds when using homochiral MOCPs for heterogeneous asymmetric catalysis [3b]:

- The use of a ditopic or polytopic chiral ligand (building blocks) bearing two or more ligating units to link adjacent, catalytically active metal centers, where the active sites are incorporated into the main backbone of the resulting homochiral MOCP (Scheme 9.3a).
- The use of chiral polyfunctional ligands as linkers and metal ions or clusters as nodes to construct homochiral MOCPs; however, the active sites do not take part

Metal-organic coordination polymer catalyst

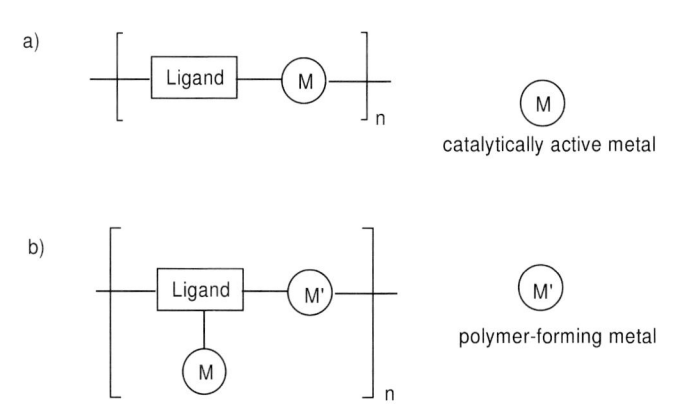

a) [—Ligand—M—]$_n$ M catalytically active metal

b) [—Ligand—M'—]$_n$ with M branch   M' polymer-forming metal

Scheme 9.3 Schematic representations of the two strategies for the construction of MOCP catalysts.

directly in the framework formation, but rather are located as pendant auxiliaries on the main chain of the MOCP (Scheme 9.3b).

In both approaches, the MOCPs are homochiral and potentially allow for a high catalyst loading. While the former approach results in immobilized chiral catalysts on the polymer backbone without using any external support, the latter approach most often uses homochiral metal phosphate hybrids or metal–organic coordination networks (MOCN) as supporting frameworks for the immobilized chiral catalysts. For the sake of convenience, we will refer to the relevant polymers as type I and II MOCPs, respectively, in later discussions. Other variants of the strategies have also been reported, as can be found in the following sections.

The type I MOCP catalysts (Scheme 9.3a) are constructed by alternating the incorporation of the chiral bridging ligands and the catalytic metals. In order for the method to be applicable to a targeted reaction, the transition state or the catalyst rest state should contain at least two ligands bridged by catalytic metal centers or clusters. Although this requirement is limiting, it offers the possibility of immobilizing multicomponent enantioselective catalysts which are challenging by using conventional methods. An important structural consideration here is that the metal centers play a dual role as both the polymer's structural binders and the catalytic active sites; hence, it is essential that they should be capable of bonding simultaneously with at least two ligand moieties (same or different) but still have vacant or labile ligating sites available for substrate and/or reagent coordination and activation. Here, the catalytically active moieties are present as repeating units of the coordination polymer, and thus the structural motif of the corresponding homogeneous catalyst should preferably be maintained during polymer synthesis if high reactivity/enantioselectivity is to be realized. In such a case, the desired catalytic performance of the resulting MOCPs can be fine-tuned by judicious modifications on the spacer or the chiral units of the polytopic ligand.

The type II MOCP catalysts (Scheme 9.3b) are usually obtained by the stepwise generation of networks and uploading of catalytically active metals (or *vice versa*). Multifunctional chiral bridging ligands and two different types of metal (if necessary) are used as the molecular building-blocks. A prerequisite for the design is the orthogonal nature of the two types of functional group in the chiral bridging ligands. Whilst the primary functional groups are responsible for connecting the network-forming metals to form extended structures, the secondary chiral groups are used to generate asymmetric catalytic sites with (or without) catalytic metals. In ideal scenarios, porous homochiral MOCPs with accessible and/or uniform catalytic sites can be obtained and used to catalyze asymmetric organic transformations in heterogeneous fashion.

When the desired chiral polytopic ligands are at hand, the synthetic protocols for accessing homochiral MOCP catalysts can be quite simple. A combination of the ligand with an appropriate metal precursor in a compatible solvent under appropriate conditions, can result in the spontaneous formation of insoluble polymers by self assembly. This polymerization process involves the successive formation of bonds between a polytopic ligand and a metal, and thus is facilitated by employing a ligand that shows high affinity towards the network metal [28].

One general requirement for any catalyst immobilization method is that the support materials (and the heterogenized catalysts) need to be stable enough during the reaction process [2b, 3, 4]. With organic components present in the structures, MOCPs are often not stable beyond 600 K [7, 20, 21]. Although this relatively unsatisfactory thermal stability would hamper their widespread use in many of the current heterogeneous catalytic processes performed by zeolites and other purely inorganic catalysts, it seem not to matter too much for most of the catalytic asymmetric synthesis procedures which are often performed under mild conditions (<373 K) for stereocontrol [1]. A primary concern in designing MOCP-type catalysts is that the chiral MOCPs should be chemically stable enough, in order for them to be recovered for reuse and to minimize/eliminate metal leachings during the heterogeneous catalytic cycles. MOCPs maintain their structural integrity by metal–ligand bonds, which encompass a broad spectrum of strengths depending on the metal-binding motifs (i.e. strength of ligation, ligand denticity, coordination number, etc.) [28]. Although sometimes nontrivial, these factors can be synergistically tuned to increase the strength of the metal–ligand interaction and thus enhance the stability of MOCPs. This can be achieved by using donor ligands of exceptionally high metal affinity (high binding constants), and/or increasing the number of donor atoms bonding to the metal by using multidentate ligands [28]. Both, strong monodentate phosphine-Rh bonding motifs and multidentate binding arrays such as BINOL-Ti(IV) have been utilized for MOCP catalyst construction, and this will discussed in the next section.

Subtle external factors such as solvent and temperature can also play a role in determining the stability of a MOCP. Generally, solvents of low polarity are beneficial for the use of MOCP catalysts, not only because of the low solubility of the extended polymeric assemblies in such solvents but also for the following reasons. The use of a coordinating solvent is often (but not always!) disadvantageous for MOCPs, as the strong donor-type solvents may compete for catalytic sites on the metal and inhibit the coordination/activation of substrate. Alternatively, they may facilitate the depolymerization/decomposition of a MOCP and drastically reduce the degree of polymerization (especially under elevated temperature). In the worst-case scenario this would introduce achiral fragments with a decreased enantioselectivity and might even lead to metal leaching. Furthermore, additives with comparable functionalities to that of the polytopic ligands may also inhibit the crosslinking of metal–ligand bonds and thus block the formation of an extended structure of a MOCP, exhibiting similar influences as described above on a heterogeneous catalysis. With these considerations in mind, a meaningful catalytic study using a MOCP as a recyclable catalyst should always include an investigation of the liquid phase supernatant over the solid-state catalyst after a catalytic cycle. This is to ensure that the MOCP does not partly dissolve and decompose in solution, so as to exclude that it is the components which act as a homogeneous catalyst in the reaction system.

The high loading with metal content alone for a MOCP solid is not a guarantee for catalytic activity, as both active site availability and space accessibility are prerequisites for an immobilized catalyst to effect catalytic transformations. On the one hand, vacant ligating sites (coordination vacancy) on the catalytic metal after the MOCP formation, should also be available for substrate activation, either by dissociation of the weakly bound labile ligands or by their exchange with the substrate molecules. On the other hand, these active sites should be well dispersed and suitably orientated on the external surface or within pores/channels of a MOCP catalyst, so that they are readily accessible to the reactants to minimize the diffusion problems. In general, a porous MOCP with a high surface area is beneficial for easy diffusion of the reactants to the active sites.

Taken together, an ideal reusable homochiral MOCP catalyst should combine the advantages of both heterogeneous catalysts (e.g. facile catalyst recovery, high stability, ease of handling) and their homogeneous counterparts (uniform active sites, high efficiency and selectivity, reproducibility). Many of these issues are often balanced during reaction development through trial-and-error. For heterogeneous asymmetric transformations, crystalline MOCP catalysts with periodically ordered chiral active sites are highly desirable, although the amorphous nature of most homochiral MOCPs developed to date does not preclude their excellent catalytic performance, nor their reusability. In the following sections, the published data relating to applications of homochiral MOCP catalysts in heterogeneous asymmetric reactions are summarized, together with an assessment of the catalytic performance and heterogeneity as well as reusability in each case. Such a catalyst immobilization technique is based on the assembly of MOCP solids by a complexation of polytopic ligands and metals without the need for an external support, and thus can be considered as a 'self-supporting strategy' [5, 6, 29, 30].

9.4
Type I Homochiral MOCP Catalysts in Heterogeneous Asymmetric Reactions

9.4.1
Enantioselective C—C Bond-Forming Reactions

9.4.1.1 Carbonyl-Ene Reaction

The ene reactions are one of the most important classes of reactions for carbon–carbon bond construction in organic synthesis, among which the asymmetric glyoxylate-ene reactions are synthetically useful in that α-hydroxyesters of biological importance can be obtained with high chiral induction [31]. Lewis acidic titanium(IV) complexes of various nonracemic 1,1′,2,2′-binaphthol (BINOL) derivatives have been developed primarily by Mikami as efficient chiral catalysts for promoting this type of reaction [32]. A remarkable positive nonlinear effect has been observed in the BINOL/Ti(IV)-catalyzed enantioselective ene reaction of glyoxylate and α-methylstyrene, which was ascribed to the dimeric nature of the titanium catalysts by Mikami and Nakai [32b,d,e]. Based on the asymmetric activation concept, Mikami and coworkers showed that the enantiopure (R)-BINOLate-Ti(OiPr)$_2$ catalyst for this reaction could be further evolved by the addition of another equivalent of (*R*)-BINOL(H_2), affording the ene-adduct in even higher enantioselectivity [32g]. Remarkably, a kinetic study also showed that the reaction catalyzed by the (*R*)-BINOLate-Ti(O*i*Pr)$_2$/(*R*)-BINOL combination was considerably faster (26-fold) than that catalyzed by the (*R*)-BINOL-Ti(O*i*Pr)$_2$ alone [32g], which suggested that a synergistic effect might exist in the resulting catalytic system. These results inspired the research groups of Sasai [29] and Ding [30] to develop, independently, a variety of recyclable homochiral Ti-bis-BINOL hybrid polymers for catalysis of the asymmetric carbonyl-ene reaction. An immobilization of the chiral titanium catalysts was achieved by the polycondensation of a rigidly bridged dimeric BINOLs (**L$_{2a}$**–**L$_{2e}$**) with Ti(O*i*Pr)$_4$ in the presence or absence of a small amount of water in dichloromethane (DCM) or chloroform, leading to the formation of insoluble Ti-bridged hybrid polymers (Scheme 9.4).

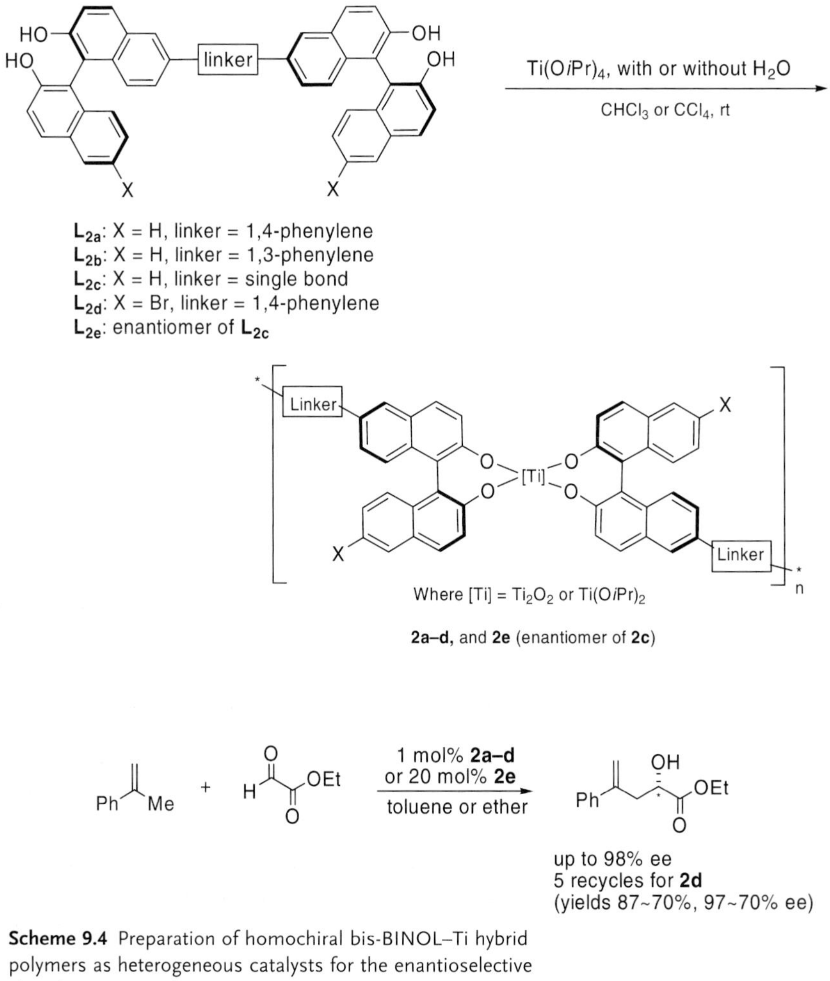
HO
HO
linker
OH
OH
X
X
Ti(O*i*Pr)$_4$, with or without H_2O
$CHCl_3$ or CCl_4, rt
L$_{2a}$: X = H, linker = 1,4-phenylene
L$_{2b}$: X = H, linker = 1,3-phenylene
L$_{2c}$: X = H, linker = single bond
L$_{2d}$: X = Br, linker = 1,4-phenylene
L$_{2e}$: enantiomer of **L$_{2c}$**
Linker
[Ti]
Linker
n
Where [Ti] = Ti_2O_2 or Ti(O*i*Pr)$_2$
2a–d, and **2e** (enantiomer of **2c**)
Ph
Me
H
OEt
1 mol% **2a–d**
or 20 mol% **2e**
toluene or ether
Ph
OH
OEt
up to 98% ee
5 recycles for **2d**
(yields 87~70%, 97~70% ee)

Scheme 9.4 Preparation of homochiral bis-BINOL–Ti hybrid polymers as heterogeneous catalysts for the enantioselective glyoxylate-ene reaction.

These assembled heterogeneous catalysts were tested in the carbonyl-ene reaction of α-methylstyrene with ethyl glyoxylate, affording the α-hydroxyester in good yield and excellent enantioselectivity (up to 98% ee) (Scheme 9.4). Remarkably, it was found that the nature of the linkers between the two BINOL units in the ditopic ligands (L_{2a}–L_{2e}) plays an important role in determining the enantioselectivity of the catalyst, with distinct Ti hybrid polymers showing dramatically different performance under identical conditions. In addition, the introduction of electron-withdrawing substituent such as Br to the backbone of BINOL units (L_{2d}) resulted in a significant improvement in the product yield (99%). It is therefore possible to fine-tune the immobilized catalysts through modifications on the polytopic ligands. The titanium hybrid catalyst **2d** could be recovered by simple filtration and reused for up to five consecutive cycles in diethyl ether, without any further activation process. However, a gradual deterioration in yields (from 87% to 70%) and enantioselectivities (from 97% to 70% ee) was observed, presumably as a result of titanium(IV) catalyst degradation during the catalysis and/or recovery.

9.4.1.2 Michael Addition

Al-Li-bis (binaphthoxide) (ALB) is a chiral heterobimetallic complex with Al and Li centers supported by two BINOLate units, capable of catalyzing a range of asymmetric reactions in high efficiency and enantioselectivity [33]. A variety of studies has revealed that synergistic cooperation between the two types of active sites is required for the reactivity of this bifunctional catalyst [33a], and therefore it is important to maintain the structural integrity of the active motif during immobilization of this multicomponent catalyst for high catalytic efficiency. This would be difficult to realize using a conventional immobilization approach, which usually involves the random introduction of ligand and functional units onto a sterically irregular polymer backbone [3, 4]. Based on these considerations, Sasai and coworkers made use of the aforementioned bis-BINOL ligands in combination with $LiAlH_4$ and BuLi to generate insoluble aluminum-bridged homochiral polymers **3a–3d**, and examined their catalytic performance in the Michael reaction between 2-cyclohexenone and dibenzyl malonate (Scheme 9.5) [29]. Whereas, all of the ALB polymers led to Michael adducts in good yields, the enantioselectivity ranged from modest to excellent, with the best ee-value (96% for **3d**) comparable to that for the homogeneous ALB catalyst (97% ee). In contrast, a polystyrene (PS)-supported ALB, generated by the random introduction of BINOL derivatives onto the PS resin, gave the same Michael adduct without enantiomeric excess [34]. These observations clearly illustrate an advantage in using the strategy for immobilization of multicomponent asymmetric catalysts.

The heterogeneity of the catalysis was confirmed by an investigation of the liquid-phase supernatants over the solid-state catalysts after a catalytic cycle. Neither the ligand nor catalyst activity in the Michael reaction were observed in the supernatant solution for the catalyst **3d**. The heterogenized ALB catalyst **3d** was recovered by syringe removal of the supernatant and rinsing with fresh solvent

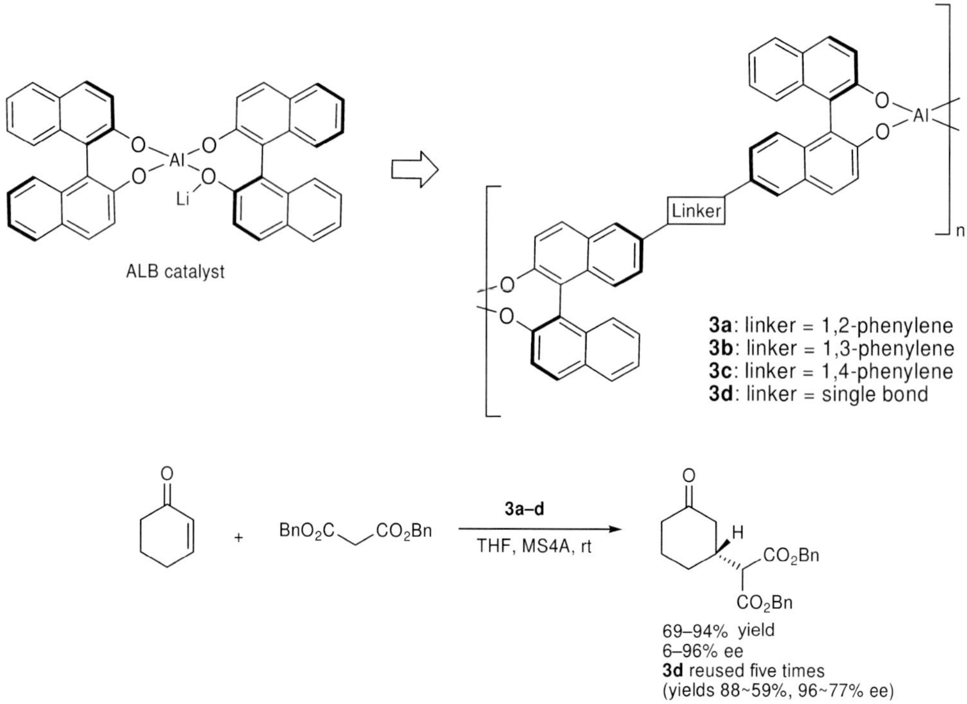

Scheme 9.5 Enantioselective Michael addition catalyzed by Al-bridged hybrid polymers.

under argon; it was then reused for five cycles with a gradual decrease in product yield (from 88% to 59%) and ee-value (from 96% to 77%).

9.4.1.3 Diethylzinc Addition to Aldehydes

Using a similar approach, Harada and Nakatsugawa immobilized a tris-BINOLate/Ti(OiPr)$_4$ catalyst for the asymmetric addition of diethylzinc to aldehydes [35]. Treatment of the rigid chiral tris-BINOL ligand **L$_3$** with Ti(OiPr)$_4$ in CH$_2$Cl$_2$/toluene gave an insoluble polymeric aggregate **L$_3$**-[Ti(OiPr)$_2$]$_3$, which was examined as a catalyst in the diethylzinc addition to benzaldehyde (Scheme 9.6). Although the reaction proceeded smoothly in toluene/hexane to give (R)-1-phenylpropanol in 74% ee and the solid catalyst was recovered quantitatively by centrifugation, control experiments indicated that the supernatant of the reaction mixture also exhibited a significant level of reactivity, albeit with lower rate and less enantioselectivity than was the case with the solid catalyst. This observation suggested that the polymer decomposition/degradation might have occurred under the reaction conditions, resulting in a leaching of some catalytic species into the solution phase. Even though the catalysis may only be regarded as partially heterogeneous in a rigorous sense, the aggregate catalyst could be isolated from the product by filtration and reused for six cycles without any significant lowering

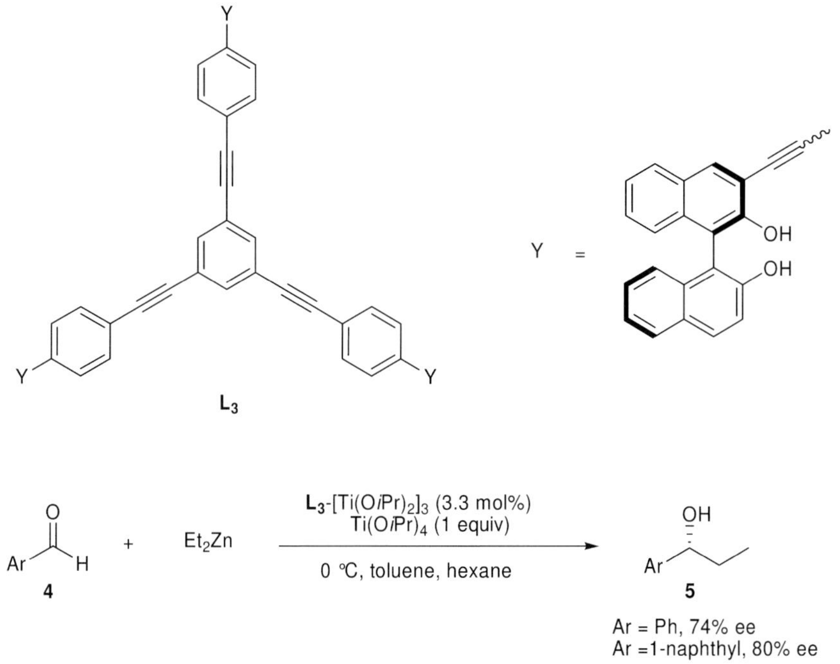

Scheme 9.6 Enantioselective diethylzinc addition to aldehydes catalyzed by a tris-BINOL Ti coordination polymer.

of the enantioselectivity (72–75% ee). Extension of the reaction to 1-naphthaldehyde gave high conversions (>95%) and up to 84% ee.

9.4.2
Enantioselective Oxidation Reactions

9.4.2.1 Epoxidation of α,β-Unsaturated Ketones

In the immobilization of homogeneous chiral catalysts using metal–organic polymers, the stereochemical features of the chiral polytopic ligand can exert a profound influence on the enantioselectivity and activity of the targeted catalytic reaction. Such an effect was studied by Ding and coworkers in the heterogenization of Shibasaki's lanthanum catalyst [36] for the enantioselective catalysis of epoxidation of α,β-unsaturated ketones [37]. The treatment of a diversity of poly-BINOL ligands ($L_{2a–2c}$ and $L_{4a–4f}$) with different linker geometries with La(OiPr)$_3$ in THF afforded heterogenized poly-BINOL/La assemblies (**6a–6i**), which were subsequently tested in the epoxidation of chalcone with cumene hydroperoxide (CMHP) as oxidant in the presence of molecular sieves and triphenylphosphine oxide as additives (Scheme 9.7; R = R′ = Ph). Although the exact reason was not clear due to a lack of structural information, the enantioselectivity for epoxidation of chalcone exhibited a significant dependence on the geometric factors such as the length or angles of the poly-BINOLs.

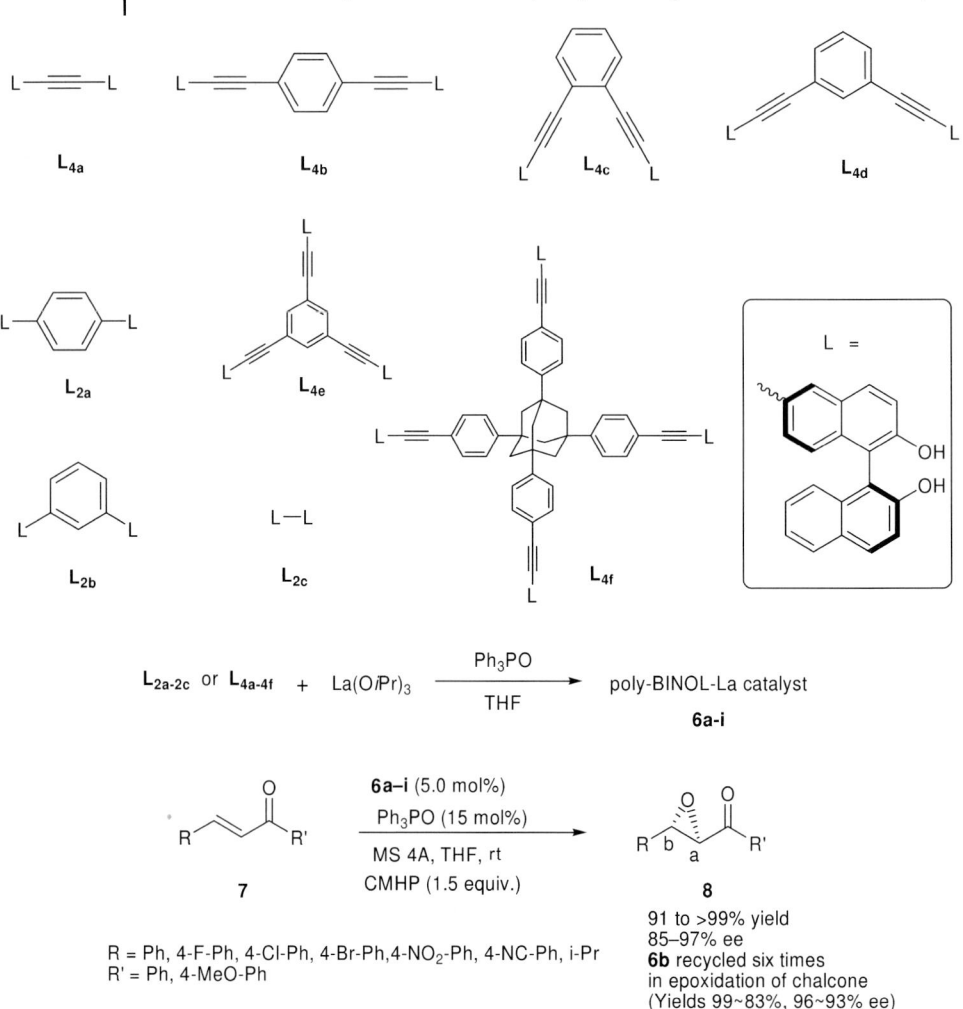

Scheme 9.7 Immobilized Shibasaki's BINOL/La catalysts for the epoxidation of α,β-unsaturated ketones.

The heterogeneous nature of the catalysis was confirmed by the inactivity of the supernatant THF solution of the catalyst composed of L_{4b} for the epoxidation of chalcone under the same experimental conditions, and the less than 0.4 ppm lanthanum leaching in consecutive runs as determined by inductively coupled plasma (ICP) spectroscopy. The catalyst with L_{4b} could be recovered by simple filtration under argon, and reused for at least six cycles in the epoxidation of chalcone without any significant loss of enantioselectivity. Furthermore, the substrate general adaptability of the immobilized catalyst was demonstrated for the enantioselective catalytic epoxidation of a variety of α,β-unsaturated ketones (**7**), affording the corresponding α,β-epoxy ketones (**8**) in excellent yields (91–99%) and enantioselectivities (84–97% ee) (Scheme 9.7).

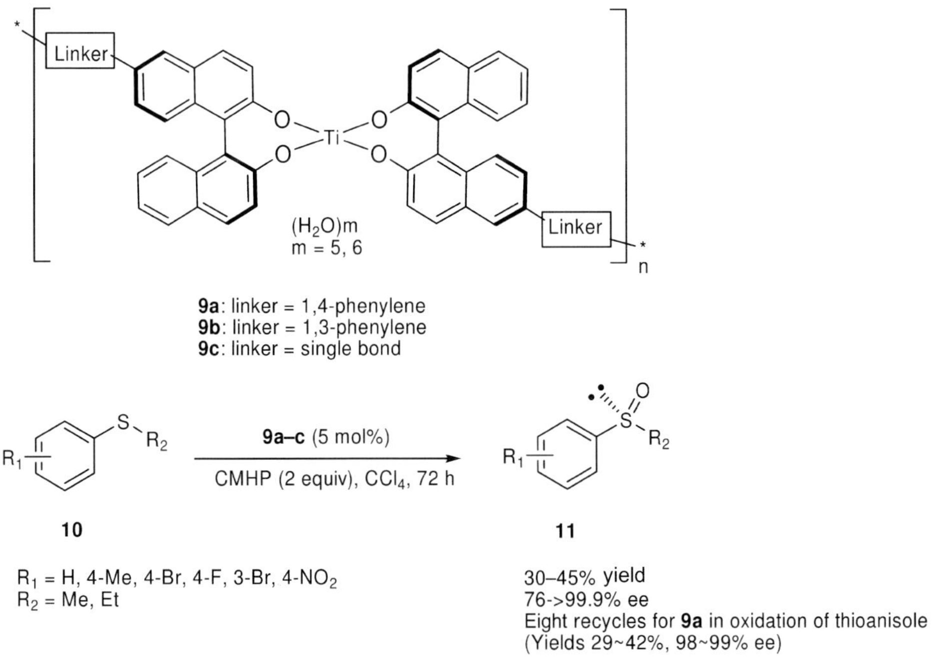
Linker
Ti
(H2O)m
m = 5, 6
Linker
n
9a: linker = 1,4-phenylene
9b: linker = 1,3-phenylene
9c: linker = single bond
9a–c (5 mol%)
CMHP (2 equiv), CCl4, 72 h
10
11
R1 = H, 4-Me, 4-Br, 4-F, 3-Br, 4-NO2
R2 = Me, Et
30–45% yield
76->99.9% ee
Eight recycles for 9a in oxidation of thioanisole
(Yields 29~42%, 98~99% ee)

Scheme 9.8 Heterogeneous catalysis of asymmetric sulfoxidations with Poly-BINOL–Ti homochiral polymers.

9.4.2.2 **Sulfoxidation of Aryl Alkyl Sulfides**

Some variants of the Ti-Poly-BINOL homochiral polymers have also been extended to the heterogeneous catalysis of asymmetric sulfoxidation reaction by Ding and coworkers [30b]. The treatment of bridged BINOL ligands $\mathbf{L}_{2a-2c}$ with $Ti(O\textit{i}Pr)_4$ in a 1 : 1 molar ratio in CCl_4, followed by the addition of 40 equivalents of water (relative to ligand), resulted in the heterogenized Uemura-type sulfoxidation catalysts **9a–c** (Scheme 9.8) [38]. It should be noted here that, in the Uemura procedure, the presence of a significant amount of water is essential to produce an effective catalyst as well as to extend the catalyst's lifetime [38]. A recent structural study conducted by Salvadori *et al.* suggested that an aggregated solution species $(BINOLate)_6Ti_4(\mu_3\text{-}OH)_4$ may act as the catalytic precursor for the homogeneous process [39].

The immobilized Uemura-type catalysts **9a–c** were subsequently examined in the heterogeneous enantioselective oxidation of a variety of aryl alkyl sulfides with CMHP as the oxidant, affording the corresponding chiral sulfoxides with extremely high enantioselectivities (96.4 to >99.9% ee) in moderate yields (*ca* 40%) (Scheme 9.8). The heterogeneous nature of the catalysis was confirmed by the control experiment showing the inactivity of the supernatant, as well as by the negligible titanium leaching into the solution phase as evidenced by ICP spectroscopic analysis. Remarkably, simple filtration in open air enabled separation of the solid catalyst **9a**, which could be reused in the reaction of thioanisole for eight cycles that covered a period of more than one month. No loss of enantioselectivity or obvious deterioration of sulfoxide yields was observed.

9.4.3
Asymmetric Hydrogenations

9.4.3.1 Hydrogenation of Dehydro-α-Amino Acids and Enamides

Recently, a variety of monodentate phosphorus ligands have been developed for the Rh-catalyzed asymmetric hydrogenation of olefins [40]. It is generally accepted that two monodentate phosphorous ligands are bonded to Rh in the catalytically active species [40e,g–i]. Thus, through the assembly of polytopic phosphorus ligands with an appropriate Rh^I precursor, a catalytically active P–Rh–P motif could be incorporated into the backbone of the resulting coordination polymer.

Ding employed this strategy to immobilize Feringa's MonoPhos/Rh catalysts [40b] for the asymmetric hydrogenation of dehydro-α-amino acid esters and enamides [41a]. Treatment of the ditopic MonoPhos ligand L_{5a-5c} with [Rh(cod)]BF$_4$ in DCM/toluene resulted in an immediate precipitation of an amorphous Rh-containing polymer **12a–c**, which were demonstrated to be effective catalysts for the asymmetric hydrogenation of β-aryl or alkyl-substituted methyl esters of dehydro-α-amino acids **13** and enamide **15**, affording the corresponding amino acid derivatives **14** and secondary amine derivative **16**, respectively, in high yields and excellent enantioselectivities (Scheme 9.9). Remarkably, all three chiral polymeric catalysts **12a–c** were shown to catalyze the asymmetric hydrogenation of enamide **15** with enantioselectivity superior to that of the corresponding homogeneous monophos/Rh catalyst (95–97% versus 89% ee). The heterogeneous nature of the catalysis was demonstrated by the inactivity of the supernatant in the catalysis, as well as the low rhodium leaching level as determined by ICP spectroscopic analysis. Last, but not least, for the hydrogenation of methyl 2-acetamidoacrylate (R = H in **13**), the catalysts could be readily recycled and reused for at least seven runs without any significant loss of activity and enantioselectivity (99% yield and 89.5% ee for the seventh run). Recently, Wong reported a chiral ditopic bis-MonoPhos-type ligand with a tetraphenylene core for the Rh(I)-catalyzed asymmetric hydrogenation of various acetamidocinnamate derivatives [41b]. With 1 mol% catalyst loading, up to 99.0% ee and 100% conversion could be achieved for reactions performed in DCM under a hydrogen pressure of 20 atmospheres. Although the *in situ*-formed complex solid is also insoluble in the reaction medium, neither catalyst recovery nor reuse were reported in this case.

In the generation of homochiral metal–organic polymers for heterogeneous asymmetric catalysis, the design and synthesis of polytopic chiral ligands is of key importance. Generally, a covalent linkage of various size and shape is used to incorporate the ligating chiral moieties into the polytopic ligand. However, the somewhat arduous synthesis of the covalently linked polytopic ligands may hamper their application. Noncovalent interactions such hydrogen bonding may also be employed as a practical alternative to covalent bond for the construction of polytopic ligands, as exemplified in a recent study by Ding and coworkers [42]. The

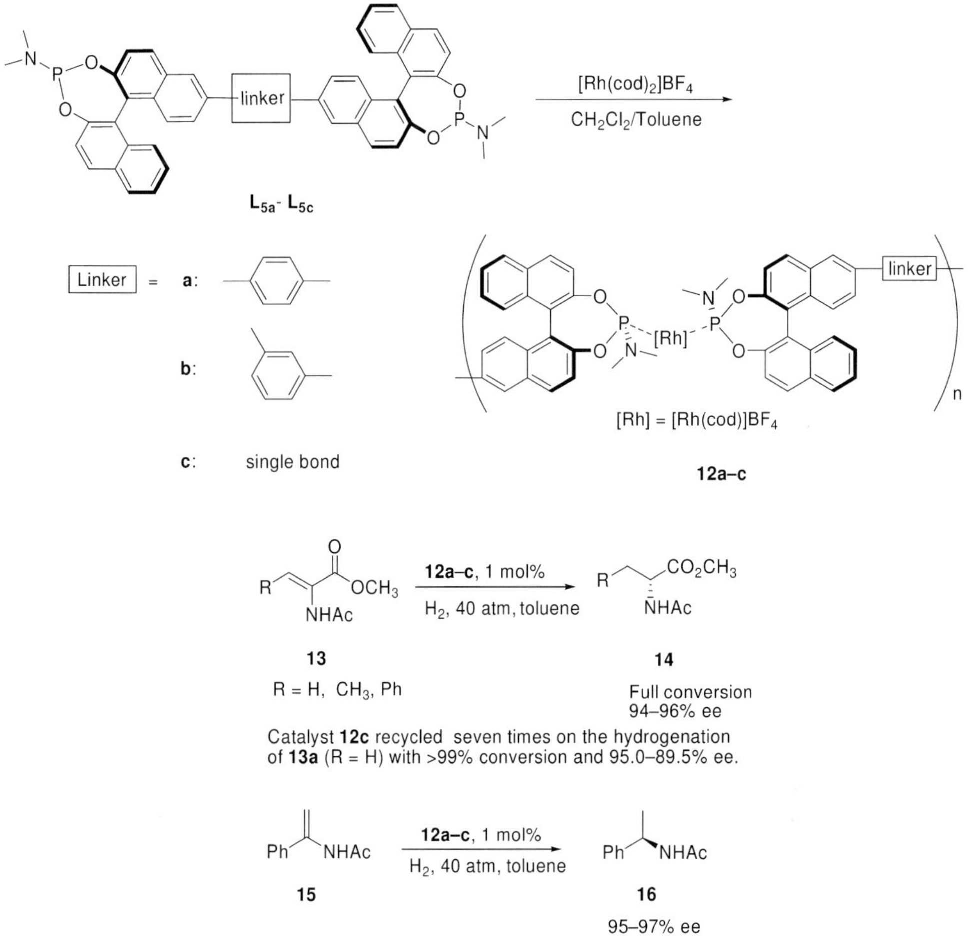

9.4 Type I Homochiral MOCP Catalysts in Heterogeneous Asymmetric Reactions

Scheme 9.9 Heterogeneous catalysis of enantioselective hydrogenations with self-supported MonoPhos/Rh catalysts.

self-complementary hydrogen-bonding unit ureido-4[1H]-ureidopyrimidone (UP) was incorporated with Feringa's MonoPhos moiety to generate a H-bonding bridged bis-MonoPhos ligand, which was assembled with [Rh(cod)$_2$]BF$_4$ to afford the supramolecular metal–organic polymer **17** insoluble in nonpolar organic solvents such as toluene (Scheme 9.10). A prerequisite for this strategy which should be noted is that the two types of interaction, namely hydrogen bonding and ligand-to-metal coordination, must be orthogonal in order to ensure an ordered alternating self-assembly and proper function. The asymmetric hydrogenation of dehydro-α-amino acid derivatives **13** and N-(1-phenylvinyl)acetamide catalyzed by **17** proceeded smoothly with 91–96% ee, which were comparable to values for the homogeneous counterpart. The heterogeneous nature of the catalysis was established in the hydrogenation of (Z)-methyl 2-acetamidobut-2-enoate (R = CH$_3$ in

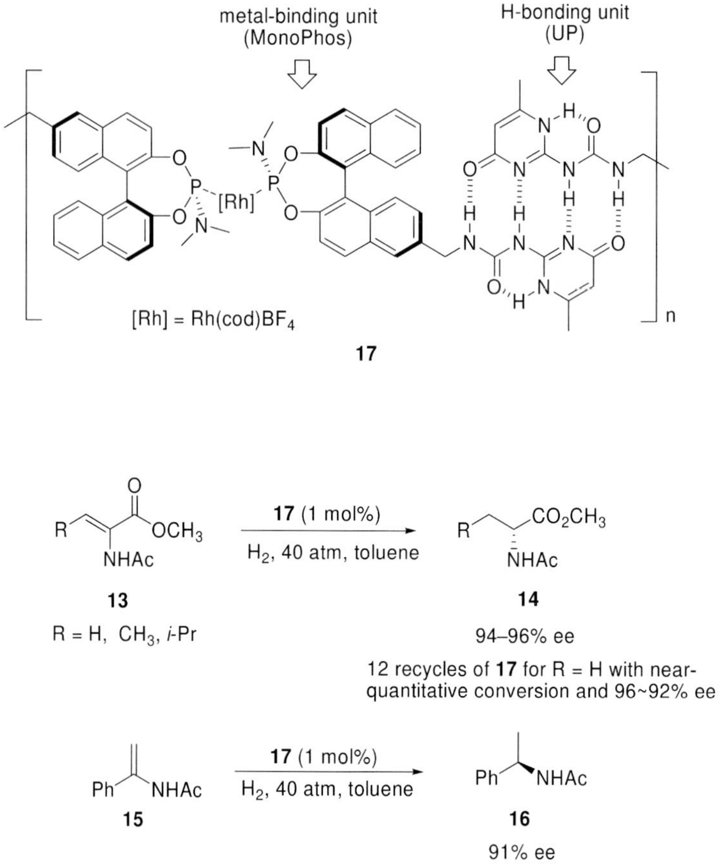

Scheme 9.10 Supramolecular metal–organic polymer **17** for the heterogeneous catalysis of asymmetric hydrogenations.

13), as evidenced by the inactivity of the supernatant and lack of metal leaching (<1 ppm) in both the filtered solution and isolated product.

The reusability of the catalyst system was also examined in the hydrogenation of (Z)-methyl 2-acetamidobut-2-enoate. Using the filtration-recovered catalyst, the hydrogenation proceeded with near-quantitative conversion and constant enantioselectivity (96–92% ee) for at least 10 cycles and with a standard reaction time of 20 h. The reaction profile of the hydrogenation showed that the catalyst reactivity declined with consecutive runs, presumably due to partial catalyst decomposition during catalyst recovery.

9.4.3.2 Hydrogenation of Ketones

In principle, the self-supporting strategy (type I MOCP) for chiral catalyst immobilization should be applicable whenever the active catalyst species contains at least two ligands coordinated to the metal (M). However, this would be challenging in practice when the two ligands are different (L_a and L_b), as a complex system containing multiple species (ML_aL_a, ML_bL_b, ML_aL_b, etc.) may form as a result of the dynamic equilibrium nature in many coordination processes. The specific formation of the hetero-ligand combinations (ML_aL_b) would require that the coordinating information stored in the ligands and metallic ion, respectively, would be sufficiently strong as to dictate their assembly process in a programmed manner [43]. The iteratively occurring structural motifs in Noyori's catalyst, [$RuCl_2${(R)-BINAP}{(R,R)-DPEN}] (BINAP: 2,2′-bis(diphenylphosphino)-1,1′-binaphthyl; DPEN: 1,2-diphenylethylenediamine), was taken by Ding and coworkers to generate [ML_aL_b]$_n$-type metal-organic polymeric catalysts [44]. Treatment of the ditopic BINAP **L$_{6a}$** or **L$_{6b}$** with [{(C_6H_6)$RuCl_2$}$_2$] in dimethylformamide (DMF), followed by reaction with ditopic DPEN **L$_7$**, selectively afforded the heterogenized Noyori-type catalysts **18a** and **18b**, which were then tested in the asymmetric hydrogenation of aromatic ketones **19** (Scheme 9.11). Excellent enantioselectivities with ee-values ranging from 94 to 98% were obtained for catalyst **18b**. Further decreasing the catalyst loading of **18b** to 0.01 mol% in the hydrogenation of acetophenone did not cause any significant deterioration in either yield or enantioselectivity (95% ee). Control experiments using the supernatant of catalyst **18b** established the heterogeneous nature of catalysis in the hydrogenation of acetophenone, and less than 0.1 ppm ruthenium was leached into the organic phase, as determined by ICP analysis. The catalyst **18b** could be recovered by simple filtration under argon, and reused for seven runs without any significant loss in enantioselectivity and conversion.

Ding and coworkers extended the programmed assembly strategy to prepare heterogenized Noyori-type catalysts by the combination of an achiral bis-BIPHEP [BIPHEP: 2,2′-bis(diphenylphosphino)-1,1′-biphenyl] and enantiopure bridged bis-DPEN **L7** with Ru ions; these authors then further examined the catalytic properties of these compounds in the asymmetric hydrogenation of aromatic ketones [45], and good ee-values were obtained. In the hydrogenation of acetophenone, the catalyst could be recovered and recycled four times without any obvious loss of selectivity and activity.

9.5
Type II Homochiral MOCP Catalysts in Heterogeneous Asymmetric Reactions

9.5.1
Enantioselective Hydrogenations

Recently, Lin and coworkers used porous zirconium phosphonate to immobilize chiral Ru–BINAP derivatives for the highly enantioselective hydrogenation of β-keto esters [46]. This strategy relies on combining the robustness of metal phosphonate frameworks with enantioselectivity of the metal complexes to obtain a highly enantioselective heterogeneous catalyst. Treatment of the rigid ligands with [{Ru(benzene)Cl_2}$_2$] in DMF gave Ru(II)-containing bisphosphonic acid

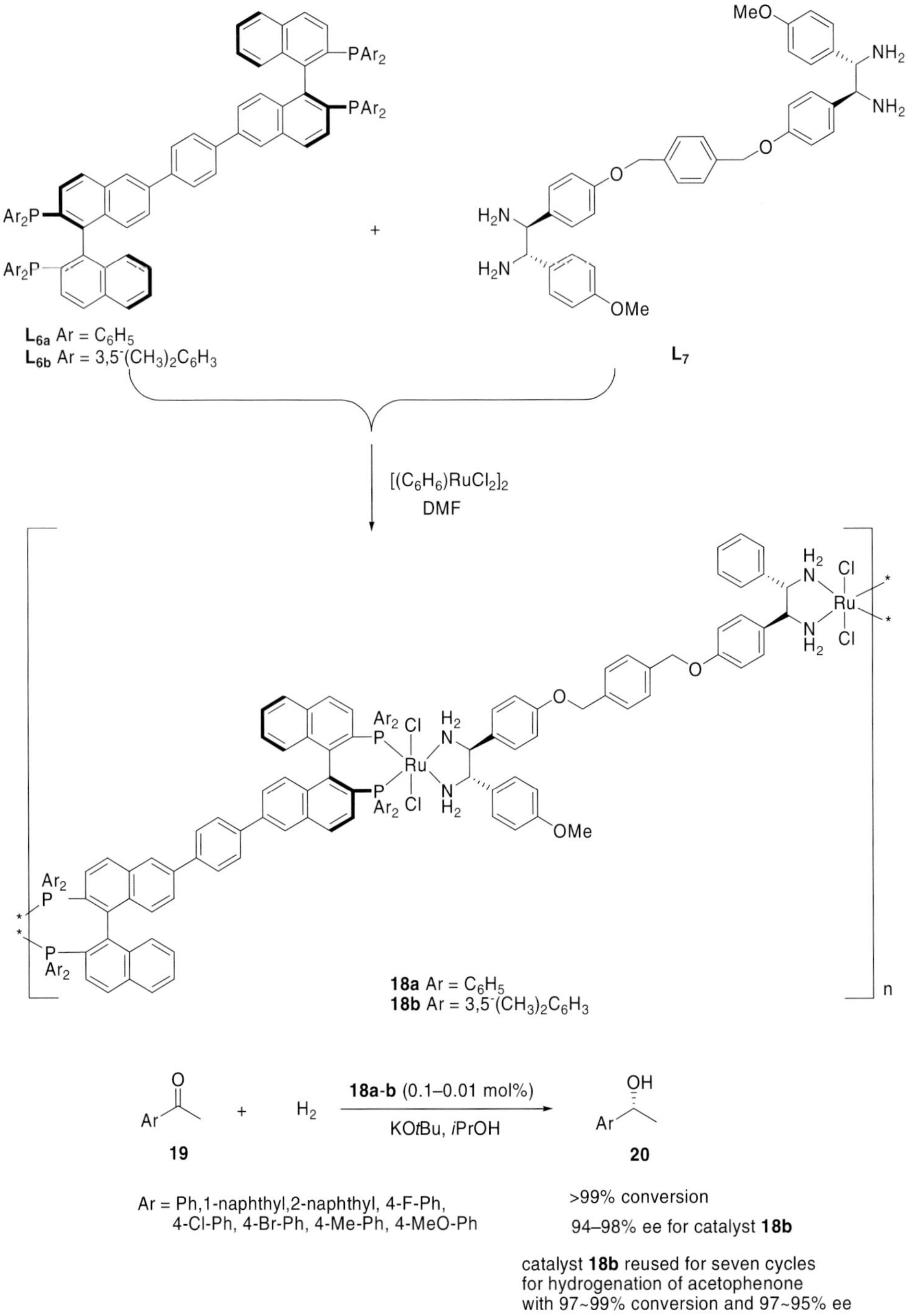

L6a Ar = C₆H₅
L6b Ar = 3,5⁻(CH₃)₂C₆H₃

L7

[(C₆H₆)RuCl₂]₂
DMF

18a Ar = C₆H₅
18b Ar = 3,5⁻(CH₃)₂C₆H₃

19 + H₂ →[18a-b (0.1–0.01 mol%)][KO*t*Bu, *i*PrOH] **20**

Ar = Ph, 1-naphthyl, 2-naphthyl, 4-F-Ph, 4-Cl-Ph, 4-Br-Ph, 4-Me-Ph, 4-MeO-Ph

>99% conversion
94–98% ee for catalyst **18b**

catalyst **18b** reused for seven cycles for hydrogenation of acetophenone with 97~99% conversion and 97~95% ee

Scheme 9.11 Programmed assembly of MOCP **18** for immobilization of Noyori's catalyst and the enantioselective hydrogenations of ketones.

intermediates **21**, which further refluxed with Zr-(O*t*Bu)₄ in methanol to yield zirconium phosphonate hybrid materials **22** containing pendant chiral bisphosphane-chelated ruthenium sites (Scheme 9.12). Here, the two metals play different roles in functioning; while zirconium is responsible for immobilization and is present in the phosphate framework backbone, ruthenium is the catalytically active

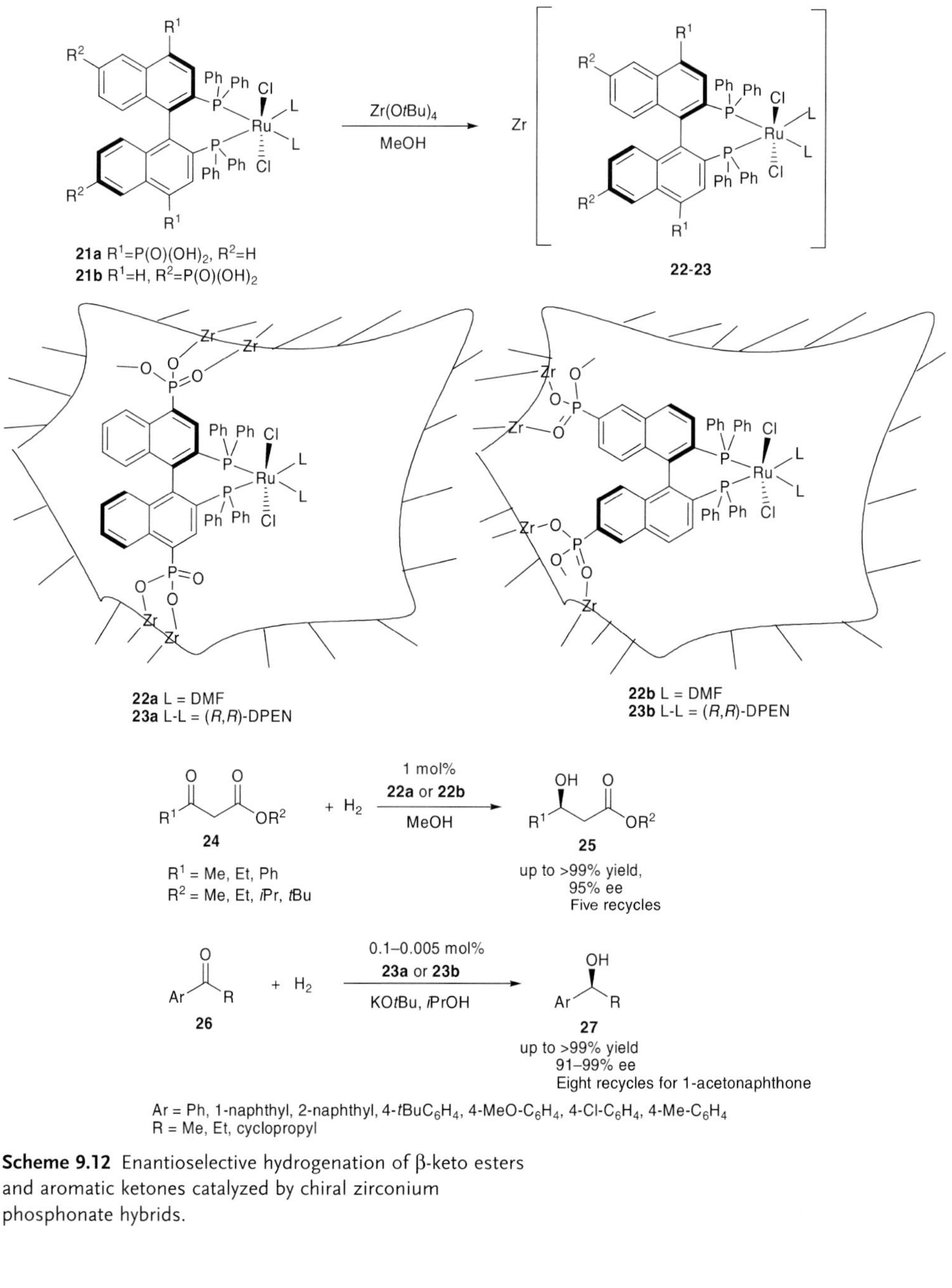

21a R¹=P(O)(OH)₂, R²=H
21b R¹=H, R²=P(O)(OH)₂

22-23

22a L = DMF
23a L-L = (*R,R*)-DPEN

22b L = DMF
23b L-L = (*R,R*)-DPEN

24 + H₂ → **25**
1 mol% **22a** or **22b**, MeOH
R¹ = Me, Et, Ph
R² = Me, Et, *i*Pr, *t*Bu
up to >99% yield, 95% ee
Five recycles

26 + H₂ → **27**
0.1–0.005 mol% **23a** or **23b**, KO*t*Bu, *i*PrOH
up to >99% yield
91–99% ee
Eight recycles for 1-acetonaphthone

Ar = Ph, 1-naphthyl, 2-naphthyl, 4-*t*BuC₆H₄, 4-MeO-C₆H₄, 4-Cl-C₆H₄, 4-Me-C₆H₄
R = Me, Et, cyclopropyl

Scheme 9.12 Enantioselective hydrogenation of β-keto esters and aromatic ketones catalyzed by chiral zirconium phosphonate hybrids.

metal in the hydrogenation and present as the pendant group. N_2 adsorption measurements indicated that these hybrid polymers are highly porous with high surface areas. Powder XRD analysis showed the solids to be amorphous, such that the possibility of any detailed structure elucidation was precluded.

Both of the hybrid polymers **22a–b** were shown to be highly efficient heterogeneous catalysts for the asymmetric hydrogenation of β-keto esters **24** (Scheme 9.12). For example, compound **22b** catalyzed the hydrogenation of β-keto esters with complete conversion in 20 h and excellent enantioselectivity (ee-values ranging from 91.7% to 95.0%). The heterogeneous nature of the catalysis was established by an absence of reactivity of the supernatants of the catalysts. Direct current plasma (DCP) spectroscopic analysis indicated that less than 0.01% leaching of the ruthenium occurred for each round of hydrogenation. Catalyst **22b** was reused for five cycles in the asymmetric hydrogenation of methyl acetoacetate with complete conversions and no significant deterioration of enantioselectivity (ee-values ranging from 93% to 88%).

Lin and coworkers further improved the chiral Zr phosphonate hybrid polymers by the incorporation of Ru–BINAP–DPEN moieties, to catalyze the heterogeneous asymmetric hydrogenation of aromatic ketones with excellent activity and extremely high enantioselectivity (**23a–b** in Scheme 9.12) [47]. With built-in Ru–BINAP–DPEN entities as active sites in the hybrid polymers, an outstanding catalytic performance was demonstrated in using these chiral solids as heterogeneous catalysts for the asymmetric hydrogenation of aromatic ketones **26**. With a 0.1 mol% catalyst loading of **23a**, all of the tested substrates showed complete conversion and high ee-values (ranging from 90.6% to 99.2%) that were superior to those for the corresponding homogeneous ruthenium complex. Furthermore, a catalyst loading of **23a** as low as 0.005 mol% could still bring about a full conversion for the hydrogenation of 1-acetonaphthone (turnover frequency (TOF) 500 h^{-1}) with an ee-value of 98.6%. A somewhat inferior enantioselectivity was observed when **23b** was used, which suggested a remarkable matrix effect and potential fine-tunability of this type of hybrid polymer. Finally, the heterogeneous catalyst **23a** could be recycled and reused in the asymmetric hydrogenation of 1-acetonaphthone for up to eight times, without any loss of enantioselectivity.

9.5.2
Asymmetric C—C Bond-Forming Reactions

Lin and coworkers prepared chiral porous Zr phosphonates with pendant chiral dihydroxy groups, and used them in combination with excess of Ti(OiPr)$_4$ in promoting asymmetric additions of diethylzinc to a range of aromatic aldehydes (Scheme 9.13) [48]. By refluxing BINOL-derived bisphosphonic acids L_{8a-c} with Zr(OnBu)$_4$ in butanol, chiral amorphous zirconium phosphonates **28** were obtained, which exhibited BET surface areas ranging from 431 to 586 $m^2 g^{-1}$, as determined from N_2 adsorption measurements. After drying at 80 °C *in vacuo*, these hybrid polymers were treated with excess Ti(OiPr)$_4$ to generate active catalysts for the addition of diethylzinc to aromatic aldehydes **4**. Under 20–50 mol%

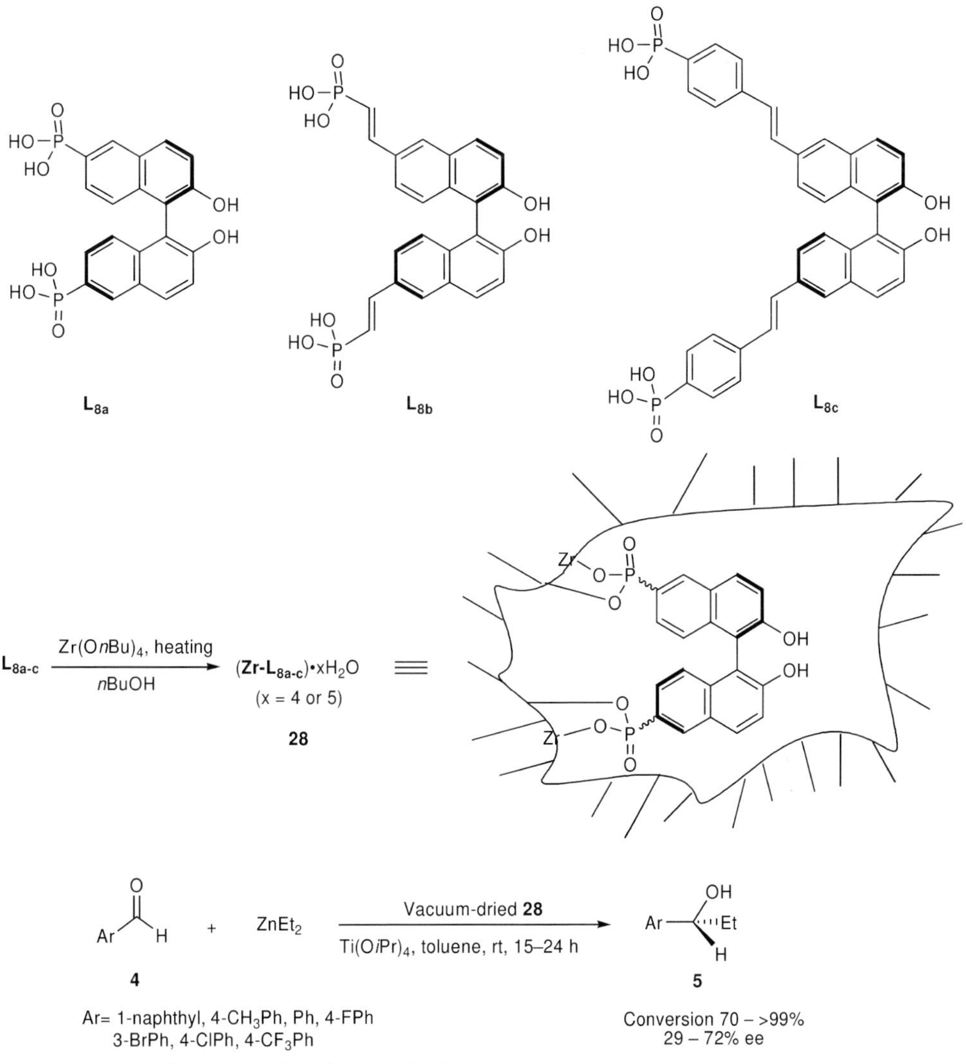

Scheme 9.13 Chiral zirconium phosphonate hybrids-Ti-catalyzed enantioselective diethylzinc addition to aldehydes.

solid loading in toluene at room temperature, the reactions afforded chiral secondary alcohols with good to high conversions and moderate enantioselectivities (ee-values up to 72%). Although the heterogeneous nature of the systems was supported by the inactivity of the supernatant in the reactions, no recycling of the catalysts was reported.

One of the most challenging tasks associated with most heterogeneous catalysts is their structure elucidation owing to their amorphous nature. This lack of detailed structure information greatly hampers the mechanistic understanding of the corresponding heterogeneous catalytic process. The immobilization of homogeneous catalysts using metal–organic coordination polymers may provide a promising solution to this problem. Since the 1990s, numerous crystalline MOCPs have been structurally characterized in their solid state by using X-ray diffraction (XRD) techniques, including some homochiral MOCPs with interesting catalytic properties. The self-assembly nature and noncovalent interactions in the synthesis of MOCP may sometimes favor the ordered stacking (crystallization) of the solids. By virtue of the single-crystallinity of some MOCP catalysts and uniformly distributed catalytic sites, an investigation of the structure–activity relationship (SAR) would be feasible, which in turn may greatly facilitate the catalytic study and application from many perspectives.

Recently, Lin and coworkers reported a crystalline homochiral MOF and its application in the heterogeneous asymmetric catalysis of diethylzinc addition to aldehydes with activity and stereoselectivity comparable to its homogeneous counterparts (Scheme 9.14) [49]. Here, the design strategy lay in utilizing chiral bridging ligands L_9 bearing bipyridyl primary functionality for network construction and the chiral BINOL unit as the secondary orthogonal functionality. By treatment of L_9 with $CdCl_2$ in MeOH/DMF, crystals of **29** were obtained. Single-crystal XRD studies revealed that in **29** the Cd centers were connected by chlorines and the pyridyl units of the bridging ligand to form a 3-D network with large chiral channels (1.6 × 1.8 nm) along the a-axis. Some of the BINOL dihydroxy groups of L_9 were found to point to the open channel, which were occupied by solvent

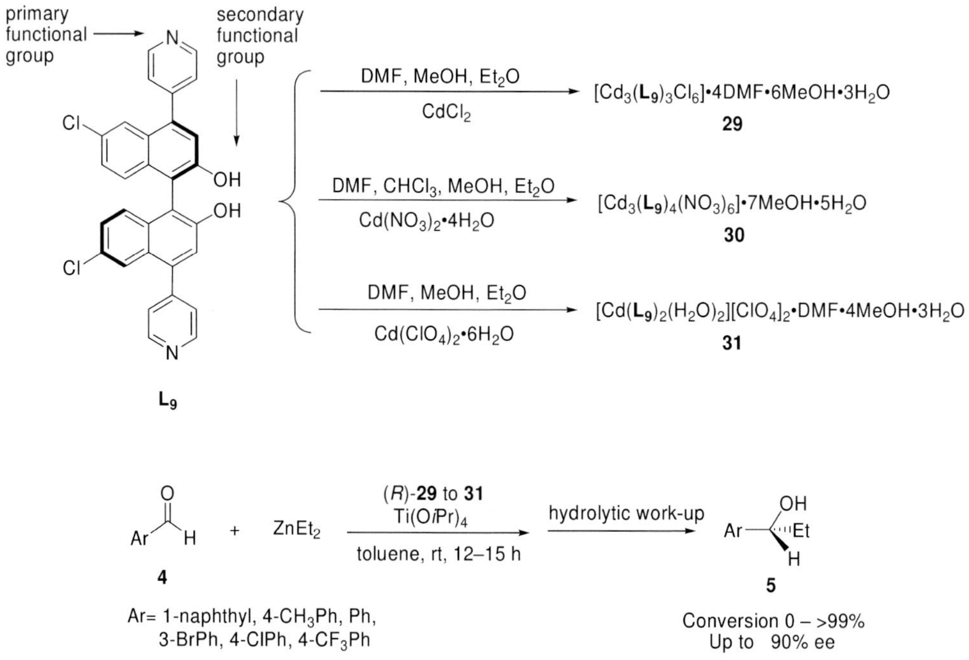

Scheme 9.14 Crystalline type II MOCP solids for titanium(IV)-catalyzed $ZnEt_2$ addition to aromatic aldehydes.

molecules in the as-synthesized sample. Upon solvent evacuation, PXRD analysis indicated that the framework structure was retained, while CO_2 adsorption measurements demonstrated its permanent porosity and high surface area (601 $m^2 g^{-1}$). Treatment of the evacuated solid of **29** with an excess of Ti(O*i*Pr)$_4$ generated the active catalyst, which catalyzed the addition of diethylzinc to a range of aromatic aldehydes with complete conversion and ee-values which were comparable with, or even superior to, those of the homogeneous analogues [49]. One remarkable feature of this catalyst system was its size selectivity. For example, when a highly sterically demanding dendritic aldehyde (size 2 nm) was used as the test substrate, the conversion was zero for the heterogeneous catalyst, which was in sharp contrast to the good conversion (95%) obtained using the homogeneous BINOL–Ti(O*i*Pr)$_4$ catalyst under the otherwise identical conditions. These results indicated that the catalytically active sites were located in the channels, a characteristic which mimicked zeolites. However, no catalyst recovery or reuse were reported.

Very interestingly, by using the same chiral bridging ligand **L$_9$** but with different Cd(II) salts, Lin's group was able to obtain another two crystalline MOFs **30** and **31** (Scheme 9.14) [50]. Single-crystal XRD studies indicated that these MOFs adopted different 3-D framework structures. PLATON calculation of the results, thermogravimetric analysis (TGA) and CO_2 adsorption measurements indicated that both **30** and **31** were porous and exhibited permanent porosity following removal of all the solvent molecules. However, when **30** and **31** were treated with Ti(O*i*Pr)$_4$ and tested as enantioselective heterogeneous catalysts for the asymmetric addition of diethylzinc to aldehydes, drastically different catalytic activities were observed. Whilst **30** catalyzed the addition of diethylzinc to a range of aromatic aldehydes with complete conversions in 15 h and moderate to high ee-values (the heterogeneous nature of the catalysis was confirmed by control experiments), **31** failed to show any activity at all under otherwise identical conditions. Based on a closer examination of the structures, the authors proposed that the steric crowdedness of the chiral dihydroxy groups in **31** might be responsible for the inactivity. It is remarkable that the structure-dependent catalytic activity study for heterogeneous asymmetric catalysis can indeed be performed using structurally well-characterized MOCP catalysts.

9.5.3
Epoxidation

Very recently, Nguyen, Hupp and their colleagues reported the immobilization of a chiral Salen–Mn complex in a crystalline microporous metal–organic framework (Scheme 9.15) [51]. A chiral Salen–Mn complex **L$_{10}$** with 4-pyridyl groups was employed as the ditopic building unit which, upon solvothermal reaction with zinc ions and biphenyldicarboxylic acid (H$_2$bpdc) in DMF, afforded a crystalline coordination network **32**. A single-crystal X-ray structure determination of **32** showed that all Mn^{III} sites were accessible to the channels. PLATON calculations and TGA showed that **32** possessed a highly porous structure with solvent molecules occupying the micropores, while PXRD measurements indicated that the evacuated

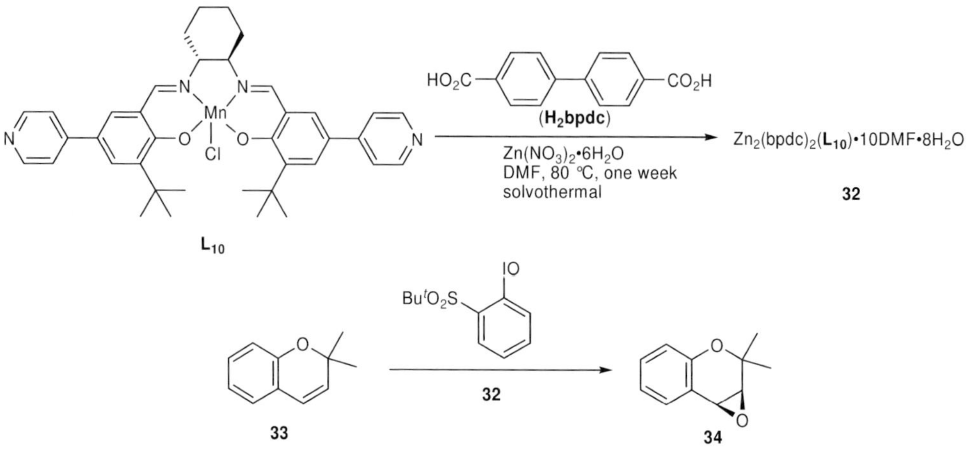

Entry	Cycle	TON	Yield (%)	ee (%)
1	1st	1430	71	82
2	2nd	1420	71	82
3	3rd	1320	66	82

Scheme 9.15 Homochiral Salen–Mn metal–organic framework for the asymmetric epoxidation of 2,2-dimethyl-2H-chromene.

sample still retained its crystallinity. The catalytic activity of **32** was examined in the asymmetric epoxidation of 2,2-dimethyl-2H-chromene **33** with 2-(tert-butylsulfonyl) iodosylbenzene as oxidant, yielding **34** with an enantiomeric excess that almost rivaled that of its homogeneous counterpart (82% for the coordination network **32** versus 88% for free L_{10}). Most interestingly, immobilization of the Salen–Mn(III) complex in this way significantly extended the catalyst's lifetime. While the homogeneous epoxidation catalysts L_{10} was found to lose much of its activity after the first few minutes, the framework-immobilized catalyst **32** exhibited close to constant reactivity during the entire catalytic process. The enhanced catalyst stability of **32** was ascribed by the authors to the confinement effect of the framework, which prevented the degrading oxidation of the Salen moiety. Finally, a recyclability study of **32** showed no loss of enantioselectivity after three cycles, albeit with a small loss in activity. Although ICP spectroscopy after removal of the MOF particles showed that between 4% and 7% of the manganese initially present in **32** was lost per cycle, measurements of reaction progress in the resulting solution phase indicated that the small quantity of dissolved manganese did not catalyze the reaction.

9.5.4
Miscellaneous

High contents of asymmetric units in a homochiral MOCP do not always guarantee excellent enantioselectivity in a reaction. Lin and coworkers have reported the

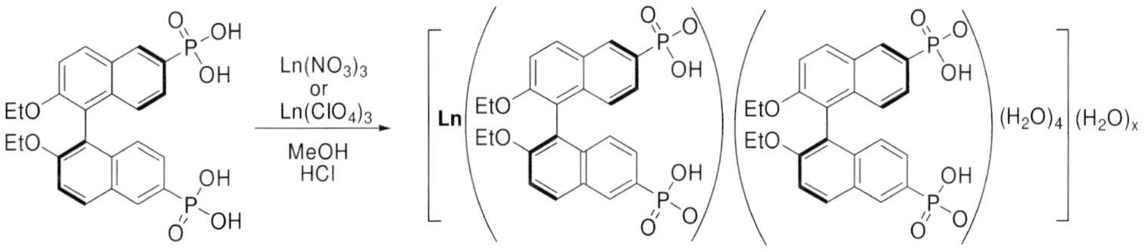

Scheme 9.16 Homochiral lanthanide bisphosphonates [52].

synthesis and catalytic applications of a series of homochiral porous lamellar lanthanide bisphosphonates **35** (Scheme 9.16) [52]. The presence of both Lewis and Brønsted acid sites in these solids rendered them capable of catalyzing several organic reactions, including the cyanosilylation of aldehydes and the ring opening of *meso*-carboxylic anhydrides; unfortunately, essentially no enantioselectivity (ee <5%) was obtained in all cases.

Very recently, Fedin, Kim and colleagues synthesized a homochiral metal–organic polymeric material, [Zn_2(bdc)(L-lac)(dmf)]·(DMF) (**36**), by using a one-pot solvothermal reaction of $Zn(NO_3)_2$, L-lactic acid (L-H_2lac) and 1,4-benzenedicarboxylic acid (H_2bdc) in DMF [53]. This 3-D homochiral microporous framework exhibited permanent porosity and enantioselective host–guest sorption properties towards several substituted thioether oxides. Although **36** could catalyze the oxidation of thioethers to sulfoxides with size and chemoselectivity, no asymmetric induction was observed.

9.6
Concluding Remarks and Outlook

In summary, new strategies to immobilize soluble chiral complexes in homochiral metal–organic hybrid polymers for heterogeneous asymmetric catalysis have been developed over the past few years. Through a modular assembly of chiral polytopic/polyfunctional ligands and (catalytic) metal ions, homochiral MOCPs with diverse structures and functionalities have been prepared, and utilized to catalyze several types of asymmetric organic reaction in a heterogeneous manner, with high efficiencies and enantioselectivities comparable with or even superior to those of their corresponding homogeneous counterparts. The molecular building units of the MOCPs can be easily fine-tuned, which allows for facile engineering of their catalytic performance. These materials are shown to promote some reactions under mild conditions, and can be easily recovered by simple filtration or decantation and reused in multiple cycles without any loss of activity or enantioselectivity. In contrast to most other heterogeneous catalysts, these MOCP catalysts possess a high density of catalytically active units and asymmetric moieties, and can be used to immobilize multicomponent chiral complexes where synergistic actions are important for the catalysis. Another salient feature which is in sharp contrast to most state-of-art heterogeneous catalysts is that, in some cases, MOCP catalysts can be highly crystalline with uniform active sites, so that their structural information can be extracted and used for a better mechanistic understanding and SAR studies. These remarkable features, coupled with the versatile methods for the rational design and synthesis of various MOCPs, suggest a broader scope for the application of these compounds in heterogeneous (asymmetric) catalysis in the future.

Although remarkable progresses have been achieved to date, many challenges remain in this emerging field of MOCP catalysis. New types of organic–inorganic hybrid materials need to be investigated for their potential applications in chemically interesting and/or industrially useful (asymmetric) catalytic reactions. Beyond activity and selectivity, significant improvements in the stability and compatibility of MOCP catalysts with reaction systems are also very desirable. Hence, further studies are required in order to provide a critical assessment of their application potential and limitations, and to evaluate if MOCPs can compete with other types of well-known heterogeneous asymmetric catalyst.

Although the amorphous nature of MOCPs does not preclude their interesting catalytic properties, the development of structurally well-defined systems is highly desirable for mechanistic and structure–activity–selectivity studies. The development of new analytical tools that can be applied to the characterization of not only the microstructure of an MOCP (e.g. to ascertain how many repeat units are actually incorporated into a polymer framework) but also the precatalyst or catalyst in its resting state, remains a formidable and important challenge.

The difficulties sometimes encountered in the synthesis of polytopic ligands might be overcome by the development of efficient synthetic protocols (such as 'click' chemistry) or the use of noncovalent bonding interactions (H bond or coordination). In addition, the development of continuous-flow reaction systems might prove advantageous for the practical applications of MOCP catalysts.

Clearly, whilst these challenges are highly interdisciplinary and multifarious, they should be met by the development of new methodologies and principles in polymer, inorganic, supramolecular, catalysis and organometallic chemistry. The strategy of using MOCPs might also be extended to a much broader scope of organic transformations and materials science. Taken together, it is expected that the use of MOCPs for catalyst immobilization will continue to attract increasing research interest from both academia and industry.

References

1 (a) Noyori, R. (1994) *Asymmetric Catalysis in Organic Synthesis*, Wiley-Interscience, New York.
(b) Ojima, I. (ed.) (2000) *Catalysis Asymmetric Synthesis*, 2nd edn, Wiley-VCH Verlag GmbH, New York.
(c) Jacobsen, E.N., Pfaltz, A. and Yamamoto, H. (eds) (1999) *Comprehensive Asymmetric Catalysis*, Vols **I–III**, Springer, Berlin.
(d) Yamamoto, H. (ed.) (2001) *Lewis Acids in Organic Synthesis*, Wiley-VCH Verlag GmbH, New York.

References | 353

2 (a) Blaser, H.-U (2003) *Chemical Communications*, 293–6.
(b) McMorn, P. and Hutchings, G.J. (2004) *Chemical Society Reviews*, **33**, 108–22.

3 (a) de Vos, D.E., Vankelecom, I.F. and Jacobs, P.A. (eds) (2000) *Chiral Catalyst Immobilization and Recycling*, Wiley-VCH Verlag GmbH, Weinheim.
(b) Heitbaum, M., Glorius, F. and Escher, I. (2006) *Angewandte Chemie – International Edition*, **45**, 4732–62.
(c) Dioos, B.M.L., Vankelecom, I.F.J. and Jacobs, P.A. (2006) *Advanced Synthesis Catalysis*, **348**, 1413–46.

4 (a) Leadbeater, N.E. and Marco, M. (2002) *Chemical Reviews*, **102**, 3217–74.
(b) Song, C.E. and Lee, S. (2002) *Chemical Reviews*, **102**, 3495–524.
(c) Fan, Q.-H., Li, Y.-M. and Chan, A.S.C. (2002) *Chemical Reviews*, **102**, 3385–466.
(d) de Vos, D.E., Dams, M., Sels, B.F. and Jacobs, P.A. (2002) *Chemical Reviews*, **102**, 3615–40.
(e) Benaglia, M., Puglisi, A. and Cozzi, F. (2003) *Chemical Reviews*, **103**, 3401–29.
(f) Cozzi, F. (2006) *Advanced Synthesis Catalysis*, **348**, 1367–90.

5 Dai, L.-X. (2004) *Angewandte Chemie – International Edition*, **43**, 5726–9.

6 (a) Ding, K.-L., Wang, Z., Wang, X.-W., Liang, Y.-X. and Wang, X.-S. (2006) *Chemistry – A European Journal*, **12**, 5188–97.
(b) Shi, L., Wang, Z., Wang, X.-W., Li, M.-X. and Ding, K.-L. (2006) *Chinese Journal of Organic Chemistry*, **26**, 1444–56.
(c) Ding, K.-L., Wang, Z. and Shi, L. (2007) *Pure and Applied Chemistry*, **79**, 1529–38.

7 Forster, P.M. and Cheetham, A.K. (2003) *Topics in Catalysis*, **24**, 79–86.

8 (a) Kesanli, B. and Lin, W.-B. (2003) *Coordination Chemical Reviews*, **246**, 305–26.
(b) Lin, W.-B. (2005) *Journal of Solid State Chemistry*, **178**, 2486–90.
(c) Ngo, H.L. and Lin, W.-B. (2005) *Topics in Catalysis*, **34**, 85–92.
(d) Lin, W.-B. (2007) *MRS Bulletin*, **32**, 544–8.

9 Cheetham, A.K., Rao, C.N.R. and Feller, R.K. (2006) *Chemical Communications*, 4780–95.

10 Janiak, C. (2003) *Dalton Transactions*, 2781–804.

11 (a) Kitagawa, S., Kitaura, R. and Noro, S. (2004) *Angewandte Chemie – International Edition*, **43**, 2334–75.
(b) Yaghi, O.M., O'Keeffe, M., Ockwig, N.W., Chae, H.K., Eddaoudi, M. and Kim, J. (2003) *Nature*, **423**, 705–14.
(c) Robin, A.Y. and Fromm, K.M. (2006) *Coordination Chemical Reviews*, **250**, 2127–57.

12 Hagrman, P.J., Hagrman, D. and Zubieta, J. (1999) *Angewandte Chemie – International Edition*, **38**, 2638–84.

13 Fujita, M., Kwon, Y.-J., Washizu, S. and Ogura, K. (1994) *Journal of the American Chemical Society*, **116**, 1151–2.

14 Ohmori, O. and Fujita, M. (2004) *Chemical Communications*, 1586–7.

15 (a) Sawaki, T., Dewa, T. and Aoyama, Y. (1998) *Journal of the American Chemical Society*, **120**, 8539–40.
(b) Sawaki, T. and Aoyama, Y. (1999) *Journal of the American Chemical Society*, **121**, 4793–8.
(c) Dewa, T. and Aoyama, Y. (2000) *Journal of Molecular Catalysis A – Chemical*, **152**, 257–60.

16 Tanski, J.M. and Wolczanski, P.T. (2001) *Inorganic Chemistry*, **40**, 2026–33.

17 (a) Saito, S., Murase, M. and Yamamoto, H. (1999) *Synlett*, 57–8.
(b) Ishida, S., Hayano, T., Furuno, H. and Inanaga, J. (2005) *Heterocycles*, **66**, 645–9.

18 Gomez-Lor, B., Gutierrez-Puebla, E., Iglesias, M., Monge, M.A., Ruiz-Valero, C. and Snejko, N. (2002) *Inorganic Chemistry*, **41**, 2429–32.

19 Perles, J., Iglesias, M., Ruiz-Valero, C. and Snejko, N. (2004) *Journal of Materials Chemistry*, **14**, 2683–9.

20 Schlichte, K., Kratzke, T. and Kaskel, S. (2004) *Microporous Mesoporous Materials*, **73**, 81–8.

21 Alaerts, L., Séguin, E., Poelman, H., Thibault-Starzyk, F., Jacobs, P.A. and De Vos, D.E. (2006) *Chemistry – A European Journal*, **12**, 7353–63.

22 Kato, C.N., Hasegawa, M., Sato, T., Yoshizawa, A., Inoue, T. and Mori, W. (2005) *Journal of Catalysis*, **230**, 226–36.

23 Mueller, U., Schubert, M., Teich, F., Puetter, H., Schierle-Arndt, K. and Pastré, J. (2006) *Journal of Materials Chemistry*, **16**, 626–36.

24 Clearfield, A. and Wang, Z.-K. (2002) *Journal of the Chemical Society–Dalton Transactions*, 2937–47.

25 Curini, M., Rosati, O. and Costantino, U. (2004) *Current Organic Chemistry*, **8**, 591–606.

26 Seo, J.S., Whang, D., Lee, H., Jun, S.I., Oh, J., Jeon, Y.J. and Kim, K. (2000) *Nature*, **404**, 982–6.

27 Holliday, B.J. and Mirkin, C.A. (2001) *Angewandte Chemie–International Edition*, **40**, 2022–43.

28 Williams, K.A., Boydston, A.J. and Bielawski, C.W. (2007) *Chemical Society Reviews*, **36**, 729–44.

29 Takizawa, S., Somei, H., Jayaprakash, D. and Sasai, H. (2003) *Angewandte Chemie–International Edition*, **42**, 5711–14.

30 (a) Guo, H.-C., Wang, X.-W. and Ding, K.-L. (2004) *Tetrahedron Letters*, **45**, 2009–12.
(b) Wang, X.-S., Wang, X.-W., Guo, H.-C., Wang, Z. and Ding, K.-L. (2005) *Chemistry–A European Journal*, **11**, 4078–88.

31 (a) Mikami, K. and Shimizu, M. (1992) *Chemical Reviews*, **92**, 1021–50.
(b) Mikami, K. and Terada, M. (1999) *Comprehensive Asymmetric Catalysis*, Vol. III (eds E.N. Jacobsen, A. Pfaltz and H. Yamamoto), Springer, Berlin, Chapter 32.
(c) Dias, L.C. (2000) *Current Organic Chemistry*, **4**, 305–42.

32 (a) Mikami, K., Terada, M. and Nakai, T. (1989) *Journal of the American Chemical Society*, **111**, 1940–1.
(b) Terada, M., Mikami, K. and Nakai, T. (1990) *Chemical Communications*, 1623–4.
(c) Mikami, K., Terada, M. and Nakai, T. (1990) *Journal of the American Chemical Society*, **112**, 3949–54.
(d) Terada, M., Mikami, K. and Nakai, T. (1994) *Chemical Communications*, 833–4.
(e) Kitamoto, D., Imma, H. and Nakai, T. (1995) *Tetrahedron Letters*, **36**, 1861–4.
(f) Mikami, K., Motoyama, Y. and Terada, M. (1994) *Inorganica Chimica Acta*, **222**, 71–5.
(g) Mikami, K. and Matsukawa, S. (1997) *Nature*, **385**, 613–15
(h) Terada, M., Matsumoto, Y., Nakamura, Y. and Mikami, K. (1997) *Chemical Communications*, 281–2.
(i) Terada, M., Matsumoto, Y., Nakamura, Y. and Mikami, K. (1999) *Inorganica Chimica Acta*, **296**, 267–72.
(j) Mikami, K. and Matsumoto, Y. (2004) *Tetrahedron*, **60**, 7715–19.

33 (a) Shibasaki, M., Sasai, H. and Arai, T. (1997) *Angewandte Chemie–International Edition*, **36**, 1236–56.
(b) Shibasaki, M. and Matsunaga, S. (2006) *Chemical Society Reviews*, **35**, 269–79.
(c) Gou, S.-H., Wang, J., Liu, X.-H., Wang, W.-T., Chen, F.-X. and Feng, X.-M. (2007) *Advanced Synthesis Catalysis*, **349**, 343–9.

34 Arai, T., Sekiguti, T., Iizuka, Y., Takizawa, S., Sakamoto, S., Yamaguchi, K. and Sasai, H. (2002) *Tetrahedron: Asymmetry*, **13**, 2083–7.

35 Harada, T. and Nakatsugawa, M. (2006) *Synlett*, 321–3.

36 (a) Bougauchi, M., Watanabe, S., Arai, T., Sasai, H. and Shibasaki, M. (1997) *Journal of the American Chemical Society*, **119**, 2329–30.
(b) Daikai, K., Kamaura, M. and Inanaga, J. (1998) *Tetrahedron Letters*, **39**, 7321–2.
(c) Kinoshita, T., Okada, S., Park, S.R., Matsunaga, S. and Shibasaki, M. (2003) *Angewandte Chemie–International Edition*, **42**, 4680–4.
(d) Matsunaga, S., Kinoshita, T., Okada, S., Harada, S. and Shibasaki, M. (2004) *Journal of the American Chemical Society*, **126**, 7559–70.

37 Wang, X.-W., Shi, L., Li, M.-X. and Ding, K.-L. (2005) *Angewandte Chemie–International Edition*, **44**, 6362–6.

38 Komatsu, N., Hashizume, M., Sugita, T. and Uemura, S. (1993) *Journal of Organic Chemistry*, **58**, 4529–33.

39 Pescitelli, G., Di Bari, L. and Salvadori, P. (2006) *Journal of Organometallic Chemistry*, **691**, 2311–18.

40 (a) Reetz, M.T. and Mehler, G. (2000) *Angewandte Chemie–International Edition*, **39**, 3889–90.
(b) van den Berg, M., Minnaard, A.J., Schudde, E.P., van Esch, J., de Vries, A.H.M., de Vries, J.G. and Feringa, B.L.

(2000) *Journal of the American Chemical Society*, **122**, 11539–40.
(c) Claver, C., Fernandez, E., Gillon, A., Heslop, K., Hyett, D.J., Martorell, A., Orpen, A.G. and Pringle, P.G. (2000) *Chemical Communications*, 961–2.
(d) Komarov, I.V. and Börner, A. (2001) *Angewandte Chemie – International Edition*, **40**, 1197–200.
(e) Hu, A.-G., Fu, Y., Xie, J.-H., Zhou, H., Wang, L.-X. and Zhou, Q.-L. (2002) *Angewandte Chemie – International Edition*, **41**, 2348–50.
(f) Wu, S., Zhang, W., Zhang, Z. and Zhang, X. (2004) *Organic Letters*, **6**, 3565–7.
(g) Liu, Y. and Ding, K.-L. (2005) *Journal of the American Chemical Society*, **127**, 10488–9.
(h) Liu, Y., Sandoval, C.A., Yamaguchi, Y., Zhang, X., Wang, Z., Kato, K. and Ding, K.-L. (2006) *Journal of the American Chemical Society*, **128**, 14212–13.
(i) Reetz, M.T., Sell, T., Meiswinkel, A. and Mehler, G. (2003) *Angewandte Chemie – International Edition*, **42**, 790–3.

41 (a) Wang, X.-W. and Ding, K.-L. (2004) *Journal of the American Chemical Society*, **126**, 10524–5.
(b) Peng, H.-Y., Lam, C.-K., Mak, T.C.W., Cai, Z., Ma, W.-T., Li, Y.-X. and Wong, H.N.C. (2005) *Journal of the American Chemical Society*, **127**, 9603–11.

42 Shi, L., Wang, X.-W., Sandoval, C.A., Li, M.-X., Qi, Q.-Y., Li, Z.-T. and Ding, K.-L. (2006) *Angewandte Chemie – International Edition*, **45**, 4108–12.

43 Lehn, J.-M. (2000) *Chemistry – A European Journal*, **6**, 2097–102.

44 Liang, Y.-X., Jing, Q., Li, X., Shi, L. and Ding, K.-L. (2005) *Journal of the American Chemical Society*, **127**, 7694–5.

45 Liang, Y.-X., Wang, Z. and Ding, K.-L. (2006) *Advanced Synthesis Catalysis*, **348**, 1533–8.

46 Hu, A.-G., Ngo, H.L. and Lin, W.-B. (2003) *Angewandte Chemie – International Edition*, **42**, 6000–3.

47 Hu, A.-G., Ngo, H.L. and Lin, W.-B. (2003) *Journal of the American Chemical Society*, **125**, 11490–1.

48 Ngo, H.-L., Hu, A.-G. and Lin, W.-B. (2004) *Journal of Molecular Catalysis A – Chemical*, **215**, 177–86.

49 Wu, C.-D., Hu, A.-G., Zhang, L. and Lin, W.-B. (2005) *Journal of the American Chemical Society*, **127**, 8940–1.

50 Wu, C.-D. and Lin, W.-B. (2007) *Angewandte Chemie – International Edition*, **46**, 1075–8.

51 Cho, S.-H., Ma, B.-Q., Nguyen, S.-B.T., Hupp, J.T. and Albrecht-Schmitt, T.E. (2006) *Chemical Communications*, 2563–5.

52 Evans, O.R., Ngo, H.L. and Lin, W.-B. (2001) *Journal of the American Chemical Society*, **123**, 10395–6.

53 Dybtsev, D.N., Nuzhdin, A.L., Chun, H., Bryliakov, K.P., Talsi, E.P., Fedin, V.P. and Kim, K. (2006) *Angewandte Chemie – International Edition*, **45**, 916–20.

10
Heterogeneous Enantioselective Hydrogenation on Metal Surface Modified by Chiral Molecules

Takashi Sugimura

10.1
Introduction

Noble metals have the ability to adsorb a variety of organic and inorganic molecules on their surface, after which the adsorbed molecules acquire special properties. Heterogeneous catalysis provides such an example, whereby a substrate and a reagent are both absorbed or adsorbed onto the catalyst surface and, when the two species meet on the surface, a reaction occurs with assistance from the surface metal. The reaction on such a noble metal catalyst consists of three steps: (i) adsorption of the substrate; (ii) reaction on the surface; and (iii) desorption of the product. Hence, the reaction selectivities are governed by both the adsorption step of the substrate and the reaction transition states. The control of reaction selectivity during the adsorption step represents one of the main characteristics of heterogeneous catalysis, and may be due not only to the adsorption rate or constant but also to the geometry of the adsorbed substrate. By applying this unique situation, a coexisting third molecule – which formally is unreactive with the reagent and reactant but may be adsorbed onto the catalyst surface – can affect the adsorption and the reaction of the substrate during the process of catalysis. If the adsorption of the third molecule is strong, then the molecule behaves as part of the catalyst. Such a molecule is referred to as a 'modifier', and has the potential to regulate, enhance – or even to cause – a certain catalytic activity. When the modifier is chiral and optically active, the catalyst has the potential to differentiate the enantioface (and/or the enantiotopos) of the substrate.

The history of heterogeneous enantioselective catalysis with chiral modification of the metal surface extends back even further than that of homogeneous molecular metal catalysis. However, successful precedents which result in a practically useful stereoselectivity (e.g. of over 80%) involve only three types, all of which involve the hydrogenation of unsaturated bonds. Initially, these reactions were realized by achieving the correct solution to address all requirements for the chiral modifier. That is, the adsorption of the modifier must occur on all of the active sites of the entire catalyst surface, and with a uniform conformation and geometry. At the same time, adsorption of the modifier should not interrupt the reaction of the substrate, either by acting as a 'poison' of the catalyst or by competitive adsorption with the substrate. In order to leave room on the surface for substrate adsorption, the degree of modification becomes critical, as a space which is greater than the substrate size can function as a non-modified site, on which hydrogenation is not affected by the chiral modifier, but rather a racemic product is created. The result is a diminution in the overall asymmetric yield, which is a common drawback of chirally modified heterogeneous catalysis, with only those systems that can overcome this problem being capable of providing high enantioselectivity. Of course, the mode of the modifier–substrate interaction is also important. Typically, the interaction on the surface must occur by a single mechanism, and should cause a sufficiently stereocontrolled adsorption and reaction of the substrate. In addition, the interaction should not *deactivate* the substrate, because the low catalytic activity at the modified site results not only in a poor turnover frequency (TOF) but also in an increase in the proportion of catalysis of the nonmodified sites. Previously, high asymmetric yields of the hydrogenations have been achieved by addressing all of these requirements.

During hydrogenation, the activated reagent of atomic hydrogen exists ubiquitously on the metal surface. One notable property of hydrogen is its size; it is so small that it can access any space and react with an adsorbed substrate–modifier complex from the narrow surface side. In contrast, if the reagent is a normal organic or inorganic molecule, the reaction requires a certain size of space at the reaction sites. In this situation the geometries of both the reagent and the substrate–modifier complex are strictly limited by the adsorption onto the surface, and close contact of the species at the reaction sites may be difficult.

Chiral modification of the catalyst surface and its application to enantioselective hydrogenation have been reviewed frequently, and consequently only those results published since 2000 [1] will be reviewed in this chapter. Details of the three enantioselective hydrogenation systems using platinum, nickel and palladium metals will be presented here; of these metals, palladium has been investigated more recently as a catalyst than either nickel or platinum, and so will be described in detail.

10.2
History of the Chiral Modification of Metal Catalysts

Historically, it took many years to achieve the almost perfect stereocontrol of an enantioselective hydrogenation reaction using a heterogeneous catalyst, with a resultant 97–99% enantiomeric excess (ee) of the product. One of the oldest such examples is that of tartaric acid-modified platinum black for the hydrogenation of an oxime to give a chiral amine of <20% ee [2]. Nonetheless, many reviews in this field have provided a clear history of the enantioselective catalysis from silk-palladium – a palladium metal supported on a chiral silk fiber to hydrogenate a

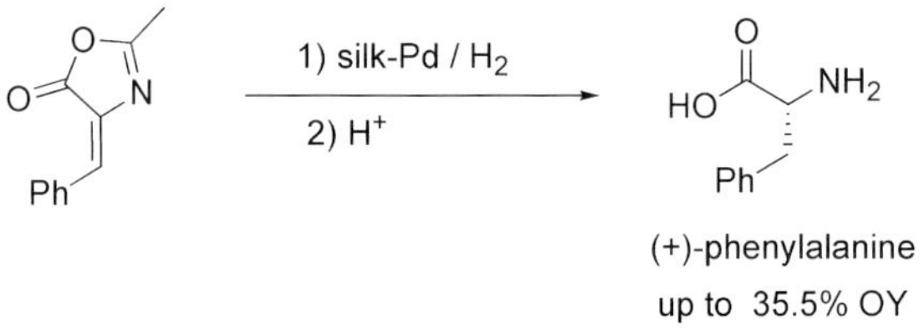

Scheme 10.1 Enantioselective hydrogenation and optical yield over the silk-palladium catalyst.

dehydrophenylalanine derivative that resulted in optical yields of up to 35.5% (Scheme 10.1) [3].

The report on silk-palladium has been well-recognized as a pioneering study demonstrating the potential of the stereocontrolling with a surface modification that opened the window of asymmetric heterogenous catalysis. The mechanism of this catalysis system can be truly regarded as a 'black box', and its poor reproducibility due to the use of a natural material caused the further development of silk-palladium to be interrupted. Through the years, several research groups have postulated that the dependency of optical yield on the individual origins of the silk may be due to the degree of decomposition of the silk protein caused during its treatment with palladium salts. If this was the case, then the amino acids produced could play a major role in the stereocontrol by being adsorbed onto the palladium metal surface. A number of studies which were conducted to identify the correct combination of a parent catalyst, chiral modifier and substrate led to the concept of two hydrogenation systems: (i) the platinum/cinchona alkaloid/α-ketoester system; and (ii) the nickel/tartaric acid/β-ketoester system. Another combination of palladium/cinchona alkaloid/functionalized olefin for the hydrogenation reaction was developed at a later date.

10.3
Cinchona Alkaloid-Modified Platinum Catalysis [1, 4]

The platinum catalyst modified by cinchonine (CN) or cinchonidine (CD) is the most widely studied system among the heterogeneous enantioselective catalysts, and has been applied to industrial processes on moderately large scales. Following the hydrogenation of ethyl benzoylformate, the product was obtained in quantitative yield with optical purities up to 80% (Scheme 10.2). This combination of metal/modifier/substrate was identified by Orito et al. in 1979 [5, 6]. As the stereoselectivity was already high in the first of these reports, the reaction conditions as well as the structure of the chiral modifiers have mostly been retained following a vast number of investigations. Today, this platinum-based hydrogenation is referred to as the 'Orito reaction'.

One notable property of the Orito reaction is that two chiral modifiers, CD and CN, are diastereomers of each other. The configurations of the functional groups which are expected to create the molecular recognition are antipodes, and only the

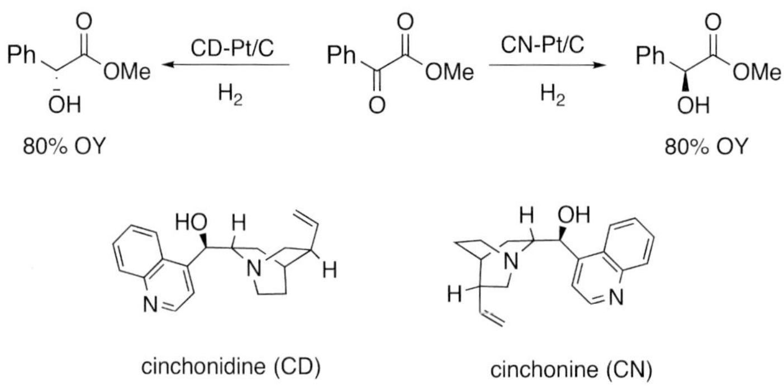

Scheme 10.2 The Orito reaction; enantioselective hydrogenation of ketone over CD- or CN-modified Pd/C.

vinylic portion (which is essentially an ethyl group during the hydrogenation) has the same stereochemistry. Hence, if both CD and CN are considered and used as pseudo-enantiomers, both enantiomers can be obtained although the stereocontrolling abilities are not the same.

During the 1990s, major progress in the Orito reaction was achieved not only to explore the mechanism but also to improve the product ee-values. Blaser et al. found that the hydrogenation of methyl pyruvate gave a better ee-value (95%) when the reaction was carried out in acetic acid and the hydroxy group in the CD was converted to a methoxy group. These improved conditions indicated that the quinuclidine nitrogen in the CD was protonated to function as a proton donor to the substrate [7]. A notable acceleration was also found in the hydrogenation rate by the CD modification [8]; such an advantageous phenomenon served not only to perform the reaction under a high TOF but also to suppress the racemic production catalyzed by the unmodified platinum surface. Following many other efforts during the 1990s, two groups achieved 97% ee by using ultrasonic radiation for the CD-modification [9] or by using a polymer stabilized nanocluster of platinum [10]. Other progress during the late 1990s included an increase in the applicable substrates and expected mechanisms for the enantiodifferentiation, which are described elsewhere [4]. Since 2000, the Orito reaction has been studied to determine the reaction mechanism in detail, to achieve further optimization of the reaction conditions, and to fully characterize the modified catalyst.

The group of Baiker also contributed significantly to the Orito reaction. For the classic combination of the Pt/CD/pyruvate ester, these authors found that a side (aldol) reaction affected the product ee-value, and that some improvement was observed following its removal [11]. Continuous hydrogenation in a fixed-bed reactor was also demonstrated in supercritical ethane [12] and in acetic acid [13]. The range of applicable substrates was also expanded to β-trifluoromethyl-β-ketoester [14], cyclohexane-1,2-dione [15], aromatic methyl ketones [16], α-hydroxyketones [17] and isatin derivatives [18]. Some selected results are shown in Scheme 10.3.

Important information regarding the stereocontrol mechanism was reported by the same group. For the interaction model of linear α-ketoesters with CD, a ther-

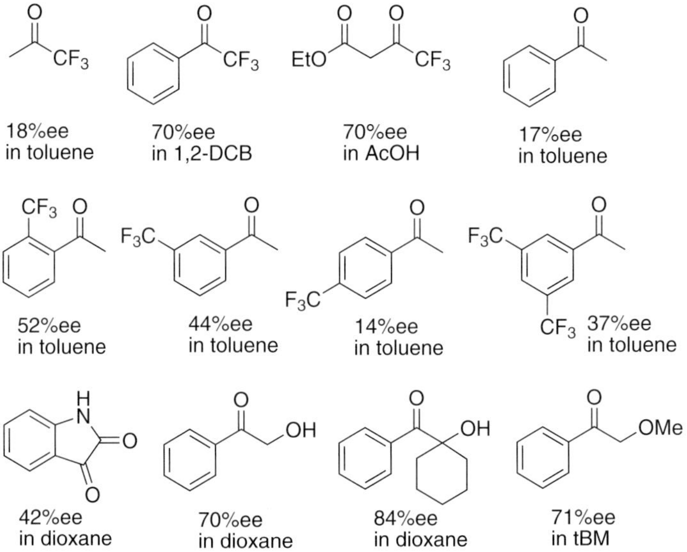

10.3 Cinchona Alkaloid-Modified Platinum Catalysis

Scheme 10.3 Selected ketones used for the enantioselective hydrogenation over the CD-modified Pt/Al$_2$O$_3$, and ee-values obtained in an optimized solvent for each reaction.

Substrate	ee
CH$_3$COCF$_3$	18% ee in toluene
PhCOCF$_3$	70% ee in 1,2-DCB
EtOCOCH$_2$COCF$_3$	70% ee in AcOH
PhCOCH$_3$ (acetophenone)	17% ee in toluene
2-CF$_3$-PhCOCH$_3$	52% ee in toluene
3-CF$_3$-PhCOCH$_3$	44% ee in toluene
4-CF$_3$-PhCOCH$_3$	14% ee in toluene
3,5-(CF$_3$)$_2$-PhCOCH$_3$	37% ee in toluene
isatin	42% ee in dioxane
PhCOCOOH	70% ee in dioxane
cyclohexyl PhCOCOOH	84% ee in dioxane
PhCOCOOMe	71% ee in tBM

modynamically stable *s-cis* conformer was initially presumed to be employed but, based on detailed calculations, this was revised to the less-stable *s-trans* conformer because of a better agreement with the experimental results [19]. Conformation of the CD adsorbed on a Pt(101) surface was reinvestigated using the first principles methods to provide a more complete picture of the modification process [20]. Similar mechanistic analyses were also demonstrated for the other substrates [21]. Among the studies with CD derivatives, the Pt catalyst modified by the O-phenyl ether of CD (phenoxy analogue) is worthy of mention as it produces an inverted stereoselectivity for the hydrogenation of pivalate up to 72% ee [22]. New chiral modifiers designed by simplification of the CD structure were tested for the hydrogenation of ketopantolactone, another representative substrate for the CD-modified platinum catalyst [23]. Selected examples are shown in Scheme 10.4. All of the modifiers contained the 1-naphthyl group, which has a weaker (but sufficient) anchoring ability against the platinum surface. The best modifier structure was found to depend significantly on the structure of the substrates [24]. For example, in the hydrogenation of ketopantolactone, a 74% ee-value for the R-configuration was obtained with CD in dichloromethane (DCM), whereas the use of N-methyl CD rather than CD provided the S-product with up to 45% ee [25].

Bartók's group also reported a number of results relating to this catalytic system. Following the report of an ee-value of 97% for the product by hydrogenation of methyl pivalate [9], the group also studied details of the ultrasound irradiation of the catalyst and found the improvement in ee-value to be due to a synergistic effect of the transformation and surface cleaning of the metal particles and enhanced cinchonidine adsorption [26]. The group also reported that the use of

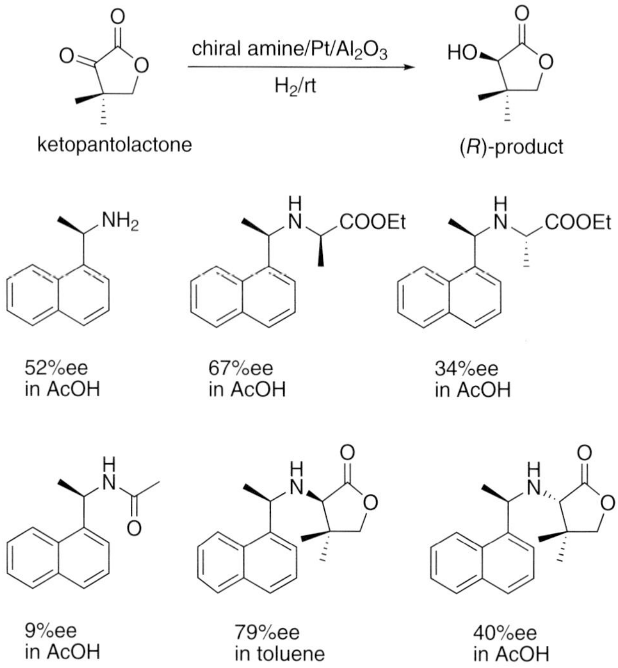

Scheme 10.4 Enantioselective hydrogenation of ketopantolactone over Pt/Al$_2$O$_3$ modified by synthetic chiral amines. Selected amines are listed with the best ee-values and the reaction solvent for the reactions.

α-isocinchonine (α-ICN; Figure 10.1) rather than CN gave similar results for the hydrogenation of ethyl pyruvate; for example, 86% ee with CN and 60% ee with α-ICN in toluene containing 6% acetic acid [27]. Under different conditions in acetic acid, CN, α-ICN and β-ICN resulted in 92, 82 and 62% ee of the product, respectively [28]. Similarly, the hydrogenation of ketopantolactone resulted in 55% and 37% ee (in acetic acid) with respective modifiers, but with β-ICN the product stereochemistry was inverted to result in up to 60% ee (in toluene) [29]. An inversion of the enantioselectivity for the hydrogenation of ketopantolactone was also reported by Baiker's group [30].

By extending the substrate design, a dynamic kinetic resolution of the α-fluoroketones [14] was demonstrated [31]. The reaction shown in Scheme 10.5 accompanied the equilibration between the enantiomers of the substrate that was faster than the hydrogenation process, to produce the *threo*-isomer with high selectivities.

Murzin *et al.* reported detailed studies of the hydrogenation of 1-phenyl-1,2-propanedione (acetyl benzoyl), which has two carbonyl groups to give four possible stereoisomers by double hydrogenation [32]. A detailed theoretical analysis has also been reported for this substrate [33].

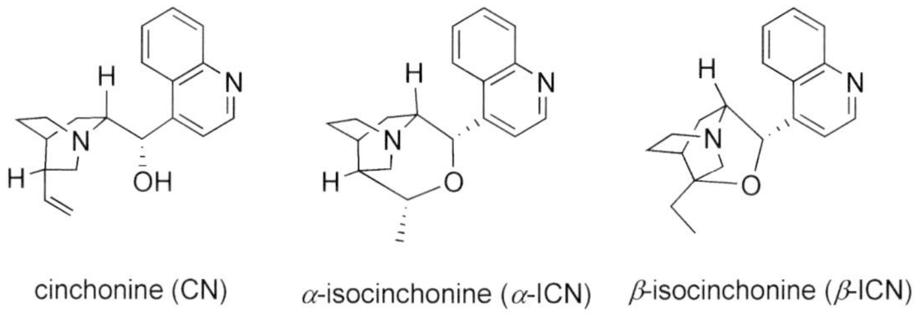

cinchonine (CN)   α-isocinchonine (α-ICN)   β-isocinchonine (β-ICN)

Figure 10.1 Structures of CN and its derivatives.

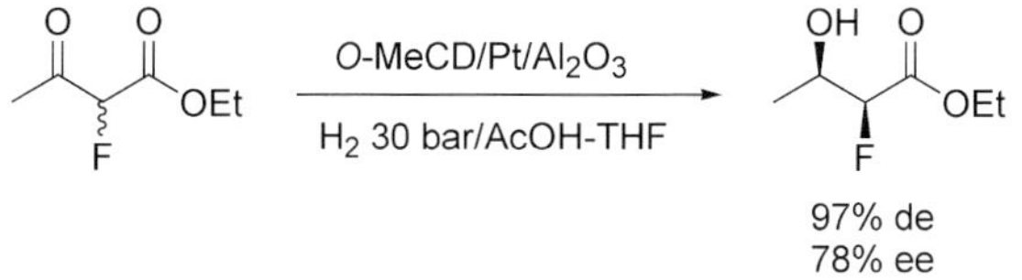

Scheme 10.5 Dynamic kinetic resolution of α-fluoro-acetoacetate ester over O-methylcinchonidine (O-MeCD)-modified Pt/Al$_2$O$_3$.

Today, the Orito reaction continues to attract much attention, and forms the largest category of heterogeneous enantioselective hydrogenation. For reasons of space limitation in this chapter, the details of many studies have been omitted, although the groups of Hutchings [34], Baltruschat [35], Richards [36], Zeara [37], McBreen [38], Wells [39], Roucoux [40], Li [41], Liu [42], Garland [43], Margitfalvi [44], Perosa [45], Williams [46], Feföldi [47] and Reyes [48] appear to be currently active in this field.

A transition-state model of the Orito reaction is presented at the end of the section; although various models have been proposed on many occasions, the structure shown in Figure 10.2 is currently the most reliable. The CD is adsorbed onto the catalyst surface with the quinoline ring facing against the surface, while the hydroxy group stays near the surface. In this conformation (the so-called 'Open-3'), the ammonium proton at the quinuclidine interacts with the carbonyl and is reduced.

10.4
Tartaric Acid-Modified Nickel Catalysis [49–52]

Enantioselective hydrogenation using a nickel catalyst is the oldest-known heterogeneous enantioselective catalysis. Izumi and Akabori began with a combination of Raney nickel/amino acid/methyl acetoacetate (MAA) [53] and identified a better combination, namely Raney nickel/tartaric acid/MAA (Scheme 10.6) [54]. Raney nickel remains a well-used base catalyst due to its high catalytic activity, while tartaric acid is the best chiral modifier for the nickel hydrogenation studied to date. MAA is the smallest β-ketoester, and is still a representative substrate of

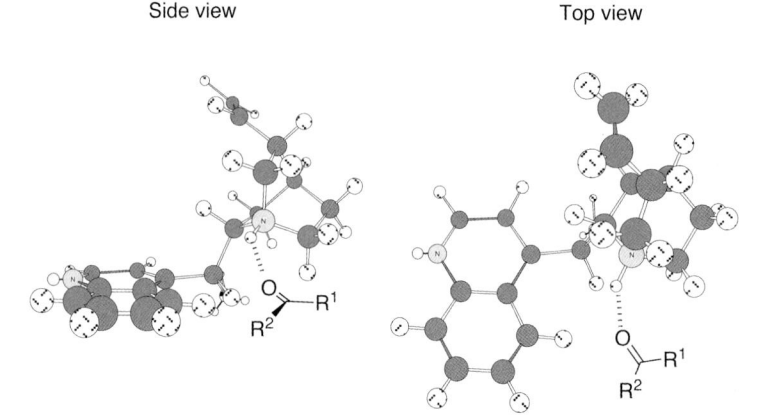

Figure 10.2 The most probable structure of the CD–ketone complex over the platinum metal surface. The structures are seen from on the surface (side view) and the upper side (top view).

Scheme 10.6 Enantioselective hydrogenation of MAA over tartaric acid-modified nickel catalyst.

β-ketoesters and their analogues. As tartaric acid is tightly adsorbed onto the catalyst surface and is stable under the hydrogenation conditions, the modified catalyst can be used for the recover/reuse process, which is different from the cinchona alkaloid-modified platinum and palladium systems.

Results identified for the nickel catalysis before 1999 are summarized as follows [1, 50, 51]:

- The nickel catalyst must consist of chemically pure crystalline having a relatively large diameter [55], although amorphous and chemically impure regions that remained in the Raney nickel may be removed by an acid treatment [56], by ultrasonic irradiation [57] or, more efficiently, by site-selective partial poisoning with sodium bromide [58].

- The substrate structure was extended from β-ketoesters not only to similar analogues, such as 1,3-diketones [59], β-alkoxy ketones [60] and α-ketosulfones [61], but also to simple alkanones such as 2-octanone [62].

- The hydrogenation conditions are preferable when a large amount of pivalic acid exists for the 2-alkanones instead of a trace amount of acetic acid for the β-ketoesters [63].
- The lifetime of the modified catalyst during its repeated use is drastically extended by the addition of a very small amount of amine [64].
- The highest product ee-value of 98.6% was reported in 1997 for the hydrogenation of methyl 3-cyclopropyl-3-oxopropanoate [65], and this value remains the highest of all heterogeneous enantioselective catalyses. Selected substrates which are applicable to nickel catalysis are shown in Scheme 10.7, together with the highest product ee-value for each substrate.

The tartaric acid-modified nickel is prepared by soaking it in a hot aqueous solution of tartaric acid (and NaBr), basically in the same way as the original procedure. This modification step necessitates a much greater amount of tartaric acid than is actually adsorbed (ca 30-fold); moreover, the modification solution must be disposed of with great care due to the high nickel salt content generated by leaching of the catalyst surface. The direct addition of tartaric acid (and NaBr) to the reaction mixture (*in situ* modification) represents the ideal approach but did not provide a sufficiently high product ee-value when Raney nickel was used. In 2000, Osawa *et al.* reported that 79% ee was attained with the *in situ* modification when a fine nickel powder was employed as the precursor catalyst for the hydrogenation of MAA [66]. The catalyst has a high durability, providing identical results for more than 20 recycles following its separation by filtration [67]. The *in situ* modification can also be applied to the hydrogenation of 2-octanone with reduced nickel which was prepared from nickel oxide [68] and could also be used repeatedly [69]. This success of the *in situ* modification with reduced nickel led to a new pre-modification method that today is carried out in an organic solvent under hydrogen pressure [70].

The hydrogenation of MAA is usually carried out in the presence of acetic acid (ca 0.1%, v/v). A careful kinetic analysis indicated that acetic acid specifically accelerates the enantioselective hydrogenation interacting with tartaric acid, but decelerates the nonenantioselective hydrogenation [71]. Kukula and Cerveny also reported the kinetic analysis of this hydrogenation system [72].

The supported nickel has notable advantages due to its easy handling and the controllability of the nickel particle size. It may also be used as a base catalyst for the tartaric acid-modification, though the supported nickels were less suitable than the unsupported nickels [73]. However, by optimizing the type of support, the precipitation method, the calcination temperature and the activation process, an ee-value of 87% was reported with Ni/Al_2O_3 by Osawa [74], and of 72% with Ni/SiO_2 by Lee [75]. Wolfson *et al.* reported that a 91% ee-value was attained with Ni/C (graphite) at 20% or less conversion of the hydrogenation [76].

Although the contribution of the nonenantioselective site to the total hydrogenation is difficult to predict, it is firmly believed that there must indeed be some degree of contribution. By comparison with the best substrate that gives the product of 98.6% ee, lower ee-values with other β-ketoesters were analyzed [77]. In a simplified model, the total product ee-value is expressed as:

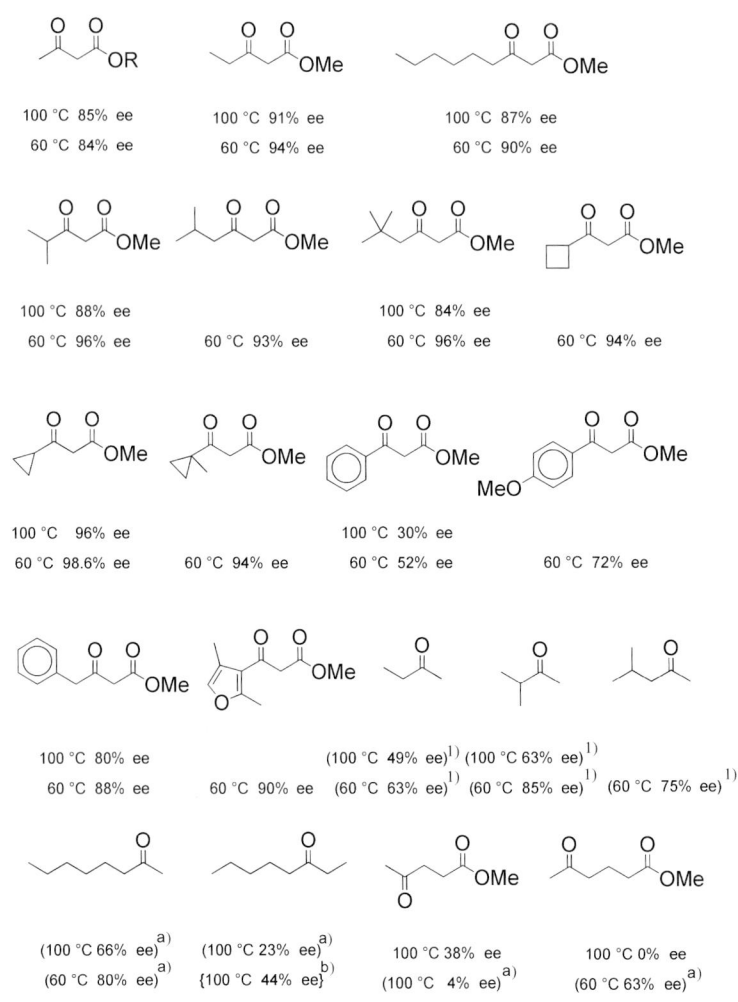

10 Heterogeneous Enantioselective Hydrogenation on Metal Surface Modified by Chiral Molecules

[Structures of β-ketoesters with ee values:]

R-O-CO-CH₂-CO-CH₃
100 °C 85% ee
60 °C 84% ee

MeO-CO-CH₂-CO-CH₃
100 °C 91% ee
60 °C 94% ee

MeO-CO-CH₂-CO-(CH₂)₄CH₃
100 °C 87% ee
60 °C 90% ee

iPr-CO-CH₂-CO-OMe
100 °C 88% ee
60 °C 96% ee

tBu-CO-CH₂-CO-OMe (isobutyl variant)
60 °C 93% ee

(CH₃)₃C-CO-CH₂-CO-OMe
100 °C 84% ee
60 °C 96% ee

cyclobutyl-CO-CH₂-CO-OMe
60 °C 94% ee

cyclopropyl-CO-CH₂-CO-OMe
100 °C 96% ee
60 °C 98.6% ee

(1-methylcyclopropyl)-CO-CH₂-CO-OMe
60 °C 94% ee

Ph-CO-CH₂-CO-OMe
100 °C 30% ee
60 °C 52% ee

(4-MeO-C₆H₄)-CO-CH₂-CO-OMe
60 °C 72% ee

PhCH₂-CO-CH₂-CO-OMe
100 °C 80% ee
60 °C 88% ee

(2,5-dimethylfuran-3-yl)-CO-CH₂-CO-OMe
60 °C 90% ee

CH₃-CO-CH₂-CH₃
(100 °C 49% ee)[1] (100 °C 63% ee)[1]
(60 °C 63% ee)[1] (60 °C 85% ee)[1] (60 °C 75% ee)[1]

CH₃(CH₂)₄-CO-CH₃
(100 °C 66% ee)[a]
(60 °C 80% ee)[a]

CH₃(CH₂)₃-CO-CH₃
(100 °C 23% ee)[a]
{100 °C 44% ee}[b]

CH₃-CO-CH₂-CO-OMe (acetoacetate)
100 °C 38% ee
(100 °C 4% ee)[a]

longer chain β-ketoester-OMe
100 °C 0% ee
(60 °C 63% ee)[a]

Scheme 10.7 Selected compounds applicable for the tartaric acid-modified Raney nickel, and the product ee-values at 60 and 100 °C. (a) The results in parentheses were obtained in a 1:1 mixture of pivalic acid and THF; (b) The result in brackets was obtained in the presence of 2 equiv. of 1-methyl-1-cyclohexanecarboxylic acid.

$$\%ee = \frac{i \times E}{(E+N)}$$

where E and N are the contributions from the enantioselective and nonenantioselective catalysis, respectively, to the total hydrogenation, and the factor-i is the product ee-value specific to the enantioselective hydrogenation. The three substrates shown in Scheme 10.8 suggested, on an experimental basis, that factor-i is

10.4 Tartaric Acid-Modified Nickel Catalysis

	(isopropyl β-ketoester, OMe)	(cyclopropyl β-ketoester, OMe)	(MAA, OR)
Product ee	96%	98.6%	86%
Factor-i	ca 1	ca 1	ca 0.9
$E/(E+N)$	0.96	0.99	0.96

Scheme 10.8 Quantitative analysis of the product ee-values obtained by the hydrogenation of β-ketoesters over the tartaric acid-NaBr-modified Raney nickel. Factor-i indicates the intrinsic enantioselective ability of the tartaric acid-modified sites, and E and N indicate the contribution of the enantioselective catalysis sites and nonenantioselective hydrogenation sites, respectively.

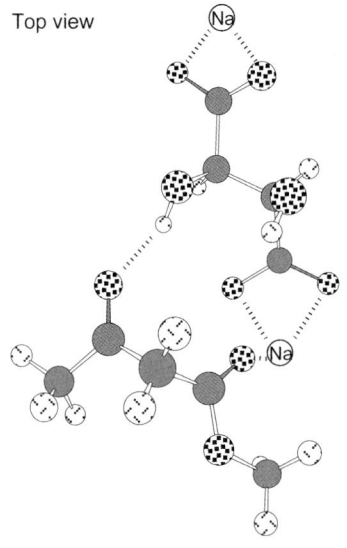

Figure 10.3 Expected structure of complex of tartaric acid and methyl acetoacetate over nickel catalyst. The figure is drawn as viewed from the upper side of the surface.

the same between the isopropyl and cyclopropyl substrates, and that $E/(E + N)$ is the same between the methyl and isopropyl substrates. The result of the analysis suggested that MAA required an improvement in factor-i to achieve the higher product ee-value; hence, a study to fine-tune the interaction between the substrate and the modifier will be a key issue.

The transition-state model for the hydrogenation of MAA was revised by Osawa after an interval of 17 years [51]. A top view of tartaric acid adsorbed onto a nickel surface, and of MAA interacted with one of the hydroxy groups of the tartaric acid and sodium of the countercation, is illustrated in Figure 10.3.

One other notable progress which has been made since 2000 is that of a surface analysis using scanning tunneling microscopy (STM). Reval and Baddeley reported several STM analyses of a tartaric acid-modified Cu(110) surface, and showed that the modifier molecules were self-assembled in rows of three, with each row stacking in parallel with the other modifiers [78]. Some years later, the details of a tartaric acid-modified Ni(111) surface was reported where, under low coverage, some molecular arrangements in either the absence [79] or presence [80] of MAA were found to suggest the existence of a stereocontrolled mechanism [81, 82].

10.5
Cinchona Alkaloid-Modified Palladium Catalysis [83]

Today, palladium catalysts can be employed for the hydrogenation of a variety of functional groups, and many types are available commercially. Due to its ease of use and well-disclosed nature, palladium-catalyzed hydrogenation has been monitored for a variety of chiral modifier/substrate combinations, although the combination presently in use is somewhat limited. Tungler et al. reported an example in which isophorone was reduced by palladium black modified by a dihydroapovincaminic acid ethyl ester, which is a saturated analogue of the natural vanilla alkaloid (Scheme 10.9) [84, 85]. The highest ee-value achieved was 55%. Except for this case, all other Pd systems employed CD/CN or their analogues as the chiral modifier.

In 1985, Perez et al. reported a hydrogenation process using a combination of Pd/CD/unsaturated acid (Scheme 10.10) [86]. Although the enantioselectivity was low (up to only 30.5% ee), this was the earliest example of a palladium-catalyzed enantioselective hydrogenation. Several years later, Nitta became interested in this dormant system and improved the product ee-value for phenylcinnamic acid (PCA) to 44% [87]. A variety of supported palladium catalysts was subsequently prepared with different mean particle sizes and size distributions, and the 5% Pd/TiO$_2$ was obtained as a result. The TiO$_2$ employed was a rutile type, with a small surface area (51 m^2 g^{-1}); thus, a 5% loading of palladium resulted in a smaller

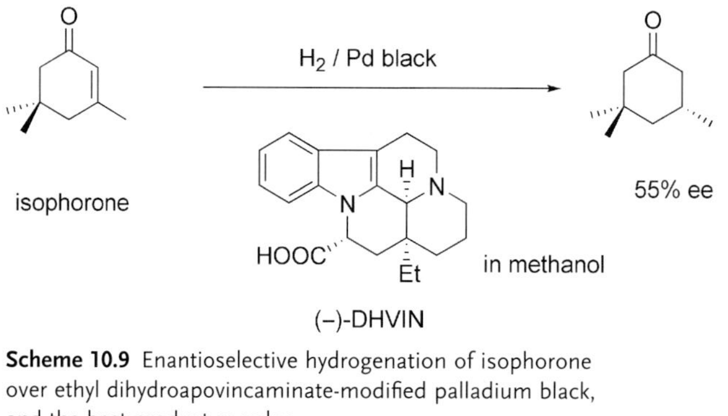

Scheme 10.9 Enantioselective hydrogenation of isophorone over ethyl dihydroapovincaminate-modified palladium black, and the best product ee-value.

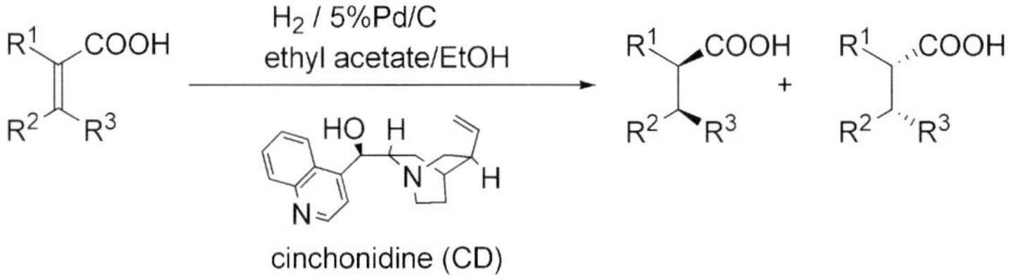

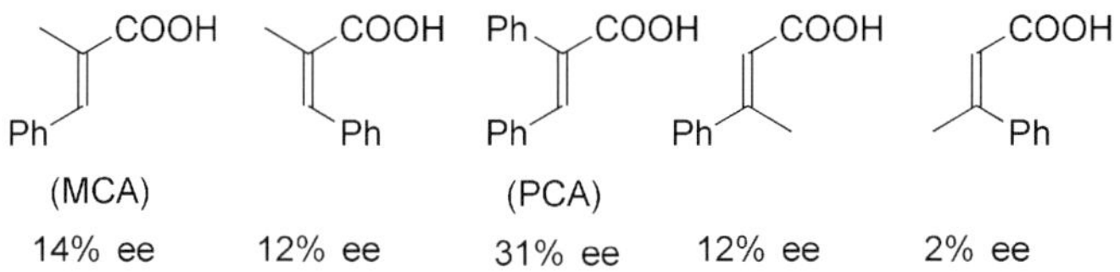

Scheme 10.10 Enantioselective hydrogenation of α,β-unsaturated carboxylic acids over Pd/C in the presence of cinchonidine, and the product ee-values under certain conditions.

dispersion of 0.52 and a larger average particle size of 2.1 nm [88]. This catalyst has long been the best available for the hydrogenation of PCA and other olefins, except for that of aliphatic unsaturated acids.

Much of the progress in this catalysis system was brought about following the report of Nitta in 1994 [87]. Since chiral modified palladium catalysts have a short history, and relevant reviews are few in number [83], all of the important findings are outlined in the present section, irrespective of the borderline of the year 2000. Investigations aimed at improving the optical yield by a combination of Pd/CD/ olefin represents a way in which to develop a heterogeneous enantioselective catalysis system. Currently, three types of olefinic substrate are known to provide good to high product ee-values, and the optimized conditions found to date will differ among the substrate types. This suggests that, whilst the stereocontrol mechanisms are also different, they too may become unified in the near future by a further optimization of the reaction conditions.

10.5.1
Aromatic α,β-Unsubstituted Carboxylic Acids

PCA is a representative of the aromatic substrates detailed in this subsection (Scheme 10.11). The hydrogenation rate and stereoselectivity when using various modifiers for the Pd/TiO$_2$ are shown in Table 10.1 [89]. The catalytic activity of the unmodified Pd/TiO$_2$ is as high so as to perform the hydrogenation in 98 h^{-1}g^{-1} of the initial rate under the same conditions in wet dimethylformamide; DMF [90]. The decrease in hydrogenation rate following the CD modification was in contrast to the acceleration in the platinum system. The stereochemistry of the product was opposite between modifications with CD and CN, as might have been expected,

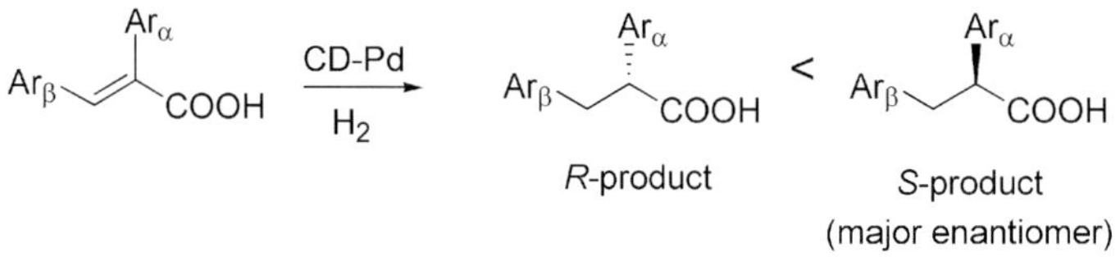

Ar$_\beta$—CH=C(Ar$_\alpha$)—COOH →[CD-Pd, H$_2$] Ar$_\beta$—CH$_2$—CH(Ar$_\alpha$)—COOH < Ar$_\beta$—CH$_2$—CH(Ar$_\alpha$)—COOH

R-product   S-product
                         (major enantiomer)

Scheme 10.11 Stereodirection during the hydrogenation of α,β-diarylpropenoic acids over a cinchonidine-modified palladium catalyst.

Table 10.1 Modifier effect on the enantioselective hydrogenation of PCA in methanol and wet DMF over a 5 wt% Pd/TiO$_2$ catalyst (conversion >99%).

Modifier	Solvent	Initial rate (mmol h^{-1}g^{-1})	ee (%)	Major enantiomer
CD	Methanol	36	49	S
	90% DMF	10	61	S
CN	Methanol	21	28	R
	90% DMF	8	29	R
Quinine	Methanol	17	10	S
	90% DMF	7	5	S
Quinidine	Methanol	18	1	S
	90% DMF	9	4	S
Norcinchol	Methanol	34	42	S
	90% DMF	8	50	S
O-Methylhydrocinchonidine	Methanol	20	12	R
	90% DMF	8	18	R
(S)-1-(1-Naphthyl)-ethylamine	Methanol	47	12	R
	90% DMF	11	4	S

although the ee-values were much different. Quinine and quinidine, each of which has a methoxy group at C6 of the quinoline ring, were not suitable modifiers. Vinyl groups in the CD and CN were converted into ethyl groups during the hydrogenation of PCA. The results with norcinchol, which has a hydroxymethyl moiety instead of a vinyl, indicates the unimportance of this part of the molecule in stereocontrol. Replacement of the hydroxy group by a methoxy group in the CD caused an inversion of stereochemistry, which again was different from the platinum system. Although the rate reduction by the modification was higher with the methoxy analogue than with the CD, a weaker absorption of the methoxy analogue was suggested by the mixed modification experiment using these two modifiers.

Many different types of support for palladium metal were tested, and those having appreciable amounts of both acidic and basic sites with a moderate specific surface area were found to be preferable [88]. Supports with a micropore structure proved not to be suitable because the active sites were situated inside the micro-

10.5 Cinchona Alkaloid-Modified Palladium Catalysis

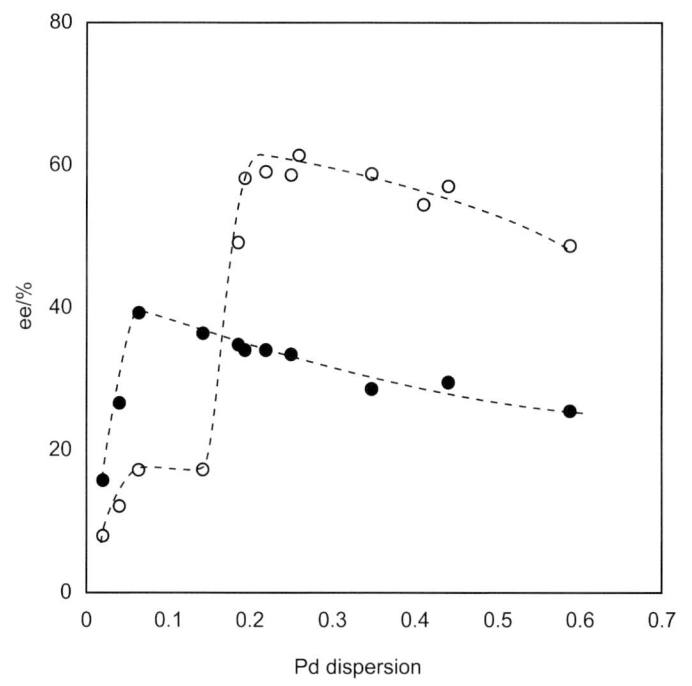

Figure 10.4 Pd-dispersion dependence of the enantioselective hydrogenation of PCA in a wet DMF (open circles) and ethyl acetate (filled circles) with the CD-modified Pd/TiO$_2$.

pores, and as a result neither the CD nor PCA could diffuse into the pores at the same time. In the micropores, the PCA molecules are converted into the racemic product catalyzed by the unmodified palladium, while the CD-modified sites in the micropores are not accessible by PCA [91].

The size of the palladium particle is also an important factor for absorption of the CD–PCA complex in a preferred structure for the hydrogenation. The product ee-values with 0.5 to 50% Pd/TiO$_2$ are shown in Figure 10.4 [92]. In a polar solvent, the ee-value gradually decreased with the increasing dispersion in the range of >0.2. As formation of the CD–PCA complex necessitates a certain space on the palladium surface, a higher ee-value with a smaller dispersion (larger particle size) is reasonable. The sudden fall in product ee to <0.2 was attributable to a larger flat (but not round) space on the palladium surface, which caused structural strain on the absorption of the CD–PCA complex. The difference between the polar and nonpolar solvents in the dispersion dependency appeared to be related to the polarity dependency of the CD conformation [93]. The solvent effect was also studied in detail [90, 94], although the relationship with the product ee-value was not straightforward. However, a polar 'wet' solvent, such as 2.5% aqueous dioxane or 2.5% aqueous DMF, consistently provided the best ee-values to date, using Pd/TiO$_2$ and Pd/Al$_2$O$_3$.

Table 10.2 The additive effect on the enantioselective hydrogenation of PCA with a CD-modified 5 wt% Pd/TiO$_2$ in 2.5% aqueous 1,4-dioxane.

Entry	Additive (equiv.)	Initial rate r_0 (mmol g^{-1} h^{-1})	ee (%)
1	None	7	59
2	Ammonia (0.5)	21	64
3	Diethylamine (0.5)	32	63
4	Triethylamine (0.5)	30	65
5	Butylamine (0.5)	28	60
6	Benzylamine (0.1)	21	64
7	Benzylamine (0.3)	33	67
8	Benzylamine (0.5)	33	71
9	Benzylamine (0.7)	34	68
10	Benzylamine (1.0)	32	71
11	None[a]	27	48
12	Benzylamine (0.5)[a]	44	60
13	None[b]	4	60
14	Benzylamine (0.5)[b]	10	69

a In methanol.
b In 2.5% aqueous DMF.

The use of an amine additive represents another important technique introduced by Nitta [94, 95]. When benzylamine (BA) was added to the hydrogenation media, the product ee of 59% increased to 64–71% (Table 10.2; entries 6–10). The change in product ee-value was accompanied by an acceleration of the reaction. Other amines having the proper basicity or the BA addition in other solvents were also effective. For the unmodified catalysis, the addition of BA decreased the hydrogenation rate by half. Based on the degree of deactivation by the CD modification (from 1/7 to 1/10), the addition of BA accelerated the modified catalysis by recovering the activity loss caused by the CD modification. Whilst BA specifically accelerated the hydrogenation of the CD–PCA complex, it decelerated the undesired hydrogenation of the free PCA to perform the nonenantioselective catalysis. As a result, the proportion of racemate formation in the total hydrogenation was decreased and this resulted in a high product ee-value. The mechanism of acceleration was assumed to be due to a quicker replacement of the product with reactant at the CD-modified reaction site. In fact, the best ee-value (72%) to be obtained without BA under sufficiently optimized conditions [96] could not be exceeded simply by the addition of BA or other amines, as shown in Table 10.2. Neither did the addition of BA improve the intrinsic stereoselectivity (factor-i).

A better base-catalyst exceeding Pd/TiO$_2$ for this type of substrate was recently identified in a study using commercially available Pd/C catalysts [97]. Some examples of Pd/C catalysts, many of which are available worldwide and of stable quality, are listed in Table 10.3. Compared with Pd/TiO$_2$, the dispersions were larger and the average diameter of the palladium particle smaller. When Pd/C was used as received for the hydrogenation of PCA, the product ee-value was low (as reported

Table 10.3 Surface area of the support (Sa) and palladium (S_M), dispersion (D_M), average particle size (D), and the performance as obtained and after the hydrogen treatment at 80°C for 30 min.

Catalyst	Sa (m²g⁻¹)	SM (m²g⁻¹)	DM	D (nm)	Product ee (%)[a]	
					As received	80°C under H²
5% Pd/TiO₂[b]	51	9.7	0.52	2.1	72	72
5% Pd/C 5R38H (eggshell)[c]		13.4	0.60	1.9	49	77
5% Pd/C 5R90 (intermediate)[c]		12.4	0.56	2.0	32	74
10% Pd/C LR385 (uniform)[d]		33.9	0.76	1.5	22	25
5% Pd/C AER (eggshell)[d]	987	14.9	0.67	1.7	55	81
5% Pd/C STD (uniform)[d]	1007	17.0	0.76	1.5	43	81

a The catalyst was employed after the CD modification, and the hydrogenation of PCA was carried out in a wet dioxane in the presence of BA at room temperature.
b Prepared by a precipitation–deposition method from JRC-TIO-3.
c Obtained from Johnson Matthey.
d Obtained from N. E. Chemcat.

previously [86]), while the eggshell-type Pd/C tended to provide a better ee-value (55%) [91]. The Pd location in the carbon support may affect the ease with which CD diffuses to the entire Pd surface. Strikingly, when Pd/C was treated with hydrogen at 80°C in a reaction solvent prior to the CD modification, the product ee-value improved up to 81% (Table 10.3, right-hand column). Hydrogen treatment under such mild conditions should not cause any change in the structure of the palladium particle, and hence the surface of the catalyst appeared to be purified. Such an improvement was not observed with Pd/TiO₂ because it was prepared by hydrogenation at 200°C.

Various PCA analogues were prepared to optimize the substrate structure for the hydrogenation over the CD-modified palladium; some selected results are shown in Scheme 10.12 [98, 99]. The product ee-value and initial hydrogenation rate with the hydrogen-treated Pd/C in wet dioxane are shown, and the reaction was carried out in the presence and absence of BA (values under the latter conditions are shown in parentheses). p-Methoxy substitution at the β-phenyl ring brought about an approximate 10% improvement in ee-value, whereas that at the α-phenyl ring did not significantly affect the ee-values of the products [98]. An ee of 92% was the highest value obtained to date for the heterogeneous enantioselective hydrogenation of unsaturated acids.

Methylcinnamic acid (MCA) is a commercially available substrate for which the product ee was only 14% under the original conditions [86], and remained low (23%) after some improvement many years later [94]. However, the application of hydrogen-treated Pd/C under optimized conditions for PCA brought about a notable improvement in the hydrogenation of MCA that resulted in a 46% ee (Scheme 10.13) [100]. The product ee-value was higher with a bulkier α-substituent, while with a β-p-methoxyphenyl group the product ee was 86%. As

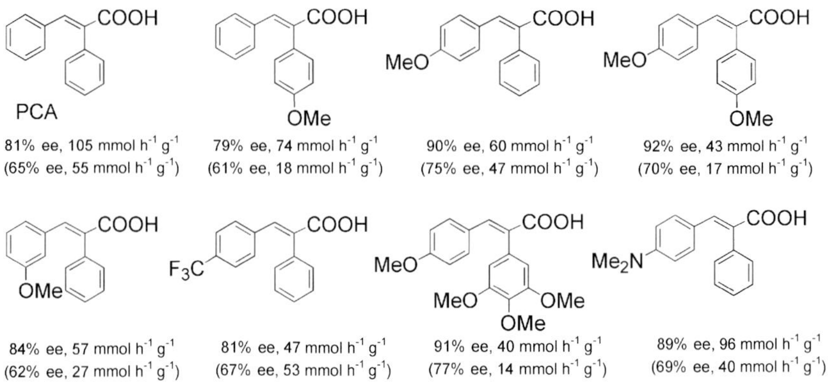

10 Heterogeneous Enantioselective Hydrogenation on Metal Surface Modified by Chiral Molecules

PCA
81% ee, 105 mmol h^{-1} g^{-1}
(65% ee, 55 mmol h^{-1} g^{-1})

79% ee, 74 mmol h^{-1} g^{-1}
(61% ee, 18 mmol h^{-1} g^{-1})

90% ee, 60 mmol h^{-1} g^{-1}
(75% ee, 47 mmol h^{-1} g^{-1})

92% ee, 43 mmol h^{-1} g^{-1}
(70% ee, 17 mmol h^{-1} g^{-1})

84% ee, 57 mmol h^{-1} g^{-1}
(62% ee, 27 mmol h^{-1} g^{-1})

81% ee, 47 mmol h^{-1} g^{-1}
(67% ee, 53 mmol h^{-1} g^{-1})

91% ee, 40 mmol h^{-1} g^{-1}
(77% ee, 14 mmol h^{-1} g^{-1})

89% ee, 96 mmol h^{-1} g^{-1}
(69% ee, 40 mmol h^{-1} g^{-1})

Scheme 10.12 Selected aromatic substrates applicable for the CD-modified Pd/C. Produce ee-values are given for the enantioselective hydrogenation in the presence and absence (shown in parentheses) of benzylamine (1.0 equiv.).

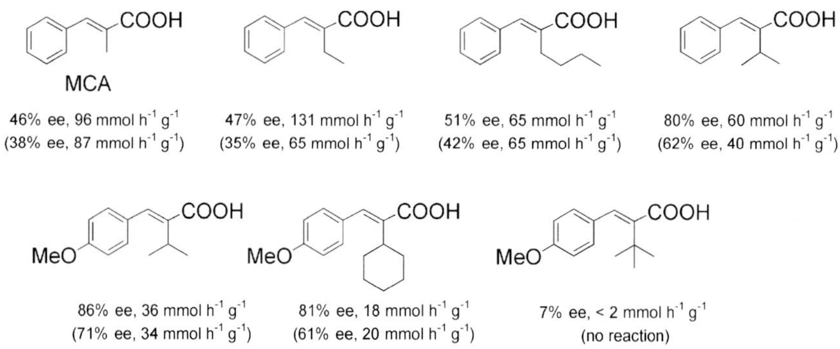

MCA
46% ee, 96 mmol h^{-1} g^{-1}
(38% ee, 87 mmol h^{-1} g^{-1})

47% ee, 131 mmol h^{-1} g^{-1}
(35% ee, 65 mmol h^{-1} g^{-1})

51% ee, 65 mmol h^{-1} g^{-1}
(42% ee, 65 mmol h^{-1} g^{-1})

80% ee, 60 mmol h^{-1} g^{-1}
(62% ee, 40 mmol h^{-1} g^{-1})

86% ee, 36 mmol h^{-1} g^{-1}
(71% ee, 34 mmol h^{-1} g^{-1})

81% ee, 18 mmol h^{-1} g^{-1}
(61% ee, 20 mmol h^{-1} g^{-1})

7% ee, < 2 mmol h^{-1} g^{-1}
(no reaction)

Scheme 10.13 Product ee-values for the enantioselective hydrogenation with α-alkyl-β-arylpropenoic acids over CD-modified Pd/C in the presence or absence (in parentheses) of benzyl amine.

the optimized conditions for the PCA family were applicable to the MCA family, they should be classified within the same substrate type, and could be enantiodifferentiated under a common stereocontrol mechanism.

10.5.2
Aliphatic α,β-Unsubstituted Carboxylic Acids

In 1996, Baiker's group began to publish results for the enantioselective hydrogenation of aliphatic α,β-unsubstituted carboxylic acids [101, 102] shortly after the initial report by Wells et al. [103]. The product ee-value was moderately high at 52% by using the CD-modified Pd/Al$_2$O$_3$ in a nonpolar solvent such as hexane under high-pressure hydrogen. One notable property of the aliphatic substrates,

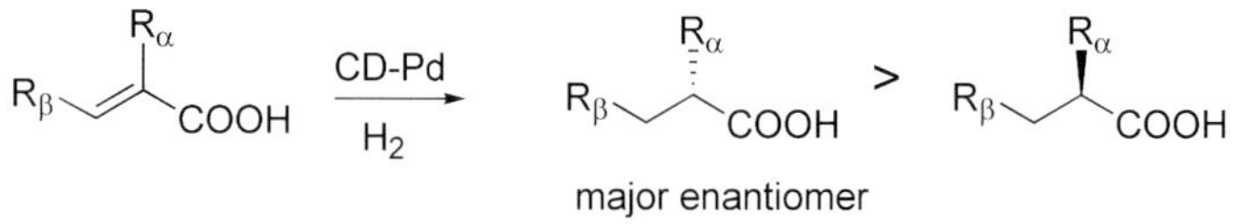

Scheme 10.14 Stereodirection during the hydrogenation of α,β-dialkylpropenoic acids over the CD-modified palladium catalyst.

Table 10.4 Product ee-values of the hydrogenation with CD-modified Pd/Al$_2$O$_3$ (conversion >99%).

Entry	R$^\alpha$	R$^\beta$	Solvent (additive)	H$_2$ pressure (atm)	Product ee (%)	Reference
1	Me	Me	Hexane	60	47	[102]
2			Toluene (BA)	50	58	[108]
3			Hexane	40	48	[109]
4	Me	Et	c-Hexane	1	28	[101]
5			c-Hexane	10	40	[101]
6			c-Hexane	60	52	[101]
7			Toluene (BA)	50	61	[108]
8			Toluene (BA)	50	66	[110]
9	Me	Pr	Toluene	50	56	[111]
10	Et	H	Hexane	60	20	[102]

represented by *(E)*-2-methyl-2-butenoic acid (tiglic acid), was formation of the *S*-product (see Scheme 10.14), which was opposite to the aromatic substrates in stereoselectivity. The dissimilarity in the optimized reaction conditions – for example, the reaction solvent, pressure and base catalyst – also indicated a difference in the stereocontrol mechanism. Based on the infra-red spectroscopic analysis of a mixture of the substrate and CD in solution, a dimer model consisting of the hydrogenation transition state with the CD and two acid molecules was proposed. A conformational analysis of the CD in solution [104], a structural study of the CD using its analogues [105], and isomerization of the substrate [106, 107] were also carried out in order to establish the current understanding of this series of hydrogenations. The function of the nonpolar solvent is to enhance interaction among the CD and two molecules of the substrate, while the high pressure suppresses the olefinic isomerization of the substrate during hydrogenation.

Selected values of the product ee, as reported by different groups, are listed in Table 10.4. The effects of the structure were not clear, but a longer alkyl chain at the β-position tended to provide a better ee-value. The addition of BA to the reaction mixture improved the stereoselectivity, but the effect was limited. The hydrogenation of a family of furan-2-carboxylic acids was also reported [112]; the best ee-value of 53% was obtained when benzofuran-2-carboxylic acid was hydrogenated in 2-propanol with the CD-modified Pd/Al$_2$O$_3$ under high-pressure conditions (30 atm).

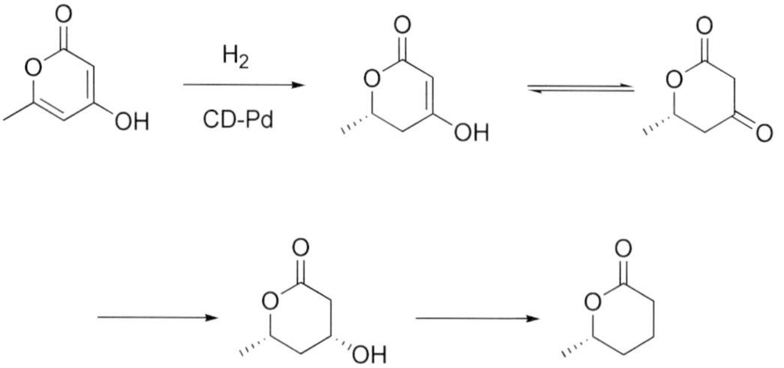

Scheme 10.15 Enantioselective hydrogenation of 4-hydroxy-6-methyl-2-pyrone over the CD-modified Pd/TiO$_2$ and succeeding diastereoselective hydrogenation.

10.5.3
2-Pyrone Derivatives

The hydrogenation of 2-pyrone derivatives is a special case of the chiral modified palladium catalysis, for multiple reasons. The enantioselective hydrogenation of 4-hydroxy-6-methyl-2-pyrone over the CD-modified palladium catalyst was first reported by Baiker's group in 2000 (Scheme 10.15) [113]. The hydrogenation was performed with Pd/TiO$_2$ in a protic solvent under atmospheric hydrogen pressure (the same as for the PCA family). The highest ee-value in the initial report was up to 85%, but this was obtained at only a 2% conversion. Such a low conversion was not due to suppression of the overreaction; rather, the main reason was the very slow hydrogenation of the substrate. Hydrogenation with the CD-modified Pd/TiO$_2$ catalyst was 25-fold slower than the corresponding reaction with the unmodified Pd/TiO$_2$. Under these conditions, hydrogenation of the CD at the quinoline ring becomes obvious and, as a result, the ee-value suddenly fell above a 5% conversion [114]. The excess addition of CD further diminished the hydrogenation rate of the substrate, and the continuous addition of small amounts of CD during the hydrogenation also proved ineffective (67% ee at 12% conversion). Both, solvent [115] and additives [116] affected the reaction, but did not essentially solve the problem. The recognition mechanism of the enantioface of 4-hydroxy-6-methyl-2-pyrone by CD on the palladium surface was proposed to be an acid–base interaction due to the high acidity of the 4-hydroxy group (pK_a = 4.73), but this was revised shortly afterwards.

The introduction of analogous substrates solved the problem of low conversion. By using the 4-methoxy analogue, the product (in 90% ee) was obtained in 80% yield accompanied by an over-reduced product of 89% ee and 98% diastereomeric excess (de) in 9% yield [117]. The 4-methyl analogue also showed a similar property. One notable finding with these substrates was that the CN modification gave a better ee-value than did the CD modification (80% versus 72% ee under certain conditions). This represented a rarely known case that CN was superior to CD as a chiral modifier among the Pt- and Pd-based enantioselective hydrogenations.

Baiker's group employed the 4-methoxypyrone as a standard substrate thereafter. Under noncatalytic conditions (the use of a stoichiometric amount of CN), a 94% ee and 95% de at an 80% conversion were achieved [118]. A stereochemical model of CN adsorption onto the palladium surface, and interaction of the pyrone derivatives with CN, were also proposed on the basis of spectroscopic investigations, product analysis and computational studies [119].

In addition to hydrogenation of the α,β-unsaturated acid and 2-pyrone derivatives, unsaturated esters were also demonstrated for the hydrogenation over the CD-modified palladium catalyst. However, at present, the product ee-value is not sufficiently high [120–122].

10.6
Conclusions

It appears that, after many years, the question remains as to whether heterogeneous enantioselective catalysis is still seen as a 'black box', and the answer is, most likely, 'yes'. Although a host of studies conducted during the past decade have provided notable progress in this field, many questions remain unanswered. Whilst initially a low enantioselectivity was a massive problem, almost perfect stereocontrol is no longer a dream. In the past the combination of metal/modifier/substrate was limited almost to accidental findings, yet today the design of chiral modifiers for specific substrates is becoming increasingly possible. Whilst a new combination remains a challenging issue, fruitful results seem to be on the horizon [123, 124]. As heterogeneous enantioselective catalysis is limited to hydrogenation, the development of a nonhydrogenation reaction is a major point. On completing this chapter, it is hoped that further developments in enantioselective heterogeneous catalysis not only employ the system's high productivity but also disclose those characteristics not possessed by molecular catalysts.

Acknowledgments

The author thanks Emeritus Professors Akira Tai, Yuriko Nitta and Tadashi Okuyama, who provided much help when they were his coworkers. He also thanks all his former and present students.

References

After the submission of the draft on June, 2007, many progress has been made in this field. Many of them will be reviewed somewhere else, but here, one good review is added. Mallat, T., Orglmeister, E., Baiker, A. (2007) *Chemical Review*, **107**, 4863–4890.

1 For results before 2000, see: (a) De Vos, D.E., Vankelecom, I.F.J. and Jacobs, P.A. (eds) (2000) *Chiral Catalyst Immobilization and Recycling*, Wiley-VCH Verlag GmbH, Weinheim.
See also: (b) Blaser, H.-U. (1991)

Tetrahedron: Asymmetry, **2**, 843–66. For other reviews, see references cited in each section.
2 Nakamura, Y. (1941) *Bulletin of the Chemical Society of Japan*, **16**, 367–70.
3 Akabori, S., Sakurai, S., Izumi, Y. and Fujii, Y. (1956) *Nature*, **178**, 323–4.
4 For recent reviews, see: (a) Baiker, A. (1997) *Journal of Molecular Catalysis A–Chemical*, **115**, 473–93.
(b) Blaser, H.-U., Jalett, H.-P., Müller, M. and Studer, M. (1997) *Catalysis Today*, **37**, 441–63.
(c) von Arx, M., Mallat, T. and Baiker, A. (2002) *Topics in Catalysis*, **19**, 75–87.
5 (a) Orito, Y., Imai, S., Niwa, S. and Nguyen, G.-H. (1979) *Journal of the Society of Synthetic Organic Chemistry of Japan*, **37**, 173–8.
(b) Orito, Y., Imai, S. and Niwa, S. (1979) *Journal of Chemical Society of Japan*, 1118–20.
6 (a) Orito, Y., Imai, S. and Niwa, S. (1980) *Journal of Chemical Society of Japan*, 670–5.
(b) Orito, Y., Imai, S. and Niwa, S. (1982) *Journal of Chemical Society of Japan*, 137–41.
7 Blaser, H.-U., Jalett, H.P. and Wiehl, J. (1991) *Journal of Molecular Catalysis*, **68**, 215–22.
8 Garland, M. and Blaser, H.-U. (1990) *Journal of the American Chemical Society*, **112**, 7048–9.
9 Török, B., Balazsik, K., Szöllösi, G., Felföldi, K. and Bartok, M. (1999) *Chirality*, **11**, 470–4.
10 Zao, X., Liu, H., Guo, C. and Yang, X. (1999) *Tetrahedron*, **55**, 7787–804.
11 Ferri, D., Bürgi, T., Borszeky, K., Mallat, T. and Baiker, A. (2000) *Journal of Catalysis*, **193**, 139–44.
12 Wandeler, R., Künzle, N., Schneider, M.S., Mallat, T. and Baiker, A. (2000) *Journal of Catalysis*, **200**, 377–88.
13 Künzle, N., Solér, J.-W. and Baiker, A. (2003) *Catalysis Today*, **79–80**, 503–9.
14 (a) von Arx, M., Mallat, T. and Baiker, A. (2000) *Journal of Catalysis*, **193**, 161–4.
(b) von Arx, M., Mallat, T. and Baiker, A. (2001) *Tetrahedron: Asymmetry*, **12**, 3089–94.
15 Sonderegger, O.J., Bürgi, T. and Baiker, A. (2003) *Journal of Catalysis*, **215**, 116–21.
16 Hess, R., Mallat, T. and Baiker, A. (2003) *Journal of Catalysis*, **218**, 453–6.
17 Sonderegger, O.J., Ho, G.M.-W., Bürgi, T. and Baiker, A. (2005) *Journal of Molecular Catalysis A–Chemical*, **229**, 19–24.
18 Sonderegger, O.J., Bürgi, T., Limbach, L.K. and Baiker, A. (2004) *Journal of Molecular Catalysis A–Chemical*, **217**, 93–101.
19 Bürgi, T. and Baiker, A. (2000) *Journal of Catalysis*, **194**, 445–51.
20 Vargas, A. and Baiker, A. (2006) *Journal of Catalysis*, **239**, 220–6.
21 (a) Diezi, S., Ferri, D., Vargas, A., Mallat, T. and Baiker, A. (2006) *Journal of the American Chemical Society*, **128**, 4048–57.
(b) Bonalumi, N., Vargas, A., Ferri, D., Bürgi, T., Mallat, T. and Baiker, A. (2005) *Journal of the American Chemical Society*, **127**, 8467–77.
22 (a) Diezi, S., Reimann, S., Bonalumi, N., Mallat, T. and Baiker, A. (2006) *Journal of Catalysis*, **239**, 255–62.
(b) See also, Diezi, S., Mallat, T., Szabo, A. and Baiker, A. (2006) *Journal of Catalysis*, **228**, 162–73.
23 Orglmeister, E., Mallat, T. and Baiker, A. (2005) *Advanced Synthesis Catalysis*, **347**, 78–86.
24 Diezi, S., Hess, M., Orglmeister, E., Mallat, T. and Baiker, A. (2005) *Journal of Molecular Catalysis A–Chemical*, **239**, 49–56.
25 Orglmeister, E., Mallat, T. and Baiker, A. (2005) *Journal of Catalysis*, **233**, 333–41.
26 Török, B., Balazsik, K., Török, M. and Felföldi, K. (2002) *Catalysis Letters*, **81**, 55–62.
27 Bartók, M., Török, B., Balazsik, K. and Bartók, T. (2001) *Catalysis Letters*, **73**, 127–31.
28 Bartók, M., Sutyinszki, M. and Felföldi, K. (2003) *Journal of Catalysis*, **220**, 207–14.
29 Bartók, M., Balazsik, K., Bucsi, I. and Szöllösi, G. (2006) *Journal of Catalysis*, **239**, 74–82.
30 Diezi, S., Szabo, A., Mallat, T. and Baiker, A. (2003) *Tetrahedron: Asymmetry*, **14**, 2573–7.

31 Szöri, K., Szöllösi, G. and Bartók, M. (2006) *Advanced Synthesis Catalysis*, **348**, 515–22.

32 (a) Toukoniitty, E., Mäki-Arvela, P., Kuzma, M., Villea, A., Neyestanaki, A.K., Salmi, T., Sjöholm, R., Leino, R., Laine, E. and Murzin, D.Y. (2001) *Journal of Catalysis*, **204**, 281–91.
(b) Toukoniitty, E., Mäki-Arvela, P., Kumar, N., Salmi, T. and Murzin, D.Y. (2003) *Catalysis Today*, **79–81**, 189–93.
(c) Toukoniitty, E., Mäki-Arvela, P., Kuusisto, J., Nieminene, V., Päivärinta, J., Hotokka, M., Salmi, T. and Murzin, D.Y. (2003) *Journal of Molecular Catalysis A–Chemical*, **192**, 135–51.
(d) Toukoniitty, E., Nieminen, V., Taskinen, A., Päivärinta, J., Hotokka, M. and Murzin, D.Y. (2004) *Journal of Catalysis*, **224**, 326–39.
(e) Toukoniitty, E., Busygin, I., Leino, R. and Murzin, D.Y. (2004) *Journal of Catalysis*, **227**, 210–16.
(f) Busygin, I., Toukoniitty, E., Leino, R. and Murzin, D.Y. (2005) *Journal of Molecular Catalysis A–Chemical*, **236**, 227–38.

33 Nieminen, V., Taskinen, A., Toukoniitty, E., Hotokka, M. and Murzin, D.Y. (2006) *Journal of Catalysis*, **237**, 131–42.

34 Dummer, N.F., Jenkins, R., Li, X., Bawaked, S.M., McMorn, P., Burrows, A., Kiely, C.J., Wells, R.P.K., Willock, D.J. and Hutchings, G.J. (2006) *Journal of Catalysis*, **243**, 165–70.

35 Fietkau, N., Bussar, R. and Baltruschat, H. (2006) *Electrochimica Acta*, **51**, 5626–35.

36 (a) Kraynov, A., Suchopar, A. and Richards, R. (2006) *Catalysis Letters*, **110**, 91–9.
(b) Kraynov, A., Suchopar, A., D'Souza, L. and Richards, R. (2006) *Physical Chemistry Chemical Physics*, **8**, 1321–8.

37 Ma, Z. and Zeara, F. (2006) *Journal of the American Chemical Society*, **128**, 1614–15.

38 Lavoie, S., Laliberté, M.-A., Temprano, I. and McBreen, P.H. (2006) *Journal of the American Chemical Society*, **128**, 7588–93.

39 Attard, G.A., Griffin, K.G., Jenkins, D.J., Johnston, P. and Wells, P.B. (2006) *Catalysis Today*, **114**, 346–52.

40 Mévellec, V., Mattioda, C., Schulz, J., Rolland, J.-P. and Roucoux, A. (2004) *Journal of Catalysis*, **225**, 1–6.

41 Li, X., You, X., Ying, P., Xiao, J. and Li, C. (2003) *Topics in Catalysis*, **25**, 1–4.

42 Yan, X., He, B., Zhang, J. and Liu, H. (2005) *Chinese Journal of Polymer Science*, **23**, 393–9.

43 Gao, F., Chen, L. and Garland, M. (2006) *Journal of Catalysis*, **238**, 402–11.

44 Margitfalvi, J.L. and Tálas, E. (2006) *Applied Catalysis A: General*, **301**, 187–95.

45 Perosa, A., Tundo, P. and Selva, M. (2002) *Journal of Molecular Catalysis A–Chemical*, **180**, 169–75.

46 LeBlanc, R.J. and Williams, C.T. (2004) *Journal of Molecular Catalysis A–Chemical*, **220**, 207–14.

47 Cserényi, S., Bucsi, I. and Felföldi, K. (2006) *Reaction Kinetics and Catalysis Letters*, **87**, 395–403.

48 Marin-Astorga, N., Pecchi, G. and Reyes, P. (2006) *Reaction Kinetics and Catalysis Letters*, **87**, 121–8.

49 Tai, A. and Harada, T. (1986) *Tailored Metal Catalyst* (ed. Y. Iwasawa), Reidel, Dordrecht, pp. 265–324.

50 Sugimura, T. (1999) *Catalysis Surveys from Japan*, **3**, 37–42.

51 Osawa, T., Harada, T. and Takayasu, O. (2000) *Topics in Catalysis*, **13**, 155–68.

52 Osawa, T., Harada, T. and Takayasu, O. (2006) *Current Organic Chemistry*, **10**, 1513–31.

53 Izumi, Y., Imaida, M., Furukawa, H. and Akabori, S. (1963) *Bulletin of the Chemical Society of Japan*, **36**, 21–5.

54 Izumi, Y. (1971) *Angewandte Chemie–International Edition in English*, **10**, 871–81.

55 (a) Nitta, Y., Sekine, F., Imanaka, T. and Teranishi, S. (1981) *Bulletin of the Chemical Society of Japan*, **54**, 980–4.
(b) Nitta, Y., Sekine, F., Imanaka, T. and Teranishi, S. (1982) *Journal of Catalysis*, **74**, 382–92.

56 Harada, T., Yamamoto, M., Ozaki, H. and Izumi, Y. (1980) *Proceedings, 7th International Congress of Catalysis*,

57 Tai, A., Kikukawa, T., Sugimura, T., Inoue, Y., Abe, S., Osawa, T. and Harada, T. (1994) *Bulletin of the Chemical Society of Japan*, **67**, 2473–7.
58 Harada, T., Yamamoto, M., Onaka, S., Imaida, M., Tai, A. and Izumi, Y. (1981) *Bulletin of the Chemical Society of Japan*, **54**, 2323–9.
59 Tai, A., Ito, K. and Harada, T. (1981) *Bulletin of the Chemical Society of Japan*, **54**, 223–7.
60 Murakami, S., Harada, T. and Tai, A. (1980) *Bulletin of the Chemical Society of Japan*, **53**, 1356–60.
61 Hiraki, Y., Ito, K., Harada, T. and Tai, A. (1981) *Chemistry Letters*, 131–4.
62 Osawa, T. (1985) *Chemistry Letters*, 1609–13.
63 Osawa, T., Harada, T. and Tai, A. (1990) *Journal of Catalysis*, **121**, 7–17.
64 Tai, A., Harada, T., Tsukioka, K., Osawa, T. and Sugimura, T. (1988) *Proceedings, 9th International Congress of Catalysis, Calgary* (eds M.J. Phillips and M. Terran), The Chemical Institute of Canada, Ottawa, p. 1092.
65 Nakagawa, S., Sugimura, T. and Tai, A. (1997) *Chemistry Letters*, 859–60.
66 (a) Osawa, T., Ozawa, A., Harada, T. and Takayasu, O. (2000) *Journal of Molecular Catalysis A–Chemical*, **154**, 271–5.
(b) Osawa, T., Hayashi, Y., Ozawa, A., Harada, T. and Takayasu, O. (2001) *Journal of Molecular Catalysis A–Chemical*, **169**, 289–93.
67 (a) Osawa, T., Sakai, S., Harada, T. and Takayasu, O. (2001) *Chemistry Letters*, 392–3.
(b) Osawa, T., Sakai, S., Deguchi, K., Harada, T. and Takayasu, O. (2002) *Journal of Molecular Catalysis A–Chemical*, **185**, 65–9.
68 Osawa, T., Ozaki, N., Harada, T. and Takayasu, O. (2002) *Bulletin of the Chemical Society of Japan*, **75**, 2695–6.
69 Osawa, T., Sawada, K., Harada, T. and Takayasu, O. (2004) *Applied Catalysis A: General*, **264**, 33–6.
70 (a) Osawa, T., Oishi, S., Yoshihisa, M., Maegawa, M., Harada, T. and Takayasu, O. (2005) *Catalysis Letters*, **102**, 261–4.
(b) Osawa, T., Maegawa, M., Yoshihisa, M., Kobayashi, M., Harada, T. and Takayasu, O. (2006) *Catalysis Letters*, **107**, 83–8.
71 (a) Osawa, T., Ando, M., Sakai, S., Harada, T. and Takayasu, O. (2005) *Catalysis Letters*, **105**, 41–5.
(b) Osawa, T., Ando, M., Harada, T. and Takayasu, O. (2005) *Bulletin of the Chemical Society of Japan*, **78**, 1371–2.
72 Kukula, P. and Cerveny, L. (2003) *Research on Chemical Intermediates*, **29**, 91–105.
73 Nitta, Y., Sekine, F., Imanaka, T. and Teranishi, S. (1982) *Journal of Catalysis*, **74**, 382.
74 Osawa, T., Mita, S., Iwai, A., Takayasu, O., Hashiba, H., Hashimoto, S., Harada, T. and Matsuura, I. (2000) *Journal of Molecular Catalysis A–Chemical*, **157**, 207–16.
75 (a) Jo, D., Lee, J.S. and Lee, K.H. (2004) *Research on Chemical Intermediates*, **30**, 889–901.
(b) Jo, D., Lee, J.S. and Lee, K.H. (2004) *Journal of Molecular Catalysis A–Chemical*, **222**, 199–205.
76 Wolfson, A., Gerech, S., Landau, M.V. and Herskowitz, M. (2001) *Applied Catalysis A: General*, **208**, 91–8.
77 Sugimura, T., Nakagawa, S. and Tai, A. (2002) *Bulletin of the Chemical Society of Japan*, **75**, 355–63.
78 (a) Lorenzo, M.O., Haq, S., Bertrams, T., Murray, P., Raval, R., Baddeley, C.J. (1999) *Journal of Physical Chemistry B*, **103**, 10661–9.
(b) Lorenzo, M.O., Baddeley, C.J., Muryn, C. and Raval, R. (2000) *Nature*, **404**, 376–9.
79 Jones, T.E. and Baddeley, C.J. (2002) *Surface Science*, **513**, 453–67.
80 Jones, T.E. and Baddeley, C.J. (2002) *Surface Science*, **519**, 237–49.
81 Reval, R. (2001) *CATTECH*, **5**, 12–28.
82 Baddeley, C.J. (2003) *Topics in Catalysis*, **25**, 17–27.
83 Nitta, Y. (2006) *Journal of the Society of Synthetic Organic Chemistry of Japan*, **64**, 827–35.
84 (a) Tungler, A., Nitta, Y., Fodor, K., Farkas, G. and Máthé, T. (1999) *Journal of*

Molecular Catalysis A–Chemical, **149**, 135–40.
(b) See also, Tungler, A. and Fogassy, G. (2001) *Journal of Molecular Catalysis A–Chemical*, **173**, 231–47.

85 For heterogeneous enantioselective catalysts there are some more candidates on the borderline of the definition. For example, when a certain α,β-unsaturated ketone was hydrogenated over Pd/C in the presence of 0.5 equiv. ephedrine, up to 36% ee of the product was obtained. It is not clear whether the demanded amount of the chiral molecule indicates a necessity of the interaction with the substrate in solution, or just a weak adsorption constant of the chiral molecule. Thorey, C., Hénin, F. and Muzart, J. (1996) *Tetrahedron: Asymmetry*, **7**, 975–6.

86 Perez, J.R.G., Malthête, J. and Jacques, J. (1985) *Comptes Rendus de l'Academie des Sciences. Paris*, **300** (II), 169–72.

87 Nitta, Y., Ueda, Y. and Imanaka, T. (1994) *Chemistry Letters*, 1095–8.

88 Nitta, Y., Kobiro, K. and Okamoto, Y. (1997) *Studies in Surface Science and Catalysis*, **108**, 191–8.

89 Nitta, Y. and Shibata, A. (1998) *Chemistry Letters*, 161–2.

90 Nitta, Y. and Kobiro, K. (1995) *Chemistry Letters*, 165–6.

91 Nitta, Y. and Okamoto, Y. (1998) *Chemistry Letters*, 1115–16.

92 Nitta, Y., Kubota, T. and Okamoto, Y. (2004) *Journal of Molecular Catalysis A–Chemical*, **212**, 155–9.

93 Bürgi, T. and Baker, A. (1998) *Journal of the American Chemical Society*, **120**, 12920–6.

94 Nitta, Y. (2000) *Topics in Catalysis*, **13**, 179–85.

95 Nitta, Y. (1999) *Chemistry Letters*, 635–6.

96 Nitta, Y. and Kobiro, K. (1995) *Chemistry Letters*, 897–8.

97 Nitta, Y., Watanabe, J., Okuyama, T. and Sugimura, T. (2005) *Journal of Catalysis*, **236**, 164–7.

98 Sugimura, T., Watanabe, J., Okuyama, T. and Nitta, Y. (2005) *Tetrahedron: Asymmetry*, **16**, 1573–5.

99 Sugimura, T., Uchida, T., Yokota, M. and Watanabe, J. (2007) The 99th Meeting of the Catalysis Society of Japan, Osaka, March 2007, Abstract, No. 2.

100 Sugimura, T., Watanabe, J., Uchida, T., Nitta, Y. and Okuyama, T. (2006) *Catalysis Letters*, **112**, 27–30.

101 Borszeky, K., Mallat, T. and Baiker, A. (1996) *Catalysis Letters*, 199–202.

102 Borszeky, K., Mallat, T. and Baiker, A. (1997) *Tetrahedron: Asymmetry*, **8**, 3745–53.

103 Hall, T.J., Johnston, P., Vermeer, W.A.H., Watson, S.R. and Wells, P.B. (1996) *Studies in Surface Science and Catalysis*, **101**, 221–30.

104 Bürgi, T. and Baiker, A. (1998) *Journal of the American Chemical Society*, **120**, 12920–6.

105 Borszeky, K., Bürgi, T., Zhaohui, Z., Mallat, T. and Baiker, A. (1999) *Journal of Catalysis*, **187**, 160–6.

106 Borszeky, K., Mallat, T. and Baiker, A. (1999) *Catalysis Letters*, **59**, 95–7.

107 Solladié-Cavallo, A., Hoernel, F., Schmitt, M. and Garin, F. (2003) *Journal of Molecular Catalysis A–Chemical*, **195**, 181–8.

108 Szöllösi, G., Hanaoka, T., Niwa, S., Mizukami, F. and Bartók, M. (2005) *Journal of Catalysis*, **231**, 480–3.

109 Bisignani, R., Franceschini, S., Piccolo, O. and Vaccari, A. (2005) *Journal of Molecular Catalysis A–Chemical*, **232**, 161–4.

110 Kun, I., Torok, B., Felföldi, K. and Bartoók, M. (2000) *Applied Catalysis A: General*, **203**, 71–9.

111 Szöllösi, G., Niwa, S., Hanaoka, T. and Mizukami, F. (2005) *Journal of Molecular Catalysis A–Chemical*, **203**, 91–5.

112 (a) Maris, M., Bürgi, T., Mallat, T. and Baiker, A. (2004) *Journal of Catalysis*, **226**, 393–400.
(b) See also: Maris, M., Mallat, T., Orglmeister, E. and Baiker, A. (2004) *Journal of Molecular Catalysis A–Chemical*, **219**, 371–67.

113 Huck, W.-R., Mallat, T. and Baiker, A. (2000) *Journal of Catalysis*, **193**, 1–4.

114 Huck, W.-R., Mallat, T. and Baiker, A. (2000) *Catalysis Letters*, **69**, 129–32.

115 Huck, W.-R., Bürgi, T., Mallat, T. and Baiker, A. (2001) *Journal of Catalysis*, **200**, 171–80.

116 Huck, W.-R., Bürgi, T., Mallat, T. and Baiker, A. (2002) *Journal of Catalysis*, **205**, 213–16.
117 Huck, W.-R., Mallat, T. and Baiker, A. (2002) *New Journal of Chemistry*, **26**, 6–8.
118 Huck, W.-R., Mallat, T. and Baiker, A. (2002) *Catalysis Letters*, **80**, 87–92.
119 (a) Bürgi, T. and Baiker, A. (2002) *The Journal of Physical Chemistry B*, **106**, 10649–58.
(b) Huck, W.-R., Bürgi, T., Mallat, T. and Baiker, A. (2003) *Journal of Catalysis*, **219**, 41–51.
(c) Huck, W.-R., Mallat, T. and Baiker, A. (2003) *Catalysis Letters*, **87**, 241–7.
(d) Vargas, A., Ferri, D. and Baiker, A. (2005) *Journal of Catalysis*, **236**, 1–8.
120 (a) Colston, N.J., Wells, R.P.K., Wells, P.B. and Hutchings, G.J. (2005) *Catalysis Letters*, **103**, 117–20.
(b) Coulston, N.J., Jeffery, E.L., Wells, R.P.K., McMorn, P., Wells, P.B., Willock, D.J. and Hutchings, G.J. (2006) *Journal of Catalysis*, **243**, 360–7.
(c) Coulston, N.J., Wells, R.P.K., Wells, P.B. and Hutchings, G.J. (2006) *Catalysis Today*, **114**, 353–6.
121 Szöri, K., Szöllösi, G., Felföldi, K., Bartók, M. and React, K. (2005) *Catalysis Letters*, **84**, 151–6.
122 Blaser, H.-U., Höning, H., Studer, S. and Wedemeyer-Exl, C. (1999) *Journal of Molecular Catalysis A–Chemical*, 253–7.
123 Sonderegger, O.J., Ho, G.M.-W., Bürgi, T. and Baiker, A. (2005) *Journal of Catalysis*, **230**, 499–506.
124 Maris, M., Ferr, D., Königsmann, L., Mallat, T. and Baiker, A. (2006) *Journal of Catalysis*, **237**, 230–6.

11
Asymmetric Phase-Transfer Catalysis
Xisheng Wang, Quan Lan, and Keiji Maruoka

11.1
Introduction

In 1971, Starks introduced the term phase-transfer catalysis (PTC) to explain the critical role of tetraalkylammonium or phosphonium salts (Q^+X^-) in the reactions between two substances located in different immiscible phases [1]. Although this was not the first observation of the catalytic activity of quaternary onium salts, the foundations of PTC were laid by Starks, together with M. Makosza and A. Brändström during the mid to late 1960s. Ever since that time, the chemical community has witnessed the exponential growth of PTC as a practical methodology for organic synthesis, featuring its simple experimental operations, mild reaction conditions, inexpensive and environmentally benign reagents and solvents, and the possibility to conduct large-scale preparations [2]. Nowadays, PTC appears to be a prime synthetic tool, being appreciated in various fields of organic chemistry, and has also found widespread industrial applications.

On the other hand, the development of asymmetric PTC based on the use of structurally well-defined chiral, nonracemic catalysts has progressed rather slowly, despite its great importance to create a new domain in modern asymmetric catalysis by taking full advantage of structurally and stereochemically modifiable tetraalkylonium cations (Q^+). However, recent enormous efforts toward this direction have resulted in notable achievements, making it feasible to perform a variety of bond- forming reactions under mild phase-transfer-catalyzed conditions [3]. The aim of this chapter is to illustrate the track of evolution of active research field, with the main focus on recent progress. Hopefully, this will provide the reader with a better understanding of the current situation and future perspectives of asymmetric PTC.

Handbook of Asymmetric Heterogeneous Catalysis. Edited by K. Ding and Y. Uozumi
Copyright © 2008 WILEY-VCH Verlag GmbH & Co. KGaA, Weinheim
ISBN: 978-3-527-31913-8

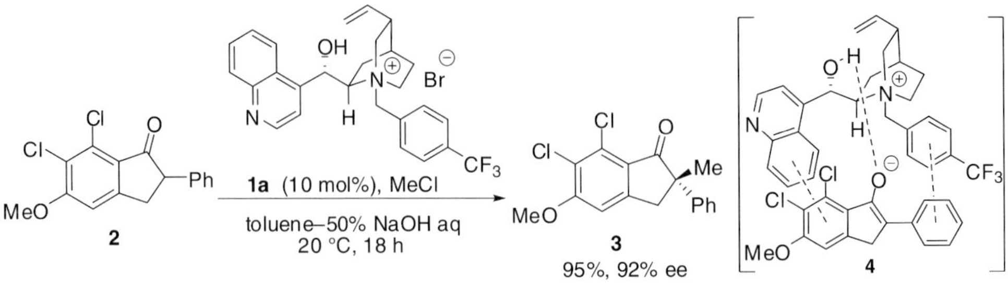

Scheme 11.1 Asymmetric phase-transfer-catalyzed alkylation of indanone derivative.

11.2
Alkylation

11.2.1
Pioneering Study

The enantioselective alkylation of active methylene compounds occupies the central place in the field of asymmetric PTC, and its development was triggered by the pioneering studies of the Merck research group in 1984 [4]. Dolling and coworkers succeeded in utilizing the quaternary ammonium salt **1a** derived from cinchonine as the catalyst for the methylation of phenylindanone derivative **2**, giving the corresponding alkylated product **3** in excellent yield with high enantiomeric excess (ee) (Scheme 11.1). These authors also proposed the intermediacy of a tight ion pair **4** fixed by an electrostatic effect and hydrogen bonding, as well as π-π stacking interactions, to account for the result.

11.2.2
Asymmetric Synthesis of α-Amino Acids and Their Derivatives

11.2.2.1 Monoalkylation of Schiff Bases Derived from Glycine
Five years after the epoch-making investigations of the Merck group, this type of catalyst was used successfully for the asymmetric synthesis of α-amino acids by O'Donnell, using glycinate Schiff base **5** as a key substrate [5]. The asymmetric alkylation of **5** by benzyl bromide proceeded smoothly under phase-transfer conditions with N-(benzyl)cinchoninium chloride [**8a** (Cl)] as a catalyst, to give the alkylation product **6** in good yield and moderate enantioselectivity (Scheme 11.2). Further optimization with hydroxyl-protected catalyst **8b** (a second-generation catalyst) enhanced the enantioselectivity to 81% ee [6]. Following this milestone reports, research in this area made little progress for some time, but a new class of cinchona alkaloid-derived catalysts bearing an N-anthracenylmethyl function (third-generation catalyst), which were developed independently by Lygo [7] and Corey [8], and showed much higher asymmetric induction, then opened a new era of asymmetric PTC (Scheme 11.2).

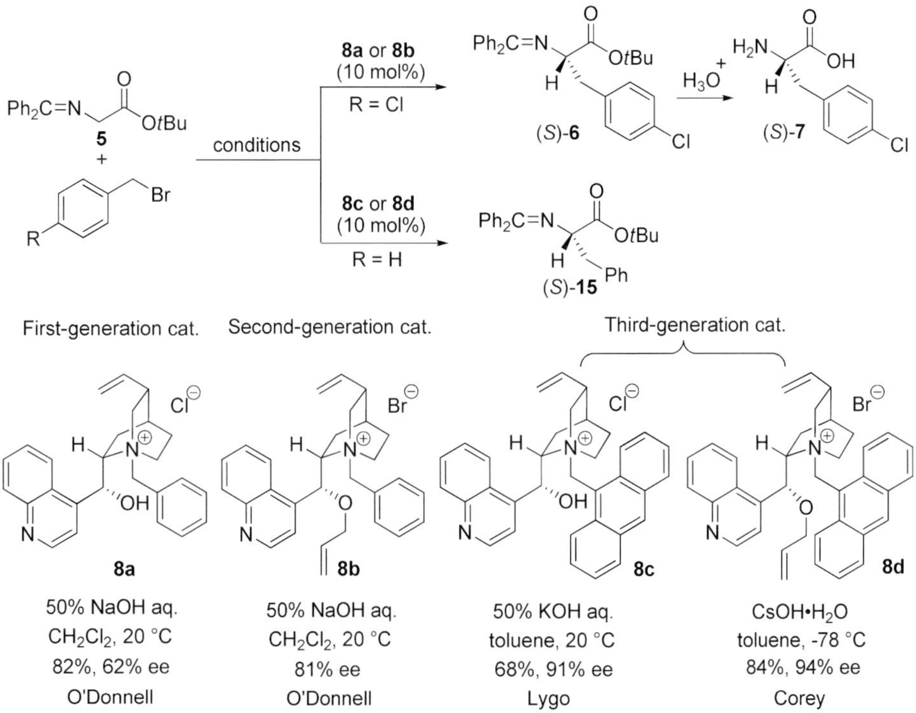

11.2 Alkylation | 385

Scheme 11.2 Development of cinchonine-derived phase-transfer catalysts.

First-generation cat. **8a**: 50% NaOH aq., CH$_2$Cl$_2$, 20 °C, 82%, 62% ee (O'Donnell)

Second-generation cat. **8b**: 50% NaOH aq., CH$_2$Cl$_2$, 20 °C, 81% ee (O'Donnell)

Third-generation cat. **8c**: 50% KOH aq., toluene, 20 °C, 68%, 91% ee (Lygo)

Third-generation cat. **8d**: CsOH·H$_2$O, toluene, −78 °C, 84%, 94% ee (Corey)

In 1999, we designed and prepared the structurally rigid, chiral spiro ammonium salts of type **9** as a new C_2-symmetric chiral phase-transfer catalyst and successfully applied them to the highly efficient, catalytic enantioselective alkylation of **5** under mild conditions [9]. The key finding was a significant effect of an aromatic (Ar) substituent at the 3,3′-position of one binaphthyl subunit of the catalyst on the enantiofacial discrimination, and (S,S)-**9e** was revealed to be the catalyst of choice for the preparation of a variety of essentially enantiopure α-amino acids by this transformation (Table 11.1). Generally, 1 mol% of **9e** is sufficient for the smooth alkylation, and the catalyst loading can be reduced to 0.2 mol% without any loss of enantiomeric excess (entry 6). In the reaction with the simple alkyl halides such as ethyl iodide, the use of aqueous cesium hydroxide (CsOH) as a basic phase at a lower reaction temperature is recommended (entry 7).

A similar electronic effect of fluoroaromatic substituents was utilized by Jew and Park to develop efficient catalysts derived from cinchona alkaloid. Evaluation of the effect of electron-withdrawing groups on the aromatic moiety of the benzylic group on the nitrogen of dihydrocinchonidinium salt **10a** revealed that the *ortho*-fluoro substituent led to a dramatic enhancement in the enantioselectivity, and catalyst **10a** having a 2′,3′,4′-trifluorobenzyl group showed the highest selectivities (Scheme 11.3) [10]. It has been proposed that hydrogen bonding interaction between C(9)-O and 2′-F in **10a** might rigidify its conformation, leading to high

Table 11.1 Effect of aromatic substituent (Ar) of **9e**-catalyzed phase-transfer alkylation of **5**.

Ar = H (**9a**), Ph (**9b**), **9c**, **9d**, **9e**

Ph$_2$C=N–CH$_2$–OtBu (**5**) + RX $\xrightarrow[\text{toluene-50\% KOH aq, 0 °C}]{(S,S)\text{-}9\ (1\ \text{mol\%})}$ Ph$_2$C=N–C(H)(R)–C(O)OtBu

Entry	Catalyst	RX	Yield (%)	ee (%) (Config)
1	9a	PhCH$_2$Br	73	79 (R)
2	9b	PhCH$_2$Br	81	89 (R)
3	9c	PhCH$_2$Br	95	96 (R)
4	9d	PhCH$_2$Br	91	98 (R)
5	9e	PhCH$_2$Br	90	99 (R)
6[a]	9e	PhCH$_2$Br	72	99 (R)
7[b]	9e	EtI	89	98 (R)
8	9e	CH$_2$=CHCH$_2$Br	80	99 (R)
9	9e	2,6-Me$_2$-PhCH$_2$Br	98	99 (R)
10	9e	4-Bz-PhCH$_2$Br	86	99 (R)

a With 0.2 mol% of (S, S)-**9e**.
b With sat. CsOH at −15 °C.

enantioselectivity. Recent studies from the same group on the evaluation of related chiral quaternary ammonium salts **10b** containing 2′-N-oxypyridine strongly support this hypothesis, suggesting the intervention of a preorganized catalyst such as **11** (Scheme 11.3) [11].

Jew and Park have also utilized the dimerization effect, as observed in the development of Sharpless asymmetric dihydroxylation, where ligands with two independent cinchona alkaloid units attached to heterocyclic spacers led to a considerable increase in both the enantioselectivity and scope of the substrates, to design dimeric and trimeric cinchona alkaloid-derived phase-transfer catalysts **12** [12] and **13** [13]. These authors investigated the ideal aromatic spacer for optimal dimeric catalysts, and found that the catalyst **14** with a 2,7-bis(bromomethyl) naphthalene spacer and two cinchona alkaloid units exhibited remarkable catalytic and chiral efficiency (Scheme 11.3) [14].

With the critical role of 3,3′-diaryl substituents of **9** in mind, we examined the effect of 4,4′- and 6,6′-substituents of one (**16**) or both (**17**) binaphthyl subunits. As shown in Scheme 11.4, the introduction of aromatic groups at 4,4′- and

11.2 Alkylation

Scheme 11.3 Development of efficient phase-transfer catalysts derived from cinchona alkaloid.

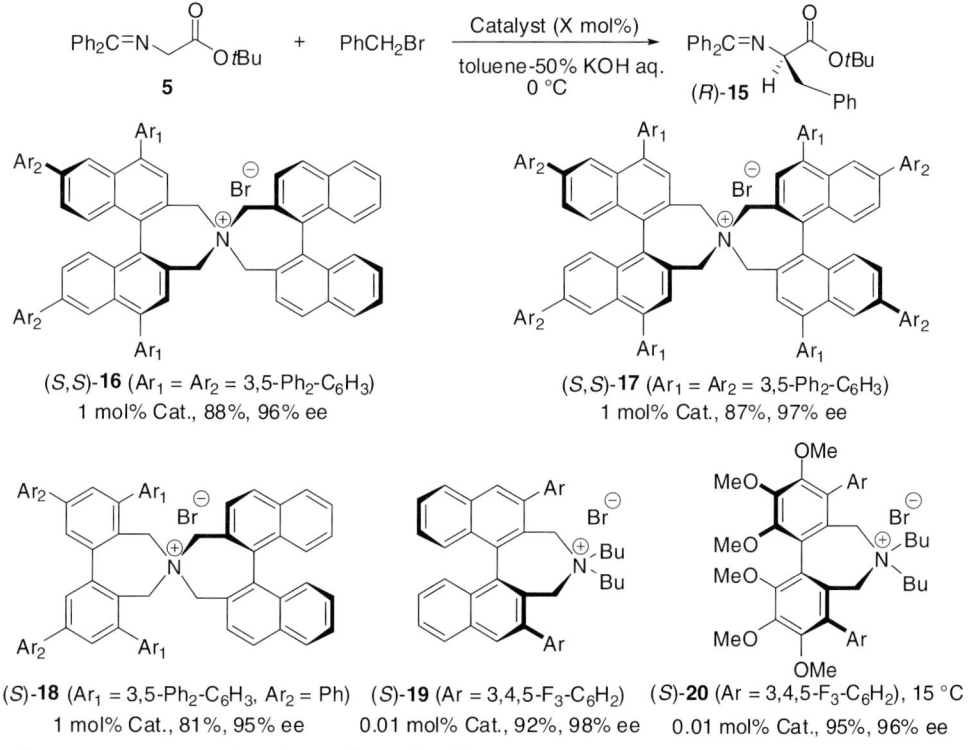

- **10a**: 90%, 96% ee
- **10b**: 94%, 96% ee
- **11**
- **12a** (X = H): 91%, 90% ee
- **12b** (X = F): 93%, 94% ee
- **13**, 3 mol%, −20 °C, 10 h, 94%, 94% ee
- **14**, 1 mol%, 95%, 97% ee

Reaction: Ph$_2$C=N-CH$_2$-CO$_2$tBu (**5**) + PhCH$_2$Br → **10** (10 mol%), 50% KOH aq, toluene–CHCl$_3$ (7:3), 0 °C, 2–7 h → (S)-**15** (Ph$_2$C=N-CH(CH$_2$Ph)-CO$_2$tBu)

Reaction: Ph$_2$C=N-CH$_2$-CO$_2$tBu (**5**) + PhCH$_2$Br → Catalyst (X mol%), toluene-50% KOH aq., 0 °C → (R)-**15**

- (S,S)-**16** (Ar$_1$ = Ar$_2$ = 3,5-Ph$_2$-C$_6$H$_3$); 1 mol% Cat., 88%, 96% ee
- (S,S)-**17** (Ar$_1$ = Ar$_2$ = 3,5-Ph$_2$-C$_6$H$_3$); 1 mol% Cat., 87%, 97% ee
- (S)-**18** (Ar$_1$ = 3,5-Ph$_2$-C$_6$H$_3$, Ar$_2$ = Ph); 1 mol% Cat., 81%, 95% ee
- (S)-**19** (Ar = 3,4,5-F$_3$-C$_6$H$_2$); 0.01 mol% Cat., 92%, 98% ee
- (S)-**20** (Ar = 3,4,5-F$_3$-C$_6$H$_2$), 15 °C; 0.01 mol% Cat., 95%, 96% ee

Scheme 11.4 Asymmetric benzylation of **5** catalyzed by C$_2$-symmetric chiral N-spiro ammonium salts.

6,6′-position led to a meaningful effect on the stereoselectivity of the phase-transfer-catalyzed alkylation of **5** [15]. Also, 4,4′,6,6′-tetraarylbinaphthyl-substituted ammonium bromide **17**, which was assembled through the reaction of aqueous ammonia with two similarly binaphthyl-modified subunits and avoided the independent synthesis of two different compounds required for **9**, showed high catalytic and chiral efficiency in the alkylation of **5** (Scheme 11.4) [16].

Although the conformationally rigid, N-spiro structure created by two chiral binaphthyl subunits represents a characteristic feature of **9** and related catalyst **16**, it also imposes limitations on the catalyst design due to an imperative use of the two different chiral binaphthyl moieties. Accordingly, we developed a new C_2-symmetric chiral quaternary ammonium bromide **18** incorporating an achiral, conformationally flexible biphenyl subunit (Scheme 11.4) [17]. The benzylation of **5** with *(S)*-**18** as a catalyst gave **15** in 95% ee. This unique phenomenon provides a powerful strategy in the molecular design of chiral catalysts; that is, the requisite chirality can be served by the simple binaphthyl moiety, while an additional structural requirement for fine-tuning of reactivity and selectivity can be fulfilled by an easily modifiable achiral biphenyl structure. This certainly obviates the use of two chiral units and should be appreciated in the synthesis of a variety of chiral catalysts with different steric and/or electronic properties.

By replacing a rigid binaphthyl moiety with flexible straight-chain alkyl groups, an unusually active chiral phase-transfer catalyst **19** was developed, which exhibited a high catalytic performance and demonstrated the remarkable efficiency and practicality of the present approach towards the enantioselective synthesis of α-alkyl-α-amino acids. Most notably, the reaction of **5** with various alkyl halides proceeded smoothly in the presence of only 0.01–0.05 mol% of *(S)*-**19** to afford the corresponding alkylation products with excellent enantioselectivities (Scheme 11.4) [18]. This strategy was also proved to be effective in successfully designing phase-transfer catalyst **20** (Scheme 11.4) [19].

These reports have accelerated research into improvements of the asymmetric alkylation of **5**, and have resulted in the emergence of a series of appropriately modified purely synthetic chiral quaternary ammonium salts. The performance of the representative catalysts in the asymmetric benzylation of **5** are summarized in Table 11.2, in order to facilitate for their preparative use.

Using glycine diphenylmethyl (Dpm) amide-derived Schiff base **28** as a key substrate, N-spiro chiral quaternary ammonium bromide **9g** was found to be an ideal catalyst to achieve high enantioselectivity, even in the alkylation with a less-reactive simple secondary alkyl halides (Scheme 11.5) [27]. Furthermore, it was found that the chiral ammonium enolate generated from **9g** and **28** had an ability to recognize the chirality of β-branched primary alkyl halides, which provides impressive levels of kinetic resolution during the alkylation with racemic halide **29** allowing for two α- and γ-stereocenters of **30** to be controlled, as exemplified in Scheme 11.5 [28].

This approach was also found to be successfully applicable to the asymmetric alkylation of Weinreb amide derivative **31**, utilizing **9f** as a catalyst (Scheme 11.6). The resultant optically active α-amino acid Weinreb amides **32** were efficiently

Table 11.2 Representative catalysts and their performance in the phase-transfer-catalyzed benzylation of **5**.

Entry	Catalyst (X mol%)	Base	Solvent	T (°C)	t (h)	Yield (%)	ee (%) (config.)	Reference
1	21 (2)	NaOH	Toluene	rt	48	>95	80 (R)	[20]
2	22 (30)	1 M KOH aq.	CH_2Cl_2	0	40	55	90 (R)	[21]
3	23 (5)	CsOH	CH_2Cl_2	0	1	55[a]	32 (R)	[22]
4	24 (1)	50% KOH aq.	Toluene	0	12	55	58 (S)	[23]
5	25a (10)	CsOH·H_2O	Toluene/CH_2Cl_2	−78	60	87	93 (R)	[24]
6	26 (2)	50% KOH aq.	CH_2Cl_2	0	20	>95	95 (R)	[25]
7	27 (1)	15 M KOH aq.	Toluene	0	1.5	89	97 (R)	[26]

a Completion (%).

converted to the corresponding amino ketone **33** by simple treatment with a Grignard reagent (Scheme 11.6) [28].

The catalytic enantioselective alkylation of **5** has also been carried out in the presence of polymer-supported cinchona alkaloid-derived ammonium salts (Table 11.3), since the enantioselective synthesis of α-amino acids employing easily available and reusable chiral catalysts presents clear advantages for large-scale synthesis. Nájera employed resin-supported ammonium salt **8e** as a chiral phase-transfer catalyst for the alkylation of glycine isopropyl ester-derived Schiff base **34** [29], leading to the formation of **35** in 90% yield with 90% ee in the benzylation (entry 1, Table 11.3). Cahard investigated the role of a flexible methylene spacer between the quaternary ammonium moiety and the polystyrene backbone in the similar benzylation of **5**, and found that catalyst **1b** anchored to the matrix through the

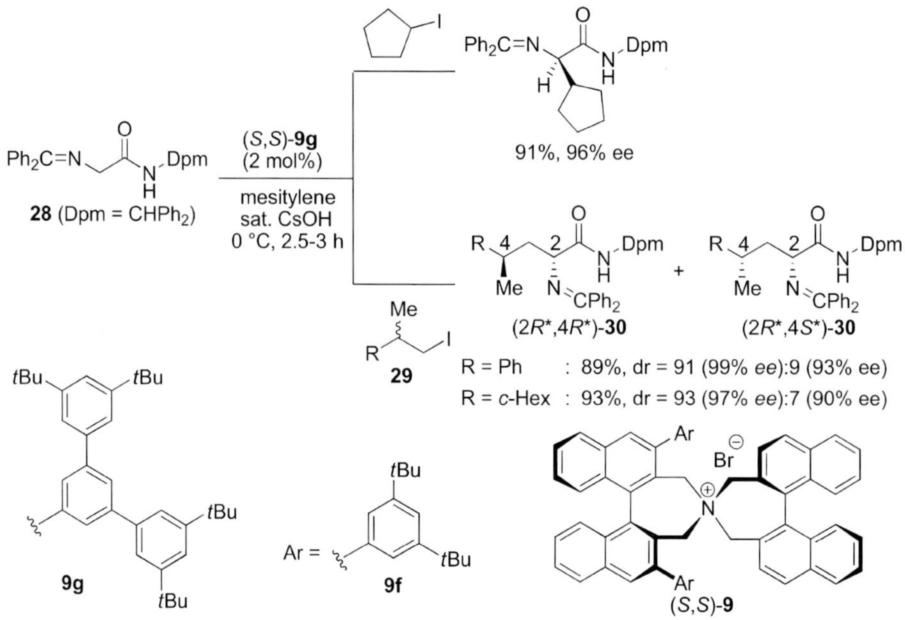

Scheme 11.5 Asymmetric alkylation of glycine Dpm amide Schiff base **28**.

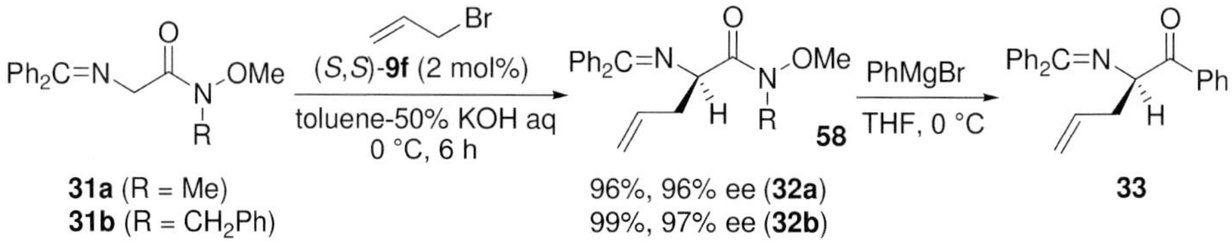

Scheme 11.6 Asymmetric alkylation of glycine Weinreb amide Schiff base **31**.

four carbon spacers was optimal, giving **15** with 81% ee (entry 2) [30]. Cahard also introduced a cinchonidine-derived quaternary ammonium salt grafted to a poly(ethylene glycol) (PEG) matrix **8f** as an efficient homogeneous catalyst for the asymmetric alkylation of **5**, and up to 81% ee was attained in the benzylation reaction (entry 3) [31]. Meanwhile, Cahard and Plaquevent succeeded in improving the enantioselectivity by attaching a Merrifield resin onto the hydroxy moiety of the cinchonidine-derived catalyst possessing a 9-anthracenylmethyl group on nitrogen (**8g**) (entry 4) [32]. Benaglia immobilized the third-generation catalyst on a modified PEG through the alkylation of C(9) hydroxy functionality. The chiral ammonium salt **8h** thus obtained acted as a homogeneous catalyst in the benzylation of **5** to afford **15** with a maximum ee-value of 64% (entry 5) [33].

A recyclable fluorous chiral phase-transfer catalyst **36** has been developed by the present authors' group and its high chiral efficiency and reusability demonstrated in the asymmetric alkylation of **5**. On completion of the reaction, **36** could be easily recovered by the simple extraction with FC-72 (perfluorohexanes) as a fluorous

Table 11.3 Asymmetric benzylation with polymer-supported phase-transfer catalysts.

Entry	Catalyst	Substrate	Base	T (°C)	t (h)	Yield (%)	ee (%) (Config.)	Reference
1	8e	34	25% NaOH aq.	0	17	90	90 (S)	[29]
2	1b	5	50% KOH aq.	0	15	60	81 (R)	[30]
3	8f	5	50% KOH aq.	0	15	84	81 (S)	[31]
4	8g	5	CsOH·H$_2$O	−50	30	67	94 (S)	[32]
5[a]	8h	5	CsOH·H$_2$O	−78	60	75	64 (S)	[33]

a Dichloromethane was used as solvent.

solvent and used for the next run, without any loss of reactivity and selectivity (Scheme 11.7) [34].

11.2.2.2 Dialkylation of Schiff Bases Derived from α-Alkyl-α-Amino Acids

Phase-transfer catalysis has made unique contributions in the development of a truly efficient method for the preparation of nonproteinogenic, chiral α,α-dialkyl-α-amino acids, which are often effective enzyme inhibitors and also indispensable for the elucidation of enzymatic mechanisms [35].

The first example of preparing optically active α,α-dialkyl-α-amino acids by chiral PTC was reported by O'Donnell in 1992 [36], giving moderate enantioselectivities for the alkylation of *p*-chlorobenzaldehyde imine of alanine *tert*-butyl ester **39a** with cinchonine-derived **1c** as catalyst (entry 1, Table 11.4). Subsequently,

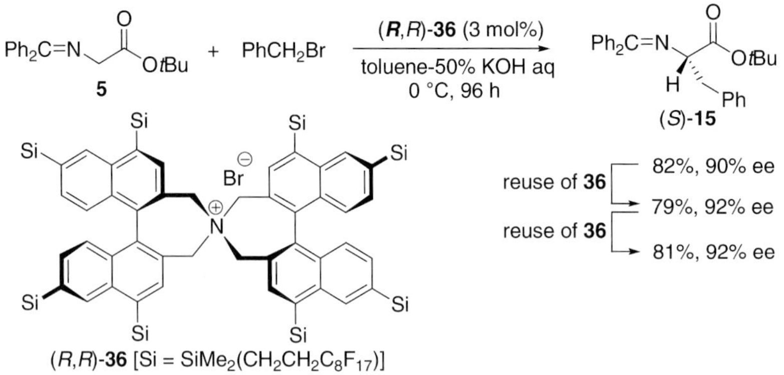

Scheme 11.7 Recyclable fluorous chiral phase-transfer catalyst.

Table 11.4 Phase-transfer-catalyzed enantioselective benzylation of aldimine Schiff bases derived from α-alkyl-α-amino acids.

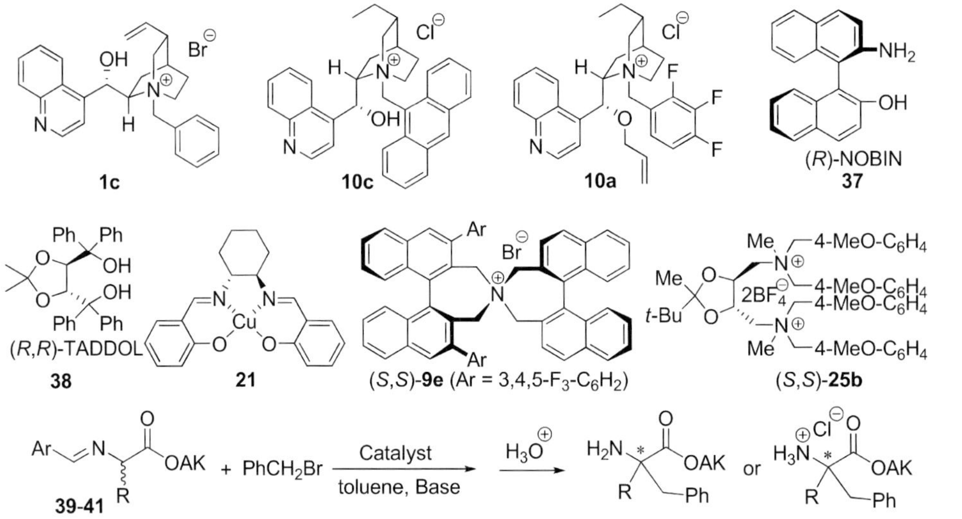

Entry	Sub.	Ar	R	AK	Cat. (X mol%)	Base	T (°C)	Yield (%)	ee (%) (config.)	Reference
1	39a	p-Cl-C$_6$H$_4$	Me	tBu	1c (10)	K$_2$CO$_3$/KOH	r.t.	80	44 (S)	[36]
2	39a	p-Cl-C$_6$H$_4$	Me	tBu	10c (10)	K$_2$CO$_3$/KOH	r.t.	95	87 (S)	[37]
3	39b	2-Naph	Me	tBu	10a (10)	RbOH	−35	91	95 (S)	[38]
4	40	Ph	Me	iPr	37 (10)	50% KOH aq.	r.t.	95	97 (S)	[39]
5	40	Ph	Me	iPr	38 (3)	50% KOH aq.	r.t.	94	94 (S)	[39]
6	41	p-Cl-C$_6$H$_4$	Et	Me	21 (2)	NaOH	r.t.	91	82 (S)	[20, 40]
7	39a	p-Cl-C$_6$H$_4$	Me	tBu	9e (1)	CsOH·H$_2$O	0	85	98 (R)	[41]
8	39c	p-Cl-C$_6$H$_4$	iBu	tBu	9e (1)	CsOH·H$_2$O	0	64	92	[41]
9[a]	39a	p-Cl-C$_6$H$_4$	Me	tBu	25b (10)	CsOH·H$_2$O	−70	83	89 (R)	[24b]

a Toluene/CH$_2$Cl$_2$ (7:3) was used as solvent.

different types of catalyst have been designed and evaluated in the alkylation of aromatic aldimine Schiff bases of α-amino acid esters (mainly alaninate). Enantiopure (4R,5R)- or (4S,5S)-TADDOL (**38**) and NOBIN (**37**), upon *in-situ* deprotonation with solid NaOH or NaH, act as chiral bases to be effective chiral phase-transfer catalysts for the enantioselective alkylation of alanine-derived imines **40** (entries 4 and 5) [39]. Their chelating ability towards the sodium cation is crucial to render the sodium enolate soluble in toluene, as well as to achieve enantiofacial differentiation in the transition state. The ability of copper(II)–salen complex **21** was used for the quaternization of aldimine Schiff base of various α-alkyl-α-amino acid methyl esters **41** which were alkylated enantioselectively in the presence of 2 mol% **21** (entry 6) [20, 40].

N-spiro chiral quaternary ammonium bromide **9e** is also applicable to the asymmetric alkylation of aldimine Schiff base **39** derived from the corresponding α-amino acids, affording the desired noncoded amino acid esters with excellent asymmetric induction (entries 7 and 8, Table 11.4) [41]. Bis-ammonium tetrafluoroborate **25b** developed by Shibasaki successfully promoted the alkylation of **39a**, even at low temperature, to give the corresponding α,α-dialkyl-α-amino ester in good yield with high enantioselectivity (entry 9) [24b].

In addition to the high efficiency and broad generality, the characteristic feature of the **9e**-catalyzed asymmetric alkylation strategy is visualized by the direct stereoselective introduction of two different side chains to glycine-derived aldimine Schiff base **42** in one-pot under phase-transfer conditions. Initial treatment of the toluene solution of **42** and *(S,S)*-**9e** (1 mol%) with allyl bromide (1 equiv.) and CsOH·H$_2$O at −10 °C and the subsequent reaction with benzyl bromide (1.2 equiv.) at 0 °C resulted in formation of the double alkylation product **43a** in 80% yield with 98% ee after hydrolysis. Notably, in the double alkylation of **42** by the addition of the halides in reverse order, the absolute configuration of the product **43a** was confirmed to be opposite, indicating the intervention of the chiral ammonium enolate **44** in the second alkylation stage (Scheme 11.8) [41].

Further, *N*-spiro chiral quaternary ammonium bromide **9e** was successfully applied by Jew and Park to the asymmetric synthesis of α-alkyl serines using

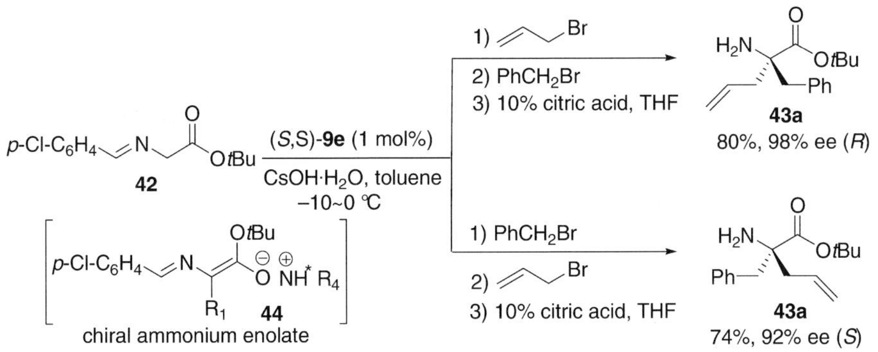

Scheme 11.8 Highly enantioselective, one-pot double alkylation.

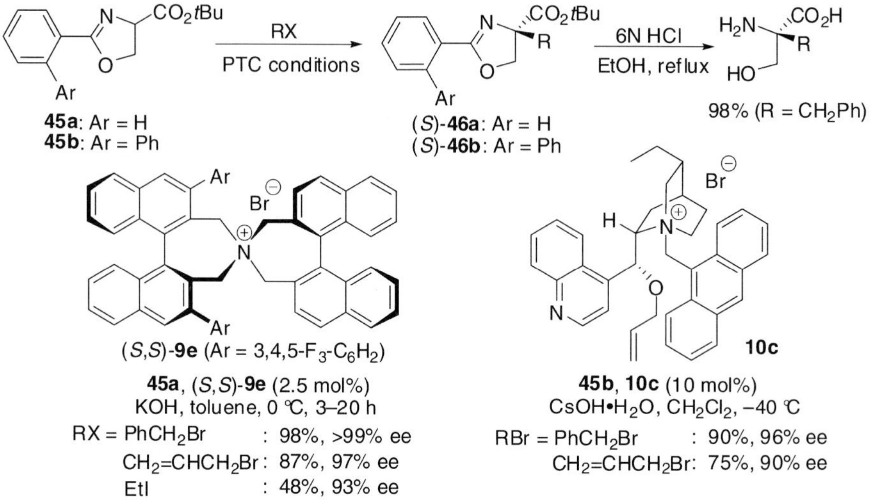

Scheme 11.9 Catalytic asymmetric synthesis of α-alkyl serines.

phenyl oxazoline derivative 45 as a requisite substrate. The reaction is general in nature and provides a practical access to a variety of optically active α-alkyl serines through acidic hydrolysis, as exemplified in Scheme 11.9 [42]. The same group also undertook the modification of the phenyl moiety of 45 to various aromatic groups and identified the o-biphenyl analogue 45b as a suitable substrate for attaining high enantioselectivity with cinchona alkaloid-derived phase-transfer catalyst 10d (Scheme 11.9) [43].

11.2.2.3 Alkylation of Schiff Base-Activated Peptides

Peptide modification is an essential, yet flexible, synthetic concept for the efficient target screening and optimization of lead structures in the application of naturally occurring peptides as pharmaceuticals. The introduction of side chains directly to a peptide backbone represents a powerful method for the preparation of unnatural peptides [44].

It was envisaged that chiral PTC should play a crucial role in achieving an efficient chirality transfer; hence, an examination was conducted of the alkylation of the dipeptide, Gly-L-Phe derivative 47 (Table 11.5). When a mixture of 47 and tetrabutylammonium bromide (TBAB; 2 mol%) in toluene was treated with 50% KOH aqueous solution and benzyl bromide at 0 °C for 4 h, the corresponding benzylation product 48 was obtained in 85% yield with the diastereomeric ratio (DL-48 : LL-48) of 54 : 46 (8% diastereomeric excess (de)) (entry 1). In contrast, the reaction with chiral quaternary ammonium bromide (S,S)-9g possessing a 3,5-bis(3,5-di-*tert*-butylphenyl)phenyl group under similar conditions realized almost complete diastereocontrol (entries 4–6) [45].

Chiral PTC with (S,S)- [or (R,R)-] 9g can be successfully extended to the stereoselective N-terminal alkylation of LL-, DL-49 and DDL-50, and the corresponding protected tri- or tetra-peptide was obtained in high yield with excellent

Table 11.5 Stereoselective N-terminal alkylation of peptides.

Entry	Substrate	Catalyst	t (h)	Yield (%)	de (%)
1	L-47	TBAB	4	85	8
2		(S,S)-9c	4	88	55
3		(R,R)-9c	6	83	20
4		(S,S)-9e	8	43	81
5		(S,S)-9f	4	98	86
6		(S,S)-9g	6	97	97
7	LL-49	(S,S)-9g	2	74	20
8	LL-49	(R,R)-9g	2	89	93
9	DL-49	(S,S)-9g	2	91	98
10	DDL-50	(S,S)-9g	2	90	94

stereochemical control when a matched catalyst for this diastereofacial differentiation was used (entries 7–10, Table 11.5) [45].

A variety of alkyl halides can be employed as an electrophile in this alkylation, and the efficiency of the transmission of stereochemical information was not affected by the side-chain structure of the pre-existing amino acid residues. Furthermore, this method allowed an asymmetric construction of noncoded α,α-dialkyl-α-amino acid residues at the peptide terminal, as exemplified by the stereoselective alkylation of the dipeptide, L-Ala-L-Phe derivative **51** (Scheme 11.10).

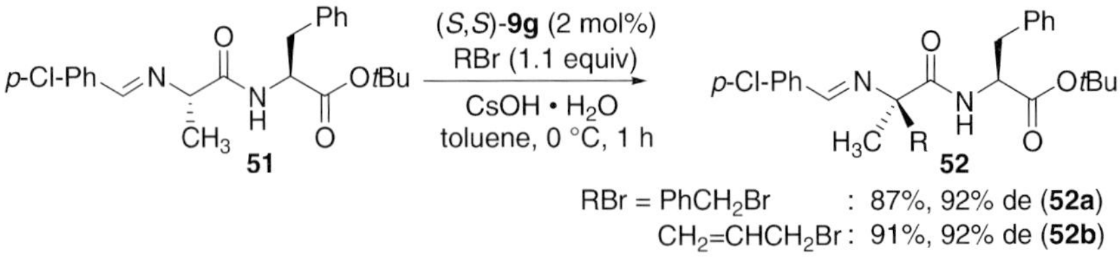

Scheme 11.10 Asymmetric construction of α,α-dialkyl-α-amino acid residues.

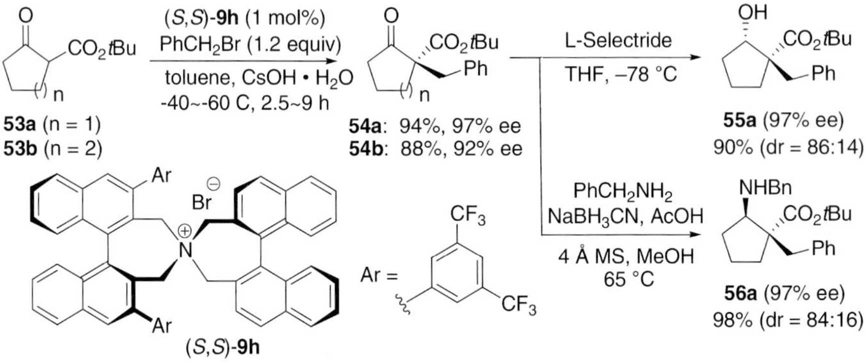

Scheme 11.11 Construction of quaternary stereocenters on β-keto esters by alkylation with the catalysis of **9h**.

11.2.3
Other Alkylations and Aromatic or Vinylic Substitutions

The efficient, highly enantioselective construction of quaternary carbon centers on β-keto esters under phase-transfer conditions has been achieved using N-spiro chiral quaternary ammonium bromide **9h** as a catalyst [46]. This system has a broad generality in terms of the structure of β-keto esters and alkyl halides (Scheme 11.11). The resultant alkylation products **54** can be readily converted into the corresponding β-hydroxy esters **55** and β-amino esters **56**, respectively.

With this alkylation strategy for the asymmetric construction of all-carbon quaternary stereocenters on 1,3-dicarbonyl compounds in mind, a practical asymmetric synthesis of functionalized aza-cyclic α-amino acid derivatives possessing quaternary stereocenters has been achieved by the phase-transfer-catalyzed alkylation of **57** in the presence of *(S,S)*-**9h** (1 mol%) (Scheme 11.12). Subsequent reduction or alkylation of the 3-keto carbonyl moiety of **58** proceeded with complete diastereochemical control to afford the corresponding β-hydroxy aza-cyclic α-amino acid derivatives having stereochemically defined consecutive quaternary carbon centers [47]. This alkylation strategy has also been successfully demonstrated to accommodate α-acyl-γ-butyrolactones nicely. In the presence of *(S,S)*-**9h** (1 mol%) and Cs_2CO_3, the reaction of α-benzoyl-γ-butyrolactones **59** with benzyl

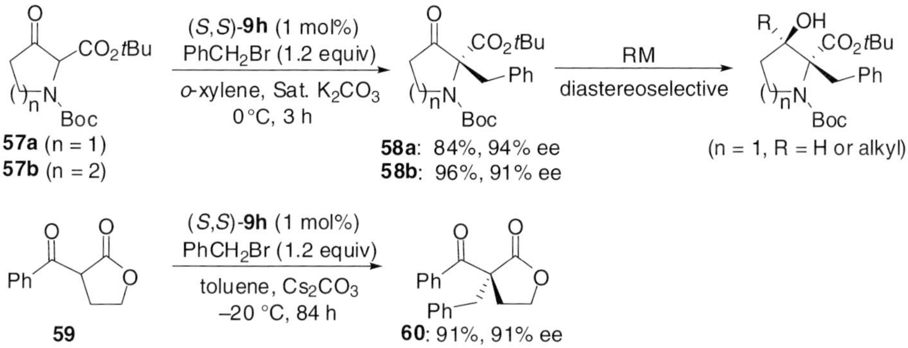

Scheme 11.12 Asymmetric alkylation of aza-cyclic α-amino acid derivatives and α-benzoyl-γ-butyrolactones.

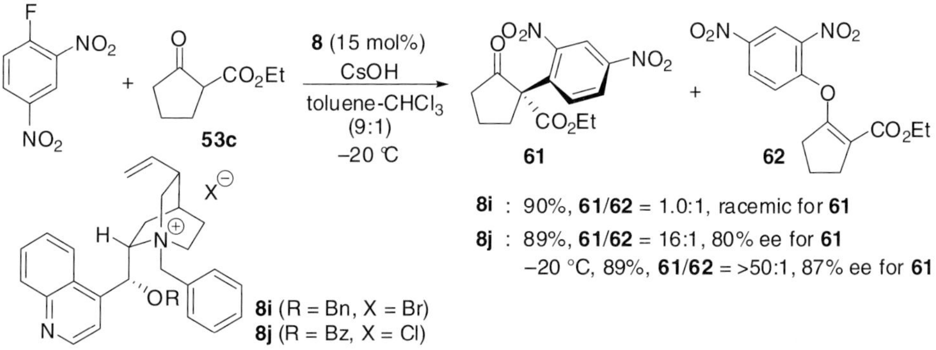

Scheme 11.13 Regioselective and enantioselective nucleophilic aromatic substitution reactions.

bromide in toluene proceeded smoothly at −20 °C, affording the adducts **60** in 91% yield with 91% ee [48].

Jørgensen developed a catalytic regioselective and enantioselective nucleophilic aromatic substitution reaction of activated aromatic compounds with 1,3-dicarbonyl compounds under phase-transfer conditions. This was crucial for obtaining the C-arylated product **61** predominantly with high enantioselectivity by replacing a benzyl with a benzoate group in the cinchona alkaloids-derived phase-transfer catalyst (Scheme 11.13) [49].

Recently, Jørgensen reported the first example of a catalytic enantioselective vinylic substitution reaction (Scheme 11.14). With a bulky 1-adamantylcarbonyl group modified phase-transfer catalyst **10d** as the catalyst, the reaction between alkyl cyclopentanone-2-carboxylates (**53a**) with (Z)-β-cholro-1-phenylpropenone (**63a**) proceeded smoothly, affording the product **64a** with Z/E > 95 : 5 and 94% ee [50]. As for the trisubstituted alkene **64b**, the α-iodine atom was tolerated in the catalytic reaction.

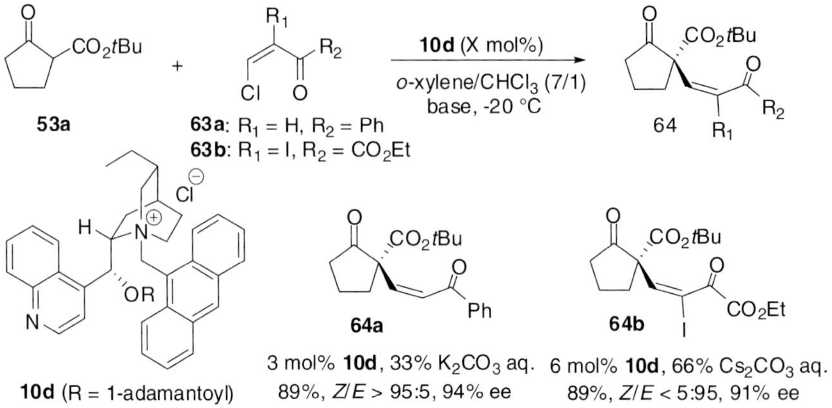

Scheme 11.14 Enantioselective nucleophilic vinylic substitution reaction.

11.3
Michael Addition

The asymmetric Michael addition of active methylene or methyne compounds to electron deficient olefins, particularly α,β-unsaturated carbonyl compounds, represents a fundamental and useful approach to construct functionalized carbon frameworks [51]. The first successful, phase-transfer-catalyzed process was based on the use of well-designed chiral crown ethers **69** and **70** as catalyst. In the presence of **69**, β-keto ester **65** was added to methyl vinyl ketone (MVK) in moderate yield but with virtually complete stereochemical control. In much the same way, crown **70** was shown to be effective for the reaction of methyl 2-phenylpropionate **67** with methyl acrylate, affording the Michael adduct **68** in 80% yield and 83% ee (Scheme 11.15) [52].

The enantioselective Michael addition of glycine derivatives by means of chiral PTC has been developed to synthesize a variety of functionalized α-alkyl amino acids. The first example was reported by Corey using **8d** as a catalyst for the asymmetric conjugate addition of glycinate Schiff base **5** to α,β-unsaturated carbonyl substrates (entries 1–2, Table 11.4) [53]. The reaction of **5** and methyl acrylate afforded the functionalized glutamic acid derivative with high enantioselectivity, which is very useful for synthetic applications because the two carbonyl groups are differentiated. To date, this type of phase-transfer-catalyzed Michael addition of **5** has been investigated with either acrylates or alkyl vinyl ketones as an acceptor, under the influence of different catalysts and bases. The representative results are summarized in Table 11.6.

Recently, we addressed the importance of a dual-functioning chiral phase-transfer catalyst such as **73** for obtaining a high level of enantioselectivity in the Michael addition of malonates to chalcone derivatives [58]. For instance, the reaction of diethyl malonate with chalcone in toluene under the influence of K_2CO_3 and **73** (3 mol%) proceeded smoothly at −20 °C with excellent enantioselectivity, while the selectivity was markedly decreased when **74** possessing no hydroxy functionality was used as catalyst (Scheme 11.16).

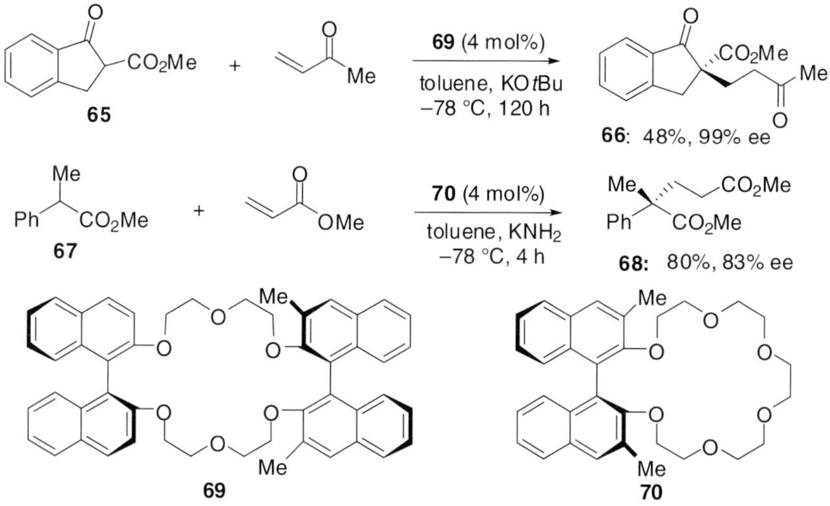

Scheme 11.15 Chiral crown ethers as phase-transfer catalysts.

Jew and Park achieved a highly enantioselective synthesis of (2S)-α-(hydroxymethyl) glutamic acid, a potent metabotropic receptor ligand, through the Michael addition of 2-naphthalene oxazoline derivative (**45c**) to ethyl acrylate under phase-transfer conditions. As shown in Scheme 11.17, the use of BEMP as a base at −60 °C with the catalysis of N-spiro chiral quaternary ammonium bromide **9e** appeared to be essential for attaining an excellent enantioselectivity [59].

The diastereoselective and enantioselective conjugate addition of nitroalkanes to alkylidenemalonates was then developed under mild phase-transfer conditions by the use of appropriately designed chiral quaternary ammonium bromide **9i** as an efficient catalyst. This new protocol offered a practical entry to optically active γ-amino acid derivatives, as shown in Scheme 11.18 [60]. The designed chiral phase-transfer catalyst **9h** (Scheme 11.18) was shown to be efficacious in the enantioselective Michael addition of β-keto esters to α,β-unsaturated carbonyl compounds, and enabled the use of α,β-unsaturated aldehydes as an acceptor, leading to the construction of a quaternary stereocenter having three different functionalities of carbonyl origin. It is of interest that the use of fluorenyl ester **53c** greatly improved the enantioselectivity. The addition of **53c** to MVK was also feasible under similar conditions, and the desired **76** was obtained quantitatively with 97% ee (Scheme 11.19) [46].

11.4
Aldol and Related Reactions

Although phase-transfer catalytic enantioselective direct aldol reactions of glycine donor with aldehyde acceptors could provide an ideal method for the simultaneous construction of the primary structure and stereochemical integrity of β-hydroxy-

Table 11.6 Asymmetric Michael addition of 5 to α,β-unsaturated carbonyl compounds.

Entry	Catalyst (X mol%)	R	Base	Solvent	T (°C)	Yield (%)	ee (%) (config.)	Reference
1	8d (10)	OMe	CsOH·H$_2$O	CH$_2$Cl$_2$	−78	85	95 (S)	[53]
2	8d (10)	Et	CsOH·H$_2$O	CH$_2$Cl$_2$	−78	85	91 (S)	[53]
3	25c (10)	OBn	CsCOO$_3$	PhCl	−30	84	81 (S)	[24]
4	70 (10)	OtBu	CsOH·H$_2$O	tBuOMe	−60	73	77 (S)	[54]
5	71 (20)	OEt	KOtBu	CH$_2$Cl$_2$	−78	65	96 (S)	[55]
6	72 (1)	Me	CsCO$_3$	PhCl	−30	100	75 (S)	[56]
7	27 (1)	Me	CsCO$_3$	iPr$_2$O	0	84	94 (S)	[57]
8	27 (1)	n-C$_5$H$_{12}$	CsCO$_3$	iPr$_2$O	0	60	94 (S)	[57]

α-amino acids, which are extremely important chiral units especially from the pharmaceutical viewpoint, the examples reported to date are very limited.

We recently developed an efficient, highly diastereoselective and enantioselective direct aldol reaction of **5** with a wide range of aliphatic aldehydes under mild phase-transfer conditions, and employing N-spiro chiral quaternary ammonium salt **9i** as a key catalyst. A mechanistic investigation revealed the intervention of highly stereoselective retro aldol reaction, which could be minimized by using a catalytic amount of 1% NaOH aqueous solution and ammonium chloride, leading to the establishment of a general and practical chemical process for the synthesis of optically active *anti*-β-hydroxy-α-amino esters **77** (Scheme 11.20) [61].

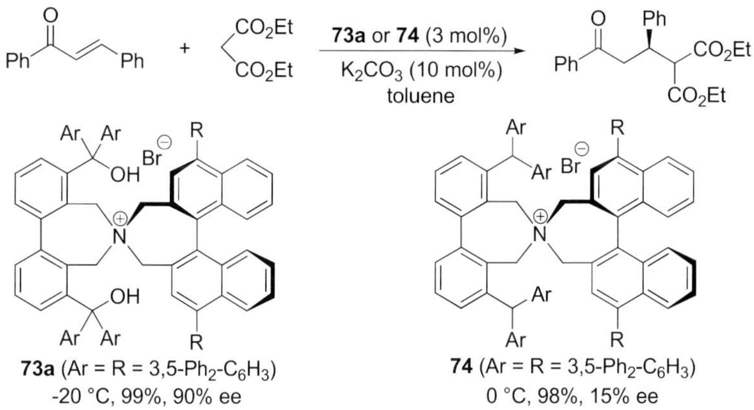

11.3 Michael Addition

Scheme 11.16 Asymmetric Michael addition of diethyl malonate to chalcone.

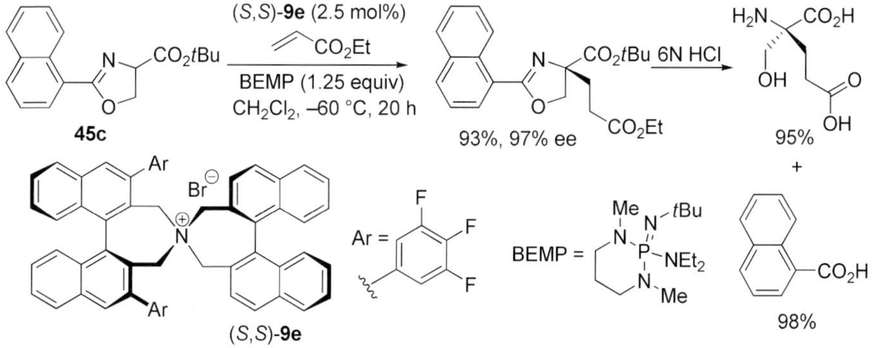

73a (Ar = R = 3,5-Ph$_2$-C$_6$H$_3$)
−20 °C, 99%, 90% ee

74 (Ar = R = 3,5-Ph$_2$-C$_6$H$_3$)
0 °C, 98%, 15% ee

Scheme 11.17 Asymmetric synthesis of (2S)-α-(hydroxymethyl)glutamic acid.

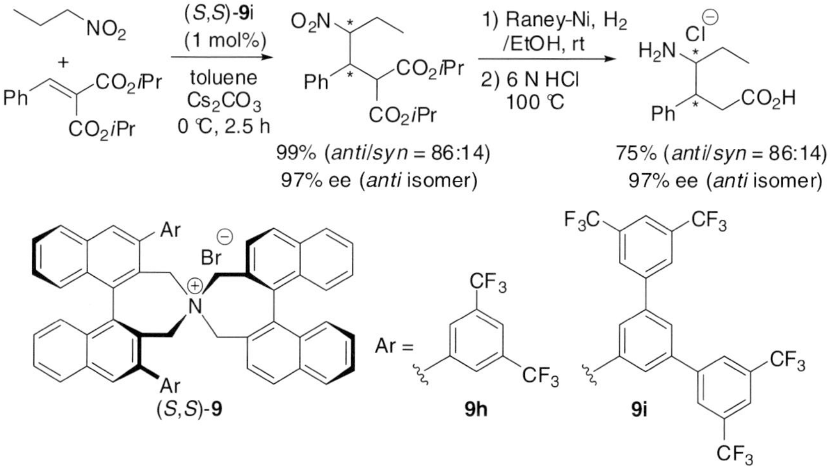

Scheme 11.18 Michael addition of nitroalkanes to alkylidenemalonates.

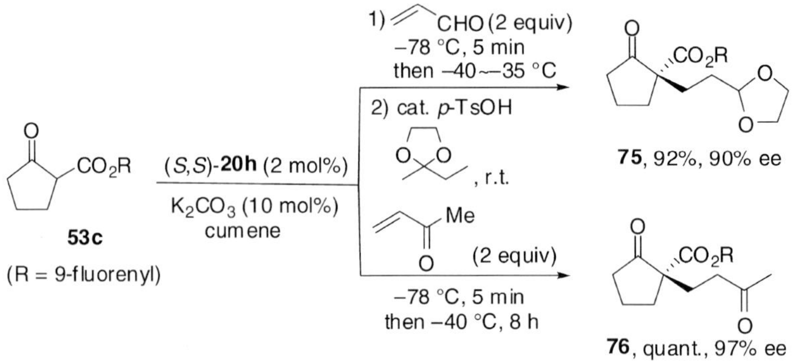

Figure 11.19 Asymmetric Michael addition of β-keto esters to acrolein and MVK.

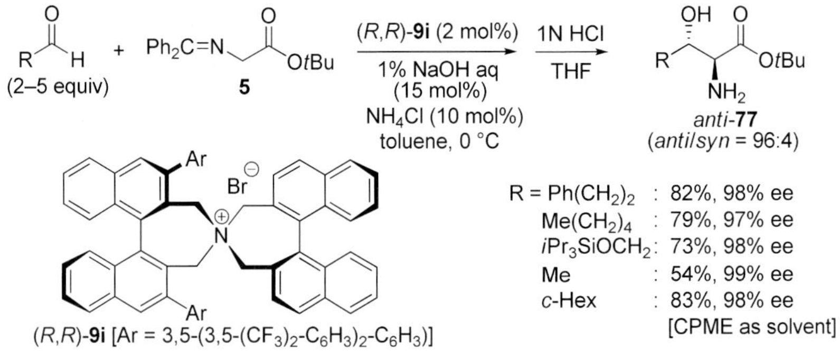

Scheme 11.20 Highly diastereoselective and enantioselective direct aldol reactions.

A phase-transfer-catalyzed direct Mannich reaction of glycinate Schiff base 5 with α-imino ester 78 was achieved with high enantioselectivity by the use of N-spiro chiral quaternary ammonium bromide 9e as catalyst (Scheme 11.21) [62]. This method enabled the catalytic asymmetric synthesis of differentially protected 3-aminoaspartate, a nitrogen analogue of dialkyl tartrate, the utility of which was demonstrated by the conversion of product (syn-79) into a precursor (80) of streptolidine lactam.

The tartrate-derived diammonium salt **25d**, which was modified by introducing an aromatic ring at the acetal side chains, was identified as an optimal catalyst for the reaction of 5 with various N-Boc imines under solid–liquid phase-transfer conditions, as exemplified in Scheme 11.22 [63]. The usefulness of the Mannich adduct **81** was further demonstrated by the straightforward synthesis of the optically pure tripeptide **82**.

Herrera and Bernardi have developed a new catalytic enantioselective approach to the asymmetric nucleophilic addition of nitromethane to N-carbamoyl imines generated from α-amido sulfones (aza-Henry reaction) [64]. The chiral phase-transfer catalyst **85a** acts in a dual fashion, first promoting the formation of the imine under mild reaction conditions and then activating the nucleophile for asymmetric addition. This new strategy for the catalytic aza-Henry reaction was

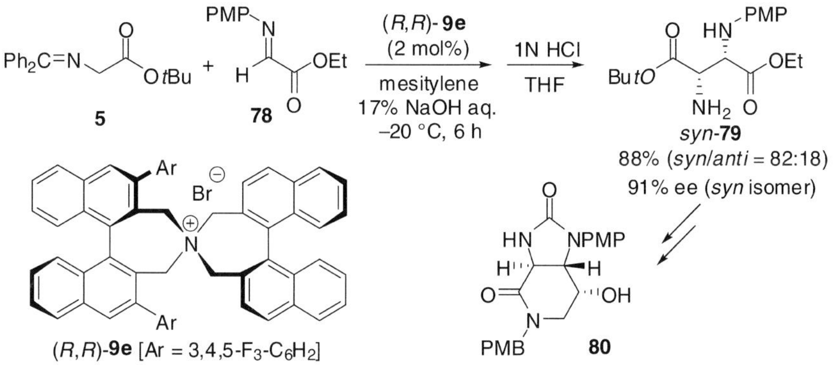

Scheme 11.21 Direct Mannich approach to the nitrogen analogue of dialkyl tartrate.

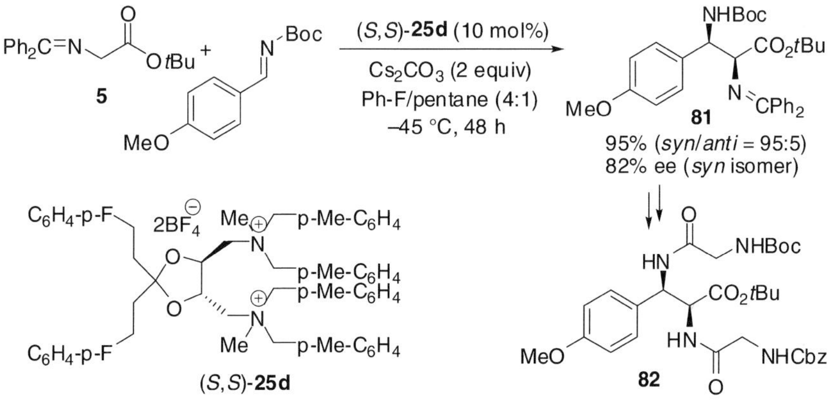

Scheme 11.22 Direct Mannich reaction with N-tert-butoxycarbonyl imines.

particularly effective for the synthesis of N-Boc α-alkyl β-nitroamines from both linear and branched aliphatic aldehyde-derived α-amido sulfones (Table 11.7).

The highly enantioselective phase-transfer-catalyzed Mannich reaction of malonates with *in situ*-generated, Cbz-protected azomethines has been developed by Ricci (Scheme 11.23) [65]. This approach gives a straightforward access to optically active β-alkyl N-carbanoyl acids, starting from readily available and stable α-amido sulfones.

11.5
Neber Rearrangement

The Neber rearrangement of oxime sulfonates has been considered to proceed via a nitrene pathway or an anion pathway. If the latter mechanism is operative, the use of a certain chiral base could result in the discrimination of two enantiotopic α-protons to furnish optically active α-amino ketones. Verification of this hypothesis was provided by realizing the asymmetric Neber rearrangement of simple oxime sulfonate **90**, generated *in situ* from the parent oxime *(Z)*-**89**, under

Table 11.7 Asymmetric catalytic Aza-Henry reaction under phase-transfer conditions.

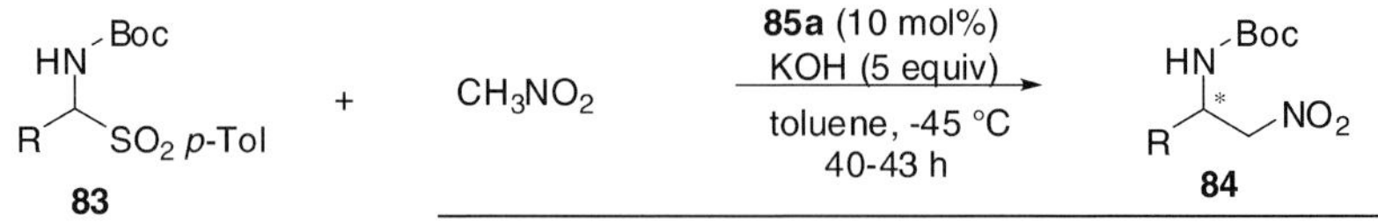

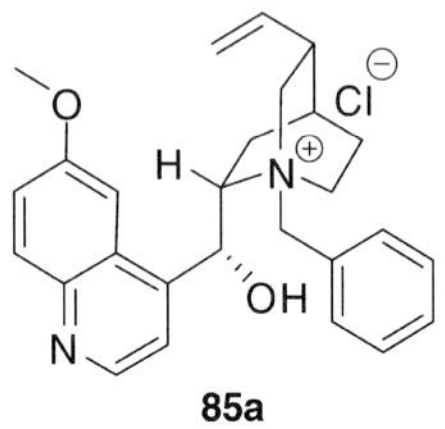

Entry	R	yield (%)	ee (%)
1	Ph	95	84
2	Ph(CH$_2$)$_2$	98	95
3	Cy	84	98
4	i-Pr	95	95
5	Et	92	94
6	Me	86	92

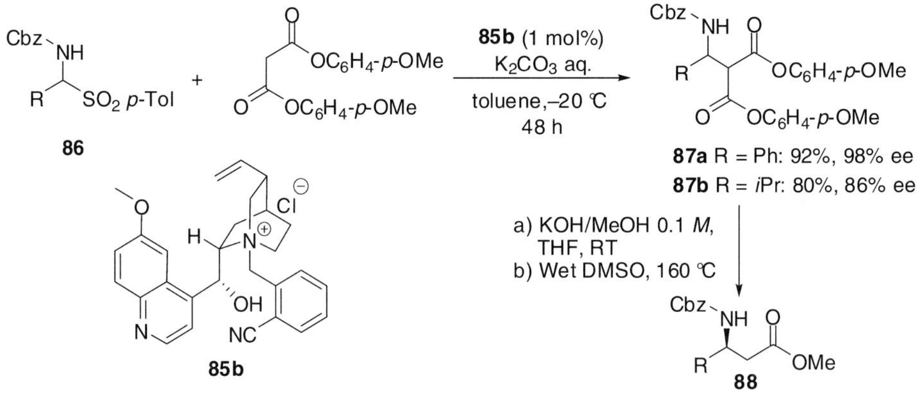

Scheme 11.23 Asymmetric Mannich reaction of malonates with α-amino sulfones under PTC conditions.

phase-transfer conditions using the structurally rigid, N-spiro-type chiral quaternary ammonium bromide **9k** as catalyst. The corresponding protected α-amino ketone **91** was isolated in high yield with notable enantiomeric excess (Scheme 11.24) [66]. The stereochemical outcome of the present asymmetric Neber rearrangement can be rationalized by postulating the transition-state model, in which the conformation of the catalyst–substrate ion pair would be fixed to appreciate the possible π–π interaction, as illustrated in Scheme 11.24.

11.6
Epoxidation

The catalytic asymmetric epoxidation of electron-deficient olefins, particularly α,β-unsaturated ketones, has been the subject of numerous investigations, and a

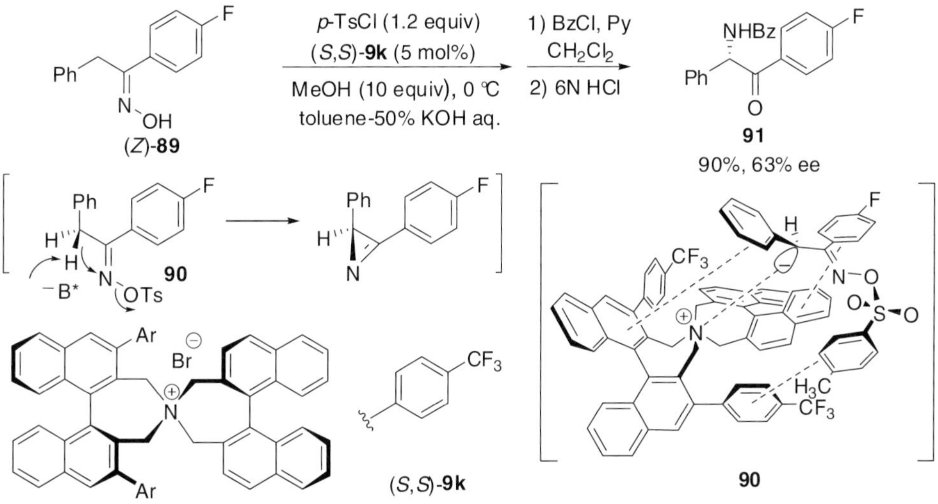

Scheme 11.24 Asymmetric Neber rearrangement of oxime sulfonate **90**.

number of useful methodologies have been elaborated [67]. Among these, the method utilizing chiral PTC occupies a unique place featuring its practical advantages, and it also allows the highly enantioselective epoxidation of *trans*-α,β-unsaturated ketones, particularly chalcone.

Shioiri and Arai carried out an asymmetric epoxidation of chalcone and its derivatives with 30% hydrogen peroxide using a chiral ammonium salt **1d** as phase-transfer catalyst (entry 1, Table 11.8) [68]. The observed enantioselectivity in epoxy chalcone was seen to be highly dependent on the *para*-substituent of **1d**. The cinchona-derived, rigid quaternary ammonium salt **10e** developed by Corey showed higher enantioselection, giving the epoxy product with 93% ee when KOCl was used as the oxidant (entry 2) [69]. Recently, Jew and Park showed that surfactants can dramatically increase both the reaction rate and enantioselectivity of phase-transfer catalytic epoxidation. Up to >99% ee was obtained with Span 20 and the amount of catalyst **12c** could be reduced to 1 mol% when using 30% hydrogen peroxide as the oxidant (entry 3) [70].

Chiral monoaza-15-crown-5 derived from D-glucose **92** was shown to be another good catalyst in the asymmetric epoxidation of chalcones with *tert*-butylhydroperoxide as the oxidant, with the highest enantioselectivity (94% ee) being reported by Bakó (entry 4, Table 11.8) [71]. The tetracyclic C_2-symmetric guanidium salt **93**, which was prepared from *(S)*-malic acid by Murphy, also showed excellent enantioselection in the asymmetric epoxidation of chalcone (entry 5) [72].

We designed a new and highly efficient chiral *N*-spiro-type quaternary ammonium salt **73** with dual functions for the asymmetric epoxidation of various enone substrates (entries 5–9, Table 11.8) [73]. The exceedingly high asymmetric induction was ascribable to the molecular recognition ability of the catalyst toward enone substrates by virtue of the appropriately aligned hydroxy functionality, as well as the

Table 11.8 Enantioselective epoxidation of enones under phase-transfer conditions.

Entry	Cat. (X mol%)	R_1	R_2	Oxidant	Solvent	T (°C)	Yield (%)	ee (%) (Config.)	Reference
1	1d (5)	Ph	Ph	30% H_2O_2, LiOH	Bu_2O	4	97	84 ($\alpha S,\beta R$)	[68]
2	10e (10)	Ph	Ph	KOCl	toluene	−40	95	93 ($\alpha S,\beta R$)	[69]
3[a]	12c (1)	Ph	Ph	30% H_2O_2, 50% KOH	toluene	r.t.	95	>99 ($\alpha S,\beta R$)	[70]
4	92 (7)	Ph	Ph	tBuOOH, 20% NaOH	toluene	−40	95	94 ($\alpha R,\beta S$)	[71]
5	93 (5)	Ph	Ph	8% NaOCl	toluene	0 – r.t.	99	93 ($\alpha S,\beta R$)	[72]
6	73a (3)	Ph	Ph	13% NaOCl	toluene	0	99	96 ($\alpha S,\beta R$)	[73]
7	73b (3)	tBu	Ph	13% NaOCl	toluene	0	99	92 ($\alpha S,\beta R$)	[73]
8	73b (3)	tBu	cHex	13% NaOCl	toluene	0	80	96 ($\alpha S,\beta R$)	[73]
9	73a (3)	O (indanone=CHPh)		13% NaOCl	toluene	0	91	99 ($\alpha S,\beta R$)	[73]

a Span 20 (1 mol%) was added as surfactant.

chiral molecular cavity. Indeed, the enantioselectivity highly was shown to depend on the steric size and the electronic factor of both Ar and R substituents in **73**.

11.7
Strecker Reaction

The catalytic asymmetric cyanation of imines – the Strecker reaction – represents one of the most direct and viable methods for the asymmetric synthesis of α-amino

11.7 Strecker Reaction

Table 11.9 Phase-transfer-catalyzed asymmetric Strecker reaction.

Entry	R	yield (%)	ee (%)
1	c-Oct	88	97
2	i-Pr	85	93
3	Ph(CH$_2$)$_2$	81	90
4	(CH$_3$)CHCH$_2$	82	88
5	t-Bu	94	94
6	Ph(CH$_3$)$_2$C	95	98
7	Ad	98	97

(R)-**96**: Ar = p-CF$_3$-Ph

Table 11.10 Asymmetric catalytic Strecker reaction with cyanohydrin under PTC conditions.

Entry	R	yield (%)	ee (%)
1	Ph(CH$_2$)$_2$	95	68
2	PhCH$_2$	95	79
3	Me	85	78
4	CH$_3$CH$_2$	88	80
5	i-Pr	92	82
6	t-Bu	85	88
7	Cy	95	50

acids and their derivatives. Recently, we disclosed the first example of a phase-transfer-catalyzed, highly enantioselective Strecker reaction of aldimines using aqueous KCN as cyanide source through the development of a new chiral quaternary ammonium salt (R)-**96** (Table 11.9). A key feature of this catalyst design was the introduction of *ortho*-aryl-substituted aromatic groups at the 3,3′-positions of the chiral binaphthyl unit. The *ortho*-aryl groups caused rotational restriction around the naphthyl–aryl biaryl axes, which provided a configurational bias to create a stereochemically defined molecular cavity over the nitrogen. Indeed, catalyst **96** bearing a stereochemically defined tetranaphthyl backbone exhibited high chiral efficiency in the reaction of **94** with 1.5 equiv KCN to afford **95** in high yield with excellent enantioselectivity [74].

Acetone cyanohydrin **97** was used as a cyanide ion source by Herrera and Ricci to accomplish the phase-transfer enantioselective cyanation of *in situ*-generated aliphatic aldimines from *N*-Boc protected α-amino sulfones (Table 11.10). Quinine-derived ammonium salt **85c** with a CF_3 at the *ortho* position of the benzyl group displayed a remarkably substrate scope, affording the α-amino nitriles with excellent yields and high enantioselectivity [75].

11.8
Conclusions

Following the breakthrough delivered by the Merck research group, a large number of naturally occurring alkaloid derivatives have been developed as powerful and readily available chiral phase-transfer catalysts. In contrast, purely synthetic chiral quaternary onium salts and chiral crown ethers have been created which feature their own characteristic advantages. This has, in turn, not only led to an attainment of considerably higher reactivity and stereoselectivity, but has also expanded the applications of asymmetric PTC in modern organic synthesis. In particular, the enantioselective functionalization of the glycinate Schiff base, as introduced by O'Donnell, has been used widely as a 'benchmark' reaction to evaluate the efficiency of the newly devised catalyst, and this has developed into a reliable and practicable method for the synthesis of optically pure α-amino acids and their derivatives. Whilst this–somewhat ironically–highlights the current limitations of existing systems, it is also driving the development of new systems to extend the concept of asymmetric PTC. It is clear that, in the immediate future, a continuous effort should be devoted to the rational design of chiral phase-transfer catalysts and their applications to synthetically valuable transformations. This surely would contribute greatly to the establishment of genuinely sustainable chemical processes targeted at the worldwide production of highly valuable substances.

Acknowledgments

We thank our colleagues at Hokkaido University and Kyoto University for the personal and scientific collaborations, and whose names appear in the references. Without their enthusiasm for organic chemistry, our research in the field of asymmetric PTC would not have been achieved.

References

1 Starks, C.M. (1971) *Journal of the American Chemical Society*, **93**, 195.
2 (a) Dehmlow, E.V. and Dehmlow, S.S. (1993) *Phase–Transfer Catalysis*, 3rd edn, Wiley-VCH Verlag GmbH, Weinheim.
(b) Starks, C.M., Liotta, C.L. and Halpern, M. (1994) *Phase–Transfer Catalysis*, Chapman & Hall, New York.
(c) Sasson, Y. and Neumann, R. (1997) *Handbook of Phase–Transfer Catalysis*, Blackie Academic & Professional, London.

(d) Halpern, M.E. (ed.) (1997) *Phase–Transfer Catalysis*, ACS Symposium Series 659, American Chemical Society, Washington, DC.

3 (a) Shioiri, T. (1997) *Handbook of Phase–Transfer Catalysis* (Y. Sasson and R. Neumann), Blackie Academic & Professional, London. Chapter 14.
(b) O'Donnell, M.J. (1998) in *Phases*, Issue 4, SACHEM, Inc., Austin, Texas, p. 5.
(c) O'Donnell, M.J. (1999) in *Phases*, Issue 5, SACHEM, Inc., Austin, Texas, p. 5.
(d) Nelson, A. (1999) *Angewandte Chemie*, **111**, 1685.
(e) Nelson, A. (1999) *Angewandte Chemie–International Edition*, **38**, 1583.
(f) Shioiri, T. and Arai, S. (2000) *Stimulating Concepts in Chemistry* (eds F. Vogtle, J.F. Stoddart and M. Shibasaki), Wiley-VCH Verlag GmbH, Weinheim, p. 123.
(g) O'Donnell, M.J. and Catalytic, I. (2000) *Asymmetric Syntheses* (ed. I. Ojima), 2nd edn, Wiley-VCH Verlag GmbH, New York. Chapter 10.
(h) O'Donnell, M.J. (2001) *Aldrichimica Acta*, **34**, 3.
(i) Maruoka, K. and Ooi, T. (2003) *Chemical Reviews*, **103**, 3013.
(j) O'Donnell, M.J. (2004) *Accounts of Chemical Research*, **37**, 506.
(k) Lygo, B. and Andrews, B.I. (2004) *Accounts of Chemical Research*, **37**, 518.
(l) Ooi, T. and Maruoka, K. (2007) *Enantioselective Organocatalysis* (ed. P.I. Dalko), Wiley-VCH Verlag GmbH, Weinheim, p. 121.

4 (a) Dolling, U.-H., Davis, P. and Grabowski, E.J.J. (1984) *Journal of the American Chemical Society*, **106**, 446.
(b) Hughes, D.L., Dolling, U.-H., Ryan, K.M., Schoenewaldt, E.F. and Grabowski, E.J.J. (1987) *Journal of Organic Chemistry*, **52**, 4745.

5 (a) O'Donnell, M.J., Bennett, W.D. and Wu, S. (1989) *Journal of the American Chemical Society*, **111**, 2353.
(b) Lipkowitz, K.B., Cavanaugh, M.W., Baker, B. and O'Donnell, M.J. (1991) *Journal of Organic Chemistry*, **56**, 5181.
(c) Esikova, I.A., Nahreini, T.S. and O'Donnell, M.J. (1997) *Phase-Transfer Catalysis* (ed. M.E. Halpern), ACS Symposium Series 659, American Chemical Society, Washington, DC. Chapter 7.
(d) O'Donnell, M.J., Esikova, I.A., Shullenberger, D.F. and Wu, S. (1997) *Phase-Transfer Catalysis* (ed. M.E. Halpern), ACS Symposium Series 659, American Chemical Society, Washington, DC, Chapter 10.

6 O'Donnell, M.J., Wu, S. and Huffman, J.C. (1994) *Tetrahedron*, **50**, 4507.

7 (a) Lygo, B. and Wainwright, P.G. (1997) *Tetrahedron Letters*, **38**, 8595.
(b) Lygo, B., Crosby, J., Lowdon, T.R. and Wainwright, P.G. (2001) *Tetrahedron*, **57**, 2391.
(c) Lygo, B., Crosby, J., Lowdon, T.R., Peterson, J.A. and Wainwright, P.G. (2001) *Tetrahedron*, **57**, 2403.

8 Corey, E.J., Xu, F. and Noe, M.C. (1997) *Journal of the American Chemical Society*, **119**, 12414.

9 (a) Ooi, T., Kameda, M. and Maruoka, K. (1999) *Journal of the American Chemical Society*, **121**, 6519.
(b) Maruoka, K. (2001) *Journal of Fluorine Chemistry*, **112**, 95.
(c) Ooi, T., Uematsu, Y. and Maruoka, K. (2002) *Advanced Synthesis Catalysis*, **344**, 288.
(d) Ooi, T., Uematsu, Y. and Maruoka, K. (2003) *Journal of Organic Chemistry*, **68**, 4576.
(e) Ooi, T., Kameda, M. and Maruoka, K. (2003) *Journal of the American Chemical Society*, **125**, 5139.

10 Yoo, S.-S., Jew, M.-S., Jeong, B.-S.II , Park, Y. and Park, H.-G. (2002) *Organic Letters*, **4**, 4245.

11 Yoo, M.-S., Jeong, B.-S., Lee, J.H., Park, H.-G. and Jew, S.-S. (2005) *Organic Letters*, **7**, 1129.

12 (a) Jew, S.-S., Jeong, B.-S., Yoo, M.-S., Huh, H. and Park, H.-G. (2001) *Chemical Communications*, 1244.
(b) Park, H.-G., Jeong, B.-S., Yoo, M.-S., Lee, J.-H., Park, B.-S., Kim, M.G. and Jew, S.-S. (2003) *Tetrahedron Letters*, **44**, 3497.

13 Park, H.-G., Jeong, B.-S., Yoo, M.-S., Park, M.-K., Huh, H. and Jew, S.-S. (2001) *Tetrahedron Letters*, **42**, 4645.

14 (a) Park, H.-G., Jeong, B.-S., Yoo, M.-S., Lee, J.-H., Park, M.-k., Lee, Y.-J., Kim,

M.-J. and Jew, S.-S. (2002) *Angewandte Chemie*, **114**, 3162.
(b) Park, H.-G., Jeong, B.-S., Yoo, M.-S., Lee, J.-H., Park, M.-K., Lee, Y.-J., Kim, M.-J. and Jew, S.-S. (2002) *Angewandte Chemie – International Edition*, **41**, 3036.

15 Hashimoto, T. and Maruoka, K. (2003) *Tetrahedron Letters*, **44**, 3313.

16 Hashimoto, T., Tanaka, Y. and Maruoka, K. (2003) *Tetrahedron: Asymmetry*, **14**, 1599.

17 (a) Ooi, T., Uematsu, Y., Kameda, M. and Maruoka, K. (2002) *Angewandte Chemie*, **114**, 1621.
(b) Ooi, T., Uematsu, Y., Kameda, M. and Maruoka, K. (2002) *Angewandte Chemie – International Edition*, **41**, 1551.

18 (a) Kitamura, M., Shirakawa, S. and Maruoka, K. (2005). *Angewandte Chemie*, **117**, 1573.
(b) Kitamura, M., Shirakawa, S. and Maruoka, K. (2005) *Angewandte Chemie – International Edition*, **44**, 1549.

19 Han, Z., Yamaguchi, Y., Kitamura, M. and Maruoka, K. (2005) *Tetrahedron Letters*, **46**, 8555.

20 Belokon, Y.N., North, M., Churkina, T.D., Ikonnikov, N.S. and Maleev, V.I. (2001) *Tetrahedron*, **57**, 2491.

21 (a) Kita, T., Georgieva, A., Hashimoto, Y., Nakata, T. and Nagasawa, K. (2002) *Angewandte Chemie*, **114**, 2956.
(b) Kita, T., Georgieva, A., Hashimoto, Y., Nakata, T. and Nagasawa, K. (2002) *Angewandte Chemie – International Edition*, **41**, 2832.

22 (a) Kowtoniuk, W.E., MacFarland, D.K. and Grover, G.N. (2005) *Tetrahedron Letters*, **46**, 5703.
(b) Rueffer, M.E., Fort, L.K. and MacFarland, D.K. See also: (2004) *Tetrahedron: Asymmetry*, **15**, 3297.

23 Mase, N., Ohno, T., Hoshikawa, N., Ohishi, K., Morimoto, H., Yoda, H. and Takabe, K. (2003) *Tetrahedron Letters*, **44**, 4073.

24 (a) Shibuguchi, T., Fukuta, Y., Akachi, Y., Sekine, A., Ohshima, T. and Shibasaki, M. (2002) *Tetrahedron Letters*, **43**, 9539.
(b) Ohshima, T., Shibuguchi, T., Fukuta, Y. and Shibasaki, M. (2004) *Tetrahedron*, **60**, 7743.

25 Sasai, H. (2003) Jpn. Kokai Tokkyo Koho, JP2003335780.

26 Lygo, B., Allbutt, B. and James, S.R. (2003) *Tetrahedron Letters*, **44**, 5629.

27 (a) Ooi, T., Sakai, D., Takeuchi, M., Tayama, E. and Maruoka, K. (2003) *Angewandte Chemie*, **115**, 2002.
(b) Ooi, T., Sakai, D., Takeuchi, M., Tayama, E. and Maruoka, K. (2003) *Angewandte Chemie – International Edition*, **42**, 5868.

28 Ooi, T., Takeuchi, M., Kato, D., Uematsu, Y., Tayama, E., Sakai, D. and Maruoka, K. (2005) *Journal of the American Chemical Society*, **127**, 5073.

29 (a) Chinchills, R., Mazón, P. and Nájera, C. (2000) *Tetrahedron: Asymmetry*, **11**, 3277.
(b) Chinchills, R., Mazón, P. and Nájera, C. (2004) *Advanced Synthesis Catalysis*, **346**, 1186.

30 Thierry, B., Plaquevent, J.-C. and Cahard, D. (2001) *Tetrahedron: Asymmetry*, **12**, 983.

31 Thierry, B., Plaquevent, J.-C. and Cahard, D. (2003) *Tetrahedron: Asymmetry*, **14**, 1671.

32 Thierry, B., Perrard, T., Audouard, C., Plaquevent, J.-C. and Cahard, D. (2001) *Synthesis*, 1742.

33 Danelli, T., Annunziata, R., Benaglia, M., Cinquini, M., Cozzi, F. and Tocco, G. (2003) *Tetrahedron: Asymmetry*, **14**, 461.

34 Shirakawa, S., Tanaka, Y. and Maruoka, K. (2004) *Organic Letters*, **6**, 1429.

35 (a) Cativiela, C. and Diaz-de-Villegas, M.D. (1998) *Tetrahedron: Asymmetry*, **9**, 3517.
(b) Schöllkopf, U. (1983) *Topics in Current Chemistry*, **109**, 65.

36 O'Donnell, M.J. and Wu, S. (1992) *Tetrahedron: Asymmetry*, **3**, 591.

37 Lygo, B., Crosby, J. and Peterson, J.A. (1999) *Tetrahedron Letters*, **40**, 8671.

38 Jew, S.-S., Jeong, B.-S., Lee, J.-H., Yoo, M.-S., Lee, Y.-J., Park, B.-S., Kim, M.G. and Park, H.-G. (2003) *Journal of Organic Chemistry*, **68**, 4514.

39 (a) Belokon, Y.N., Kochetkov, K.A., Churkina, T.D., Ikonnikov, N.S., Chesnokov, A.A., Larionov, O.V., Parmar, V.S., Kumar, R. and Kagan, H.B. (1998) *Tetrahedron: Asymmetry*, **9**, 851.
(b) Belokon, Y.N., Kochetkov, K.A., Churkina, T.D., Ikonnikov, N.S., Vyskocil,

S. and Kagan, H.B. (1999) *Tetrahedron: Asymmetry*, **10**, 1723.
(c) Belokon, Y.N., Kochetkov, K.A., Churkina, T.D., Ikonnikov, N.S., Chesnokov, A.A., Larionov, O.V., Singh, I., Parmar, V.S., Vyskocil, S. and Kagan, H.B. (2000) *Journal of Organic Chemistry*, **65**, 7041.

40 (a) Belokon, Y.N., North, M., Kublitski, V.S., Ikonnikov, N.S., Krasik, P.E. and Maleev, V.I. (1999) *Tetrahedron Letters*, **40**, 6105.
(b) Belokon, Y.N., Davies, R.G. and North, M. (2000) *Tetrahedron Letters*, **41**, 7245.
(c) Belokon, Y.N., Davies, R.G., Fuentes, J.A., North, M. and Parsons, T. (2001) *Tetrahedron Letters*, **42**, 8093.
(d) Belokon, Y.N., Bhave, D., D'Addario, D., Groaz, E., North, M. and Tagliazucca, V. (2004) *Tetrahedron*, **60**, 1849.
(e) Belokon, Y.N., Fuentes, J.A., North, M. and Steed, J.W. (2004) *Tetrahedron*, **60**, 3191.

41 Ooi, T., Takeuchi, M., Kameda, M. and Maruoka, K. (2000) *Journal of the American Chemical Society*, **122**, 5228.

42 (a) Jew, S.-S., Lee, Y.-J., Lee, J., Kang, M.J., Jeong, B.-S., Lee, J.-H., Yoo, M.-S., Kim, M.-J., Choi, S.-H., Ku, J.-M. and Park, H.-G. (2004) *Angewandte Chemie*, **116**, 2436.
(b) Jew, S.-S., Lee, Y.-J., Lee, J., Kang, M.J., Jeong, B.-S., Lee, J.-H., Yoo, M.-S., Kim, M.-J., Choi, S.-H., Ku, J.-M. and Park, H.-G. (2004) *Angewandte Chemie – International Edition*, **43**, 2382.

43 Lee, Y.-J., Lee, J., Kim, M.-J., Kim, T.-S., Park, H.-G. and Jew, S.-S. (2005) *Organic Letters*, **7**, 1557.

44 Seebach, D., Beck, A.K. and Studer, A. For a comprehensive review on the use of peptide enolates, see: (1995) *Modern Synthetic Methods*, Vol. 7 (eds B. Ernst and C. Leumann), Wiley-VCH Verlag GmbH, Weinheim, p. 1.

45 (a) Ooi, T., Tayama, E. and Maruoka, K. (2003) *Angewandte Chemie*, **115**, 599.
(b) Ooi, T., Tayama, E. and Maruoka, K. (2003) *Angewandte Chemie – International Edition*, **42**, 579.
(c) Maruoka, K., Tayama, E. and Ooi, T. (2004) *Proceedings of the National Academy of Sciences of the United States of America*, **101**, 5824.

46 (a) Ooi, T., Miki, T., Taniguchi, M., Shiraishi, M., Takeuchi, M. and Maruoka, K. (2003) *Angewandte Chemie*, **115**, 3926.
(b) Ooi, T., Miki, T., Taniguchi, M., Shiraishi, M., Takeuchi, M. and Maruoka, K. (2003) *Angewandte Chemie – International Edition*, **42**, 3796.

47 Ooi, T., Miki, T. and Maruoka, K. (2005) *Organic Letters*, **7**, 191.

48 Ooi, T., Miki, T. and Maruoka, K. (2006) *Advanced Synthesis Catalysis*, **348**, 1539.

49 (a) Bella, M., Kobbelgaard, S. and Jørgensen, K.A. (2005) *Journal of the American Chemical Society*, **127**, 3670.
(b) Kobbelgaard, S., Bella, M. and Jørgensen, K.A. (2006) *Journal of Organic Chemistry*, **71**, 4980.

50 (a) Poulsen, T.B., Bernardi, L., Bell, M. and Jørgensen, K.A. (2006) *Angewandte Chemie*, **118**, 6701.
(b) Poulsen, T.B., Bernardi, L., Bell, M. and Jørgensen, K.A. (2005) *Angewandte Chemie – International Edition*, **45**, 6551.

51 For recent reviews on catalytic asymmetric Michael reaction, see: (a) Krause, N. and Hoffman-Röder, A. (2001) *Synthesis*, 171.
(b) Sibi, M. and Manyem, S. (2000) *Tetrahedron*, **56**, 8033.
(c) Kanai, M. and Shibasaki, M. (2000) *Catalytic Asymmetric Synthesis*, 2nd edn (ed. I. Ojima), Wiley-VCH Verlag GmbH & Co. KGaA, New York, p. 569.
(d) Tomioka, K. and Nagaoka, Y. (1999) *Comprehensive Asymmetric Catalysis*, Vol. 3 (eds E.N. Jacobsen, A. Pfaltz and H. Yamamoto), Springer, Berlin. Chapter 31.1.

52 Cram, D.J. and Sogah, G.D.Y. (1981) *Journal of the Chemical Society D – Chemical Communications*, 625.

53 Corey, E.J., Noe, M.C. and Xu, F. (1998) *Tetrahedron Letters*, **39**, 5347.

54 Arai, S., Tsuji, R. and Nishida, A. (2002) *Tetrahedron Letters*, **43**, 9535.

55 Akiyama, T., Hara, M., Fuchibe, K., Sakamoto, S. and Yamaguchi, K. (2003) *Chemical Communications*, 1734.

56 Arai, S., Tokumaru, K. and Aoyama, T. (2004) *Chemical & Pharmaceutical Bulletin*, **52**, 646.

57 Lygo, B., Allbutt, B. and Kirton, E.H.M. (2005) *Tetrahedron Letters*, **46**, 4461.
58 Ooi, T., Ohara, D., Fukumoto, K. and Maruoka, K. (2005) *Organic Letters*, **7**, 3195.
59 Lee, Y.-J., Lee, J., Kim, M.-J., Jeong, B.-S., Lee, J.-H., Kim, T.-S., Lee, J., Ku, J.-M., Jew, S.-S. and Park, H.-G. (2005) *Organic Letters*, **7**, 3207.
60 Ooi, T., Fujioka, S. and Maruoka, K. (2004) *Journal of the American Chemical Society*, **126**, 11790.
61 (a) Ooi, T., Taniguchi, M., Kameda, M. and Maruoka, K. (2002) *Angewandte Chemie*, **114**, 4724.
(b) Ooi, T., Taniguchi, M., Kameda, M. and Maruoka, K. (2002) *Angewandte Chemie – International Edition*, **41**, 4542.
(c) Ooi, T., Kameda, M., Taniguchi, M. and Maruoka, K. (2004) *Journal of the American Chemical Society*, **126**, 9685.
62 Ooi, T., Kameda, M., Fujii, J. and Maruoka, K. (2004) *Organic Letters*, **6**, 2397.
63 (a) Okada, A., Shibuguchi, T., Ohshima, T., Masu, H., Yamaguchi, K. and Shibasaki, M. (2005) *Angewandte Chemie*, **117**, 4640.
(b) Okada, A., Shibuguchi, T., Ohshima, T., Masu, H., Yamaguchi, K. and Shibasaki, M. (2005) *Angewandte Chemie – International Edition*, **44**, 4564.
64 (a) Fini, F., Sgarzani, V., Pettersen, D., Herrera, R.P., Bernardi, L. and Ricci, A. (2005) *Angewandte Chemie*, **117**, 8189.
(b) Fini, F., Sgarzani, V., Pettersen, D., Herrera, R.P., Bernardi, L. and Ricci, A. (2005) *Angewandte Chemie – International Edition*, **44**, 7975.
65 Fini, F., Bernardi, L., Herrera, R.P., Pettersen, D., Ricci, A. and Sgarzani, V. (2006) *Advanced Synthesis Catalysis*, **348**, 2043.
66 Ooi, T., Takahashi, M., Doda, K. and Maruoka, K. (2002) *Journal of the American Chemical Society*, **124**, 7640.
67 (a) Porter, M.J. and Skidmore, J. (2000) *Chemical Communications*, 1215.
(b) Nemoto, T., Ohshima, T. and Shibasaki, M. (2002) *Journal of the Society of Synthetic Organic Chemistry of Japan*, **60**, 94.
68 (a) Arai, S., Tsuge, H. and Shioiri, T. (1998) *Tetrahedron Letters*, **39**, 7563.
(b) Arai, S., Tsuge, H., Oku, M., Miura, M. and Shioiri, T. (2002) *Tetrahedron*, **58**, 1623.
69 Corey, E.J. and Zhang, F.-Y. (1999) *Organic Letters*, **1**, 1287.
70 (a) Jew, S.-S., Lee, J.-H., Jeong, B.-S., Yoo, M.-S., Kim, M.-J., Lee, Y.-J., Lee, J., Choi, S.-H., Lee, K., Lah, M.S. and Park, H.-G. (2005) *Angewandte Chemie*, **117**, 1407.
(b) Jew, S.-S., Lee, J.-H., Jeong, B.-S., Yoo, M.-S., Kim, M.-J., Lee, Y.-J., Lee, J., Choi, S.-H., Lee, K., Lah, M.S. and Park, H.-G. (2005) *Angewandte Chemie – International Edition*, **44**, 1383.
71 (a) Bakó, T., Bakó, P., Keglevich, G., Bombicz, P., Kubinyi, M., Pál, K., Bodor, S., Makó, A. and Tőke, L. (2004) *Tetrahedron: Asymmetry*, **15**, 1589.
(b) Bakó, P., Bakó, T., Mészáros, A., Keglevich, G., Szöllősy, A., Bodor, S., Makó, A. and Tőke, L. (2004) *Synlett*, 643.
72 Allingham, M.T., Howard-Jones, A., Murphy, P.J., Thomas, D.A. and Caulkett, P.W.R. (2003) *Tetrahedron Letters*, **44**, 8677.
73 Ooi, T., Ohara, D., Tamura, M. and Maruoka, K. (2004) *Journal of the American Chemical Society*, **126**, 6844.
74 Ooi, T., Uematsu, Y. and Maruoka, K. (2006) *Journal of the American Chemical Society*, **128**, 2548.
75 Herrera, R.P., Sgarzani, V., Bernardi, L., Fini, F., Pettersen, D. and Ricci, A. (2006) *Journal of Organic Chemistry*, **71**, 9869.

12
The Industrial Application of Heterogeneous Enantioselective Catalysts

Hans-Ulrich Blaser and Benoît Pugin

12.1
Introduction

There is no doubt that the trend towards the application of single enantiomers of chiral compounds is increasing. Whilst this is especially the case for pharmaceuticals, it is also applicable to agrochemicals, flavors and fragrances [1, 2]. Among the various methods used to produce, selectively, a single enantiomer of a chiral compound, enantioselective catalysis is arguably the most attractive. Starting with a minute quantity of a (usually expensive) chiral auxiliary, large amounts of the desired product can be produced. At present, biocatalysts [3] and homogeneous metal complexes with chiral ligands [4] represent the most versatile enantioselective catalysts. However, it must be stressed that most enantiomerically enriched compounds are currently manufactured using racemate resolution or chiral pool methodologies [1]. From an industrial point of view, those catalysts which are not soluble in the same phase as the organic reactant have the inherent advantage of easy separation, and sometimes also of better handling properties. Although catalyst separation is often not a critical point for the industrial application of homogeneous catalysts [5], in some cases the heterogeneous nature might be decisive with regards to which catalytic method is used.

In this chapter we will review the industrial applications of heterogeneous enantioselective catalysts. Such catalysts may be insoluble solids or they may be soluble in a second phase that is not miscible with the organic one. Of practical relevance here are the chirally modified heterogeneous catalysts – that is, classical metal catalysts modified with a low-molecular-weight chiral compound, and chiral complexes anchored to a solid support or a vehicle which renders the catalyst insoluble in the liquid organic reaction medium. Initially, we will briefly discuss the conditions that a catalyst must meet to make it viable for the industrial manufacture of chiral products. Then, we will describe the scope and limitations of the most important practically useful types of heterogeneous chiral catalyst. Finally, applications to the synthesis of industrially relevant products will be summarized and the most important processes described in some detail.

Handbook of Asymmetric Heterogeneous Catalysis. Edited by K. Ding and Y. Uozumi
Copyright © 2008 WILEY-VCH Verlag GmbH & Co. KGaA, Weinheim
ISBN: 978-3-527-31913-8

12.2
Industrial Requirements for Applying Catalysts

In order to understand the challenges facing the application of chiral catalysts in the fine chemicals industry, it is important not only to understand the essential industrial requirements but also how process development is carried out and which criteria determine the suitability of a catalyst.

12.2.1
Characteristics of the Manufacture of Enantiomerically Pure Products

The manufacture of chiral fine chemicals such as pharmaceuticals or agrochemicals can be characterized as follows (note: the numbers given in parentheses reflect the experience of the authors):

- Multifunctional molecules produced via multistep syntheses (five to 10 or more 10 steps for pharmaceuticals, and three to seven steps for agrochemicals) with short product lives (often <20 years).

- Relatively small-scale products (1–1000 tons y^{-1} for pharmaceuticals, 500–10000 tons y^{-1} for agrochemicals), usually produced in multipurpose batch equipment (continuous processes are very rare and catalyst recycling is cumbersome, especially under a Good Manufacturing Practice (GMP) regime).

- High purity requirements (usually >99% and <10 ppm metal residue in pharmaceuticals).

- High added values and therefore tolerant to relative high process costs (especially for very effective, small-scale products).

- Short development time for the production process (less than a few months to 1–2 years), as the time to market affects the profitability of the product. In addition, the development costs for a specific compound must be kept low as process development starts at an early phase when the chance of the product to ever 'make it' is about 10% [6].

- At least in European companies, chemical development is carried out by all-round organic chemists, sometimes in collaboration with technology specialists. For this reason, catalysts must be available commercially as there is little expertise (and time) for preparing complex catalysts.

12.2.2
Process Development: Critical Factors for the Application of (Heterogeneous) Enantioselective Catalysts

It is useful to divide the development of a manufacturing process into different phases:

- Phase 1: Outlining and assessing possible synthetic routes on paper. Here, the decision is made whether to use chemocatalytic steps for making the enantioenriched product or use other methodology (chiral pool starting materials, enzymes, resolution, etc.). This decision will depend on a number of considerations such as the goal of the development, the know-how of the investigators, the time frame, or the available manpower and equipment.

- Phase 2: Demonstrating the chemical feasibility of the key step, often the enantioselective catalytic reaction and show that the catalytic step fits into the overall synthetic scheme.

- Phase 3: Optimizing the key catalytic reaction as well as the other steps. Show the technical feasibility (catalyst separation, metal impurities, etc.)

In Phases 2 and 3, not only the results of the catalyst tests (selectivity, activity, productivity, catalyst costs, etc.) but also the total product costs decide whether the catalytic route will be further developed or abandoned.

- Phase 4: Optimizing and scale-up of the catalytic step as well as the over-all process.

In the final analysis, the choice whether a synthesis with an enantioselective catalytic step is chosen depends very often on the answers to two questions:

- Can the costs for the overall manufacturing process compete with alternative routes?
- Can the catalytic step be developed in the given time frame?

If we assume that enantioselective catalysis is the method of choice, the next question in the context of our treatise is: Homogeneous or heterogeneous? Table 12.1 provides a very condensed comparison of the two classes of catalyst. Here, two comments should be made: First, this is a somewhat subjective view of the authors and mirrors their personal experiences. Second, the importance of the various factors may change for specific catalytic transformations and applications (and in some cases the opposite might well be true!). Catalysts for biphasic systems are similar to homogeneous catalysts with respect to scope, handling and understanding, while separation, availability and diffusion issues are comparable to those of heterogeneous catalysts.

12.2.3
Requirements for Practically Useful Heterogeneous Catalysts

Based on the published results and our own experience, we consider that several requirements should be met in order to make enantioselective catalysts feasible for industrial applications.

12.2.3.1 Preparation Methods
- Generally applicable: It is not really possible to predict what type of catalyst will be suitable for a given substrate and process. Therefore, flexible methods that

Table 12.1 Comparison of homogeneous and heterogeneous catalysts.

	Homogeneous	Heterogeneous
Scope	Rather broad scope and well-known limitations, also for commercially interesting targets.	Often inferior catalyst performance (ee, TON, TOF) compared to an optimized homogeneous analogue.
Separation, recycling	Separation via distillation, filtration or extraction. Catalyst recycling difficult (sometimes insufficient activity and/or productivity would require recycling to achieve acceptable costs).	Relatively easy separation via filtration. Recovery and recycling of catalyst possible (BUT: metal leaching can still be problematic).
Stability, handling	Some ligands/catalysts must be handled under an inert atmosphere (glovebox).	Often better stability and handling due to stabilizing effect of solid support.
Availability	Many ligands are commercially available also on a technical scale.	Only very few catalysts are readily available; most must be prepared as part of the process development.
Understanding, control	Relatively well understood on a molecular level (close to organic chemistry).	Often very complex structure (active sites not uniform) and difficult synthesis (reproducibility). Opportunity to control the active sites (isolation or cooperation effects; size / shape exclusion). Diffusion to and within catalyst can be a problem for large, multifunctional molecules.

allow different combinations of the most important catalyst components such as metal, chiral auxiliary or support are vitally important.

- Simple and efficient: For immobilized chiral complexes, the additional costs of immobilizing a ligand or complex must be lower than the cost for the separation of the homogeneous catalyst, or must be outweighed by other advantages such as catalyst recycling. Furthermore, the 'molecular weight' of a catalyst immobilized to a support should not exceed 10 kDa per mole of active sites, which makes a high surface loading capacity important. For a modified heterogeneous catalyst, a simple *in situ* modification is a major advantage; otherwise, collaboration with an experienced manufacturer of heterogeneous catalysts is essential.

12.2.3.2 Catalysts
- Catalyst costs: The catalyst costs will only be important later when the costs of goods of the desired product are compared. For homogeneous catalysts, the chiral ligand often is the most expensive component (typical prices for the most important chiral phosphines are 100–500 $ g^{-1} for laboratory quantities and 5000

to >40 000 $ kg^{-1} on a larger scale). For immobilized complexes, the dominant cost elements depend on the type of catalyst, but in our estimation a heterogenized catalyst is about 1.5- to 3-fold more expensive than its homogeneous analogue. For chirally modified catalysts the costs of the modifier and the modification procedure is important.

- Availability of the catalysts: If a chiral catalyst is not available at the right time and in the appropriate quantity, it will not be applied due to the time limitations for process development. Today, a sizable number of homogeneous catalysts and ligands (especially for hydrogenation) are available commercially in technical quantities. This is not the case for most immobilized complexes systems because very few are commercially available (to our knowledge none is available on a technical scale) so that their (large-scale) synthesis must form part of the process development. Most chirally modified catalytic systems are based on classical heterogeneous catalysts and modifiers based on chiral pool products. For this reason, availability should not be a problem unless complex pretreatment and/or modification procedures are required. An important point here is that the preparation and characterization of any heterogeneous catalyst requires know-how and equipment that usually is not on hand in a standard development laboratory of a fine-chemicals or pharmaceutical company.

12.2.3.3 Catalytic Properties and Handling

- Catalytic performance: The performance (selectivity, activity, productivity) of any heterogeneous catalyst should be comparable or better than that of a competing homogeneous catalyst.

- Separation: Separation should be achieved by a simple filtration using conventional multipurpose equipment, and at least 95% of the catalyst should be recovered.

- Metal leaching: Leaching must be minimal, not only because the metal is very often expensive but, even more importantly, because the limit for metal contamination in biologically active compounds is in the ppm region. An additional purification step will significantly diminish the advantage of heterogeneity.

- Re-use: Although not mandatory, and sometimes very difficult to accomplish (especially under a GMP regime), under some circumstances this would be a major advantage from an economics point of view.

12.2.4
Practically Useful Types of Heterogeneous Enantioselective Catalyst

Three types of chiral heterogeneous catalyst which we consider to be practically useful for synthetic purposes are listed and characterized in Figure 12.1. In the following sections, the scope and limitations with respect to the industrial application of these three types of heterogeneous catalyst will be discussed, and the most important industrial applications described in some detail. It should be

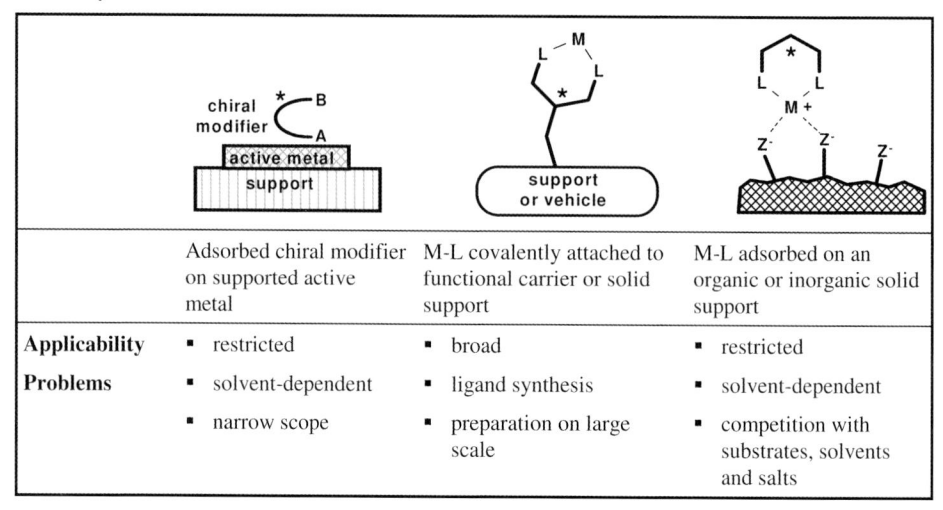

	Adsorbed chiral modifier on supported active metal	M-L covalently attached to functional carrier or solid support	M-L adsorbed on an organic or inorganic solid support
Applicability	• restricted	• broad	• restricted
Problems	• solvent-dependent	• ligand synthesis	• solvent-dependent
	• narrow scope	• preparation on large scale	• competition with substrates, solvents and salts

Figure 12.1 A schematic representation and indication of the important properties of chiral heterogeneous catalysts.

emphasized at this point, however, that other methodologies such as chiral polymers (as catalysts [7a, 8] as well as supports [7a]), entrapped metal complexes [9], chiral metal–organic assemblies [10] or porous metal–organic frameworks [11] which do not fit into this scheme are also currently under development. However, within our assessment these catalysts – with very few exceptions – are not (yet) sufficiently mature for industrial application.

12.3
Chirally Modified Heterogeneous Hydrogenation Catalysts

Historically, the chiral modification of classical heterogeneous hydrogenation catalysts was the first successful approach to practically useful enantioselective catalysts [7]. (For a detailed update, see Chapter 11.) However, despite considerable efforts only two types of catalytic system are of practical importance for preparative purposes, namely Ni catalysts modified with tartaric acid, and cinchona-modified Pt (and to some extent Pd) catalysts.

12.3.1
Nickel Catalysts Modified with Tartaric Acid

12.3.1.1 Background
Two recent reviews by Tai and Sugimura [12a] and Osawa *et al.* [12b] have provided an historical account as well as an extensive description of the important parameters affecting catalyst performance. As summarized in Scheme 12.1, suitable sub-

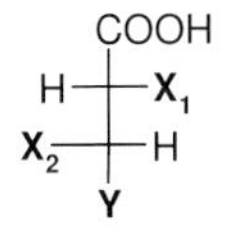

X_1	X_2	Y	ee (%)
OH	OH	COOH	83
OH	H/OR	COOH	61–65
OR	OR	COOH	0–8
OH	OH	H/Me	0–1

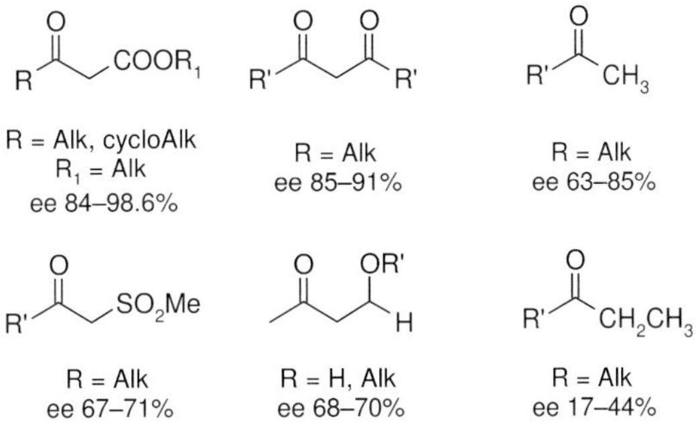

Scheme 12.1 Effect of modifier and substrate structure on enantioselectivity for Ra-Ni-catalyzed reactions of ketones. Standard reaction: Methyl acetoacetate with Ra-Ni/tartrate/NaBr. For 2- and 3-alkanones a bulky carboxylic acid was used as the co-modifier.

strates are β-functionalized ketones and bulky methyl ketones; tartaric acid in the presence of NaBr is the only modifier leading to high values of enantiomeric excess (ee). For industrial applications there are two problems: (i) In most cases, the catalyst activity is rather low and high pressures and temperatures are required for reasonable reaction times; and (ii) freshly prepared Raney nickel is the catalyst of choice and its preparation procedure is quite cumbersome. The composition of the Ni–Al alloy, the impregnation procedure (tartaric acid and NaBr concentrations, pH, temperature, time, ultrasound treatment) as well as the reaction conditions (solvent, pressure, temperature, acid comodifiers) are crucial to achieve good results. Furthermore, the modification procedure leads to a significant leaching of Ni into the wash water, thereby rendering large-scale production problematic. Some of these difficulties were addressed recently by Osawa et al. [12b], who described catalysts prepared by *in situ* modification with tartaric acid/NaBr of a prehydrogenated metallic Ni powder with reasonable activity and enantioselectivity. At present, none of these catalysts is commercially available, which makes synthetic applications difficult for organic chemists.

12.3.1.2 Synthetic and Industrial Applications

Very few synthetic applications of the Ni/tartrate system have been reported. Multistep syntheses of various isomers of the sex pheromone of the pine sawfly were carried out with the Ni/tartrate-catalyzed hydrogenation of various β-keto esters as the key step [13]. The same catalyst system was also reported to lead to biologically active C_{10}–C_{16} 3-hydroxyacids starting from the corresponding ketoesters with 83–87% ee, which can be increased to >99% with a simple crystallization [14]. A convenient and efficient ligand synthesis for homogeneous enantioselective hydrogenation was described by Bakos et al. [15], starting with the stereoselective hydrogenation of acetyl acetone to give 2,4-pentanediol with very good stereoselectivities (see Scheme 12.2). This transformation was originally developed by Izumi's group [16] and seems to have been commercialized by Wako Pure

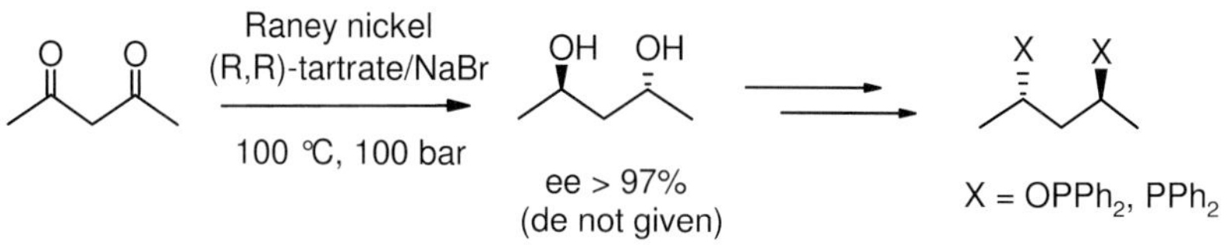

Scheme 12.2 Synthesis of diphosphine ligands.

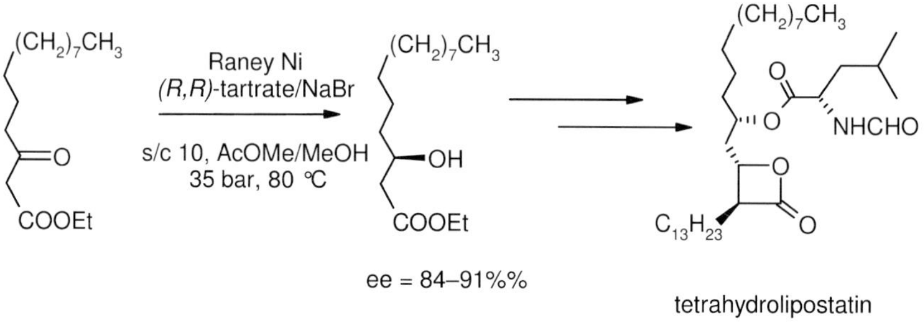

Scheme 12.3 Synthesis of tetrahydrolipostatin.

Chemicals Industries [17]. A pilot process was developed for the asymmetric hydrogenation of an intermediate in the synthesis of tetrahydrolipostatin, a pancreatic lipase inhibitor developed by Roche [18] (Scheme 12.3). Raney nickel modified by (R,R)-tartaric acid/NaBr proved to be an efficient catalyst which, under optimized conditions, gave 99% chemical yield at 35 bar and 80 °C. High catalyst loadings were required but the catalyst could be recycled up to 15 times, with ee-values falling from 91% to 84%. The process was scaled-up and ton quantities of the β-hydroxyester were produced [18a]. Interestingly, this heterogeneous variant was favored due to its simpler procedure over a Rh–biphep-catalyzed homogeneous reaction, despite the fact that the Ru catalyst gave up to 97% ee and a turnover number (TON) of 50 000.

12.3.2
Catalysts Modified with Cinchona Alkaloids

Since these are the only heterogeneous enantioselective catalysts with significant synthetic and industrial application, a somewhat more detailed description of their scope and limitations is provided. For recent updates, the reader is referred to Ref. [19].

12.3.2.1 Background
Significant progress in the substrate scope of cinchona-modified catalysts has been made during the past few years [19a]. In addition to α-keto acids and esters, some trifluoromethyl ketones and α-keto acetals and α-keto ethers have also been found to give high ee-values with Pt-based catalysts. Pd-based catalysts give moderate enantioselectivities for α,β-unsaturated acids and up to 94% ee for selected pyrones (see Scheme 12.4). Nevertheless, for the synthetic chemist the substrate scope is

Scheme 12.4 Structures of 'good' (upper), 'medium' (center) and 'bad' (lower) substrates for cinchona-modified catalysts.

still relatively narrow, and it is not expected that new important substrate classes will be found quickly. On the other hand, the chemoselectivity of this system has not yet been fully exploited, and this might represent a potential for future synthetically useful applications.

12.3.2.1.1 **Catalysts** Platinum is the metal of choice for the hydrogenation of functionalized ketones, although recently Rh was also shown to be effective for aromatic ketones, with up to 80% ee [20]. Whilst different supports are suitable, Al_2O_3, SiO_2, TiO_2 and zeolites provide the best performance. Results obtained with colloids show that the support plays only an indirect role. In general, high metal dispersions seem to be detrimental for high enantioselectivities; however, Pt colloids with 1–2 nm particle sizes also give very high ee-values. Pd is preferred for the enantioselective hydrogenation of C=C bonds but, with few exceptions, the enantioselectivities are rather low.

Two commercial 5% Pt/Al_2O_3 catalysts have shown superior performance: E 4759 from Engelhard and JMC 94 from Johnson Matthey. While both catalysts have dispersions around 0.2–0.3, E 4759 has rather small pores and a low pore volume, while JMC 94 is a wide-pore catalyst with a large pore volume. Recently, E 4759 has emerged as the 'standard' catalyst for many groups working with the Pt–cinchona system. For most catalysts a reductive pretreatment in flowing hydrogen at 400 °C just before catalyst use significantly increases the enantioselectivity and reaction rate. While this is not a major problem on the small scale, it complicates process applications. CatASium F214, which does not require any pretreatment for optimal performance, was developed recently by Degussa in collaboration with Solvias [21]. The reason for the better performance of the catASium F214 catalyst was attributed to a lower reducible residue level and a narrower Pt crystal size distribution.

		R	Z		
Cinchonine	(Cn)	Vinyl	H	(Cd)	Cinchonidine
10,11-Dihydrocinchonine	(HCn)	Ethyl	H	(HCd)	10,11-Dihydrocinchonidine
Quinidine	(Qd)	Vinyl	OMe	(Qn)	Quinine
10,11-Dihydroquinidine	(HQd)	Ethyl	OMe	(HQn)	10,11-Dihydroquinine

Scheme 12.5 Structures of the parent cinchona alkaloid families.

12.3.2.1.2 Modifiers and Solvents Apart from the parent cinchona derivatives described by Orito (see Scheme 12.5), a large number of other cinchona derivatives as well as cinchona mimics have been prepared and tested for the hydrogenation of various activated ketones. From these studies the following conclusions can be drawn:

- The best results are usually obtained with cinchonidine or slightly altered derivatives such as HCd or 9-methoxy-HCd (MeOHCd) to give the *(R)*-alcohol and with cinchonine derivatives to give the *(S)*-enantiomer, albeit with somewhat lower enantioselectivity. The most effective cinchona modifiers are commercially available or can easily be prepared from these.

- Three structural elements in the cinchona molecule were identified to affect the rate and ee-value of the enantioselective hydrogenation of α-keto acid derivatives. These included: (i) an extended aromatic moiety; (ii) the substitution pattern of the quinuclidine (the absolute configuration at C_8 controls the sense of induction, N-alkylation yields the racemate); and (iii) the substituents at C_9 (OH or MeO is optimal, larger groups reduce enantioselectivity).

- The choice of solvent had a significant effect on enantioselectivity and rate. MeOHCd in acetic acid and HCd/toluene are often the most effective modifier/solvent combinations.

12.3.2.2 Industrial Applications

The first technical application of a cinchona-modified Pt catalyst was reported by Ciba-Geigy as early as 1986 for the synthesis of methyl *(R)*-2-hydroxy-4-phenyl butyrate (*(R)*-HPB ester), an intermediate for the angiotensin-converting enzyme (ACE) inhibitor benazepril (see Scheme 12.6) [22].

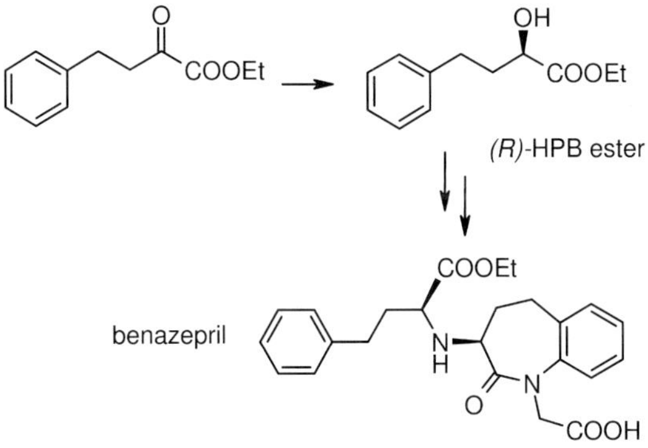

Scheme 12.6 Preparation of (R)-HPB ester via Pt-cinchona-catalyzed hydrogenation of an α-keto ester.

The development of a viable process for the HPB ester took more than a year. Even before the age of high-throughput screening, the obvious strategy was first, to screen for the best catalyst, modifier and solvent, second, to optimize relevant reaction parameters (pressure, temperature, concentrations, etc.) and, finally, to scale-up and solve relevant technical questions. Indeed, during the course of the process development more than 200 hydrogenation reactions were carried out. The most important results of this development work may be summarized as follows:

- Catalyst: 5% Pt/Al_2O_3 catalysts gave the best overall performance, E 4759 from Engelhard was the final choice.
- Modifier: About 20 modifiers were tested; HCd (in toluene) and MeOHCd (in AcOH) gave the best results and were chosen for further development.
- Solvent: Acetic acid was found to be far superior to all classical solvents, allowing up to 92% ee for the HPB ester and 95% for ethyl pyruvate (then a new world record!) [23]. For technical reasons toluene was chosen as the solvent for the production process.
- Reaction conditions: The best results (full conversion after 3–5 h, high yield, 80% ee) were obtained at 70 bar, room temperature, with 0.5% (w/w) 5% Pt/Al_2O_3 (pretreated in H_2 at 400 °C and 0.03% (w/w) modifier).
- Substrate quality: The enantioselective hydrogenation of α-keto esters proved to be exceptionally sensitive to the origin of the substrate [24].

After about two years, the production process was developed, patented and scaled-up, and in 1987 several hundred kilograms were successfully produced in a 500-liter autoclave. The progress of the optimization can best be demonstrated by the variations in ee-value versus the experiment number in the different development phases (see Figure 12.2). The effect of various measures can be seen that led to improved enantioselectivities and a stabile process. Despite this success, the

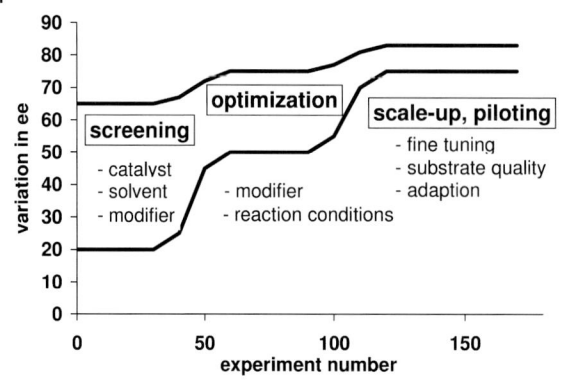

Figure 12.2 Variation in enantioselectivity during different phases of (R)-HPB ester process development.

Scheme 12.7 Second-generation synthesis of (R)-HPB ester.

pharma production section eventually decided to buy (R)-HPB ester from an external supplier.

A few years later, a new process for the (R)-HPB ester was developed by Solvias in collaboration with Ciba SC. After assessing a variety of synthetic routes, attention was focused on that depicted in Scheme 12.7. This incorporated a Claisen condensation of cheap acetophenone and diethyl oxalate, followed by chemoselective and enantioselective hydrogenation of the resulting diketo ester and hydrogenolysis to the HPB ester [25]. Although the 2,4-dioxo ester was a new substrate type, it took only a few months to develop, scale-up and implement the new process. The following aspects were the keys to the success:

- The low price of the diketo ester prepared via Claisen condensation of acetophenone and diethyl oxalate.
- The high chemoselectivity in the Pt–cinchona hydrogenation.
- The possibility of enriching the hydroxy ketone intermediate with ee-values from as low as 70% to >99% in a one-crystallization step.

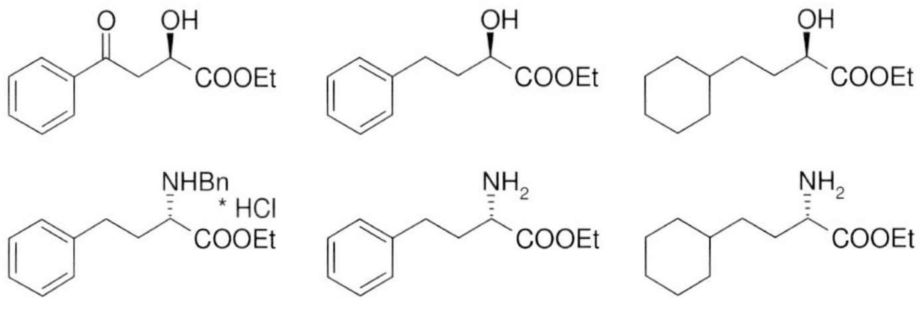

Scheme 12.8 HPB ester-related compounds available from Fluka.

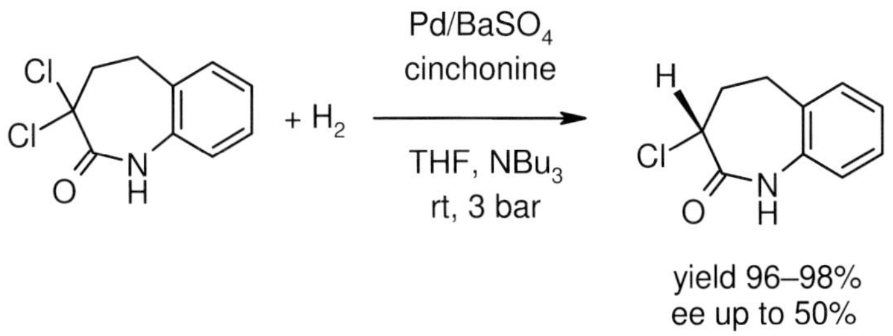

Scheme 12.9 Enantioselective dehalogenation of a dichlorobenzazepinone.

Removal of the second keto group via a Pd-catalyzed hydrogenolysis did not lead to any racemization. Derived from the keto-hydroxy intermediate, a whole range of chiral building blocks is now available in laboratory quantities from Fluka, both in the *(R)*- and *(S)*-forms (see Scheme 12.8) [26].

Attempts were also made to develop an enantioselective process for the benzazepinone building block of benazepril by selective dehalogenation of the dichloro precursor, as depicted in Scheme 12.9. However, despite an extensive screening of various modifiers, the catalyst and bases enantioselectivities never exceeded 50%, which was insufficient for a technical application even though excellent yields of the mono-dehalogenated product were achieved [27].

Two biocatalytic routes were also developed to the pilot stage by Ciba-Geigy, namely the enantioselective reduction of the corresponding α-keto acid with immobilized *Proteus vulgaris* (route **A** in Scheme 12.10) and with D-LDH in a membrane reactor (route **B**), respectively. It was therefore of interest to compare the four approaches. The EATOS (Environmental Assessment Tool for Organic Syntheses) program was used to compare the mass consumption (kg input of raw materials for 1 kg of product) as well as other parameters [28].

As shown in Figure 12.3, the new route **D** had the lowest overall mass consumption, even though the ee-value and yield for the reduction step were the lowest. This was compensated by there being fewer steps, a higher atom efficiency and a lower solvent consumption for the synthesis and extraction.

Two bench-scale processes were developed for the enantioselective hydrogenation of *p*-chlorophenylglyoxylic acid derivatives using both homogeneous and heterogeneous catalysts (Scheme 12.11) [29]. For the production of kilogram amounts of *(S)-p*-chloro mandelic acid a Ru/MeO-biphep catalyst achieved 90–93% ee

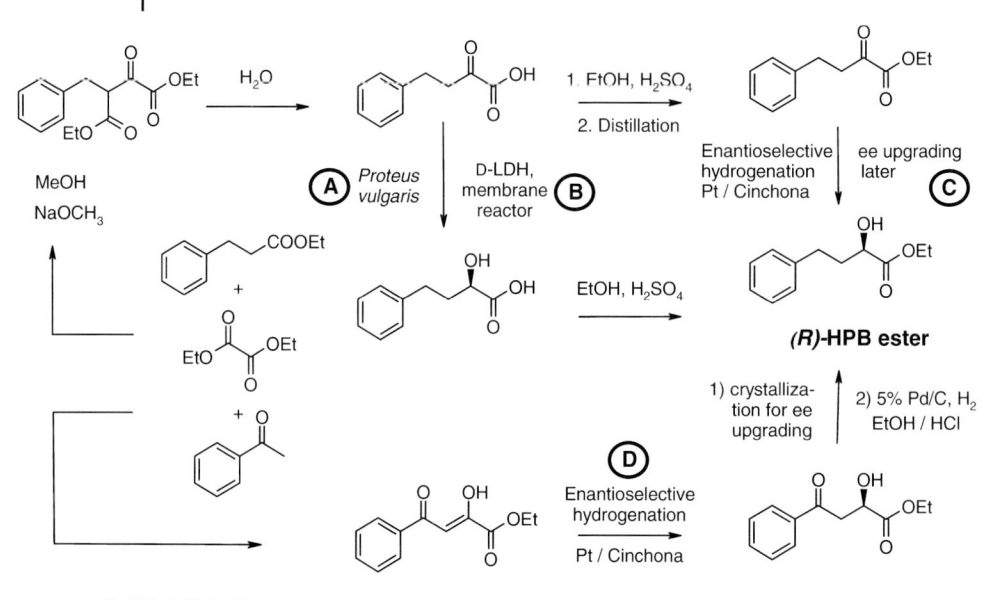

Scheme 12.10 Four routes to the (R)-HPB ester.

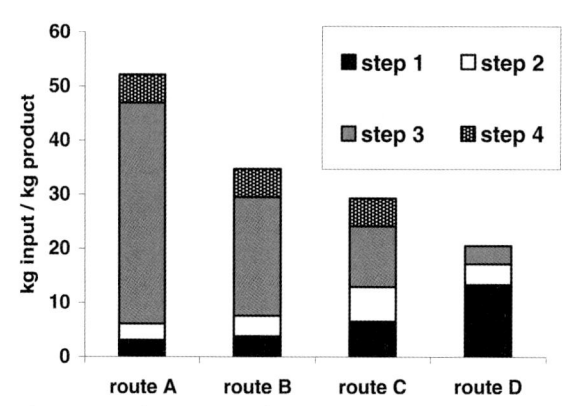

Figure 12.3 Pilot processes for (R)-HPB ester: Mass consumption (without water) for routes **A–D**.

(substrate/catalyst ratio 4000 and turnover frequency (TOF) up to $210\,h^{-1}$). A modified Pt catalyst achieved 93% ee for the *(R)*- and 86% ee for the *(S)*-methyl *p*-chloromandelate using HCd and *iso*-Cn as modifier, respectively. For the HCd-Pt system, a scale-up from 100 mg to 15 g presented no problems, indicating that the Pt-cinchona system might represent a viable alternative to the homogeneous catalyst for production of the *(R)*-enantiomer.

The hydrogenation of an α-ketolactone depicted in Scheme 12.12 is the key step for an enantioselective synthesis of pantothenic acid, which is produced by Roche

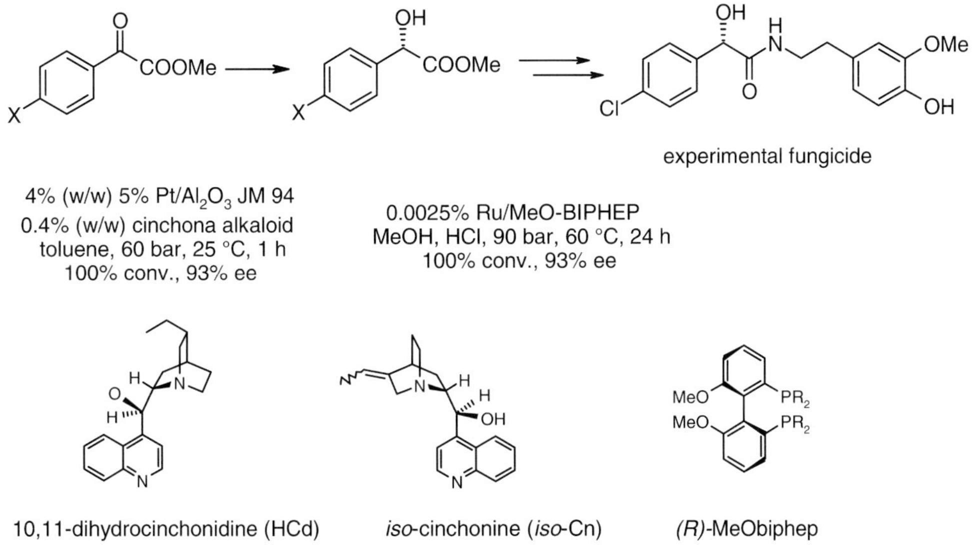

Scheme 12.11 Hydrogenation of p-chlorophenylglyoxylic acid derivatives.

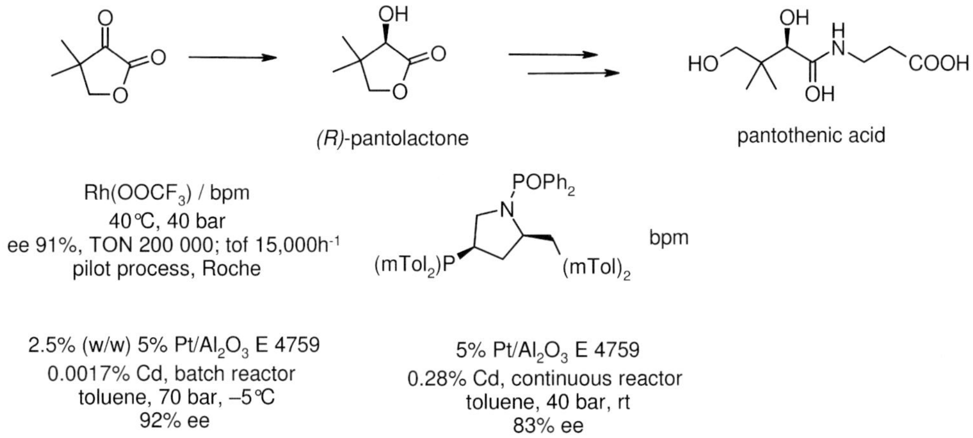

10,11-dihydrocinchonidine (HCd)

iso-cinchonine (iso-Cn)

(R)-MeObiphep

4% (w/w) 5% Pt/Al$_2$O$_3$ JM 94
0.4% (w/w) cinchona alkaloid
toluene, 60 bar, 25 °C, 1 h
100% conv., 93% ee

0.0025% Ru/MeO-BIPHEP
MeOH, HCl, 90 bar, 60 °C, 24 h
100% conv., 93% ee

(R)-pantolactone

pantothenic acid

Rh(OOCF$_3$) / bpm
40 °C, 40 bar
ee 91%, TON 200 000; tof 15,000 h^{-1}
pilot process, Roche

bpm

2.5% (w/w) 5% Pt/Al$_2$O$_3$ E 4759
0.0017% Cd, batch reactor
toluene, 70 bar, −5 °C
92% ee

5% Pt/Al$_2$O$_3$ E 4759
0.28% Cd, continuous reactor
toluene, 40 bar, rt
83% ee

Scheme 12.12 Hydrogenation of ketopantolactone.

via the racemate separation of pantoic acid. A homogeneous pilot process was developed by Roche [18b] and *(R)*-pantolactone was produced in multi-100 kg quantities. A Rh/bpm catalyst proved to be highly active with satisfactory selectivity of 91% ee. As described previously by Niwa *et al.* [30], α-keto lactones are also suitable substrates for Pt/cinchona catalysts. This was confirmed by the Baiker group, who developed a continuous bench-scale process for the hydrogenation of keto-pantolactone [31a]. The reaction was carried out in a fixed-bed reactor using the standard E 4759 catalyst at 40 bar pressure. Cinchonidine was added continuously with the feed, otherwise the ee-value fell very rapidly. The productivity was

94 mmol g catalyst^{-1} h^{-1}, and up to 83% ee was obtained. Compared to the heterogeneous batch process with ee-values of up to 92% at much lower Cd concentrations [31b], or to the homogeneous process, the continuous variant was probably not competitive. At present, pantothenic acid is still produced using the established racemate separation route.

12.4
Immobilized Chiral Metal Complexes

12.4.1
Background

Since chiral metal complexes have an extraordinarily broad scope for synthetic application [32], their immobilization was attempted at a very early stage [33]. However, it was soon observed that most immobilized complexes had a lower activity and selectivity compared to their homogeneous counterparts. Yet, during recent years this situation has changed and increasing numbers of immobilized catalysts with comparable catalytic performances have been developed (at least for model reactions). A list of immobilized catalysts, together with details of their catalytic performances (ee-values, TOF) has been complied in Table 12.2. These catalysts are, in principle, sufficient for synthetic and in some cases even industrial application. It is no accident that hydrogenation catalysts appear so prominently in the table as they represent the most versatile family of transformations for synthetic (and technical) applications. On the other hand, these data also show that almost any type of catalyst can somehow be anchored to a solid support, with good results. For the majority of research groups in this field the main motivation when immobilizing a successful homogeneous catalyst is to allow catalyst separation by filtration and to recycle the catalyst as many times as possible. Indeed, most reports provide such data, although the significance of these findings is often questionable as no comparison with homogeneous analogues has been made because the recycling was carried out with too much catalyst and/or the metal leaching was not controlled.

Despite these promising results, we do not yet know of any industrial process where an immobilized catalyst is currently applied to the commercial production of a chiral intermediate. The reasons for this are manifold, not the least being that the majority of chiral compounds are still produced via resolution. In our opinion, another important reason is that, with the exceptions described below, no immobilized catalysts have been developed from a technical point of view and/or are available commercially. As it is not possible to predict the optimal catalyst for any given transformation, the screening of as many candidates as possible is necessary. For obvious reasons this must be carried out with homogeneous catalysts which (at least for hydrogenation reactions) are available commercially in large numbers. The most promising approaches for implementing an immobilized catalyst in a process are, therefore, those which allow a rapid adaptation of the

Table 12.2 Catalytic performance of immobilized catalysts sorted according to chemical transformation.

Entry	Reaction / catalyst	Support	Method	ee[a]	TOF[b]	Reference(s)
Hydrogenation of C=C						
1	Rh–pyrphos	Silica gel	Covalent	>99	400	[34]
3	Rh–phosphinite	SO_3H on silica or resins	Adsorbed	95	>1000	[35, 36]
3	Rh–various diphosphines	Silica gel	Covalent	95	15 000	[37]
4	Rh–phosphoramidite	Aluminosilicate	Ion pair	97	>5000	[38]
5	Rh–duphos	MCM-41	Adsorbed	99	5000	[39]
6	Rh–duphos	Heteropolyacid on support	Adsorbed	96	6500	[40]
7	Ru–sulfonated binap	Controlled pore glass	SAP	96	40	[41]
Hydrogenation of C=O						
8	Ru–binap-diamine	Crosslinked polystyrene[c]	Covalent	97	~200	[42]
9	Ru–tosyldiamine[d]	Silica, MCM-41, SBA-15	Covalent	97	~100	[43]
10	Ru–binap-diamine	Polydimethylsiloxane	Entrapped	92	42	[44]
Hydrogenation of C=N						
11	Ir–josiphos	Silica gel	Covalent	79	20 000	[45]
Hydroformylation						
12	Rh–phosphine-phosphinite	Crosslinked polystyrene	Covalent	93	~200	[46]
Hydrolytic kinetic resolution of epoxides						
13	Co–salen	Crosslinked polystyrene	Covalent	99	~70	[47]
14		Silica gel	Covalent	96	~70	[47]
Dihydroxylation of alkenes						
15	Os-1,4-bis (9-*O*-quinyl)phthalazine	Silica gel	Covalent	99	6	[48a]
16		Polyacrylate	Covalent	99	6	[48b]
Epoxidation of alkenes						
17	Mn–salen	Crosslinked PS	Covalent	95	~6	[49]
18	Ti–tartrate	Silica, MCM	Covalent	86	<1	[50]
Formation of cyanohydrins with TMS-CN						
19	V–salen	Silica	Covalent	85	~8	[51]
Wacker-type cyclization						
20	Pd–oxazoline	Crosslinked PS-PEG	Covalent	96	<1	[52]
Olefin metathesis						
21	Mo–biphenol	Polystyrene	Covalent	98	40	[53]
Diels–Alder reaction						
22	Cu–oxazoline	Mesocellular silica	Covalent	>90	~10	[54]

a Best ee-value reported.
b Turnover frequency (TOF) for complete conversion.
c Polymer-bound binap commercially available from Oxford Asymmetry.
d Reducing agent $HCOOH/NEt_3$.

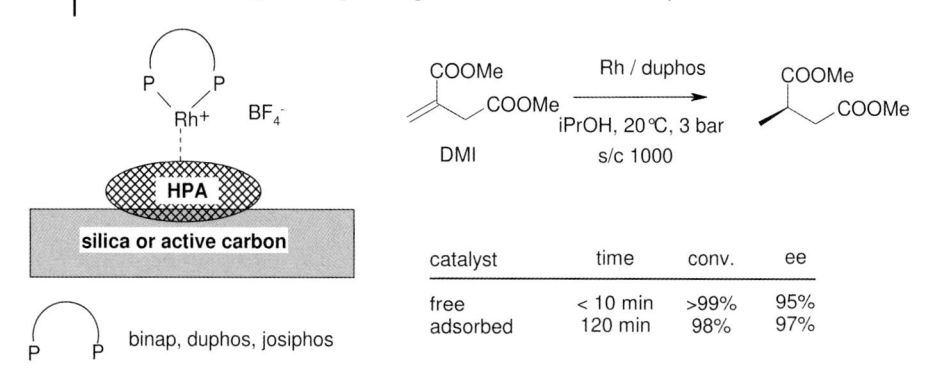

Scheme 12.13 Rh–diphosphine complexes adsorbed onto a heteropolyacid (HPA) supported on silica or active carbon.

catalyst	time	conv.	ee
free	< 10 min	>99%	95%
adsorbed	120 min	98%	97%

Reaction conditions: DMI, Rh/duphos, iPrOH, 20 °C, 3 bar, s/c 1000. Ligands: binap, duphos, josiphos.

homogeneous results. In our view this can be done best either by the adsorption strategy or by using a 'toolbox approach', as described in the following two sections.

12.4.2
Complexes Adsorbed on Solid Supports

Two somewhat different approaches have been described in the literature which are either currently being commercialized or at least have the potential for industrial application. An innovative modular method was developed by Augustine, who used supported heteropolyacids as anchoring agents to attach a large variety of metal complexes to different supports (see Scheme 12.13) [40]. Very high TOFs and good ee-values have been achieved for hydrogenation reactions with anchored Rh–diphosphine complexes of this type (for examples, see Scheme 12.13 and Table 12.2, entry 6). A Rh–Josiphos complex, when adsorbed onto a heteropolyacid-modified alumina, was shown to be active in the continuous hydrogenation of dimethyl itaconate with reasonable ee-values in supercritical CO_2 [55]. It is not quite clear how the complexes are adsorbed, since not only cationic but also neutral species can be immobilized efficiently. This methodology is currently undergoing further development, among others by the Engelhard Corporation [56] (also P. Baumeister, Chemgo, personal communication) and by Johnson Matthey [40b], and is currently being commercialized by these companies.

The second approach is to use commercially available solid supports which are able to strongly adsorb selected metal–phosphine complexes. At present, MCM-41 [39] (Scheme 12.14) and active carbon [57] have each been described as giving stable catalysts with catalytic performances for industrially relevant ligands such as duphos (see Table 12.2, entry 5), josiphos, bophoz or phanephos for the hydrogenation of dimethyl itaconate with performances similar to those of the homogeneous analogues. Leaching rates were shown to be in the ppm range for several of these adsorbed catalysts. Furthermore, it was demonstrated that recycling is possible without any significant loss in enantioselectivity.

Scheme 12.14 Rh–duphos complex adsorbed onto the inner surface of MCM-41.

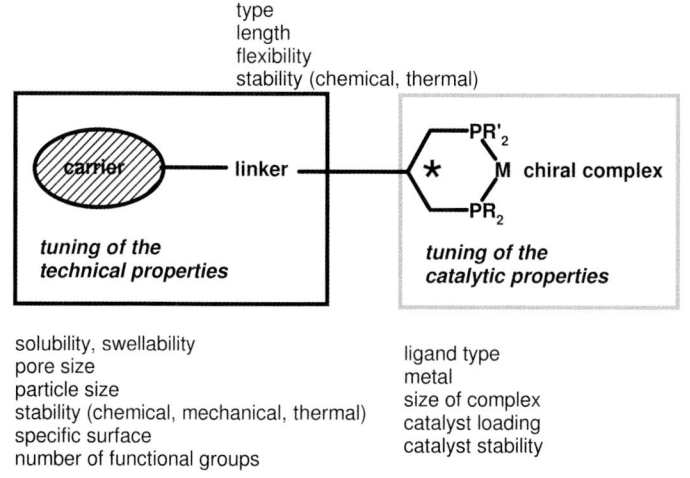

Figure 12.4 Elements and parameters of the modular toolbox.

12.4.3
Toolbox for Covalent Immobilization

Inspired by the seminal report of Nagel [34], which described a very active and selective Rh–pyrphos catalyst attached covalently to silica gel, Pugin and colleagues have developed the modular toolbox which his depicted schematically in Figure 12.4. The main elements of their system are functionalized chiral diphosphines, where three different linkers are based on isocyanate chemistry and various carriers [37, 45, 58]. This approach allows for a systematic and rapid access to a variety of immobilized chiral catalysts, with the possibility of adapting their catalytic and technical properties to specific needs.

Here, we will illustrate the concept with the josiphos-based toolbox, which has been applied to several project studies involving industrially relevant target molecules. Josiphos was selected as the modular ligands can be tuned by choice of the correct R and R′ groups, which renders this ligand class exceptionally versatile; furthermore, several technical applications have already been established [59]. The basis of the toolbox is that the Josiphos ligands are selectively functionalized at

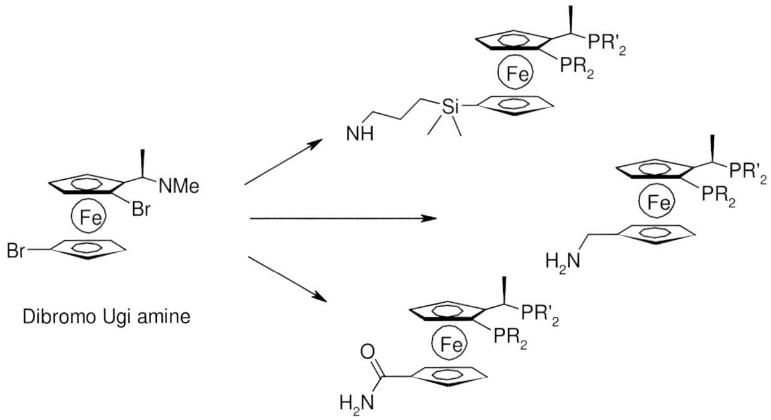

Scheme 12.15 Functionalized Josiphos ligands.

Table 12.3 Hydrogenation of MEA imine using Ir–xyliphos catalysts.

Ligand	s/c ratio	TOF (h^{-1})	ee (%)	Comment
Free Josiphos	120 000	55 385	80	
Water-soluble Josiphos	120 000	36 000	79	
Josiphos-SiO$_2$	120 000	12 000	78	
Josiphos-SiO$_2$	250 000	9750	75	50 °C, 78% conv.
Josiphos-polystyrene	50 000	1140	73	95% conv.

Reaction conditions: 19.5 g MEA-imine, 2 ml acetic acid, 10 mg tetrabutylammonium iodide, 80 bar, 25–30 °C, 100% conversion unless otherwise noted.

the unsubstituted chiral diphosphine ring (Scheme 12.15). As the synthesis of all derivatives begins with the dibromo Ugi amine and involves straightforward, established chemistry, any combination of R, R′ and functional group can be prepared within a short period of time.

As shown in Scheme 12.16, Josiphos ligands have been covalently attached to dendrimers [60], inorganic supports or organic polymers [45], as well as to vehicles which renders them soluble in ionic liquids [58] or water [61].

The functionalized ligands were tested for various hydrogenation reactions (see Scheme 12.17). Ir–Josiphos bound to silica gel as well as to a water-soluble complex produced TONs in excess of 100 000 and TOFs up to 20 000 h^{-1} for the Ir-catalyzed hydrogenation of 2-methyl-6-ethyl aniline (MEA) imine to give an intermediate for (S)-metolachlor [45a]. The polymer-bound Ir complex was much less active, and in all cases the ee-values were comparable to those for the homogeneous catalyst. Selected results are summarized in Table 12.3. However, no immobilized system could compete with the homogeneous catalyst which is used to produce >10 000 tons y^{-1} of enantioenriched (S)-metolachlor and which, under optimized condi-

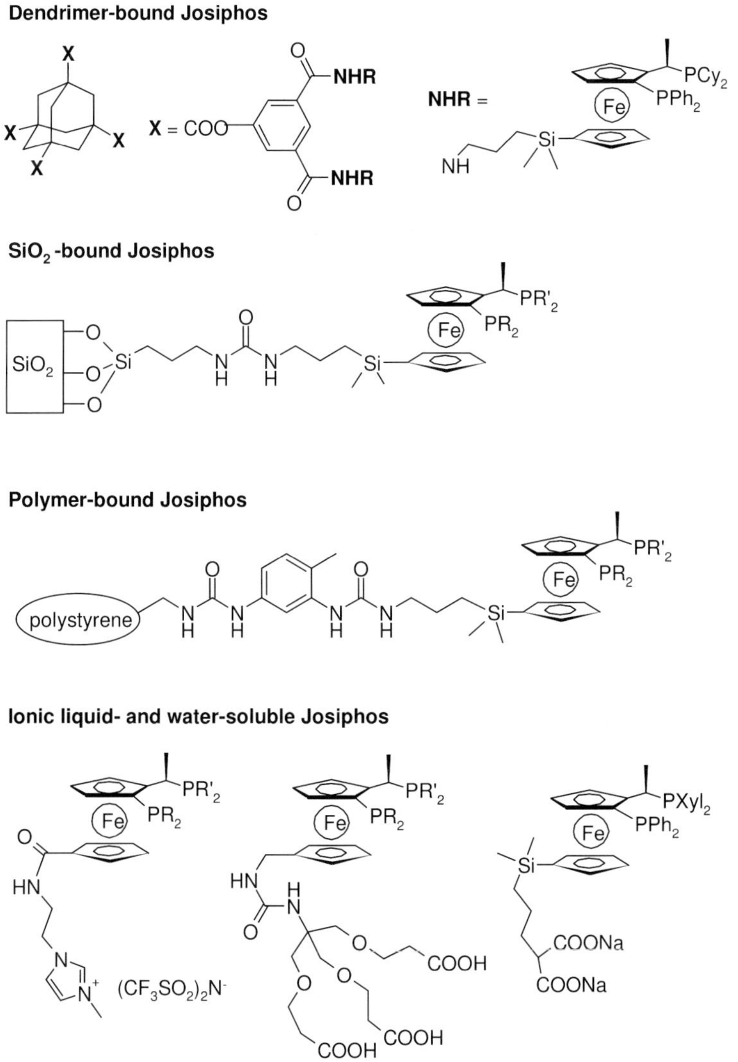

12.4 Immobilized Chiral Metal Complexes

Dendrimer-bound Josiphos

SiO$_2$-bound Josiphos

Polymer-bound Josiphos

Ionic liquid- and water-soluble Josiphos

Scheme 12.16 Josiphos ligands attached to various supports and vehicles.

tions, achieves 2 000 000 turnovers within 3 h [45b]. The dendrimer-supported Rh–Josiphos complexes hydrogenated DMIT with ee-values up to 98.6% and with similar activities as the mononuclear catalyst [60]. Similarly, ligands with an imidazolium tag had a comparable catalytic performance to the nonfunctionalized ligands for the Rh-catalyzed hydrogenation of methyl acetamino acrylate (MAA) and dimethyl itaconate (DMI), both in classical solvents and under two-phase conditions in ionic liquids. However, both catalysts were easy to separate and to recycle [58].

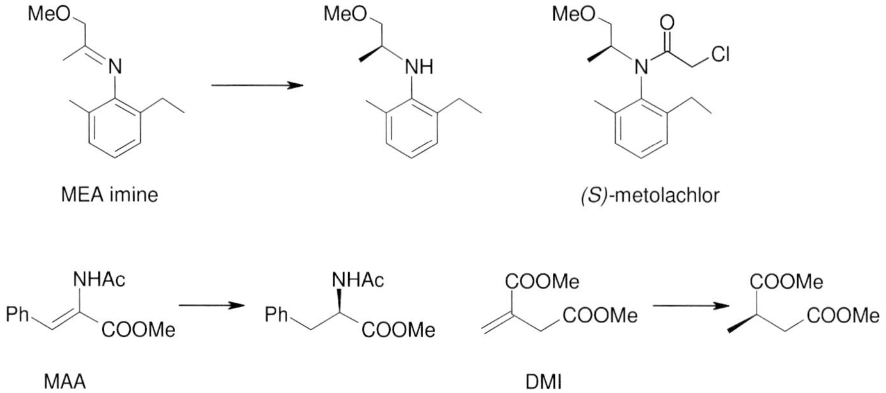

Scheme 12.17 Hydrogenation of MEA imine, MAA and DMI.

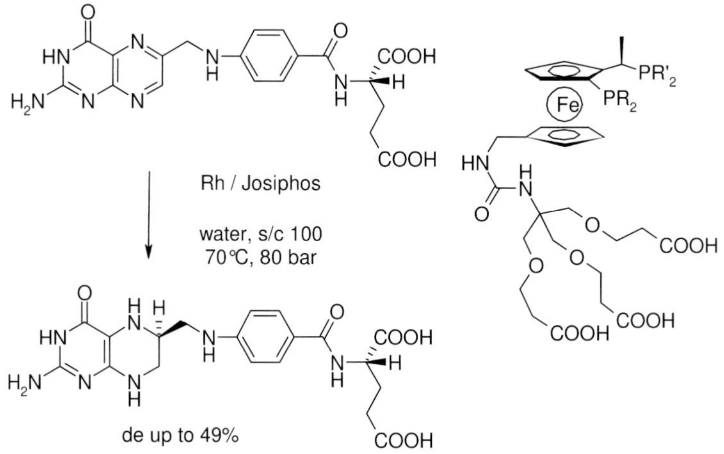

Scheme 12.18 Hydrogenation of folic acid.

The stereoselective hydrogenation of folic acid to tetrahydrofolic acid (Scheme 12.18) might represent an attractive alternative to the present unselective process which uses heterogeneous catalysts (diastereomeric excess (de) = 0%). One major problem here is the insolubility of folic acid in most organic solvents. Initial results with up to 42% de but very low TONs and TOFs were obtained by Brunner et al. using a Rh/BPPM complex adsorbed onto silica gel in aqueous buffer solutions [62]. Unfortunately, however, the originally claimed 90% de [62a] turned out to be an analytical artifact [62c].

Functionalized catalysts offer the opportunity to perform this reaction in water, and it was subsequently shown that a Rh complex of the water-soluble josiphos could achieve diastereoselectivities of up to 49% [61]. Although substrate/catalyst ratios of up to 1000 were possible, the de-values and TONs were insufficient for technical applications.

12.5
Conclusions and Outlook

Although heterogeneous enantioselective catalysts undoubtedly have operational advantages over their homogeneous counterparts with regards to handling and separation, very few of them can compete with alternative synthetic methods, either homogeneous catalysis or, even more frequently, with enantiomer separation or chiral pool approaches. Until now, only cinchona-modified Pt catalysts have been used on a ton scale for the manufacture of chiral intermediates, although several feasibility studies have shown that other types of activated ketone can be hydrogenated competitively with this catalyst. Ni catalysts modified with tartaric acid are probably not sufficiently active to compete with Noyori's catalyst for the hydrogenation of β-keto esters and related ketones.

Whilst immobilized metal complexes have not yet been applied to industrially relevant targets on a larger scale, it might be considered that several types of such heterogenized catalysts are, in principle, feasible for technical applications in the pharmaceutical and agrochemical industries. However, convincing evidence is also available that the application of anchored chiral catalysts will remain a niche application for use only in rare cases where competing technologies fail.

It is the view of the present authors that the most important problem here is the high complexity of many heterogeneous systems, because this leads to poor predictability of the catalytic properties. Additional problems with immobilized chiral catalysts include not only their lack of commercially availability but also the very high costs involved in their preparation.

References

1 Blaser, H.U. and Schmidt, E. (eds) (2003) *Large Scale Asymmetric Catalysis*, Wiley-VCH Verlag GmbH, Weinheim.

2 Ager, D.J. (ed.) (2005) *Handbook of Chiral Chemicals*, 2nd edn, CRC Press, Boca Raton.

3 For e recent overview, see Pantaleone, D.P. (2005) *Handbook of Chiral Chemicals*, 2nd edn (ed. D.J. Ager), CRC Press, Boca Raton, p. 359.

4 Jacobsen, E.N., Hayashi, T. and Pfaltz, A. (eds) (1999) *Comprehensive Asymmetric Catalysis*, Springer, Berlin.

5 Blaser, H.U. and Schmidt, E. (eds) (2003) *Large-Scale Asymmetric Catalysis*, Wiley-VCH Verlag GmbH, Weinheim, p. 1.

6 Federsel, H.-J. (2005) *Nature Reviews Drug Discovery*, 685.

7 For historical overviews, see (a) Blaser, H.U. (1991) *Tetrahedron: Asymmetry*, **2**, 843.
(b) Tai, A. and Harada, T. (1986) *Tailored Metal Catalysts* (ed. Y. Iwasawa), D. Reidel, Dordrecht, p. 265.

8 Carrea, G., Colonna, S., Kelly, D.R., Lazcano, A., Ottolina, G. and Roberts, S.M. (2005) *Trends in Biotechnology*, **23**, 507.

9 McMorn, P. and Hutchings, G.J. (2004) *Chemical Society Reviews*, **33**, 108.

10 Dai, L.X. (2004) *Angewandte Chemie – International Edition*, **43**, 5726.

11 Lin, W. (2005) *Journal of Solid State Chemistry*, **178**, 2486.

12 (a) Tai, A. and Sugimura, T. (2000) *Chiral Catalyst Immobilization and Recycling* (eds D.E. De Vos, I.F.J. Vankelecom and P.A. Jacobs), Wiley-VCH Verlag GmbH, Weinheim, p. 173.

(b) Osawa, T., Harada, T. and Takayasu, O. (2006) *Current Organic Chemistry*, **10**, 1513.

13 (a) Tai, A., Imaida, M., Oda, T. and Watanabe, H. (1978) *Chemistry Letters*, 61.
(b) Tai, A., Morimoto, N., Yoshikawa, M., Uehara, K., Sugimure, T. and Kikukawa, T. (1990) *Agricultural and Biological Chemistry*, **54**, 1753.

14 Nakahata, M., Imaida, M., Ozaki, H., Harada, T. and Tai, A. (1982) *Bulletin of the Chemical Society of Japan*, **55**, 2186.

15 Bakos, J., Toth, I. and Marko, L. (1981) *Journal of Organic Chemistry*, **46**, 5427.

16 Tai, A. and Harada, T. (1986) *Tailored Metal Catalysts* (ed. Y. Iwasawa), D. Reidel, Dordrecht, p. 265.

17 *Catalogue of Wako Pure Chemicals Industries*, Osaka, 22nd edn, pp. 471 and 547 (cited in Ref. [16], p. 269).

18 (a) Schmid, R. and Broger, E.A. (1994) Proceedings of the ChiralEurope '94 Symposium, Spring Innovation, Stockport, UK, p. 79.
(b) Schmid, R. and Scalone, M. (1999) *Comprehensive Asymmetric Catalysis* (eds E.N. Jacobsen, H. Yamamoto and A. Pfaltz), Springer, Berlin, p. 1439.

19 (a) Studer, M., Blaser, H.U. and Exner, C. (2003) *Advanced Synthesis Catalysis*, **345**, 45.
(b) Murzin, D., Maki-Arvela, P., Toukoniitty, E. and Salmi, T. (2005) *Catalysis Reviews – Science and Engineering*, **47**, 175.
(c) Baiker, A. (2005) *Catalysis Today*, **100**, 159.
(d) Bartók, M. (2006) *Current Organic Chemistry*, **10**, 1533.

20 Sonderegger, O.J., Ho, G.M.-W., Bürgi, T. and Baiker, A. (2005) *Journal of Catalysis*, **230**, 499.

21 Ostgard, D.J., Hartung, R., Krauter, J.G.E., Seebald, S., Kukula, P., Nettekoven, U., Studer, M. and Blaser, H.U. (2005) *Chemical Industries (Catalysis of Organic Reactions)*, **104**, 553.

22 Sedelmeier, G.H., Blaser, H.U. and Jalett, H.P. (1986) EP 206'993, assigned to Ciba-Geigy AG.

23 Blaser, H.U., Jalett, H.P. and Wiehl, J. (1991) *Journal of Molecular Catalysis*, **68**, 215.

24 Blaser, H.U., Jalett, H.P. and Spindler, F. (1996) *Journal of Molecular Catalysis A – Chemical*, **107**, 85.

25 (a) Studer, M., Burkhardt, S., Indolese, A.F. and Blaser, H.U. (2000) *Chemical Communications*, 1327.
(b) Herold, P., Indolese, A.F., Studer, M., Jalett, H.P. and Blaser, H.U. (2000) *Tetrahedron*, **56**, 6497.

26 Blaser, H.U., Burkhardt, S., Kirner, H.J., Mössner, T. and Studer, M. (2003) *Synthesis*, 1679.

27 Blaser, H.U., Boyer, S.K. and Pittelkow, U. (1991) *Tetrahedron: Asymmetry*, **2**, 721.

28 Blaser, H.U., Eissen, M., Fauquex, P.F., Hungerbühler, K., Schmidt, E., Sedelmeier, G. and Studer, M. (2003) *Large-Scale Asymmetric Catalysis* (eds H.U. Blaser and E. Schmidt), Wiley-VCH Verlag GmbH, Weinheim, p. 91.

29 Cederbaum, F., Lamberth, C., Malan, C., Naud, F., Spindler, F., Studer, M. and Blaser, H.U. (2004) *Advanced Synthesis Catalysis*, **346**, 842.

30 Niwa, S., Imamura, J. and Otsuka, K. (1987) JP 62158268, (*CAN 108*, 128815).

31 (a) Künzle, N., Hess, R., Mallat, T. and Baiker, A. (1999) *Journal of Catalysis*, **186**, 239.
(b) Schürch, M., Künzle, N., Mallat, T. and Baiker, A. (1998) *Journal of Catalysis*, **176**, 569.

32 Blaser, H.U., Pugin, B. and Spindler, F. (2005) *Journal of Molecular Catalysis A – Chemical*, **231**, 1.

33 One of the earliest examples: Dumont, W., Poulin, J.C., Dang, P.T. and Kagan, H.B. (1973) *Journal of the American Chemical Society*, **95**, 8295.

34 (a) Nagel, U. and Kinzel, E. (1986) *Journal of the Chemical Society D – Chemical Communications*, 1098.
(b) Nagel, U. and Leipold, J. (1996) *Chemische Berichte*, **129**, 815.

35 Selke, R., Häupke, K. and Krause, H.W. (1989) *Journal of Molecular Catalysis*, **56**, 315.

36 Canali, L. and Sherrington, D.C. (1999) *Chemical Society Reviews*, **28**, 85.

37 (a) Pugin, B. (1996) *Journal of Molecular Catalysis*, **107**, 273.
(b) Pugin, B. (1996) EP 728,768, assigned to Ciba-Geigy AG, 1996.

38 Simons, C., Hanefeld, U., Minnaard, I.W.C.E., Arends, A.J., Maschmeyer, T. and Sheldon, R.A. (2004) *Chemical Communications*, 2830.

39 Hems, W.P., McMorn, P., Riddel, S., Watson, S., Hancock, F.E. and Hutchings, G.J. (2005) *Organic and Biomolecular Chemistry*, 1547.

40 (a) Augustine, R., Tanielyan, S., Anderson, S. and Yang, H. (1999) *Chemical Communications*, 1257.
(b) Augustine, R. (2004) *Speciality Chemicals Magazine*, **24** (4), p. 19 and references cited therein.

41 Wan, K.T. and Davis, M.E. (1995) *Journal of Catalysis*, **152**, 25.

42 Ohkuma, T., Takeno, H., Honda, Y. and Noyori, R. (2001) *Advanced Synthesis Catalysis*, **343**, 369.

43 Liu, P.-N., Gu, P.-M., Deng, J.-G., Tu, Y.-Q. and Ma, Y.-P. (2005) *European Journal of Organic Chemistry*, 3221.

44 (a) Vankelecom, I.F.J., Tas, D., Parton, R.F., Van de Vyver, V. and Jacobs, P.A. (1996) *Angewandte Chemie – International Edition*, **35**, p. 1346.
(b) Janssen, K.B.M., Laquire, I., Dehaen, W., Parton, R.F., Vankelekom, I.F.J. and Jacobs, P.A. (1997) *Tetrahedron: Asymmetry*, **8**, 3481.
(c) Tas, D., Thoelen, C., Vankelecom, I. and Jacobs, P.A. (1997) *Chemical Communications*, 2323.

45 (a) Pugin, B., Landert, H., Spindler, F. and Blaser, H.U. (2002) *Advanced Synthesis Catalysis*, **344**, 974.
(b) Blaser, H.U. (2002) *Advanced Synthesis Catalysis*, **344**, 17.

46 Nozaki, K., Itoi, Y. and Shibahara, F. (1998) *Journal of the American Chemical Society*, **120**, 4051.

47 Annis, D.A. and Jacobsen, E.N. (1999) *Journal of the American Chemical Society*, **121**, 4147.

48 (a) Song, C.E., Yang, J.W. and Ha, H.J. (1997) *Tetrahedron, Asymmetry*, **8**, 841.
(b) Song, C.E., Yang, J.W., Ha, H.J. and Lee, S. (1996) *Tetrahedron, Asymmetry*, **7**, 645.

49 Song, C.E., Roh, E.J., Yu, B.M., Chi, D.Y., Kim, S.C. and Lee, K-J. (2000) *Chemical Communications*, 615.

50 Xiang, S., Zhang, Y., Xin, Q. and Li, C. (2002) *Angewandte Chemie – International Edition*, **41**, 821.

51 Baleizão, C., Gigante, B., Garcia, H. and Corma, A. (2003) *Journal of Catalysis* **215**, 199.

52 Hocke, H. and Uozumi, Y. (2002) *Synlett*, 2049.

53 (a) Hultsch, K.C., Jernelius, J.A., Hovyeda, A.H. and Schrock, R.R. (2002) *Angewandte Chemie – International Edition*, **41**, 589.
(b) Dolman, S.J., Hultzsch, K.C., Pezet, F., Teng, X., Hovyeda, A.H. and Schrock, R.R. (2004) *Journal of the American Chemical Society*, **126**, 10945.

54 Lancaster, T.M., Lee, S.S. and Ying, J.Y. (2005) *Chemical Communications*, 3577.

55 Stephenson, P., Kondor, B., Licence, P., Scovell, K., Ross, S.K. and Poliakoff, M. (2006) *Advanced Synthesis Catalysis*, **348**, 1605.

56 Brandts, J.A.M. and Berben, P.H. (2003) *Organic Process Research and Development*, **7**, 393.

57 Barnard, C.F.J., Rouzaud, J. and Stevenson, S.H. (2005) *Organic Process Research and Development*, **9**, 164.

58 Feng, X., Pugin, B., Küsters, E., Sedelmeier, G. and Blaser, H.U. (2007) *Advanced Synthesis Catalysis*, **349**, 1803.

59 Blaser, H.U., Brieden, W., Pugin, B., Spindler, F., Studer, M. and Togni, A. (2002) *Topics in Catalysis*, **19**, 3.

60 Köllner, C., Pugin, B. and Togni, A. (1998) *Journal of the American Chemical Society*, **120**, 10274.

61 Pugin, B., Groehn, V., Moser, R. and Blaser, H.U. (2006) *Tetrahedron: Asymmetry*, **17**, 544.

62 (a) Brunner, H. and Huber, C. (1992) *Chemische Berichte*, 1252085.
(b) Brunner, H., Bublak, P. and Helget, M. (1997) *Chemische Berichte Recueil*, **130**, 55.
(c) Brunner, H. and Rosenboem, S. (2000) *Monatshefte für Chemie*, **131**, 1371.

Index

a

acetophenone 102–103, 190, 224, 243–245, 343
addition, *see also* cycloaddition; Michael addition
– conjugate 56–57, 61–63
– dialkylzincs to aldehydes 55, 58–59, 162–163, 262, 336–337, 346–349
– to imines 81, 162–163
– in ionic liquids 258–259
– organometallic compounds to aldehydes 149–151, 193–197
– phenylacetylene 80–81
additive effect 372
adsorption 4, 25, 357–358, 430–431
– chirally modified 17–18
agrochemicals 414
AITUD 227, 229
ALB complex 88–89, 166–167, 335–336
aldehydes
– addition of organometallic compounds 149–151, 193–197
– dialkylzincs addition 55, 58–59, 162–163, 262, 336–337, 346–349
– α-hydroxyaldehydes 271
– silylcyanation 81–82, 255–256, 257
aldimines 392–393, 407
aldol reactions 57, 63–64, 84–87
– core-functionalized dendrimers 153–155
– in ionic liquids 265–269
– in phase-transfer catalysis 399–402
– peripherally modified dendrimers 168–169
aliphatic α,β-unsubstituted carboxylic acids 374–376
alkenes
– epoxidation 248–250, 317–318
– hydroformylation 279–281
α-alkyl-α-amino acids 391–394

alkylation
– allylic 111–115, 163–166, 214–217
– α-amino acids synthesis 384–396
– carbonyl compounds 74–79
– Friedel–Crafts hydroxyalkylation 57, 65
– in phase-transfer catalysis 384–398
allylic addition 258–259
allylic alcohols 43, 49–51
allylic alkylation 111–115, 163–166, 214–217
allylic amination 163–166, 217–218, 259
allylic esters 216–218, 220
allylic etherification 218–220
allylic nitromethylation 115–116, 221–222
allylic oxidation 187–188
allylic substitution 54–55, 58, 152–153, 256, 258
amination, allylic 163–166, 217–218, 259
amino acids
– α-amino acids 82–84, 384–396, 408
– hydrogenation of dehydro-α-amino acids 340–342
– organocatalysts derived from 307–317
amino alcohols 74–75, 77, 101, 150, 194–195, 252
aminohydroxylation
– α-aminoxylation 270–274
– olefins 44, 52–53
amphiphilic
– additives 209
– dendrimers 160–161
– PS-PEG 210–212
anti-inflammatory agents 279
aquacatalytic process
– combinatorial approach 210–214
– imidazoindole phosphine 214–222
aqueous media 86, 209–210
– chiral-switching of aquacatalytic process 210–222

– heterogeneous- and aqueous-switching of 222–229
aromatic α,β-unsubstituted carboxylic acids   369–374
aryl alkyl sulfides   339–340
availability, catalysts   417
aza-Henry reaction   402, 404
aziridination, olefins   44, 53–54

b

batch equipment, multipurpose   414
benazepril   422–426
benzaldehyde   63–64, 74–77, 194–197, 257, 269
benzylamine   372
benzylation   299, 302, 389, 391–392
BICOL-type dendrimers   141
BINAPHOS   123, 279–281
BINAPs   107–108
– in aqueous media   222–224
– in core-functionalized dendrimers 135–138, 142–144
– in ionic liquids   239–241
– in MOCPs   343–346
– in supercritical carbon dioxide   276–278
BINOLs
– in core-functionalized dendrimers 149–150, 171–173
– fluorous   195–196
– functionalized dendrimers   166–168
– lanthanum catalysts   337–338
– titanium ligands   333–337, 339
biocatalysts   2, 413
BIPHEP-cored dendrimers   140
bisoxazolines   259–561
bisphosphine ligands   236
borane   101–104, 147–149, 160–161
boron   81–82

c

carbon dioxide, *see* supercritical carbon dioxide
carbon–carbon bonds   151–152
– formation using fluorous catalysis 193–195
– formation using inorganic supports 54–66
– formation using ionic liquids   250–264
– formation using MOCPs   333–337, 346–349
– formation using supercritical carbon dioxide   282–283
– hydrogenation using inorganic supports   28–29

carbon–carbon double bonds
– hydrogenation   110–111, 235–239
– reduction   191–193
carbon–heteroatom bond formation
– using inorganic supports   54–66
– using ionic liquids   250–264
carbon–nitrogen double bonds
– hydrogenation   29, 111
– reduction   191–193
carbon–oxygen bonds, fluorous catalysis 184–193
carbon–oxygen double bonds
– hydrogenation   241–244
– reduction   29–30
carbonyl compounds, *see also* aldehydes; ketones
– alkylation   74–79
– Michael addition   400
carbonyl-ene reaction   87, 333–335
Carbosil   58
carbosilanes   169–170
carboxylic acids
– aliphatic α,β-unsubstituted   374–376
– aromatic α,β-unsubstituted   369–374
catalyst heterogenization   2
catalyst immobilization, *see* immobilization
chalcones   95–96, 316–317, 337–338
– in phase-transfer catalysis   399, 401, 405
chinchona alkaloids   82–83
– PTC epoxidation   95–96
chiral modifiers, *see* modifiers
chromium–salen   252–254
cinchona alkaloids   16, 82, 95, 302–305
– in phase-transfer catalysis   299–300, 384–387
– industrial applications   420–428
– palladium-modified   368–377
– platinum-modified   359–363
cinchonidine   359–361, 364, 422
cinchonine   359–363
Claisen–Schmidt condensation   51, 424
'click' chemistry   175, 310, 352
cobalt–salen complexes   101, 161–162, 186–187, 254–255
combinatorial approach   297
– aquacatalysis   210–214
confinement effects   26, 32–33, 53, 66
conjugate addition   56–57, 61–63
continuous-flow conditions   304, 319
continuous-flow membrane reactor   176
convergent synthesis   131–132, 176
copper   53–54, 200–202, 259–562
core-functionalized dendrimers   9–10, 133–135

- addition of organometallic compounds 149–151
- aldol reaction 153–155
- allylic substitution 152–153
- borane reduction of ketones 147–149
- hetero-Diels–Alder reaction 155–156
- hydrogenation 135–145
- Michael addition 151–152
- solid-supported 170–173
- transfer hydrogenation 145–147
costs, *see* economics
covalent immobilization 4–5, 25
- toolbox 431–434
- using inorganic supports 31–34, 45
crown ethers 398–399, 408
cyanation
- imines 406–408
- silylcyanation 81–82, 255–257, 326
cyanohydrins 255, 407
cyanosilylation 81–82, 255–257, 326
cycloaddition
- Diels–Alder cycloadditions 92–94
- 1,3-dipolar cycloaddition 94
cyclopropanation 57, 64–65
- in ionic liquids 259–262
- styrene 200–202, 260–261
- using organic polymer supports 116–118

d

DACH dendrimer 145
dehydro-α-amino acids 340–342
'dendrimer effect' 132
dendrimer supports 8–10, 131–134, 175–177, *see also* core-functionalized dendrimers; peripherally modified dendrimers; solid-supported dendrimers
dendritic structures 9, 132
dendrizyme 132, 135
dendronized polymers 142–143, 176
deprotonation 91–92
designing
- fluorous catalysts 182–184
- imidazoindole phosphine 214
- MOCPs 330–333
development time 414
(DHQ)PHAL 246
dialkylation, Schiff bases 391–394
dialkylzincs
- addition to aldehydes 55, 58–59, 162–163, 262, 336–337, 346–349
- addition to imines 162–163
diamines 107–109, 146
diarylmethanols 79, 150
diastereoselectivities 265, 267–268, 270–271

dichloromethane 251
Diels–Alder reactions 56, 59–61
- core-functionalized dendrimers 155–156
- cycloadditions 92–94
- using fluorous catalysis 203–204
- using ionic liquids 250–252, 270–274
- using MOCPs 326–327
- peripherally modified dendrimers 167–168
- using supercritical carbon dioxide 282
diethylzinc, *see* dialkylzincs
dihydronaphthalene 185–186
dihydroxylation
- in ionic liquids 245–248
- olefins 43–44, 52–53, 94–95, 245–246
- using ionic liquids and supercritical CO_2 285
DIOP, rhodium–DIOP 235–236
dioxiranes 317
diphosphines 135, 137–139, 157–158, 238
1,3-dipolar cycloaddition 94
divergent synthesis 131–132, 176
p-divinylbenzene (DVB) 7, 77
DPENDS 242–243
DPENs 106, 159–160
- cored dendrimers 137–138, 145–146
- functionalized dendrimers 159–160
- in aqueous media 224–227
- in ionic liquids 242

e

E 4759 421
EATOS program 425
economics
- catalysts 416–417
- industrial applications 1, 293, 415
electrostatic interactions 4–6
enamines 110, 268
enantioselectivities 26
- immobilised catalysts 429
- in fluorous catalysis 194
- in ionic liquids 236–237, 239–243, 249, 267–268, 270–271
- in production 423–424
ene reactions 56, 61, 87, 333–335
enones, epoxidation 43, 51–52, 406
entrapment 5–6, 25
environmental factors 293
- 'green' chemistry 12, 14, 73, 233, 294, 297–298
- nonconventional media 12, 14–15
enzymes 2, 391
ephedrine derivatives 163–164

epoxidation
- alkenes   248–250, 317–318
- allylic alcohols   43, 49–51
- amino acid derivatives   314, 316–317
- enones   43, 51–52, 406
- fluorous catalysis   184–186
- in ionic liquids   248–250
- in MOCPs   337–339, 349–350
- Sharpless   49, 99–100
- using organic polymer supports   95–100
- using phase-transfer catalysis   95–96, 404–406
- unfunctionalized olefins   40–49
- α,β-unsaturated ketones   337–339, 404–406
epoxides
- hydrolytic kinetic resolution   100–101, 186–188, 254–255
- ring opening   161–162, 173, 252–254
esters
- allylic esters   216–218, 220
- HPB ester   422–426
- β-ketoesters   241, 345–346, 363–368, 396, 402
etherification, allylic   218–220

f

ferrocenyl phosphines   166
fluorination   262–264
fluoroalkylated BINAPs   276–278
fluorophilicity   183–184
fluorous biphase system   181–182
fluorous catalysis   13–14, 181–182, 205–206
- allylic oxidation   187–188
- carbon–carbon bond formation   193–195
- carbon–oxygen bond formation   184–193
- designing fluorous catalysts   182–184
- Diels–Alder reaction   203–204
- 'flourous ponytails'   182–184
formylation, see hydroformylation
Fréchet-type dendrons   145, 147
Friedel–Craft hydroxyalkylation   57, 65
functionalized dendrimers   131, 133, see also core-functionalized dendrimers; peripherally modified dendrimers; solid-supported dendrimers

g

'generation' of growth   131
glycinate Schiff base   384, 398, 402, 408
glycines   203–204, 384–391

glyoxylate-ene reactions   333–335
'green' chemistry   12, 14, 73, 233, 294, 297–298

h

'heavy fluorous'   184, 187, 205
hetero-Diels–Alder reaction   155–156
heterogeneous catalysts   2
- compared to homogeneous   25, 415–416
- industrial applications   18–19, 415–418
homochiral MOCPs, see MOCPs
homogeneous catalysis   235
- dendrimer supports   175
- drawbacks   25, 323
- economics   1
homogeneous catalysts
- compared to heterogeneous   25, 415–416
- costs and availability   416–417
- immobilization   2–3
- dendrimer-supported   8–10
- on inorganic materials   4–6
- self-supported   10–11
- using nonconventional media   12–16
- using organic polymers   7–8
- properties and handling   417
HPB ester   422–426
hybrid dendrimers   159–160
hybrid metal oxides   328
hydrindane   219–221
hydroformylation   123, 181, 279–281
hydrogen   358
hydrogen bonding, immobilization   34–39
hydrogen concentration and pressure   236–237
hydrogenation, see also reduction; transfer hydrogenation
- carbon–carbon bond   28–29
- carbon–carbon double bond   110–111, 235–239
- carbon–nitrogen double bond   29, 111
- carbon–oxygen double bond   241–244
- chirally modified metal surface see metal surface catalysts
- core-functionalized dendrimers   135
- in supercritical carbon dioxide   274–279
- industrial applications   418–428
- ketones   107–110, 342–343, 360–361, 420–428
- peripherally modified dendrimers   157–159
- using ionic liquids and supercritical CO_2 283–284
- using MOCPs   340–346

hydrogen-transfer reduction, *see* transfer hydrogenation
hydrolysis 121–123
hydrolytic kinetic resolution 100–101, 186–188, 254–255
hydrophilic modification 209
hydrovinylation 169–170
α-hydroxyaldehydes 271
hydroxyalkylation, Friedel–Crafts 57, 65
α-hydroxyketones 271
hydroxylation, *see* aminohydroxylation; dihydroxylation
hyperbranched polymers 176

i

imidazoindole phosphine
– allylic alkylation 214–217
– allylic amination 217–218
– allylic etherification 218–220
– design and preparation 214
– synthetic application 219–222
imidazolinones 312–314
imidazolium-based ionic liquids 235, 238, 244, 264, 266, 270
imines
– addition to 81, 162–163
– aldimines 392–393, 407
– cyanation 406–408
– reduction 31, 193
iminophosphines 165
immobilization 2–3
– common techniques 4–17
– covalent 31–34, 45, 431–434
– general requirements 332
– in ionic liquids 234
– poly(amino acids) 316
– problems 323
– process 295–299, 319
– ship-in-a-bottle approach 47–48
– on solid supports 28–30, 41–44, 55–57, 294
– via coordination, ionic and other interactions 45–47
– via hydrogen bonding, ionic and other interactions 34–39
immobilized chiral metal complexes 413, 428–434
indene 184–185
industrial applications 18–19, 293, 320, 413, 435, *see also* pharmaceuticals
– economics 1, 293, 415
– hydrogenation catalysts 418–428
– immobilized chiral metal complexes 428–434
– industrial requirements 414–418
inorganic supports 25–27, 66–67
– asymmetric oxidation 40–54
– asymmetric reduction 27–39
– carbon–carbon and carbon–heteroatom bond formation 54–66
– immobilization techniques 4–6
insoluble polymer-supported catalysts 299–300, 303, 319
insulating groups 183
internally functionalized dendrimers, *see* core-functionalized dendrimers
ionic bonding 34–39, 45–47
ionic liquids 233–235, 285–286
– asymmetric oxidations in 245–250
– asymmetric reductions in 235–245
– carbon–carbon and carbon–heteroatom bond formation 250–264
– catalytic reactions in 14–15
– organocatalysis and 264–274
– supported catalysis 48–49, 63
– used with supercritical CO_2 283–285
ionic tag 266–267, 270
ion-pair formation 25

j

Jacobsen's catalyst 47, 254, 297
Janus dendrimers 176
JMC 94 421
Josiphos ligands 157–158, 431–433
Josiphos units 157–158
Juliá–Colonna epoxidation 314, 316

k

β-ketoesters 241, 345–346, 363–368, 396, 402
ketones
– borane reduction 101–104, 147–149, 160–161
– epoxidation of α,β-unsaturated 337–339, 404–406
– hydrogenation 107–110, 342–343, 345–346, 360–361, 420–428
– reduction 189–191
ketopantolactone 427
kinetic resolution, terminal epoxides 100–101, 186–188, 254–255

l

β-lactam 304
lanthanum 337–338, 351

latent biphasic system 134, 143, 176
leaching 1, 294, 419
– in ionic liquids 238–240, 286
– rhodium 280
Lewis acids 84, 92, 326–327
Lewis bases 305–306
'light fluorous' 184, 205
lithium amides 91–92

m

magnesium oxide 51–52, 62–63
malonates 197–199, 214–216, 399, 401, 404
manganese–salen complexes 40–42, 45–46, 48–49, 96–99
– for asymmetric epoxidation 248–249, 349–350
– in fluorous catalysis 184–186
Mannich reaction 270–274, 402–404
MCF 64
MCM-22 47–48
MCM-41 27–29, 41, 58, 65
– confinement effects 32–34
– in organocatalysis 308–309
MCM-48 48–49, 249–250
membrane filtration 133, 134, 176
metal complexes, immobilized 413, 428–434
metal surface catalysts 17–18, 357–358, 377
– cinchona alkaloid-modified palladium 368–377
– cinchona alkaloid-modified platinum 359–363
– history of modification 358–359
– tartaric acid-modified nickel 363–368, 418–420
metalated salen ligands 95–97
metal-free catalysis 203–205, 294, 297
metal–organic coordination networks (MOCNs) 324, 331
metal–organic coordination polymers (MOCPs), see MOCPs
metal–organic frameworks (MOFs) 59, 324, 328
metal–organic hybrid materials (MOHMs) 324
metathesis
– olefins 118–120
– ring-closing 120
methyl acetoacetate 190, 363–368
methylation, allylic nitromethylation 115–116, 221–222
methylcinnamic acid 373–374

metolachlor 31, 432
Michael addition 87–91
– core-functionalized dendrimers 151–152
– fluorous catalysis 204–205
– in ionic liquids 269–271
– in phase-transfer catalysis 398–402
– peripherally modified dendrimers 166–167
– using MOCPs 335–336
MOCNs 324, 331
MOCPs 323–325, 351–352
– carbon–carbon bond formation 333–337, 346–349
– design and construction 330–333
– historical account of applications 326–330
– hydrogenation 340–346
– oxidation reactions 337–340, 349–350
modifiers 17–18, 357–359, 361
– in cinchona alkaloid palladium catalysts 369–370, 422–423
– nickel in tartaric acid 369
modular toolbox 431–434
MOFs 59, 324, 328
MOHMs 324
molybdenum 50, 118–120
monoalkylation 384–391
monodendate phosphines, see MOPs
monodendate phosphoramidites 140–142
MonoPhos 340–342
monophosphines 169–170
MOPs 197–198, 211, 213, 226, 228
Mukaiyama aldol reactions 84
multipurpose batch equipment 414

n

nanosized membrane filtration 133–134, 176
Neber rearrangement 403–405
'negative' dendrimer effect 157
nickel 363–368, 418–420
nitroaldol reaction 57, 63–64
nitroalkanes 151, 401
nitromethylation 115–116, 221–222
NOBIN-cored dendrimers 156
noble metals 357
nonconventional media 2, see also aqueous media; fluorous catalysis; ionic liquids; supercritical carbon dioxide
– catalyst immobilization 12–16
nonionic cinchona-derived catalysts 302–305
Noyori-type catalyst 137–138, 242, 342–343
nucleophilic aromatic substitution 397–398

o

olefins, *see also* alkenes
- aminohydroxylation 44, 52–53
- aziridination 44, 53–54
- dihydroxylation 43–44, 52–53, 94–95, 245–246
- epoxidation 40–49
- metathesis 118–120

optimization, catalyst structure 296–297

organic polymer supports 7–8, 73, 123–124
- addition reactions 80–81
- aldol reactions 84–87
- alkylation reactions 74–79, 111–115
- α-amino acid synthesis 82–84
- borane reduction of ketones 101–104
- carbonyl-ene reaction 87
- cycloadditions 92–94
- cyclopropanation 116–118
- deprotonation 91–92
- epoxidation 95–100
- hydroformylation 123
- hydrogenation reactions 110–115
- hydrolysis 121–123
- hydrolytic kinetic resolution 100–101
- Michael reaction 87–91
- nitromethylation 115–116
- olefin metathesis 118–120
- phenylation 79–80
- Reissert-type reaction 120–121
- ring-closing metathesis 120
- Sharpless dihydroxylation 94–95
- silylcyanation of aldehyde 81–82
- transfer hydrogenation 104–107
- Wacker-type cyclization 121

organic solvent filtration 239
organic solvents, drawbacks to 233
organic–inorganic hybrid polymers 324–325, 328, 352

organocatalysis, *see also* aldol reactions; Michael addition
- in ionic liquids 264–274

organocatalysts 293–295, 319–320
- derived from amino acids 307–317
- immobilization process 295–299
- Lewis bases 305–306
- miscellaneous 317–318
- mobilization 16–17
- nonionic cinchona-derived 302–305
- phase-transfer 299–303

organometallic compounds, addition to aldehydes 149–151, 193–197

Orito reaction 359–360, 363, 422
osmium 52–53
osmium tetroxide 94–95, 246–248

oxazaborolidines 101–102, 147
oxazolines 154, 226, 228
- bisoxazolines 259–261

oxidation, *see also* dihydroxylation; epoxidation
- allylic oxidation in fluorous catalysis 187–188
- aziridination of olefin 44, 53–54
- in ionic liquids 245–250
- using inorganic supports 40–54
- using MOCPs 337–340, 349–350
- using organocatalysts 297–298

oxime sulfonates 403–404, 405

p

palladium
- in allylic aminations 164–165, 259
- in allylic substitutions 54–55, 58, 152–153, 256, 258
- cinchona alkaloid-modified 368–377, 420, 425
- in fluorinations 263–264
- in fluorous catalysis 197–200
- PS-PEG catalysts 210–217
- silk-palladium 358–359

PAMAM dendrimers 160–165
pantothenic acid 426–427
peptides 86–87, 394–396
perfluorocarbons 181–182, 276–280
peripherally modified dendrimers 9–10, 133, 156–157
- addition reactions 162–163
- aldol reaction 168–169
- allylic amination and alkylation 163–166
- Diels–Alder reaction 167–168
- hydrogenation 157–160
- hydrovinylation 169–170
- Michael addition 166–167
- reduction reaction 160–161
- ring opening of epoxides 161–162
- solid-supported 174–175

pharmaceuticals 279, 323
- dendrimer supports 148, 169
- industrial applications 422–428
- manufacture 414
- peptides as 394

phase separation 133–134, 143–145
phase-transfer catalysis 3, 16, 383, 408
- aldol reactions 399–402
- alkylation 384–398
- epoxidation 95–96, 404–406
- Michael addition 398–399, 400–402
- Neber rearrangement 403–405
- Strecker reaction 406–408

phase-transfer catalysts 299–303
phenylacetylene 80–81
phenylation 79–80
phenylcinnamic acid 368–374
phosphines 210–213, 226, 228, *see also* diphosphines; imidazoindole phosphine; MOPs
– bisphosphines 236
– ferrocenyl phosphines 166
– iminophosphines 165
– monophosphines 169–170
phosphoramides 305
photooxygenation 295–296
platinum 250–251, 359–363, 420–424
poly(amino acids) 315–316
poly(ethylene glycol) (PEG) 8, 73, 295–296, *see also* PS-PEG
– supported imidazolidinone 314
– supported phase-transfer catalysts 302–303
– supported proline derivatives 308
– supported TEMPO 297–298
polymers, *see* MOCPs; organic polymer supports
polystyrene 7–8, 73, 75, 281, 311
polystyrene-poly(ethylene glycol) (PS-PEG), *see* PS-PEG
polystyrene-supported dendrimers 175
polytopic ligands 329–332, 337, 340, 352
pore sizes, tunable 27
porphyrin 295–296
'positive' dendrimer effect 157, 161
PPI dendrimers 164, 168–169
process development 414–415
proline 85–87, 307–312
– in aldol reaction 168–169
– in the Michael reaction 89–90
– L-proline 265–266, 268–270
prolinols 74–75, 102–104, 147–148
propanation, *see* cyclopropanation
propargylamines 81
PS-PEG 86, 210–217, 226–228
– supported BINAP 222–224
PTC, *see* phase-transfer catalysis
Pybox 81, 116
2-pyrone derivatives 376–377
Pyrphos units 138–139, 157–158, 163–164

q

quaternary ammonium salts 82, 388, 390, 393–394, 396, 405

r

Raney nickel 363–368, 419
recycling 209, 304, 319, 416
– dendrimer catalysts 133, 143–145, 176
– fluorous catalysts 205–206
– in homogeneous catalysis 5
– in ionic liquids 249, 252–254, 285–286
– MOCPs 351
– organocatalysts 73, 264, 298
– requirements 3
reduction, *see also* hydrogenation
– borane reduction of ketones 101–104, 147–149, 160–161
– carbon–nitrogen and carbon–carbon double bonds 191–193
– in inorganic supports 27–39
– in ionic liquids 235–245
– ketones in fluorous catalysis 189–191
Reissert-type reaction 120–121
rhodium 13, 192, 202, 222–224, 229
– DIOP complex 235–236
– in hydrogenation of C–C 28
– in ionic liquids 237–239, 243
– in supercritical carbon dioxide 280–281
– MonoPhos catalysts 340–342
ring opening, epoxides 161–162, 173, 252–254
ring-closing metathesis 120
ruthenium 106–108, 116
– BINAP analogues 137–138, 142–143, 239–241, 276–278, 343–346
– DPEN catalysts 137–138, 224, 226
– in hydrogenation 145–147, 244–245
– in ionic liquids 242

s

salen ligands, *see also* manganese–salen complexes
– chromium–salen 252–254
– cobalt–salen 101, 161–162, 186–187, 254–255
– dendrimer-cored 172–173
– metalated 95–97
– vanadium–salen 256–257
SAPC 235
scanning tunneling microscopy 368
Schiff bases 402, 408
– activated peptides 394–396
– α-alkyl-α-amino acid derived 391–394
– glycine-derived 384–391
self-supported chiral catalysts 10–11, 333, 341–342
Sharpless dihydroxylation 52, 94–95, 245–246, 386

Sharpless epoxidation   49, 99–100
Shibasaki's catalyst   337–338
ship-in-a-bottle   25, 47–48
Si–H insertion   57, 65–66
SILC   48–49, 63
SILF   261–262
silica
– in inorganic supports   28–29, 41
– supported dendrimers   174
– supported DPEN   224–226
silk-palladium catalyst   358–359
SILP, *see* supported ionic liquid phase (SILP) catalyst
silylcyanation   81-2, 255–257
site isolation   26, 31, 66
small-scale products   414
solid supports, *see also* inorganic supports
– adsorption on   430–431
– immobilised catalysts   429
– immobilization with   4
– organocatalysts   294, 319
solid-supported dendrimers   133, 170
– internally functionalized   170–173
– peripherally functionalized   174–175
soluble polymers   295
soluble polymer-supported catalysts   8, 301, 307, 319
solvent precipitation   133–134, 143, 176
solvents
– choice   233, 422–423
– for MOCPs   332
– ionic liquids   14
stability
– in catalysis   3
– thermal stability   328, 332
stereoselectivities, in ionic liquids   250–251
Strecker reaction   406–408
structural characterization   347–348, 352
structure–activity relationship   348, 352
styrene, *see also* polystyrene
– copolymerization with   171–172
– cyclopropanation   200–202, 260–261
– derivatives   94
– formylation   279–281
substitution
– nucleophilic aromatic   397–398
– palladium-catalyzed   54–55, 58, 152–153, 256, 258
sulfides   339–340
sulfimidation   262
sulfoxidation   44, 339–340
supercage   47–48
supercritical carbon dioxide   15–16, 234, 274, 286
– carbon–carbon bond formation   282–283
– fluorous catalysis in   191–192
– hydroformylation in   279–281
– hydrogenation in   274–279
– used with ionic liquids   283–385
supercritical fluids   15
supported aqueous-phase catalysis (SAPC)   235
supported ionic liquid catalysis (SILC)   48–49, 63
supported ionic liquid films (SILF)   261–262
supported ionic liquid phase (SILP) catalyst   235, 249–250, 253–254, 268–269
surface metal, *see* metal surface catalysts

t
TADDOL-cored dendrimers   171–173
tartaric acid   363–368, 418–420
TEMPO   297–298
terminal epoxides   100–101, 186–188, 254–255
thermal stability   328, 332
thermotropic solubility   13
thiourea   297, 318
titanium   87
– BINOL ligands   333–337, 339
– in fluorous catalysis   196
– Ti-TADDOLates   75–76, 92, 171
titanium dioxide   368–373, 376
toolbox, modular   431–434
transfer hydrogenation   104–107, 189–190
– core-functionalized dendrimers   145–147
– DPEN catalysts   224, 226
– in ionic liquids   244–245
– peripherally modified dendrimers   159–160
transition-metal catalysis   14–15, 135, *see also* individual metals
TsDPENs, *see* DPENs
turnover frequency (TOF)   137
– immobilised catalysts   429
– in ionic liquids   237–238
turnover number (TON)   137, 264
type I MOCPs   331, 333–343
type II MOCPs   331, 343–351

u
Uemura-type catalysts   339
ultrasound irradiation   361
α,β-unsubstituted carboxylic acids
– aliphatic   374–376
– aromatic   369–374

v

vanadium(salen)   256–257
vinylation   169–170
vinylic substitution   397–398

w

Wacker-type cyclization   121
water, *see also* aqueous media
– catalyst immobilization in   12–13, 209
'wet ionic liquid'   237

x

X-ray diffraction   348

z

zeolites   5–6, 32, 47
zinc
– alkylation of carbonyl compounds   74–75
– dialkylzinc addition   55, 58–59, 162–163, 262, 336–337, 346–349
– organozinc addition   149–151, 193–195
zirconium phosphanates   343, 345–347